Lecture Notes in Computer Science 14639

Founding Editors

Gerhard Goos
Juris Hartmanis

Editorial Board Members

Elisa Bertino, *Purdue University, West Lafayette, IN, USA*
Wen Gao, *Peking University, Beijing, China*
Bernhard Steffen, *TU Dortmund University, Dortmund, Germany*
Moti Yung, *Columbia University, New York, NY, USA*

AF755295

The series Lecture Notes in Computer Science (LNCS), including its subseries Lecture Notes in Artificial Intelligence (LNAI) and Lecture Notes in Bioinformatics (LNBI), has established itself as a medium for the publication of new developments in computer science and information technology research, teaching, and education.

LNCS enjoys close cooperation with the computer science R & D community, the series counts many renowned academics among its volume editors and paper authors, and collaborates with prestigious societies. Its mission is to serve this international community by providing an invaluable service, mainly focused on the publication of conference and workshop proceedings and postproceedings. LNCS commenced publication in 1973.

Thomas Noll · Mariana Montiel ·
Francisco Gómez · Omar Costa Hamido ·
José Luis Besada · José Oliveira Martins
Editors

Mathematics and Computation in Music

9th International Conference, MCM 2024
Coimbra, Portugal, June 18–21, 2024
Proceedings

Springer

Editors
Thomas Noll
Escola Superior de Música de Catalunya
Barcelona, Spain

Mariana Montiel
Georgia State University
Atlanta, Georgia

Francisco Gómez
Universidad Politécnica de Madrid
Madrid, Spain

Omar Costa Hamido
University of Coimbra
Coimbra, Portugal

José Luis Besada
Complutense University of Madrid
Madrid, Spain

José Oliveira Martins
University of Coimbra
Coimbra, Portugal

ISSN 0302-9743 ISSN 1611-3349 (electronic)
Lecture Notes in Computer Science
ISBN 978-3-031-60637-3 ISBN 978-3-031-60638-0 (eBook)
https://doi.org/10.1007/978-3-031-60638-0

© The Editor(s) (if applicable) and The Author(s), under exclusive license
to Springer Nature Switzerland AG 2024

This work is subject to copyright. All rights are solely and exclusively licensed by the Publisher, whether the whole or part of the material is concerned, specifically the rights of translation, reprinting, reuse of illustrations, recitation, broadcasting, reproduction on microfilms or in any other physical way, and transmission or information storage and retrieval, electronic adaptation, computer software, or by similar or dissimilar methodology now known or hereafter developed.
The use of general descriptive names, registered names, trademarks, service marks, etc. in this publication does not imply, even in the absence of a specific statement, that such names are exempt from the relevant protective laws and regulations and therefore free for general use.
The publisher, the authors and the editors are safe to assume that the advice and information in this book are believed to be true and accurate at the date of publication. Neither the publisher nor the authors or the editors give a warranty, expressed or implied, with respect to the material contained herein or for any errors or omissions that may have been made. The publisher remains neutral with regard to jurisdictional claims in published maps and institutional affiliations.

This Springer imprint is published by the registered company Springer Nature Switzerland AG
The registered company address is: Gewerbestrasse 11, 6330 Cham, Switzerland

If disposing of this product, please recycle the paper.

Preface

The 9th Biennial International Conference for Mathematics and Computation in Music (MCM 2024) took place during June 18–21, 2024, at the Centre for Interdisciplinary Studies and the Centre for Informatics and Systems of the University of Coimbra, Portugal. MCM 2024 continued the pattern, initiated in 2007 at the first MCM meeting, of biennial international conferences held on alternating sides of the Atlantic: Berlin in 2007, New Haven in 2009, Paris in 2011, Montreal in 2013, London in 2015, Mexico City in 2017, Madrid in 2019 and then in Atlanta, in 2022, with the one-year delay due to the Covid epidemic.

As the main conference of the Society for Mathematics and Computation in Music (SMCM), MCM 2024 provided a platform for the communication and exchange of ideas among researchers in mathematics, informatics, music theory, musicology, and related disciplines. It brought together researchers from around the world who combine mathematics or computation with music analysis, music cognition, composition, and performance.

The schedule is available at https://www.uc.pt/ceis20/mcm24/. The scientific program featured 28 talks, 5 posters, 5 communication-performance talks, a new modality that we have included in our MCM conference, and a tribute paper. The presentations were grouped around the following subjects: Mathematical Scale Theory and Tuning; Rhythm Analysis and Rhythm Generation; Categorical and Algebraic Approaches to Music; Quantum Music; Theory and Algorithms for Melodic/Harmonic Analysis and Generation; Geometric Approaches to Musical Algorithms and Microtonality; Fourier Analysis for Music; and Similarity and Distance Measures for Music. Additionally, Dmitri Tymozcko presented a workshop on grupoids in music.

There were two plenary talks. Norman Carey and David Clampitt presented *35 Years of Well-Formed Scale Theory: Origins and Development* and William Sethares presented *The Many Tunings of Balinese Gamalean.*

There were three concerts presented: a concert by MCM participants as composers and performers, a concert where Moreno Andreatta presented a selection of his "Music and Maths" repertoire, and a concert of Coimbra Fado music by the group 'Quarteto de Coimbra' (Fábio Almeida, Tiago Rodruigues, Simão Mota Vasco Rodrigues).

We received 42 submissions of which 30 long papers and 9 short papers were accepted. All papers were peer reviewed. The submissions came from researchers in 15 countries in North and South America, Europe, Africa, Asia, and Australia.

We thank the following institutions for providing their infrastructure and human resources for the organization and promotion of MCM 2024:

- Society for Mathematics and Computation in Music
- University of Coimbra
- Escola Superior de Música de Catalunya
- Department of Mathematics and Statistics, Georgia State University
- Universidad Politécnica de Madrid

The MCM2024 conference is partially funded by FCT – Fundação para a Ciência e a Tecnologia, I.P., in the scope of the project UIDB/00460/2020.

June 2024

Thomas Noll
Mariana Montiel
Francisco Gómez
Omar Costa Hamido
José Luis Besada
José Oliveira Martins

Organization

Steering Committee

Society for Mathematics and Computation in Music

Mariana Montiel	Georgia State University, Georgia, USA
Thomas Noll	Escola Superior de Musica de Catalunya, Spain
Tom Fiore	University of Michigan-Dearborn, USA

Editorial Committee (Proceedings)

Thomas Noll	Escola Superior de Musica de Catalunya, Spain
Mariana Montiel	Georgia State University, Georgia, USA
Francisco (Paco) Gómez	Universidad Politécnica de Madrid, Spain
Omar Costa Hamido	CEIS20, University of Coimbra, Portugal
José L. Besada	Universidad Complutense de Madrid, Spain
José Oliveira Martins	CEIS20, FLUC, and University of Coimbra, Portugal

Scientific Program Chairs (Conference)

The Scientific Program Committee was responsible for the scientific content of MCM 2024. They coordinated and supervised the reviewing process for the submitted papers, and communication-performances as well as prepared the final list of oral and poster presentations. The committee is formed by the following members.

José Oliveira Martins	CEIS20, FLUC, and University of Coimbra, Portugal
Mariana Montiel	Georgia State University, Georgia, USA
Omar Costa Hamido	CEIS20, University of Coimbra, Portugal
Amílcar Cardoso	CISUC and DEI, University of Coimbra, Portugal
Pedro Martins	CISUC Universidade de Coimbra, Portugal
Jônatas Manzoli	Unicamp, Brazil and Univ. Coimbra, Portugal
Riccardo D. Wanke	CEIS20, University of Coimbra, Portugal
José L. Besada	Universidad Complutense de Madrid, Spain

Scientific Committee (Reviewers)

Adrian Kassel	ENS de Lyon, France
Alexandre Popoff Philips	Research Paris - Medisys, France
Amílcar Cardoso	CISUC, Universidade de Coimbra, Portugal
Andreé Ehresmann	LAMFA, Université de Picardie Jules Verne, France
Andreas Nestke	Hochschule für Technik und Wirtschaft Berlin, Germany
Andrew Milne	The MARCS Institute for Brain, Behaviour and Development, Sydney, Australia
Brian Martinez-Rodríguez	Conservatori Superior de Música Joaquín Rodríguez de València, Spain
Carlos Agon	IRCAM, France
Christian Kassel	Institut de Recherche Mathématique Avancée, CNRS et Université de Strasbourg, France
Christopher Dobrian	University of California, Irvine, USA
Clifton Callender	Florida State University, USA
Costas Tsougras	Aristotle University of Thessaloniki, Greece
Daniel Harasim	École polytechnique fédérale de Lausanne, France
Darrell Conklin	Department of Computer Science and Artificial Intelligence Facultad de Informática University of the Basque Country UPV/EHU Donostia - San Sebastián, Spain
David Meredith	Aalborg University, Denmark
David Temperley	Eastman School of Music, USA
Emilio Lluis-Puebla	Universidad Nacional Autónoma de México, México
Emmanuel Amiot	LAMPS, Université de Perpignan, France
Fabian Moss	Julius-Maximilians-Universität Würzburg, Germany
Filippo Carnovalini	Università degli Studi di Padov, Italy
Francisco (Paco) Gómez Martín	Universidad Politécnica de Madrid, Spain
Franck Jedrzejewski	Université de Paris-Sud, France
Harald Fripertinger	Universität Graz, Austria
Ingolf Max	Universität Leipzig, Germany
Jason Yust	Boston University, USA
Jay Hook	Indiana University Bloomington, USA
Jean-Paul Allouche	CNRS Centre National de la Recherche Scientifique, Paris, France
Jeremy Kastine	Life University, Marietta, Georgia, USA

Jônatas Manzolli	Music Department, Institute of Arts Interdisciplinary, Nucleus for Sound Studies, University of Campinas - Unicamp, Brazil
José Luis Besada	Universidad Complutense de Madrid, Spain
Julian Hook	University of Indiana, USA
Luis Nuño	ITACA Institute, Universitat Politècnica de València, Spain
Marek Zabka	Department of Musicology of the Comenius University, Bratislava, Slovak
Maria Mannone	Institute of Physics and Astronomy, University of Potsdam, Germany; DSMN - ECLT, Ca'Foscari University of Venice, Italy
Mariana Montiel	Department of Mathematics and Statistics Georgia State University, USA
Mattia Bergomi	Independent Researcher, Milan, Italy
Maximos Kaliakatsos-Papakostas	Hellenic Mediterranean University, Greece
Moreno Andreatta	CREAA, Université de Strasbourg, France
Norman Carey	CUNY Graduate Center, SUA
Octavio Agustín-Aquino	Universidad Tecnológica de la Mixteca, México
Omar Costa Hamido	CEIS20, Universidade de Coimbra, Portugal
Pedro Martins	CISUC, Universidade de Coimbra, Portugal
Peter beim Graben	Bernstein Center for Computational Neuroscience, Berlin, Germany
Pierre Sylvain Iloga	University of Maroua, Cameroon
Reinhard Blutner	University of Amsterdam, the Netherlands
Riccardo Wanke	CEIS20, Universidade de Coimbra, Portugal
Richard Cohn	Yale University, USA
Richard Parncutt	Universität Graz, Germany
Robert Peck	Louisiana State University, USA
Rodrigo Castro	The Jackson Laboratory, USA
Thomas Noll	Escola Superior de Música de Catalunya, Spain
Tom Collins	University of Miami, USA
Tom Fiore	University of Michigan-Dearborn, USA
Vicente Liern	Universidad de Valencia, Spain
William Sethares	University of Wisconsin-Madison, USA

Local Organizing Committee

The Local Organizing Committee consisted of students and other local participants at the conference who helped with logistical issues. At CEIS20 Marlene Taveira, Ângela Lopes, Sofia Melim, Cláudia Morais were the local organizers.

Collaborating Institutions

- Foundation for Science and Technology (Portugal), funding under program IUDB 00460/2020
- At the University of Coimbra:

- Centre for Interdisciplinary Studies (CEIS20)
- Centre for Informatics and Systems (CISUC)
- Faculty of Arts and Humanities
- Faculty of Science and Technology, Department of Informatics Engineering
- Laboratory of Research in Artistic Practices (LIPA)

Contents

Mathematical Scale Theory and Tuning

Quarter-Tone Music: A Tuning System Rooted in Natural Harmonic Series

Aziz Ifzarne(✉)

LIPIM, Sultan Moulay Slimane University, ENSA, Khouribga, Morocco
a.ifzarne@usms.ma

Abstract. The objective of this work is to study quarter-tone music through the lens of harmonic series. The 11th harmonic (11/8), which corresponds to the major fourth interval, offers a remarkably precise approximation of the quarter tone. By employing the "circle of major fourths," we may encompass the complete spectrum of 24 notes and establish a tuning system of quarter-tones that is based on overtones. We analyze its distinct attributes and perform a detailed examination of its impact, specifically on maqam music. To solve issues specifically related to modulation, a non-octave tuning is proposed by dividing the major fourth interval into 11 quarter-tones.

Keywords: Tuning system · microtonal music · harmonic series · just intonation · quarter-tone · 24-TET · 11th harmonic · circle of fifths · non-octave tuning

1 Introduction

Quarter-tone music is a fascinating approach to convey musical expression. This kind of microtonal music differs from western music in that it employs intervals that are half the size of a semitone, resulting in 24 pitches per octave instead of 12. In contrast to the more common 12-tone equal temperament (12-TET), the 24-tone equal temperament (24-TET) divides the octave into 24 equidistant steps. The tuning system is also referred to as 24-EDO, (24 Equal Divisions of the Octave). In this tuning system, the smallest step is the quarter tone, which is 50 cents. Quarter-tone music has a long history and is widely practiced in several regions of the world. It continues to be prevalent in Turkish, Persian and Arabic music [4]. To an uninitiated listener, the quarter tones might be initially jarring and dissonant, although this initial discomfort usually disappears after a period of listening. During the 20th century, western composers, like Ivan Wyschnegradsky [13], Charles Ives [5], Alois Hába and Don Ellis [3], conducted thorough investigations into this unique form of music. Due to the inclusion of all the pitches found in 12-TET, musicians can utilize the additional tonalities in 24-TET without sacrificing the techniques available in 12-TET. The fact that 24 is divisible by 12 makes it convenient to perform 24-TET instrumentally by using two typical 12-TET instruments tuned a quarter-tone apart, such as two pianos.

© The Author(s), under exclusive license to Springer Nature Switzerland AG 2024
T. Noll et al. (Eds.): MCM 2024, LNCS 14639, pp. 3–14, 2024.
https://doi.org/10.1007/978-3-031-60638-0_1

Western music is deeply linked to the occurrence of overtones, also referred to as harmonics or partials, which are supplementary frequencies generated along the fundamental frequency when a musical note is produced. As soon as a tone sounds, overtones vibrate. Overtones come directly from the rules of natural vibration. Just intonation tunings provide intervals that are pure, as they are derived from simple ratios contained in the harmonics. For instance, the octave is derived from the ratio of 2:1, specifically the 2nd harmonic, and the perfect fifth (ratio of 3:2) can be obtained from the 3rd harmonic. By utilizing the circle of fifths, one can get the 12 chromatic pitches. The corresponding tune is the 12-tone Pythagorean temperament. In his preface to "24 Préludes op. 22", Wyschnegradsky underlined the importance of the 11th harmonic in quarter-tone music [13].

The aim of this paper is to study quarter-tone music from the perspective of overtones. The 11th harmonic provides a highly accurate approximation of the quarter tone. The interval in question is the major fourth, which is composed of a perfect fourth plus a quarter tone. We contemplate the "circle of major fourths" like the concept of the "circle of fifths." This will allow us to access the entire range of 24 notes produced by quarter-tone music. We construct a tuning system that relies on overtones, with the focus being on the 11th harmonic (11/8). We investigate the specific characteristics of this quarter-tone tuning system and thoroughly examine its impact, specifically on maqam music. To deal with problems specifically related to modulation, a non-octave tuning is suggested, achieved by dividing the major fourth interval into 11 quarter-tones.

2 Quarter-Tones Hidden in Overtones

In quarter-tone music, 12 tones are part of the conventional semitone system, while the remaining 12 tones are new intermediary tones. The 24-tone equal temperament (24-TET) is the tuning system that divides the octave into 24 equal parts, with each part being exactly 50 cents. Each step corresponds to the frequency ratio of 21/24, or the 24th root of 2. The 24-TET is also known as quarter-tone tuning or 24 Equal Divisions of the Octave (24 EDO).

In order to represent all the 24 notes, it is imperative to provide additional symbols for ascending and descending alterations, consisting of two symbols for each type of alteration. The notation of quarter tones is still not completely standardized. We'll utilize this quite common notation (Table 1).

Table 1. Signs of alteration often used in quarter-tone music.

Half sharp (hs or +): raised ¼ tone	𝄲
Three quarter sharp: raised ¾ tone	𝄰
Half flat (hf or -): lowered ¼ tone	𝄳
Three quarter flat: lowered ¾ tone	𝄳♭

The chromatic quarter-tone scale can be represented in the following manner:

Now, let us identify the quarter tone inside the compilation of the initial 20 harmonics. If the harmonics are transposed into the span of one octave (by suitably multiplying or dividing the ratios by a power of two), harmonics n and 2n will correspond to the identical interval. Therefore, it is sufficient to consider only the odd harmonics (Table 2).

Table 2. The first 20 odd harmonics and the corresponding intervals. Intervals are compressed into the span of one octave and compared with the 12-tone equal temperament.

Harmonic	Note	Interval	Ratio	Cents	12-TET cents	Diff in cents
1	C	Unison	1	0	0	0
3	G	Fifth	3/2	702	700	2
5	E	Major third	5/4	386	400	-14
7	≈ B♭	Minor seventh	7/4	969	1000	-31
9	D	Major second	9/8	204	200	4
11	F+	Major fourth	11/8	551	600	**-49**
13	≈ A-	Neutral sixth	13/8	841	800	**41**
15	B	Major seventh	15/8	1088	1100	-12
17	C♯	Minor second	17/16	105	100	5
19	E♭	Minor third	19/16	298	300	-2

It is interesting to note that certain harmonics can be accurately represented by the notes of the chromatic scale, which is widely used in the Western music tradition. However, there are certain harmonics, such as the 7th harmonic, that pose difficulties in this regard. An interesting fact is that the 11th and 13th harmonics approximate specific quarter-tone intervals of the 24-tone equal-tempered scale. Starting with C as the fundamental, the 11th harmonic is 49 cents below F# while the 13th harmonic is 41 cents over A♭. It's clear that the 11th harmonic (11/8) gives the best approximation of the quarter tone. It is approximately 551 cents, which is just a cent away from the value of the major fourth in 24-TET. Thus, the 11th harmonic approaches almost perfectly the major fourth, which is found, for example, between C and F half-sharp. Other names for the major fourth include superfourth, up-fourth, mid-fourth, wide fourth, and semiaugmented fourth.

A sound spectrum can be employed to display the harmonics that are present in a musical note. In this case, the note is C3 (130 Hz) (Fig. 1).

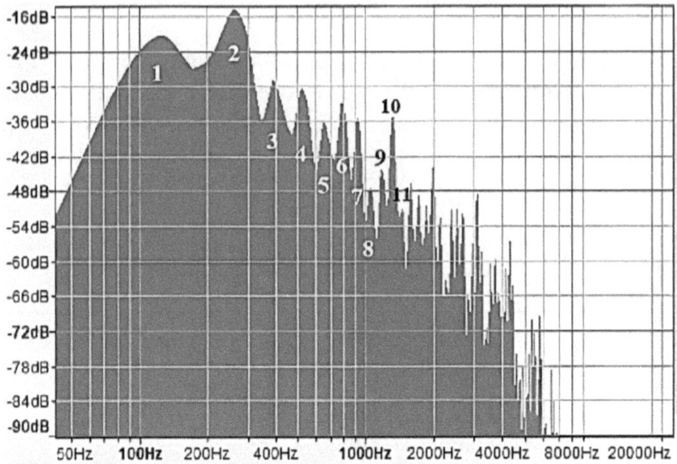

Fig. 1. A spectrum showing the sound produced by playing the note C3 (130 Hz) on a piano. Theoretical calculations indicate that the frequency of the 11th harmonic is $11 \times 130 = 1430$ Hz, which closely aligns with the observed frequency in the spectrum. Spectrum made by *Audacity*.

Remarks.

1. There is a curious point worth mentioning regarding the major fourth. Within a 24-tone equal temperament, its value is equal to $2^{\frac{11}{24}} \cong 1.3739$. This number is irrational. An excellent method for obtaining a rational approximation is the continued fraction representation:

$$1 + \cfrac{1}{2 + \cfrac{1}{1 + \cfrac{1}{2 + \cfrac{1}{13 + \cfrac{1}{1 + \cfrac{1}{4 + \cdots}}}}}}$$

Here are the three first rational approximations:

Contiued fraction	[1; 2]	[1; 2; 1]	[1; 2; 1; 2]
Rational approximation	3/2	4/3	11/8
Decimal equivalent	1.5	1.3333	1.375
Error (%)	−9.18%	2.96%	−0.08%

The fraction 11/8 appears as the 3rd approximation. It provides a highly accurate estimation with an error rate close to zero (−0.08%).

2. Considering the role of the major fourth in the 24-tone system as well as the usage of continued fractions, it is possible to create an interesting connection with well-formed scale theory. In the 12-tone system, pentatonic, diatonic, and chromatic scales share the same underlying structure, that of the well-formed scale [7]. In the same way as the semi-convergents $\left\{\frac{1}{2}, \frac{2}{3}, \frac{3}{5}, \frac{4}{7}, \frac{7}{12}, \frac{10}{17}, \ldots\right\}$ of the perfect fifth $Log_2\left(\frac{3}{2}\right)$ correspond to a hierarchy of fifth-generated scales, which play an essential role in the music theory of traditional western music, here is the list of semi-convergents of the major fourth $Log_2\left(\frac{11}{8}\right)$:

$$\left\{\frac{1}{2}, \frac{1}{3}, \frac{2}{5}, \frac{3}{7}, \frac{4}{9}, \frac{5}{11}, \frac{6}{13}, \frac{11}{24}, \ldots\right\}$$

Using the ratio $\frac{6}{13}\frac{6}{13}$, we recover Ivan Wyschnegradsky's 13-tone scale that he called diatonized chromatism [13]. The ratio $\frac{5}{11}\frac{5}{11}$ corresponds to an 11-tone scale, which at the same time forms the complement of Wyschnegradsky's 13-tone scale. And there is also the semi-convergent $\frac{3}{7}\frac{3}{7}$ which corresponds to a well-formed 7-note scale. By ordering the notes on this scale, we obtain {C, E−, E+, F+, A, Bb, B}. Major fourth between C and F+ is actually a fourth in the literal scale-theoretic sense that spans three scale steps. And there it would deserve an attribute like "neutral" or "perfect" because of its role as the scale generator.

3 A Tuning System Rooted in the 11th Harmonic

Pythagorean tuning derives the frequency ratios of all intervals from the pure perfect fifth, which has a ratio of 3:2 (3rd harmonic). The circle of fifths enables the construction of the 12 notes in the chromatic scale, despite its inability to close the circle by an excess of around 23.46 cents, which is nearly a quarter of a semitone. This surplus is referred to as the Pythagorean comma. The circle of fourths serves the same purpose as it is essentially a reversed circle of fifths. Except for the ascending and descending semitone circles, only the two previous circles manage to reveal all 12 notes of the chromatic scale.

Now, let us move on to the quarter-tone music. The circle of fifths fails to enable the construction of all 24 notes found in quarter-tone music. How can we identify the intervals for which the corresponding circles satisfy this condition?

The cyclic group of integers modulo 24, $\mathbb{Z}/24\mathbb{Z}$, can represent the 24-TET pitch classes: $C = 0$; $C+ = 1; \ldots; B+ = 23$.

Identifying the circles that surround the 24 notes can be resolved mathematically by searching the generators of the additive group $\mathbb{Z}/24\mathbb{Z}$. It is well known that:

$$\overline{a}\,\text{is a generator of}\,\mathbb{Z}/n\mathbb{Z} \Leftrightarrow a\,\text{is relatively prime to}\,n.$$

We deduce the list of the eight generators of $\mathbb{Z}/24\mathbb{Z}$: 1, 5, 7, 11, 13, 17, 19, 23.

The circles derived from the numbers 13, 17, 19, and 23 are the inverse of the circles derived from 11, 7, 5, and 1, respectively. Excluding 1 (representing the circle of quarter-tones), the remaining numbers of 5, 7, and 11 are the most intriguing:

- 5 (D+): Major second quarter-high.
- 7 (E−): Neutral third.
- 11 (F+): Major fourth.

While the initial two intervals may be interesting, it is the final one that maintains a strong connection to the harmonic series, as it represents the 11th harmonic (11/8). The circle of major fourths reveals all 24 notes found in quarter-tone music. This fact was noticed by I. Wyschnegradsky [10] who diatonicized the 24 chromatic scale to 13 tones by descending through the circle of major fourths.

To construct a tuning system based on the 11th harmonic, we utilize the cycle of major fourths rather than the cycle of fifths (Fig. 2).

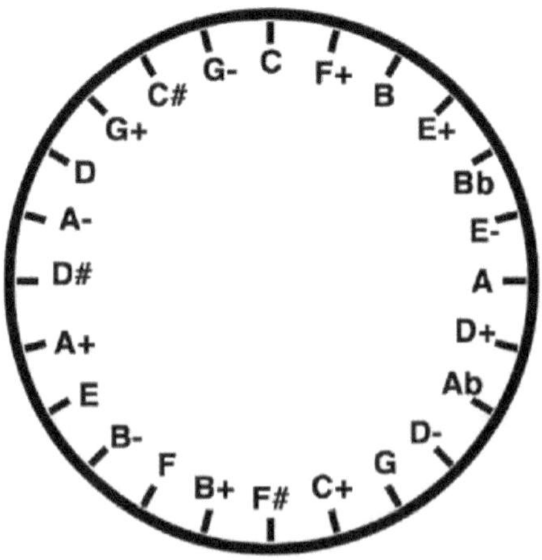

Fig. 2. The circle of major fourths

These formulas are used to calculate the ratios and the interval's sizes (in cents):

$$Ratio = \frac{\left(\frac{11}{8}\right)^{Major\,fourth\,degree}}{2^{Octave\,degree}}; \quad Cents = 1200\,Log_2(Ratio)$$

This tuning system will be designated as the “11th Harmonic Tuning” (11th HT) or as the “Circle of Major Fourths Tuning” (CMFT).

Here are some observations regarding this “11th Harmonic Tuning”:

Table 3. Comparison of the 11th Harmonic Tuning (11th HT), 24-Tone Equal Temperament (24-TET) and Just Scale. Just scale ratios are obtained from irrational 24-TET intervals using suitable continued fraction approximations.

Degree	Note	11th Harmonic Tuning				24-TET		Just Scale		
		Ratio	Cents	Major fourth degree	Octave degree	Cents	Diff (cents)	Ratio	Cents	Diff (cents)
0	C	1.0000	0	0	0	0	0	1	0	0
1	C+	1.0380	64	11	5	50	14	35/34	50	14
2	C#/Db	1.0774	129	22	10	100	29	17/16	105	24
3	D-	1.0980	162	9	4	150	12	12/11	151	11
4	D	1.1397	226	20	9	200	26	9/8	204	22
5	D+	1.1615	259	7	3	250	9	15/13	248	11
6	D#/Eb	1.2056	324	18	8	300	24	6/5	316	8
7	E-	1.2287	357	5	2	350	7	11/9	347	9
8	E	1.2754	421	16	7	400	21	5/4	386	35
9	E+	1.2998	454	3	1	450	4	9/7	435	19
10	F	1.3491	518	14	6	500	18	4/3	498	20
11	F+	1.3750	551	1	0	550	1	11/8	551	0
12	F#/Gb	1.4272	616	12	5	600	16	7/5	583	33
13	G-	1.4814	680	23	10	650	30	16/11	649	32
14	G	1.5098	713	10	4	700	13	3/2	702	11
15	G+	1.5671	778	21	9	750	28	17/11	754	24
16	G#/Ab	1.5971	811	8	3	800	11	8/5	814	-3
17	A-	1.6577	875	19	8	850	25	13/8	841	35
18	A	1.6895	908	6	2	900	8	5/3	884	24
19	A+	1.7536	972	17	7	950	22	19/11	946	26
20	A#/Bb	1.7872	1005	4	1	1000	5	9/5	1018	-12
21	B-	1.8551	1070	15	6	1050	20	11/6	1049	20
22	B	1.8906	1103	2	0	1100	3	15/8	1088	14
23	B+	1.9624	1167	13	5	1150	17	33/17	1148	19
24	C	2.0000	1200	-	-	1200	0	2	1200	0

1. A complete cycle is formed in the space of 11 octaves. The difference between 24 major fourths and 11 octaves is equal to the frequency ratio:

$$\frac{\left(\frac{11}{8}\right)^{24}}{2^{11}} \cong 1.018438 \equiv 31.63\,cents$$

The difference with the Pythagorean comma (23.46 cents) is 8.17 cents. Like the observation of a wolf fifth in the Pythagorean tuning, one of the 24 major fourths (namely the final one between G- and C) exhibits a distinct size compared to the rest. This is the cost paid for seeking just intonation. Indeed, the completion of the cycle is not possible when tuning musical intervals are based on whole number ratios $\frac{p}{q}$ (just intonation). The Rational Root Theorem yields this outcome:

$$\left(\frac{p}{q}\right)^n \neq 2^m, \forall p, q, n, m \in \mathbb{N}^* \text{ with } \frac{p}{q} < 2.$$

2. For this construction, we chose C as the fundamental note. One had the option to select another note, such as A. In this instance, the ratios will be adjusted to the new fundamental note. The wolf's major fourth will likewise get changed. For instance, if the fundamental note is A, the wolf major fourth will be between E− and A.
3. To get the ratios of intervals in different octaves, multiply (or divide) the ratios in Table 3 by powers of 2 for higher (or lower) octaves.
4. The intervals of "11th harmonic tuning" closely approach some intervals found in the 24-TET system, for instance, the neutral third, which is a very important interval in quarter-tone music. Nevertheless, the intervals of the just scale, particularly the perfect fifth and major third, reveal greater distance.

4 Applications to Maqam Music

Maqam music is a traditional system of melodic modes that is employed in the classical music of the Middle East, North Africa, and certain regions of Central Asia. The term pertains to the precise organization of pitches and intervals that constitute a musical scale or mode. The notion is predominantly present in Arabic music [4], where the term "maqam" originates from the Arabic root that signifies "place" or "location." The term "makam" is used in Turkish music, whereas in Azerbaijan it is referred to as "mugam." Both terms are akin to the Persian music concept known as "dastgāh." The maqām employs several intervals, including half tones and whole tones, which are also present in Western music. Additionally, it incorporates three-quarter and five-quarter tones. There are around 40 maqāmāt in Arabic music; however, only a few of them are commonly employed. We shall focus on these four highly popular maqams featuring quarter-tone intervals: Rast, Bayati, Saba, and Huzam. Except for the Huzam, the first three maqams contain the major fourth. This interval is particularly found between:

- E− and A
- B♭ and E−
- B− and E

In each of the following tables, the 2nd column indicates the difference (in cents) between the 11th Harmonic Tuning and 24-TET, while the 4th column indicates the steps (in cents) between two consecutive notes in the 11th Harmonic Tuning (Tables 4, 5, 6 and 7).

Table 4. Maqam Rast on C

Note	C	D	E-	F	G	A	B-	C
Diff 24-TET (cents)	0	26	7	18	13	8	20	0
Steps	-	C D	D E-	E- F	F G	G A	A B-	B- C
11th HT Steps (cents)	-	226	131	161	195	195	162	130

Table 5. Maqam Bayati on D

Note	D	E-	F	G	A	B♭	C	D
Diff 24-TET (cents)	26	7	18	13	8	5	0	26
Steps	-	D E-	E- F	F G	G A	A B♭	B♭ C	C D
11th HT Steps (cents)	-	131	161	195	195	97	195	226

Table 6. Maqam Saba on D

Note	D	E-	F	G♭	A	B♭	C	D♭
Diff 24-TET (cents)	26	7	18	16	8	5	0	29
Steps	-	D E-	E- F	F G♭	G♭ A	A B♭	B♭ C	C D♭
11th HT Steps (cents)	-	131	161	98	292	97	195	129

5 Another Alternative: Non-octave Tuning

An unavoidable disadvantage of tuning systems that rely on a circle of a predetermined interval is the presence of the wolf interval, which is some type of wolf fifth. This limits the ability to easily change keys. An alternative approach is to explore non-octave tunings.

An equal division of the octave (EDO) is a tuning system that divides the octave into a specific number of equal steps. This implies that the interval between any two successive pitches is equal. A typical instance is the common tuning 12-EDO (or 12-TET) used worldwide now. Non-octave tuning refers to a tuning system where the interval

Table 7. Maqam Huzam on E-

Note	E-	F	G	A♭	B	C	D	E-
Diff 24-TET (cents)	7	18	13	11	3	0	26	7
Steps	-	E- F	F G	G A♭	A♭ B	B C	C D	D E-
11th HT Steps (cents)	-	161	195	98	292	97	226	131

of equivalency is not the octave (2:1). A pseudo-octave (or impure octave) refers to an interval that is not exactly in a 2:1 frequency ratio but is still treated as if it were equivalent to this ratio. When tuning a piano, it is often required to deal with stretched octaves, which occur due to the inharmonicity generated by the thickness and tension of the strings. This requires slightly widening every interval. One example is the Perfect Fifths Tempered Scale (PFTS), initially proposed by S. Cordier [2]. In this temperament, all fifths are tuned to the frequency ratio of the just scale (3:2), which implies that each fifth is made up of seven equal semitones $(1.5)^{\frac{1}{7}} \cong 1.059634$. The corresponding octave is then $(1.5)^{\frac{12}{7}} \cong 2.003875$ which is 3.4 cents sharper than the normal octave.

In a general framework, let's consider a frequency ratio, denoted as r, that is equally divided into m steps. The tuning system resulting from dividing the interval r into m equal divisions is denoted as "m ED r". Considering the context, it is natural to think about an equal division of the major fourth (the 11th harmonic). In our particular situation, the step we are referring to is specifically the quarter-tone. Therefore, the value of m is 11, as the major fourth includes 11 quarter-tones, and the value of r is 11/8. This tuning system can be denoted "11 ED 11/8". To create an equal division with major fourths, the frequency ratio 11/8 needs to be split into 11 equal quarter-tones. This may be achieved by taking its 11th root $\left(\frac{11}{8}\right)^{\frac{1}{11}} \cong 1.02937 \equiv 50.12$ cents.

This value is only 0.12 cents greater than the standard quarter-tone given by 24-TET. The resulting octave can be calculated as $\left(\frac{11}{8}\right)^{\frac{24}{11}} \cong 2.00332 \equiv 1202.87$cents, which is 2.87 cents higher than the pure octave. Table 8 presents all the octave intervals expressed in decimals and cents. These values remain the same, regardless of the choice of fundamental note. Since all the intervals (for instance major fourths) possess identical values, the issue of modulation no longer arises within this tuning system.

Even though the values given are approximate and in decimal form, theoretically, this is not the case. Contrary to the Circle of Major Fourths Tuning, all intervals except the major fourth (11/8) are irrational.

Table 8. Equal division of the major fourth 11/8 into 11 quarter-tones (11 ED 11/8).

Degree	Note	Decimal	Cents
0	C	1.0000	0.00
1	C+	1.0294	50.12
2	C#/D♭	1.0596	100.24
3	D-	1.0907	150.36
4	D	1.1228	200.48
5	D+	1.1558	250.60
6	D#/E♭	1.1897	300.72
7	E-	1.2246	350.84
8	E	1.2606	400.96
9	E+	1.2976	451.08
10	F	1.3358	501.20
11	F+	1.3750	551.32
12	F#/G♭	1.4154	601.44
13	G-	1.4570	651.56
14	G	1.4998	701.68
15	G+	1.5438	751.80
16	G#/A♭	1.5892	801.92
17	A-	1.6358	852.04
18	A	1.6839	902.16
19	A+	1.7334	952.28
20	A#/B♭	1.7843	1002.40
21	B-	1.8367	1052.52
22	B	1.8906	1102.64
23	B+	1.9462	1152.76
24	C	2.0033	1202.88

6 Conclusion

When the harmonics are condensed inside one octave, some of them can be approximated by the notes of the chromatic scale, which consists of twelve equally spaced semitones (12-TET). However, this tuning is slightly inaccurate for several of the harmonics, particularly the 7th, 11th, and 13th harmonics. Nevertheless, to quote Carl Dahlhaus, "the interval-distance of the natural-tone-row (overtones), counting up to 20, includes everything from the octave to the quarter tone, (and) useful and useless musical tones. The natural-tone-row (harmonic series) justifies everything, that means, nothing." [9]. David Cope introduces the notion of interval strength, where the strength, consonance, or stability of an interval is measured by how closely it approximates a lower and stronger, or higher and weaker, position in the harmonic series [1].

The focus of our study was to examine the 11th harmonic, which is rather close in ranking to the fundamental frequency. An intriguing observation is that the 11th harmonic (11/8) provides a highly accurate approximation of a quarter-tone interval, specifically the major fourth. This interval is frequently encountered in quarter-tone music. Undoubtedly, 24-tone equal temperament remains the most practical tuning system due to its numerous advantages. Nevertheless, the idea of investigating tuning systems based on natural harmonics remains exciting and holds symbolic significance.

Acknowledgements. The author would like to express his sincere gratitude to the referee for his/her valuable observations and suggestions.

References

1. Cope, D.: Techniques of the Contemporary Composer, New York. Schirmer Books, New York (1997)
2. Cordier, S.: Piano bien tempéré et justesse orchestrale: le tempérament égal à quintes justes. Buchet et Chastel, Paris (1982)
3. Ellis, D.: Quarter Tones: A Text with Musical Examples, Exercises and Etudes. Harold Branch Publishing, Plainview NY (1975)
4. Farraj, J., Abu Shumays, S.: Inside Arabic Music: Arabic Maqam Performance and Theory in the 20th Century. Oxford University Press, Oxford (2019)
5. Ives, C.: Some Quarter-Tone Impressions, Franco-American Music Society Bulletin vol. 25 no. 3, 1925: reprinted in Essays Before a Sonata and Other Writings, Howard Boatwright (ed.), W.W. Norton, New York, 1962 (1970)
6. Jedrzejewski, F.: Dictionnaire des musiques microtonales (2004)
7. Norman, C., Clampitt, D.: Aspects of Well-Formed Scales, Music Theory Spectrum, vol. 11 no. 2, autumn (1989)
8. Partch, H.: Genesis of a Music, The University of Wisconsin Press (1949)
9. Sabbagh, P.: The Development of Harmony in Scriabin's Works, p. 12 (2003). Universal. ISBN 9781581125955. Cites: Dahlhaus, Carl "Struktur und Expression bei Alexander Skrjabin", Musik des Ostens, vol. 6, p. 229 (1972)
10. Wyschnegradsky, I.: Manual of Quarter-Tone Harmony. (1932). ed. Noah Kaplan. Brooklyn, NY: Underwolf (2017)
11. Wyschnegradsky, I.: Manuel d'harmonie à quarts de ton, La Sirène musicale, Paris, 1932, p. 24. Reprint Max Eschig, Paris (1980)
12. Wyschnegradsky, I.: La loi de la pansonorité. Paris, 1946. Reedition of version year 1953, by Franck Jedrzejewski (ed.), Editions Contrechamps, Genève (1996)
13. Wyschnegradsky, I.: 24 Préludes op. 22, en quarts de ton dans l'échelle chromatique diatonisé à 13 sons, (musical score with Preface by the author), Frankfurt, M.P. Belaieff (1979)

An Exploration of the Discontinuous-Continuous Fusion in *Yuunohui'tlapoa* for Keyboard

Julio Estrada(✉)

Institute for Research in Aesthetics, National Autonomous University of Mexico, UNAM, Ciudad de México, Mexico
ejulioestrada@gmail.com

Abstract. This work arises from the compositional and theoretical research on music scales through *d1-theory,* and on the continuum; it resorts to a terminology useful for composers, interpreters, or musicologists without mathematical training, as the author himself. The musical topic of the continuum emerged in the 20th century, introducing ideas like rhythm as frequency, microintervals, timbre, noise, glissandi, and spatialization. A unified study to comprehend the musical continuum as a global field needs to focus on a blend of components of rhythm (pulse, attack, microdurations) and sound (pitch, dynamics, color). This idea is understood as a *macrotimbre.* When these components in a continuous macrotimbre are intended to be asynchronous, the audible result is an amalgam of disruptions, interferences or crossings that are perceived as an elastic musical matter. By combining sequential pitch with continuous transitions of chords density, harmonic content, dynamics, and speed of pulse, such as *yuunohui'tlapoa* for a keyboard, asynchrony produces a discontinuous-continuous macrotimbre. The synthesis of this last is perceived as internal collisions of rhythm, melody, and harmony within a *melo-harmonic* texture or as a macro-timbre texture. The substantial differences between discontinuity and continuity in terms of calculation, memory, perception, or imagination, prompted the need for a distinct compositional approach to allow the notation of its constant spatial-temporal evolution, this being fulfilled by chronographic recording of rhythm and sound.

Keywords: Continuum · Discontinuity · Macrotimbre · *d1-theory*

1 Discontinuous and Continuum

The discontinuous-continuous intersection is a seductive musical theme to be considered, because of the opening it offers towards new forms of musical creation, a subject that invites the development of new theoretical research or creative quest. The preeminence of the discontinuous in music can be seen from different angles, such as the perceptual division between the discontinuity of rhythm and the continuity of sound, or the discontinuity of the sound within a scale – as well as synchrony as a coordinating element of rhythm and sound. As we will observe in this article, the intersection between the discontinuous and the continuous deserves to be considered as a space with a favorable future in the field of musical creation:

© The Author(s), under exclusive license to Springer Nature Switzerland AG 2024
T. Noll et al. (Eds.): MCM 2024, LNCS 14639, pp. 15–30, 2024.
https://doi.org/10.1007/978-3-031-60638-0_2

- Let us imagine a universe of free circulation between the different physical states of matter, where the fluidity of the gaseous combines and even merges with that of the magmatic or solid, where the transition between materials could erase the sensation of boundaries between contiguous states. One way of interpreting this idea in terms of the ear, for example, would be to suggest that the molecules of the ethereal and dispersed timbre of the wind tend to concentrate in an extremely elastic nucleus, hardening progressively until it becomes as rigid as a stone. The same idea could be validated as a musical matter formed by highly diffuse movements in each of its various rhythmic and tonal physical components, gradually losing mobility in some of them until reaching a whole in a stationary state.

Although it is certainly possible to hum the above descriptions by ear, or even to project them intuitively into music, access to the theme and its concretization in a work will be more effective and precise by searching for possible intersections between the continuous and the discontinuous. If these two fields are complementary, the different forms of transition from one to the other allow us to observe the extent of their relations and forms of cohabitation. Such forms of cohabitation are innumerable – such as the transition from a fixed duration to its maximum acceleration – and are observed in the traditional mixtures of a continuous articulation in a discontinuous matter.[1] The evidence of their relationship appears at the basis of the frequency relations between rhythm and sound – as indicated by Henry D. Cowell (Cowell 1969). Each epoch and each musical imaginary have brought original proposals; today, the subject is supported by scientific tools that, from the process of musical research and creation, may allow the advancement of our answers. This work requires us to address a specific research-creative project in which the composition is based on methods to approach the continuum and theories to delve into the combinatorics of the discontinuous material of the scale.

The discontinuous and the continuum constitute two major territories of the musical universe, which are constantly manifested in the perceptive, the physical, and in the theoretical and methodological options that can be proposed for their respective forms of organization. Although, from a structural point of view, the division between the discontinuous and the continuum in no way opposes the passage from rhythm to sound through their respective ambituses of frequency, amplitude, or harmonic content. The differences in the handling of these two universes are important for understanding the characteristics adopted by musical matter in each case. If, by any chance, such a division leads to a radical split between the two universes, the specific character of each is useful in promoting an exchange of methods in the treatment of the other, so that it is possible to aspire to a more fluid model of the discontinuous universe as well as greater precision in the continuum.

The physical-acoustic and mathematical method of generating intervals of equal dimension in the scales is of great use, among other things, in guiding perception towards a more precise distinction between the discontinuous and the continuum in the musical matter. Although this method is not the subject of a study in this text, it is at least worth considering some of the mental aspects involved in the observation of intervals of

[1] Among all these examples, let's look at earlier musical times, where some of the rhythmic and tonal components were already flexibly combined with other components - *accelerando, ritardando, portando,* or *crescendo, decrescendo.*

greater or lesser magnitude in whatever ambitus the rhythmic or sonic components of the musical matter may be:

- If we take *roughly* four different scales whose respective resolutions would range from a minimum to a maximum of partitions, the auditory experience in front of these scales would attempt to bring the spectator into relation with the accessibility of this musical material in terms of mental calculation, perceptive impressions, memory, or imagination:

  1. Divisions into a reduced number of terms would only require small calculation operations, and perception, memory and imagination would not find obstacles to operate with their partitions.
  2. An average increase in division adding new terms would require an increase in calculation operations, while at the same time allowing us to observe a certain reduction in both perceptual discrimination and memorization, or free-action facilities for the imaginary.
  3. A much greater increase in division would then lead to greater difficulties in calculation operations, perceptual discrimination, memorization, and the freedom of the imagination.
  4. Finally, a considerable increase in division would imply that the discontinuity inherent in the scales would no longer be understood as such, since it would imply a large number of contiguous points, which would prompt the listener to perceive an ambitus of a continuous type. In the latter case, calculation operations could involve the handling of a very large set of terms, which would increase the complexity of dealing with such a domain. Therefore, the reduction of perceptual discrimination of microunits would remove obstacles to memory, while imagination might have the advantage of flexibility to exert free will.

The existing relationships between the four cases allow a broader approach to the problem of theoretical organization:

- -The extreme cases, 1 and 4, offer a similarly simple configuration: the few terms present in 1 could offer an operational simplicity close to the extreme condensation in 4, provided there is a sense of continuity in the latter. In the most elementary discontinuity, 1, information decreases as the freedom of mental resources increases, just as in the continuity, 4, when the sense of high or maximum resolution of information is reached. In a way, it's as if, in arriving at the impression of continuity, we rediscover in a new perspective the opportunities offered by the initial discontinuity. On a theoretical level, the task requires a minimum of organization of the material, freeing the necessary criteria for any structural manipulation.
- At the same time, the cases within the two limits, 2 and 3 – in the vast middle range of discontinuity – share the problem of the increase of information as well as the decrease of mental resources. Theoretical elaboration and criteria for treatment of structures become most demanding in 2 and 3.

As numerical information decreases and mental resources increase, the theoretical task in both cases would require a minimum of organization of the material, opening to freer criteria for the manipulation of structures. As the density of terms is resolved, the free exercise of constructive manipulation, listening, and imagining would be operationally facilitated. The pattern of simplicity observed in the extreme cases, 1 and 4, invites us to approach the problem of cases 2 and 3 from a perspective whose theoretical goal is to reduce information and calculation operations to elementary expressions, thereby facilitating the action of the mental factors. The search for a scientific methodology for organizing musical material in the case of scales is a twofold task: to make it possible to exploit the totality of information and to simplify the problems posed by the abundance of data.

2 Chromatic Continuum of Scales

Starting from the discontinuous-continuous interaction it will be useful in the following pages to approach first the theory of the combinatorial potential of intervals, *Theory dl* (ESTRADA 1994e), presently developed as a *chromatic continuum of scales.* The term "chromatic" suggests a continuous division of the *frequency duplication interval, fdi* – the "octave" in a heptatonic scale –,[2] and is used to represent a *chromatic continuum of scales* by gradually dividing the *fdi.* A scale "S" as that of whole tones will be a useful simple start to get into the subject. Let's consider a scale as S6, stated as a cyclic series of terms (0, 1, 2, 3, 4, 5) or as a set of adjacent intervals, in which the distance between the last term and the first of the scale in turn constitutes an adjacent interval [1 1 1 1 1 1], 1 being the smallest distance between two terms. The first of these two options operates with *pitch classes,* while the second with *interval classes,* a mode of representation that reduces interval combinatorics to a minimum. The total set of possible groupings of adjacent intervals in the S6 scale is synthesized into a minimum of 11 cases, produced by the successive, ordered partition of its dimension. Those cases are denominated *interval identities, Ii,* equivalent 'to identifiers'[3] that group intervals in ascending order and divide them into as many levels of density "N" as there are adjacent terms or intervals included in each range. [Fig. 1]. The identities contain an ordered set of intervals, from 1 to 12 levels, serving as a reference to generate all their permutations.

[2] In contrast to the pitch division employed by Julián Carrillo and Alois Hába, Augusto Novaro proposed the division of the *fdi* (Novaro, 1951). The term "scale" is used here in the broadest sense: it applies to pitch ranges as well as to any form of scalar organization of the various components of rhythm and sound. The division of the *fdi* into equal intervals is useful for generalization purposes; however, the method is useful for knowing and exploring the combinatorics of scales with unequal intervals, i.e., as in the shō, 17 non-'octaviated' pitches.

[3] The present notion of identity is different from that of 'identity' in finite group theory, a previous research. Estrada, Julio / Gil, Jorge, *Música y teoría de grupos finitos. 3 variables booleanas. Con un resumen en inglés,* Instituto de Investigaciones Estéticas, UNAM, México, 1984.

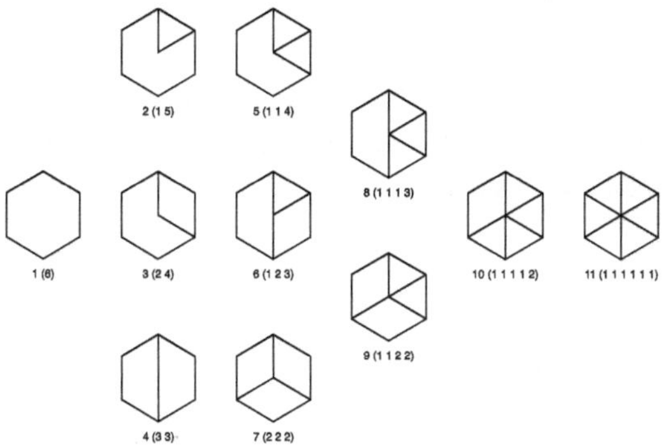

Fig. 1. REPRESENTATION IN A HEXAGON OF THE ORDERED PARTITIONS OF S6. *The total of ordered partitions of S6, represented by the different radii dividing each hexagon, constitute the 11 possible interval identities of that scale; they are ordered from left to right in 6 levels of density, N, from one interval [6] to the division between 6 adjacent intervals [1 1 1 1 1 1].*[4]

The *d1 theory* is based on minimal distance transitions, *d1*, which operate continuous-type changes within discontinuous-type materials. A transition to *d1* can be the result of several operations with interval identities, as shown by the following rapports – four of them are between the intervals of a given identity, and two more between identities of contiguous scales –:

- Permutation within the identity: permute the position between two intervals of the same interval identity: [1 2 3] d1 [2 1 3]
- Permutation between identities: permute a pair of intervals of one identity for another pair of intervals of an identity with the same density level: [1 2 2] d1 [1 1 3][5]

[4] The images of this article, produced by Christian Morales, and Adelina Amparán (A.A., Figure 5) belong to the book in preparation at UNAM, *Theory d1. The Chromatic Continuum of Scales.*

[5] This type of transition involves the merging and dissolution of two intervals whose sum is identical: [2 2] *d1* [1 3].

- Fusion: fuse two intervals into one when moving from a higher to a lower density level: [2 2 2] d1 [2 4]
- Division: divide an interval in two when moving from a lower to a higher density level: [2 2 2] d1 [1 1 2 2]
- Subtraction: eliminate a gap when moving to the next smaller scale range: S6 [1 1 2 2] d1 S5 [1 2 2]
- Addition: increase an interval when moving to the next larger range: S6 [1 1 2 2] d1 S7, N4 [1 1 2 3], or S7, N5 [1 1 1 2 2].

The notion of *combinatorial potential of the scale* designates the total set of interval identities; in these are generated all possible permutations of the intervals they contain, as well as all their relationships at *d1,* within the same level or between adjacent levels. [Fig. 2].

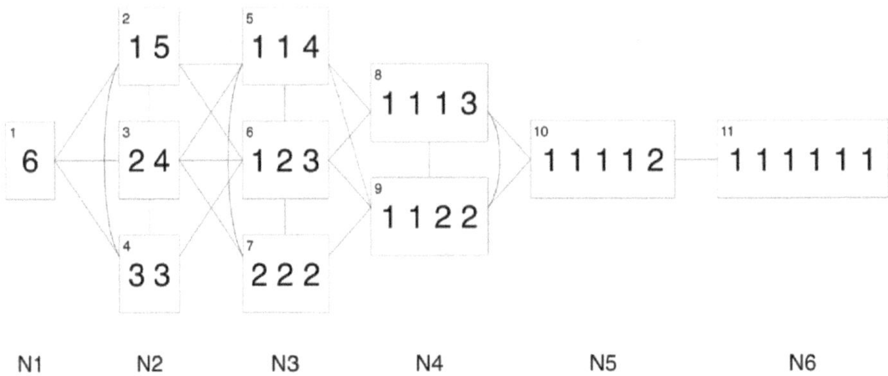

Fig. 2. COMBINATORIAL POTENTIAL NETWORK FOR S6. *In each small rectangle appears the content in intervals of each identity; the lines connecting two different identities indicate the transition from a given identity to another(s) found either within the same level N or between contiguous levels; such proximity is equivalent to a minimum distance, d1, an operation that occurs by permutation, fusion, division, subtraction or addition, as indicated by the 6 cases mentioned above.*

The identities consisting of a single interval [6] or of equal intervals [3 3], [2 2 2 2], [1 1 1 1] are invariant, while the identities consisting of different intervals allow to obtain their cyclic permutations in which *d1* is maintained from one operation to the next – a *d1* cycle –, as for example the identity 10 [1 1 1 1 2]:

[1 1 1 1 2] *d*1 [1 1 1 2 1] *d*1 [1 1 2 1 1] *d*1 [1 2 1 1 1] *d*1 [2 1 1 1 1] *d*1

As seen in Fig. 2, the projection of the combinatorial potential of S6 can be traversed continuously through transitions at *d1* between identities in two possible directions:

- Distribution of interval dimensions in identities: this offers a progressive vertical evolution in each column, from the smallest divisions – upper part – to divisions between a larger number of ample segments – lower part in columns N2, N3, N4.

- Distribution of interval identities: this offers a continuous progression on the horizontal by density levels, from the minimum of the division – N1, on the left of the diagram – to the maximum – N6, on the right.

If we take up the initial idea of relating the discontinuous to the continuum, the two previous distributions can be adopted as reference scales for creating continuum-like transitions in which, by means of curves, changes in content or changes in density in interval identities will be represented respectively. These two changes can be projected with two different orientations with respect to the combinatorial potential space; for example, an increase in energy – as the one produced by contiguous scale terms – can be represented, in the vertical, as a continuous trend towards smaller intervals, while, in the horizontal, towards an increase in density in the number of intervals. Figure 3 illustrates, with a single trajectory, these two types of continuous transition in S6: the vertical evolution will represent the amount of energy changes – from small to large intervals – in interval content, while the horizontal evolution will represent harmonic density, between one and six tones. [Fig. 3].

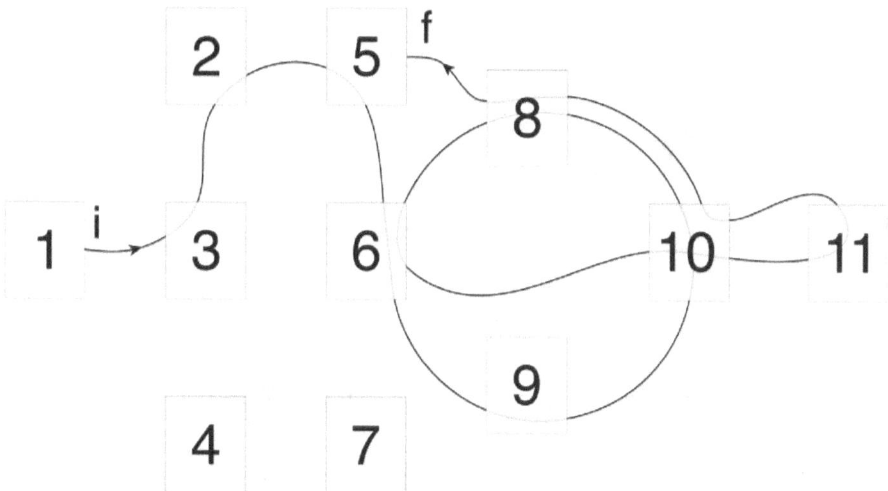

Fig. 3. S6'S COMBINATORY POTENTIAL PROJECTION, AND CONTINUOUS TRAJECTORY BETWEEN IDENTITIES. *The path of the combinatorial potential shows the continuity at the minimum distance of the transitions, d1, which at the auditory level implies a flexible mode of change along the whole trajectory; it starts from the initial point i, identity 1 [6], and reaches the final point f, identity 5 [1 1 4], with an evolution that includes 9 of the 11 identities, except for 4 [3 3] and 7 [2 2 2], covering all the levels, from N1 to N6.*

The range of possible resolutions of the chromatic continuum of scales would range from 2 to about 72 terms (1/12 of a tone), a scale dimension bordering on the continuum, a hypothetical perceptual limit to distinguish the difference between a scalar term and its two adjacent terms. The continuum of the scales is based on the series of prime numbers (2,... 71), their multiples and the combinations they produce; this allows to observe the numeric relationships between different scales and between the identities of several scales.

The S12 scale, which we will often refer to, is composed of 77 identities, a combinatorial reduction that offers a new access to traditional and contemporary harmony. For example, the basic interval identity in traditional harmony, Ii 18 [3 4 5], contains 3 main intervals ordered from smallest to largest: minor third [3], major third [4], fourth [5]. The permutation of its three different intervals – combinatoric type *abc*, 3! – gives rise to six basic cases, which ultimately correspond to the intervals contained in the minor and major chords in their respective fundamental, first and second inversion. The ordered permutations of these intervals – one interval is maintained while two intervals are permuted – create a *d1* cycle. This *d1* cycle will appear in 7 identities of S12 having 3 intervals different between them: S12 (1 2 9) (1 3 8) (1 4 7) (1 5 6) (2 3 7) (2 4 6) (3 4 5). The choice of three of those identities according to their intervallic content allows

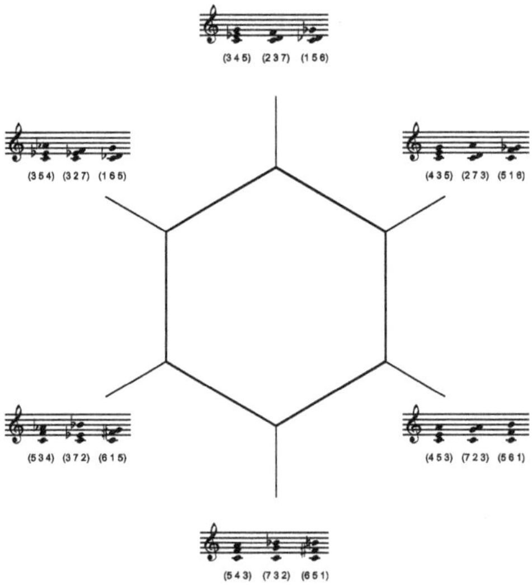

p. 186: exemple II 040515

Fig. 4. PERMUTATION CYCLE OF 3 IDENTITIES, II 18, II 14, II 12, TYPE *ABC*, LEVEL L3. *The d1 cycle creates a hexagon, which is denominated the* permutahedron *of the identity.*[6] *All the permutations are based on the note C.*

[6] The term *permutahedron* was taken from the idea of the *combinahedron,* discussed by mathematician David Romero, UNAM, co-author with J. L. Ramírez Alfonsín of a text on that topic (Ramírez Alfonsín/Romero 2002).

to observe their strong combinatoric relationship in despite of being or not consonant [Fig. 4]:

- 3 consonant intervals, Ii 18 (3 4 5)
- 1 dissonant interval, Ii 14 (2 3 7)
- 2 dissonant intervals, Ii 12 (1 5 6)

The previous sequence of permutations starting from the identity Ii 18 [3 4 5] can be read from the top right:

- minor fundamental
- major fundamental
- minor first inversion
- major second inversion
- minor second inversion
- major first inversion.[7]

The global control of interval combinatorics opens a new space for the exploration of the vast chromatic continuum of scales. An example illustrates the new possibilities opened by the interaction of pitch ranges whose divisions gradually progress from discontinuous to continuum:

- Consider the structural affinity which, in the case of scales of different dimensions, can share a set of 3 different intervals between them. The continuous transition of this interval distribution maintains a ratio close to the previous one at each new scale maintaining the minimal distance *d1* in each move. Starting with S6 in Ii [1 2 3] – first identity of the chromatic continuum of scales containing three different intervals – and ending with S24, we obtain a sequence of three different intervals each time, a sequence that sometimes allows two or more alternatives within the same scale, as for example starting from S8: [1 3 4] followed by S9: [1 3 5] or [2 3 4], both at *d1:*

S6 [1 2 3]
S7 [1 2 4]
S8 [1 3 4]
S9 [1 3 5] or [2 3 4]
S10 [1 3 6] or [2 3 5]
S11 [2 3 6] or [2 4 5]
S12 [2 4 6] or [3 4 5]
S13 [2 4 7] or [3 4 6]
S14 [3 4 7] or [3 5 6]
S15, [3 5 7] or [4 5 6]
S16 [3 5 8] or [4 5 7]
S17 [3 5 9], [4 5 8] or [4 6 7]
S18 [3 6 9], [4 5 9] or [5 6 7]
S19 [3 6 10], [4 6 9] or [5 6 8]
S20 [3 6 11], [3 7 10], [4 7 9] [5 6 9] or [5 7 8]
S21 [3 7 11], [4 7 10], [5 7 9] or [6 7 8]
S22 [3 7 12], [4 7 11], [5 7 10] or [6 7 9]
S23 [3 8 12], [4 8 11], [5 8 10] or [6 8 9]
S24 [3 8 13], [4 8 12], [5 8 11], [6 8 10] or [7 8 9].

Musically, the interest in the sequence of these various interval identities is to allow a continuous transition at *d1* from one scale to another, by changing only one pitch interval at a time. For example, the continuous transition from semitones to quartertones – from

[7] The traditional naming – major, minor, fundamental and their two inversions – could prevail for those results established by traditional harmony; however, at higher density levels such a choice would be complex. Hence, our approach is to maintain the notion of identity and to count the minimal distances *d1* between a permutation and its source.

S12 to S24 – could result in a sequence of chords, arpeggio melodies or other forms of presentation, like this one:

$$[3\,4\,5]d1[3\,4\,6]d1[3\,4\,7]d1[3\,5\,7]d1[4\,5\,7]d1[4\,5\,8]d1[4\,5\,9]d1[4\,6\,9]$$
$$d1[5\,6\,9]d1[5\,7\,9]d1[6\,7\,9]d1[6\,8\,9]d1[7\,8\,9]$$

Aurally, the previous sequence brings at each step a new opening that preserves the same combinatorial potential, a very similar proportion, and, to the ear, a sonority whose closeness to the initial one is perceivable enough.[8]

An appropriate methodology has yet to be defined for the treatment of continuum-type materials, as in case 4. From a mathematical and physic acoustical perspective, the continuum's constitution will always retain the same scalar structure, although maximum resolution indices lead us to perceive it as information that is impossible to discern in detail. At the perceptual level, the notion of *resolution* is associated with sensations of discontinuity or continuity in musical material – as the progressive succession of the number of scale terms within a broad spectrum, the kind of continuous transition from S2 to S72 –: in the presence of greater resolution, the tendency to perceive continuous material will increase. For example, excluding the individual attacks of each tone, the glissando sensation arises from scales above 64 terms.[9] In more general terms, such as auditory, visual, or tactile, resolution is equivalent to the "grain" or number of dots per field offered by the perceived image, on which will depend the sense of fidelity that can be obtained with what is reproduced.

3 Creation of a Discontinuous-Continuous Macrotimbre

The blend of components of rhythm (pulse, attack, microdurations) and sound (pitch, dynamics, color) is understood as a *macrotimbre* – an extension of the idea of timbre. The following pages will look at a concrete application arising from the transformation of a continuous macrotimbre into a *discontinuous-continuous macrotimbre.* When these components in a continuous macrotimbre are intended to be asynchronous – i.e., pitch is not coordinated by pulse – the audible result is an amalgam of disruptions, interferences or crossings that are perceived as an elastic musical matter. In the case of a mixture of discontinuous and continuous components, the result is a discontinuous-continuous macro-timbre, such as *yuunohui'tlapoa* (*Zapotec: yuunohui,* fresh clay, without stones; Nahuatl: *tlapoa,* cristals), for a keyboard – piano, organ or harpsichord – (Estrada 1999), in which melodies and harmonies seem to join and separate in a smooth and rough way, like a stone dragged by the flow of a current.

According to Henry D. Cowell around 1930 (Cowell 1969), the multiplication of a given duration, such as 1", will become a pitch, such as C. This makes it possible to observe a physical continuum linking rhythm and sound, a principle that allows the scales

[8] The software *Theory d1. MuSIIC-Win* (ESTRADA / DÍAZ / ADÁN, 1997–2003) allows to listen via MIDI this continuous transit between scales of different magnitudes. https://drive.google.com/file/d/1182BlxPWN9KHxXO2SmdwfZ4OianLLltW/view?usp=drive_link.

[9] For example, in terms of the d1 theory, this scale reaches a total of 1,741,630 intervallic identities, which already indicates a combinatorial richness beyond the limit of the controllable.

of pitch and duration intervals to be integrated into a unified field. This idea requires the consideration of the 3 main components of the waveform: frequency, amplitude, and harmonic content. In the case of rhythm and sound, these components are as follows[10]:

Frequency:	Rhythm: pulse	Sound: pitch
Amplitude:	Rhythm: attack	Sound: dynamics
Harmonic content:	Rhythm: microdurations[10]	Sound: color

The mixture of these six components does not exclude others amplifying the extension of the macro-timbre, such as the physical space of performance – height, depth, laterality – or other approaches such as noise, pressure, etc.

The elaboration of *yuunohui'tlapoa* starts from the five original trajectories utilized in the continuous macrotimbre of *yuunohui'yei,* for cello (1983), basic material for seven other versions for strings, winds, or voices of the *yuunohui* series. Each of these initial trajectories retains its independence from the others to create a non-synchronous relationship between them. In the specific case of *yuunohui'tlapoa,* its transformation retains the continuous intensity and pulse rate components, while three of these trajectories are converted into the resolution of the S12 scale – in terms of pitch identities, their density levels, and their intervallic content.

The creation of a macrotimbre that facilitates the combination of the discontinuous and the continuous requires considering the *reference scale* of each component as a *scale of resolution,* in which the number of terms or their combinatorics will be less important than if they were treated by calculating the evolution point by point. Since the dimensions of the reference scales are relatively small in the case of keyboard attack or color, the superposition of different resolutions specific to the different components of the discontinuous-continuous macrotimbre will partially diminish the typical asynchronous character maintained between the components in the *yuunohui* series. In turn, the predominance of transitions at *d1* in the case of pitch will offer a continuous becoming of microinstants with a rare spatio-temporal order.

The theoretical basis briefly outlined above is applicable to any scale that can be formed with discrete values in any component of the macro-timbre, provided that the number of terms or intervals does not exceed the capacities of perception, calculation, or memory in the creative process. In the specific case of *d1 theory,* the current application adopts the S12 scale, whose identities offer an attractive combination of content and density that can be ordered by *d1* transitions. This idea is one of the foundations of *yuunohui'tlapoa,* whose trajectories emerge from the chronographic record of *yuunohui'yei,* cello, whose continuous mode of transformation remains almost identical to the original piece, an aspect that allows the superimposition of any version in the collective versions of *ensemble'yuunohui* (1985–2019).

In *yuunohui'tlapoa,* the assumption of the combinatorial potential of the S12 scale tends to cover it by the fluidity of the trajectories and a connectivity based on *d1* transitions, within an attempt to operate outside the characteristic limits proposed by the systems, ancient or modern [Fig. 5].

---

[10] Microdurations become evident to the ear when a strong blow on the snare drum produces a sequence of small pulses, the product of the harmonic content of the rhythm, a kind of *rhythmic timbre,* a notion parallel to that of *sound color.*

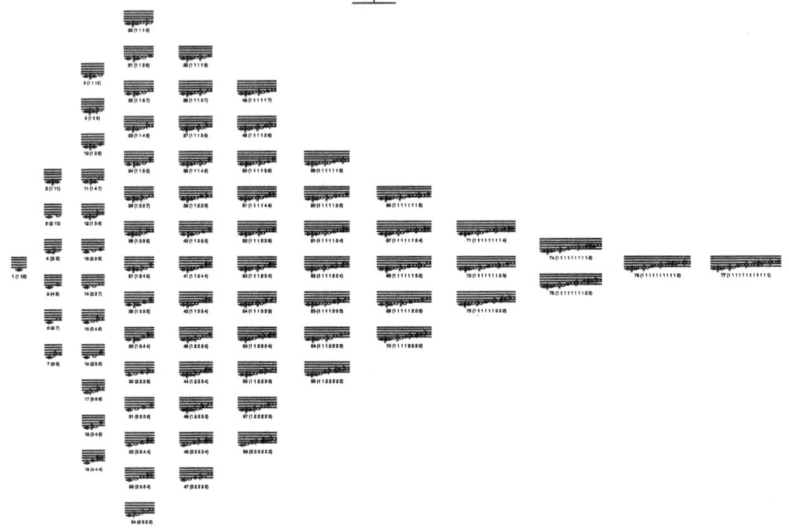

Fig. 5. S12, 77 INTERVAL IDENTITIES, INTERVALLIC CONTENT, 12 DENSITY LEVELS. *The diagram shows the pitch conversion of each identity in the frequency duplication interval, fdi.*

Despite the timbral registers of the organ and harpsichord, and their unfeasibility of vibrato, the work retains keyboard writing, so that its fixed color means that in this version only three of the five components used in the original version for cello are shared: pitch, intensity, and rhythmic pulsation. Thus, the five components of the first version are given the following assignments in *yuunohui'tlapoa*:

a) Pitch: monodic line of the lowest voice.
b) Dynamics: *crescendi-decrescendi,* only in the piano.
c) Color: here equivalent to the intervals contained by the identities in each vertical cluster. In the projection of the combinatorial potential, all changes in the intervallic content of the identities of the same density level tend to progressively reduce the smaller intervals – for example, N4, composed of 15 identities, goes from a small cluster of semitones [1 1 1 9] to a cumulation of minor thirds [3 3 3 3]. At the auditory level, the accumulation of small intervals resembles a maximum of energy or distortion, since the auditory sensation of intelligibility is reduced.
d) Rhythmic pulse: continuous rhythmic articulation – acceleration or deceleration of independent attacks that interfere with the evolution of the other components.
e) Rhythmic accent: intensity of the attack.

The addition of a new component became useful to substitute for the vibrato – understood by affinity, though not by equivalence, like vibrato in the case of strings–:

f) Rhythmic harmonic content: the micro-durations within a main duration appear here as a resource for modifying the synchronous, periodic, or stable order of the chords, which makes possible to obtain micro-differences of verticality in one or more voices; this aspect presents two alternatives:

- Evolution by continuous changes of duration: acceleration or deceleration and ascending or descending arpeggios.
- Discontinuous evolution: superposition of several divisors according to the Pythagorean method of harmonic division of the unit.

The transformation of the six components into their more precise equivalents in the discontinuous is not sufficient to create a discontinuous-continuous macrotimbre capable of articulating with sufficient fluidity the intervals contained by the identities. Hence the addition of two new independent components:

- Density in the number of intervals, which depends directly on the number of terms contained in each resolution scale. In S12, the heterogeneous distribution includes the following number of identities per level

N1	N2	N3	N4	N5	N6	N7	N8	N9	N10	N11	N12
1	6	12	15	13	11	7	5	3	2	1	1

- The order within the identity –which becomes a second level of harmonic variation–: in addition to the color of each identity itself, there is the permutation of the order of its intervals, which in turn requires a reference scale to measure the number of *d1* separating one permutation from another. These minimum distances can be observed through the connective relations produced by the permutation of the intervals contained in each identity. The total set of permutations of each identity generates a lattice structure in the form of a virtual polyhedron, the permutahedron [Fig. 6].

The combinatorial potential of S12 is the space in which seven components of the macrotimbre converge, here characteristically differentiated:

- Discontinuous:
  - Pitch, the lowest pitch of each vertical aggregate
  - Color, the intervals contained by the identities
  - Density, the number of intervals comprising the chord
  - Order of the permutations in the permutahedron
- Continuous:
  - Pulse, acceleration or deceleration
  - Rhythmic attack, acceleration or deceleration
  - Dynamics, crescendo or decrescendo.

As far as the discontinuous is concerned, the transitions within *d1* facilitate the perception of each vertical change, which in turn facilitates the perception of the confluence of the components as a "melo-harmonic" synthesis of the macro-timbre. The melodic and the harmonic are always compressed within a mobile register that does not exceed

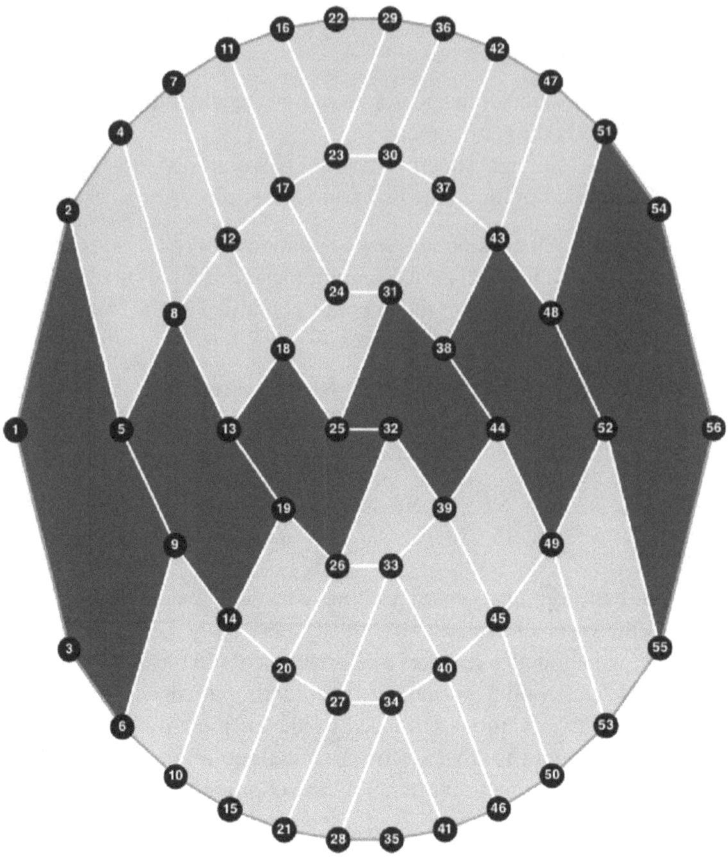

Fig. 6. PERMUTAHEDRON 27 OF S12: 30 SQUARES AND 6 HEXAGONS, 56 PERMUTATIONS, G12, N8, IDENTITY 67 [1 1 1 1 1 1 2 4], AAAAAABC. *Each permutation is enciphered: the identity, 1, is centered on the left; the lines between the operations indicate d1 links.* IMAGE: A. A., CH. M.[11]

the 12 semitones of the *fdi*, moving across the keyboard according to the changes in the bass pitch of each chord.

A fragment of the score of *yuunohui'tlapoa*'s discontinuous-continuous macrotimbre, Section S2–3, allows to observe the result. [Fig. 7].

[11] For example, the shortest path between the identity and its opposite – permutation 56, centered on the right – requires a minimum of 13 contiguous steps to get there, *d13* – traversing the peripheries at the top or bottom.

Fig. 7. YUUNOHUI'TLAPOA, KEYBOARD. *Continuous-discontinuous macrotimbre made up of six components: extract from score, Sect. 2, page 3, 10 duration units (equivalent to micro-measures).* Manuscript score, JULIUSEDIMUS

The preceding score extract shows six different modes of continuous articulation of the discontinuous macrotimbre:

- *Crescendi-decrescendi*
- Numerous changes at *d1* in the evolution of vertical intervals: duration units 1 to 3 and 7 to 9
- Cluster glissandi with modified ambitus – duration units 4 and 5, lower section –
- Arpeggio in whole-tone clusters – duration unit 7, upper part, groups of white notes –
- Acceleration and deceleration of the rate of change of pitch – signs of speed between each pair of staves and non-periodic distribution of the moments of change of speed –
- Interference of duration pulsation on pitch change rhythms – duration units 7 to 10, vertical signs in the additional staff above.

The topic of discontinuous-continuous fusion discussed so far is far from closed, and the case of *yuunohui'tlapoa* is only one of many. The attempt to make fixed-pitch instruments – such as keyboards – compatible with flexible instruments – such as strings, some wind instruments, or the human voice – has as its underlying goal the search for a way to approximate different conceptions or, given the specific characteristics of the discontinuous and the continuous, opposite methods. One of the main problems seems to be the difficulty of producing intervals as unusual as those of the chromatic continuum of scales, although today it can be overcome thanks to the resources of new technologies focused on the design of new instruments. The wide spectrum of the chromatic continuum of scales is still to be explored in many directions, from the use of non-tempered scales to the promising combination of several superimposed scales, as has been done for centuries in the field of rhythm. Like a river of crystals, the various dimensions of the vast universe of chromatic scales beg to be conquered by new research, opening further gaps in the creative process.

Computer Sources. Estrada, Julio / Díaz, Max / Adán, Víctor, *Theory d1. MuSIIC-Win,* Escuela Nacional de Música, UNAM, México, 2006, programa de instalación y manual trilingüe (Español, Francés, Inglés), 2006. https://drive.google.com/file/d/1l82BlxPWN9KHxXO2SmdwfZ4OianLL ltW/view?usp=drive_link.

Scores and Audio Sources. Estrada, Julio, *yuunohui'tlapoa,* keyboard (piano, harpsichord, organ), *Juliusedimius,* Mexico, 1999.

Estrada, Julio, *ensemble'yuunohui,* instructions for all *yuunohui* ensemble versions, *Juliusedimius,* Mexico, 1998.

Estrada, Julio, https://soundcloud.com/user-335833664.

Estrada, Julio, *yuunohui'tlapoa,* piano version, Christopher W. Jones, MP3 format.

References

Cowell, H.D.: New Musical Resources, with a preface and notes by Joscelyn Godwin, Something Else Press, Inc., EE UU, 1964

Estrada, J.: Théorie de la composition: discontinuum – continuum, tesis doctoral, Université de Strasbourg, France (1994)

Estrada, J.: Theory d1. The Chromatic Continuum of Scales, book in preparation, Institut of Research on Aesthetics, UNAM, México

Estrada, J., Adán, V.: Transformación continua de la forma de onda por medio de la permutación de sus intervalos de tiempo", ponencia, International Society of Musical Acoustics, Escuela Nacional de Música, UNAM, México, 2002

Estrada, J., Gil, J.: Música y teoría de grupos finitos. 3 variables booleanas. Con un resumen en inglés, Instituto de Investigaciones Estéticas, UNAM, México, 1984

Novaro, Augusto, Teoría natural de la música, México, 1951

Ramírez Alfonsín, J.L., Romero, D.: Embeddability of the combinahedron. Discret. Math. **254**, 473–483 (2002). Elsevier Science B.V. 5

Rhythm Analysis and Rhythm Generation

On Brazilian Drum Claves and Generating Rhythm Patterns Out of Them

Quan Duong[1(✉)], Kjell Lemström[1], and Olli Reijonen[2]

[1] Department of Computer Science, University of Helsinki, Helsinki, Finland
quan.duong@helsinki.fi
[2] Helsinki, Finland

Abstract. Rhythmic patterns, or claves, are fundamental to many Latin American and Caribbean music styles. Clave rhythms are used as drum patterns to provide rhythmic frameworks and are often used as the basis for improvisation. This paper introduces the theory of Brazilian claves and how they can be understood in the Western music framework based on measures. This is the first systematic ethnomusicological model ever presented for the Brazilian clave system. Moreover, we present a simple but effective genetic algorithm that generates random Brazilian-style rhythms based on the clave theory introduced.

Keywords: Brazilian Drum · Rhythm generation · Claves

1 Introduction

The use of the clave in Brazilian music has a long and varied history. Clave originates in Sub-Saharan Africa, from where it was brought to Cuba and Brazil by enslaved people as a part of their musical heritage [7]. Thus, clave has been a feature of Brazilian music for centuries and is used in many different contexts. Clave has become an essential part of Brazilian music culture and is a focal element of the Brazilian sound.

In this paper, we first introduce the theory behind Brazilian claves and a systematic ethnomusicological framework to build and understand the rhythmic patterns in Brazilian-style music. To our knowledge, this is the very first model to this end and is based on the study of [9]. When generating a Brazilian rhythm, the clave gives the baritone line from which the tenor and bass lines are generated by following possible rhythmical developments. Having described the theory behind claves, we report our genetic algorithm-based system capable of generating Brazilian-style rhythms based on the defined framework. In the algorithm, we represent the clave as a binary mark that tells when the beats will be played on a percussion instrument. The marks are then exposed to the genetic algorithm operators to create random but Brazilian-style-sounding rhythmic patterns.

O. Reijonen—Independent Researcher.

© The Author(s), under exclusive license to Springer Nature Switzerland AG 2024
T. Noll et al. (Eds.): MCM 2024, LNCS 14639, pp. 33–44, 2024.
https://doi.org/10.1007/978-3-031-60638-0_3

2 Related Work

More recently, deep learning approaches have been used widely for generation tasks, including music. Huang et al. [5] proposed MusicTransformer, an autoregressive generative model using a self-attention mechanism [12] for symbolic music generation. Their results show significant improvement in generated samples' sequence length and quality consistency. Another direction in music generation is using Generative Adversarial Networks (GANs), exemplified by Dong et al.'s work [2]. Their MuseGAN model can create polyphonic music incorporating harmonic, rhythmic, and multi-track dependencies along with temporal structure. However, the requirement for using a deep learning approach like GAN or Transformer is to have large datasets available for training.

Narrowing down the problem from music in general to rhythm, Gibson and Byrne [3] combine neural networks and genetic algorithms to train a rhythm generator. In their study, the training data is first labelled as "good" or "bad" by humans. Then, the neural model is trained to classify the rhythms generated by the genetic algorithm as "good" or "bad". This classification model serves as the judge of the fitness function. Hence, it helps the evolution process to converge. A similar approach was used in another study (see [4]), where an interactive interface allows the user to rate the output of the genetic algorithm as the fitness function. With this approach, we can have a good outcome without necessarily understanding the features of rhythm and considering the explainability of the model. However, at the time of writing this paper, the modelling of Brazilian rhythms has not been seen in so many studies. One of the mathematical models for Afro-Brazil-Cuban rhythms was analysed with six 4×4 fundamental claves (Shiko, Son, Soukous, Rumba, Bossa-Nova, and Gahu) [11]. In that study, various techniques were applied, from geometric representation to complexity analysis, to find relationships between rhythms. The study pointed out that the new rhythms can be generated from the combination and permutation of the root rhythms.

3 Clave

Unlike the humanities in general, this study of music is based on quantitative sciences from the point of view of scientific hypothesis and theory. A hypothesis is an assumption based on which different possibilities are being considered. A theory is a rule that can be used to study an object. It can also be used to make predictions about the subject's behaviour. Brazilian percussion music is mainly based on three factors: harmony, divisive structure and clave.

Harmony in Brazilian music, especially batucada, has been measured and found to exist [9]. The basis of harmony has remained unexplored so far. The instrument physics shows the foundation. According to the Brazilian view, the tuning of the drums is based on the tuning levels called by the names frouxo, medio, aperto, muito aperto: loose, medium, tight and very tight. The basic structure of the drum is a tube. When a suitable stimulus is given, standing wave motion, i.e. sound is formed inside the tube. It reflects from the closed end of the tube just as well as from the open end. At the open end, there is an air mass

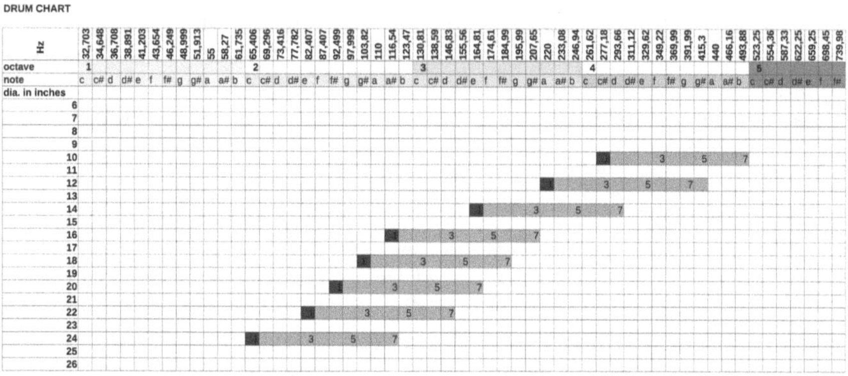

Fig. 1. Drum chart: tuning table for melodic tuning of drums.

opposite. Such pipes are also all wind instruments and organs. Their physics is the same as a drum. The lowest frequency formed in a tube closed at one or both ends is 1/4 of the fundamental frequency resonance, the fourth (4th) overtone. It is two octaves higher than the fundamental frequency. It's a drum keynote. In addition to that, only odd top notes 5, 7, 9, etc. [8] can be generated in such a tube. In practice, this means the prime and the odd notes of the natural tone harmony thirds, fifths or sevenths, etc. This requires that the drum is tuned high enough so that the air inside the drum begins to resonate. The drum is tuned by adjusting the drum membrane or both membranes together to the resonance of the harmonic frequency produced by the tube. In this case, the membranes of the drum and the air inside the tube vibrate harmoniously together and mutually reinforce each other. The tension of the membranes determines which harmonic frequency they wake up. The result is a clear and strong sound with a measurable stable pitch. The drum produces the note.

Since the frequencies produced by the drum's air space depend directly on the diameter of the drum's membrane, the tuning chart for the drums is simple. The tunings of batucadas surdo drums are shown in the Fig. 1. The drums' basic pitches and harmonic frequencies have been measured from real drums.

The lowest pitch of each diameter is marked in red. The measurement accuracy is about half note because the drum bodies' diameters and the membranes' free-sounding areas are not precisely according to the figures indicated. The Brazilian tuning levels from loose very tense are expressed in musical terms: first, third, fifth and seventh. Viewed in this way, the harmony of the drum group is based on choosing drums of appropriate size and tuning them into a musical set, which is based on natural tone harmony.

Divisiveness is a clear fact in the European popular and classical tradition in general. It is also proven to be the basis of the structure and rhythms of the Afro-Brazilian batucada [9]. Brazilian conductor Oscar Pereira de Souza explained the structural levels of music and how they came about by the term cortar (to cut) [9]. Faster rhythms are created by cutting longer tones of slower rhythms into shorter parts. Cutting can be continued until the last layer is a continuous shortest possible stroke. His point of view and the event he describes is purely divisive. Divisiveness can be seen as a theory that can be used to analyse and make predictions on European and Afro-Brazilian music.

The African core of Brazilian percussion music is based on three different rhythmic categories: Marcar, cortar and ripicar [9]. The marcar is evenly spaced, with four 1/16 stroke repeated series. The cortar rhythm category is based on the three-part tresillo clave [10]. Repicar, on the other hand, means divisive rhythms based on Brazilian 4/4 long claves. Tresillo clave is the basis of all samba claves except one. Of all possible tresillo-based theoretical claves constructed, only a few are commonly used. The clave is a rhythmic and synchronic framework for rhythms cut from it (diminished). First, the clave determines the base from which the cutting arises. Secondly, it defines cutted rhythms that have mutual synchrony. Because denser rhythms are created by cutting the long sounds of the base rhythm into shorter parts, the clave is always present and recognisable in the rhythm created by cutting. If the mutual synchrony of the cut rhythms is not in accordance with the clave, i.e. some rhythm shifts and the synchrony changes, it is a mistake called ritmo cruzado, crossed rhythm.

Let us denote by capital "X", small "x", and by "|" the beat, cut beat and bar separator, respectively. A commonly known clave is as follows:

$$X\ o\ o\ \mid X\ o\ o\ \mid X\ o\ o\ o\ \mid X\ o\ \mid X\ o\ o\ o \tag{1}$$

Now, to produce a diminished rhythm, the long beats of clave (1) are cut into smaller units, for example, as in clave (2):

$$X\ x\ o\ \mid X\ o\ x\ \mid X\ o\ x\ o\ \mid X\ o\ \mid X\ o\ o\ x \tag{2}$$

In Afro-Brazilian terms, the result is a rhythmic pattern known as p'e-chato (lame leg). Clave is the base rhythm of p'e-chato, which is recognisable from its rhythm of p'e-chato. According to European musical practice and understanding, the first beat of the bar is the strongest and supports the music until the end of the bar. This is a European concept. The Afro-Brazilian perception is different. According to it, the first beat of the clave does not need to be emphasised on Brazilian drum claves and generate rhythm patterns out of them. That is why the first beat is played as upbeat, prematurely. This is standard practice in clave-based music. When this practice is taken into account, the p'e-chato rhythm changes slightly but is then perfect in clave (3):

$$o\ x\ o\ \mid X\ o\ x\ \mid X\ o\ x\ o\ \mid X\ o\ \mid X\ o\ o\ x \tag{3}$$

The emphasis on the rhythm pattern changes according to Afro-Brazilian musical practice. All claves work according to these principles. They are base rhythms, from which a denser rhythm of the next level is produced by cutting, and the beginning of the clave is not emphasised, but the first beat of the cut rhythm is played as upbeat.

Let us now consider the formation of a simple rhythmic composition based on the clave presented above. The following example uses a slightly different but commonly known cut rhythm that is close relative to p'e-chato. The bass line is composed by subtracting beats from the cut rhythm. In the example below, the rhythmic composition has three levels, tenor (4), baritone (5) and bass (6), that are played by tamborim, caixa/atabaque and 3rd surdo, respectively.

$$o\ x\ o\ \mid X\ x\ o\ \mid X\ o\ x\ o\ \mid X\ o\ \mid X\ o\ o\ x \tag{4}$$
$$X\ o\ o\ \mid X\ o\ o\ \mid X\ o\ o\ o\ \mid X\ o\ \mid X\ o\ o\ o \tag{5}$$
$$o\ o\ o\ \mid o\ x\ o\ \mid X\ o\ o\ o\ \mid o\ o\ \mid X\ o\ o\ x \tag{6}$$

At this point, it's just three separate rhythmic lines. The lines must be in musical harmony with each other to form a musical composition. Now, it's time to choose the drums of the right size that can be tuned harmoniously. Harmony combines sounds into a musical whole.

4 Generating Method

4.1 Genetic Algorithm

Genetic algorithm (GA) is a metaheuristic algorithm commonly used in optimization and search problems. The process of natural selection of evolution inspires the idea behind genetics. In the process, genetic algorithms rely on biological operators such as selection, crossover and mutation [6]. In this section, we explain the general concept of genetic algorithm and how it could be adapted to the problem of generating Brazilian-style rhythms out of given claves.

The process begins with an initial population of candidate solutions, usually represented as binary digit (gene) strings. Each candidate solution in the population is evaluated for its fitness, which measures how well it solves a specified optimization problem. The fitter a candidate solution, the more likely it is to be selected for reproduction [6].

The selection process involves choosing a subset of the candidate solutions from the population to be parents for the next generation; the better the parents' fitness, the better the chance of their offspring to survive. This process keeps on iterating, and a fit generation of individuals has been found in the end. After parent selection, they're mated to create offspring for the next generation. A crossover operator, swapping parts of parents' genes, is commonly used to form a new candidate solution. The generated offspring might undergo a mutation that introduces random changes to maintain genetic diversity and avoid premature convergence to sub-optimal solutions.

Algorithm 1 sketches the evolution process. Five phases are considered in a genetic algorithm: **population initialization**, **fitness function calculation**, **parent selection**, **cross over** and **mutating**. Here, the population initialization is simplified by a random process. This notion can be applied to searching for the best candidate for generating rhythm from a given clave, as shown next.

Algorithm 1. Genetic Algorithm

```
procedure RUN_EVOLUTION(n_gen, n_population)
    generation = 1
    Initial random population
    while generation < n_gen do
        compute_fitness(population) -> scores
        select_parent(scores) -> parent1, parent2
        add parent to new population: new_population = [p1, p2]
        while new_population < n_population do
            make_crossover(parent1, parent2) -> child1, child2
            make_mutation(child1) -> child1
            make_mutation(child2) -> child2
            extend_population(child1, child2)-> new_population
        end while
        Set new population for next generation:
        population = new_population
        generation ++
    end while
end procedure
```

4.2 Experiment

Let us apply a genetic algorithm to generate rhythm using the clave framework. The chosen clave can be represented straightforwardly by a binary sequence with 1 for a beat that is played (*on-pulse*) and 0 for a beat that is not played (*off-pulse*). For now, we consider only 4×4 claves. For example, the format of the Bossa nova clave could be represented as a string: [1 0 0 1 0 0 1 0 0 0 1 0 0 1 0 0].

The chosen clave gives the baritone line the eventual Brazilian-style rhythm. For the complete rhythm, the tenor and bass lines will be generated using a genetic algorithm simulating the idea presented in Sect. 3 to have the complete rhythm. These generations differ only in their fitness functions and inputs: while the tenor line uses clave (baritone line) as its input for fitness function, the bass line is generated from the tenor line instead. Thus, the tenor line must be generated before the bass line.

We implemented the genetic algorithm by adapting it to this specific rhythm generation issue. In our setup, the number of generations was set to 100, the population number to 50, and the parent pool of the best candidate to 4. Furthermore, we implemented an acceptance threshold as an early stop before the end of the evolution process. This way, we could stop the evolution loop whenever the best candidate had a fitness score smaller than this threshold. The threshold was set to 10 points for both tenor and bass generators.

Tenor Line Generation: To guide the algorithm to generate a valid tenor line from the chosen clave, we need a set of fundamental principles. As described above, the genetic algorithm must calculate each generation's fitness score. To this end, we created violation criteria with penalty points. The more severe the violation, the more penalty points should be given. Below, we list the violations and the penalty points.

- 1st beat is an on-pulse: 1000 pts
- 2nd beat is an off-pulse: 1000 pts
- The last beat is off-pulse: 1000 pts
- No common on-pulses with clave except the 1st beat: 1000 pts
- Three consecutive on-pulses: 100 pts
- On-pulses less than 40%: 10 pts
- On-pulses more than 60%: 10 pts

In our penalty system, the severity of a violation expresses itself in the order of magnitude of the penalty points. We decide the points and their scales based on the feature's importance of clave rhythms. Though the range from 10 to 1000, subjectively, is clear enough to emphasize the essential features. The first four criteria are critical and thus given 1000 penalty points (pts). Violating these rules would break the Brazilian-style structure of the rhythm. On the other hand, the violation "Three continuous on-pulses" is not severe, but such density would not typically appear in Brazilian rhythms. The two last criteria give the usual density of tenor lines of Brazilian rhythms.

Bass Line Generation: A similar process is also applied to generating a bass line. While the tenor line compares with the clave for fitness function input, the bass line depends on the tenor line for this step. Inherit from the tenor, the baseline has its own criteria for calculating fitness scores. After generating the tenor line as a candidate, we use it as input to create the bass line with the penalty criteria listed below.

- On-pulses on 1st and 3rd measure: 1000 pts
- No on-pulse on 2nd and 4th measure: 1000 pts
- Two continuous on-pulses: 100 pts
- No common on-pulses with the tenor line: 10 pts
- On-pulse more than 40%: 10 pts

The strict rule for the bass line does not allow on-pulses in the 1st and 3rd measures. On the other hand, there should be on-pulses in the 2nd and 4th measures of the Western music system. An essential rule of the bass line favours alignments of the bass and tenor lines. However, this is not compulsory, as in the associated penalty score 100. Similarly to the tenor line, we do not prefer two consecutive bass on-pulses in the bass line. Lastly, the rhythm would be too dense if the proportion of on-pulses exceeds 40% of the total beats. Violating any of these latter rules would generate a punishment of 10 points in the fitness function calculation.

Parent Selection: Parent selection is made through a ranking method. All the individual fitness scores are sorted in ascending order. Instead of selecting the two best candidates as the parents, we chose top-K candidates and then used a random selection to get a pair of parents. With this method, the candidate can have a chance to mate with more variants.

Crossover: In this operation, we use a single-point crossover to combine the genes of the selected parent. We choose a random cutting point from 16 beats from a 4×4 clave. Thus, any two parents produce two children with novelty in their genes by swapping their beat strings.

Mutation: The mutation is an essential operation for genetic algorithms. Mutating genes randomly creates a new offspring variant that may boost the individual's fitness. This process helps the optimizer to avoid stacking at a local minimum. In our experiment, we set the mutation probability to 0.1. We observed this setting to give a fast convergence, still avoiding the local minima.

5 Results

We ran multiple experiments on different 4×4 claves. In our experiments, we observed quick convergences in all the tenor and bass line generations cases: The best candidates meet the acceptance threshold already before reaching the end of the evolution process (100 generations). As an example of a generated Brazilian rhythm, the below rhythm is generated starting with the Bossa nova clave, which is the one appearing in the baritone line (8).

$$o\ x\ o\ X \mid o\ o\ X\ x \mid o\ x\ X\ o \mid X\ x\ o\ x \tag{7}$$

$$X\ o\ o\ X \mid o\ o\ X\ o \mid o\ o\ X\ o \mid X\ o\ o\ o \tag{8}$$

$$o\ o\ o\ o \mid o\ o\ X\ o \mid o\ o\ o\ o \mid X\ o\ x\ o \tag{9}$$

From the presented clave theory perspective, the tenor line (7) meets all the fundamental rules of Afro-Brazilian music. The penalty from the fitness function for this tenor is 0. The bass line (9) also satisfies the strict principle we defined. However, the fitness score of this bass line is **10** because it violates the "generated beats aligned with the tenor line" criteria (beat three on measure 4). This, however, is an interesting point of the algorithm. As the algorithm tries to optimize the preference, the priority is set for more critical aspects and is more relaxed for minor issues, as in this example.

Nevertheless, the soft constraint depends on how the penalties, acceptance threshold and evolution stopping are set. The example observation can be archived in our experiment with the acceptance threshold 0 and the max evolution number 100. Given different scenarios, we can control the outcome by changing those parameters. Hence, sometimes it is good to allow violating the non-strict rules for having creativity in the artefact.

In addition to the auditing result by marking format as shown in 7, 8, and 9, we also converted them into the midi format. Therefore, we can also assess the rhythms generated by listening. We use Music21 library [1] for sound synthesizer with unpitched percussion instruments like Bass Drums, Bongo Drums and Woodblock for bass, baritone and tenor, respectively. The randomly selected samples can be listened to via the following anonymous link[1].

[1] https://bit.ly/3o9v0Cp.

6 Evaluation

Music is a non-semantic form of communication compared to spoken language, which is very challenging for evaluation. An objective evaluation reflects only one objective of the output, not presenting an overall assessment, for example, in terms of aesthetics, emotion and creativity. In music generation, which uses a supervised learning approach, objective evaluation is usually done by calculating the loss score. Please note that the score is also used for model optimisation. As in research from Huang et al., [5], a negative log-likelihood (NLL) score was used for benchmarking the result. However, this evaluation also creates a risk because of the originality of the output. A study pointed out that music generation algorithms can overfit, producing copies of excerpts of input pieces [13]. This issue also exists in a more extensive data set, which raises concern about the reported success of music generation-based deep learning.

Accompanying some objective evaluation methods is often subjective assessment. The most popular approach is a listening survey in which the audience will give a rating. There are different scenarios for rating; for example, the audiences are either asked to provide scores for a music piece (Likert scale) or select the better one between the two (pairwise comparisons) [5]. Even with a subjective assessment, there are diverging results due to variations in domain knowledge, expertise, and human perception bias.

This section will discuss the effectiveness of Genetic Algorithms compared to the Random approach. As the rhythms are generated without training samples, we cannot conduct a benchmark based on loss against ground truth. However, the genetic algorithm is an optimisation process that indicates the performance based on fitness score. We conduct a simple assessment by measuring an average fitness score for tenor and bass lines. This average fitness score is applied to the Genetic Algorithm and the Random method. We design a random generator based on the templates, which respects the introduced clave theory in Sect. 3. Eqs. 10 and 12 illustrate the template for the tenor and bass line corresponding to the baritone line (given clave) of Eq. 11.

$$o\ x\ \_\ X \mid \_\ \_\ X\ \_ \mid \_\ \_\ X\ \_ \mid X\ \_\ \_\ x \tag{10}$$

$$X\ o\ o\ X \mid o\ o\ X\ o \mid o\ o\ X\ o \mid X\ o\ o\ o \tag{11}$$

$$o\ o\ o\ o \mid \_\ \_\ \_\ \_ \mid o\ o\ o\ o \mid \_\ \_\ \_\ \_ \tag{12}$$

From the figures above, the underscore "_" symbol indicates a possible position where one could either have a play note "x" or a silent note "o". These templates are designed for random methods to fill in the missing notes. We set the random threshold for each blank position as 0.5, and the number of samples is 50. This number is consistent with the population number in genetic algorithms (Table 1).

We conducted 100 experiments each for the genetic algorithm and a random method, collecting fitness scores upon completion. Utilising a new clave for each experiment and generating populations of 50 candidates, the genetic algorithms

Table 1. Average of fitness penalty scores, lower is better.

	Genetic algorithm	Random method
Tenor line	0.0	4.2
Bass line	0.2	182

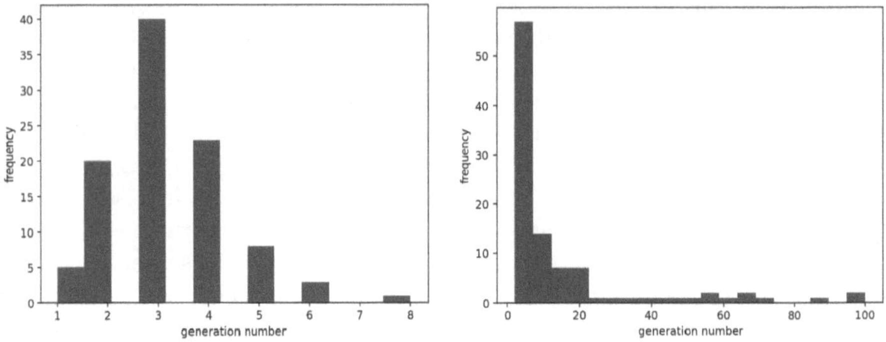

Fig. 2. Distribution generations. On the left, the genetic algorithm rapidly optimised the tenor line in fewer than ten generations, while the bass line (to the right) mostly converged before reaching 20 generations.

consistently produced superior fitness scores for both tenor and bass lines. The tenor line exhibited no criteria violations, with a minimal penalty score of 0.2 in the random method, accounting for 2 per cent of observations. Although non-perfect samples were allowed, the random method surprisingly resulted in relatively low penalties for the tenor generator. Conversely, the bass line generator showed a higher penalty score due to a higher likelihood of violating fitness criteria, particularly concentrated around two closely played notes, which might be tolerable for tenor but not ideal for bass.

Next, we evaluate the convergence of the evolution process from the genetic algorithm. Figure 2 demonstrates the tenor's generation number distribution before reaching the acceptable fitness score. As we set the score for acceptance as 0, all the tenor samples have satisfied the criteria. The number of generations converged quickly after less than ten generations. In the case of the bass line (see Fig. 2), it took longer to have the fitness score meet the acceptance score of 0. In particular, two observations have not converged to the minimum point after 100 generations, though most of the evolution process is done after 20 generations. After analysing those two observations, we discovered that the infraction is due to non-aligned play notes between the bass line and the tenor line with a penalty score of 10. This issue, as mentioned in Sect. 5, is acceptable and does not break the clave theory.

Nonetheless, we were unable to conduct a subjective assessment. A survey without knowing the personal background of users is not reliable. As in music judgment, the differences between experts and ordinary people could produce different results. Also, a small number of rating samples might not reflect the actual distribution accurately. On the development path for further research, we will design a subjective evaluation method suitable for non-trivial music.

7 Discussion

This research delves into a comprehensive examination of Afro-Brazilian clave theory to provide a generalized understanding of its properties. The genre exhibits unique characteristics that necessitate using distinct analysis methods compared to Western music systems. To lay the groundwork for future exploration, we present an overview of the fundamental concepts underlying this genre. Additionally, our study introduces a user-friendly tool for modelling Afro-Brazilian music, wherein we conducted experiments with random and rule-based methods to generate rhythm based on a given clave. However, these methods exhibit limitations in terms of flexibility and control over the resulting output.

While advanced machine learning techniques have shown promise in music generation, they heavily rely on extensive and diverse datasets—a resource not readily available for clave music. Moreover, utilizing complex algorithms like deep learning may prove excessive for the relatively simple structure of the clave. As an alternative, we propose the utilization of genetic algorithms, a heuristic optimization process that holds the potential to yield satisfactory Afro-Brazilian rhythm music.

In the initial stage of our study, we deliberately limited our experimentation to a 4×4 clave pattern. It is important to note that the purpose of this model was not to evaluate the aesthetic quality of the generated rhythm but rather to assess the feasibility of developing a model tailored to this genre of music. By establishing this foundation, it becomes possible to generate more intricate and complex rhythmic patterns that can be effectively incorporated into advanced music generation techniques.

8 Conclusions

In this study, we introduced the Brazilian Claves system and its fundamental characteristics, highlighting its differences from Western music. Using a simple algorithm, we demonstrated the ability to generate rhythm from a given clave. This approach offers a flexible framework for controlling the outcome through an approximation algorithm rather than strictly complex coding rules. The generated artefact's variation is achieved through an optimization process rather than a random method, resulting in exciting results that adhere to established criteria.

In future, we will give a more profound description of the clave framework and its analytical capabilities. We are also keen on developing more sophisticated rhythm generative algorithms based on the framework. As the primary objective of this study does not involve benchmarking generative algorithms, our evaluation focuses solely on comparing the efficiency of a basic heuristic genetic model against a random approach. To conduct a reasonable evaluation, we plan to design a comprehensive survey system that includes user background and parameter-based assessment. Additionally, other objective evaluation aspects are also considered for further research. Moreover, it would be interesting to study how, for instance, Western music could be automatically turned into a Brazilian-style cover version using the framework.

References

1. Cuthbert, M.S., Ariza, C.: music21: a toolkit for computer-aided musicology and symbolic music data. In: Downie, J.S., Veltkamp, R.C. (eds.) ISMIR, pp. 637–642. International Society for Music Information Retrieval (2010)
2. Dong, H.W., Hsiao, W.Y., Yang, L.C., Yang, Y.H.: MuseGAN: multi-track sequential generative adversarial networks for symbolic music generation and accompaniment. AAAI Press (2018)
3. Gibson, P., Byrne, J.: Neurogen, musical composition using genetic algorithms and cooperating neural networks. In: 1991 Second International Conference on Artificial Neural Networks, pp. 309–313. IET (1991)
4. Horowitz, D.: Generating rhythms with genetic algorithms. In: Proceedings of the Twelfth AAAI National Conference on Artificial Intelligence, AAAI'94, p. 1459. AAAI Press (1994)
5. Huang, C.Z.A., et al.: Music transformer: generating music with long-term structure. In: International Conference on Learning Representations (2018)
6. Mitchell, M.: An Introduction to Genetic Algorithms. MIT press, Cambridge (1998)
7. Peñalosa, D.: The Clave Matrix. Bembe Books, California (2012)
8. Pohjola, M.: The physics of instruments (soitinten fysiikkaa). MSc thesis (2016). https://oulurepo.oulu.fi/handle/10024/8197
9. Reijonen, O.: Lost Batucada. The Art of Deixa Falar, Portela and Mestre Oscar Bigode. Ph.D. thesis, University of Helsinki, Finland (2017)
10. Sandroni, C.: Feitiço decente: transformações do samba no Rio de Janeiro, 1917-1933. Jorge Zahar Editor (2001)
11. Toussaint, G.: A mathematical analysis of African, Brazilian, and Cuban clave rhythms. In: Sarhangi, R. (ed.) Bridges: Mathematical Connections in Art, Music, and Science, pp. 157–168. Bridges Conference, Southwestern College, Winfield, Kansas (2002). http://archive.bridgesmathart.org/2002/bridges2002-157.html
12. Vaswani, A., et al.: Attention is all you need. In: Proceedings 31st International Conference on Neural Information Processing Systems, NIPS'17, p. 6000-6010 (2017)
13. Yin, Z., Reuben, F., Stepney, S., Collins, T.: Measuring when a music generation algorithm copies too much: the originality report, cardinality score, and symbolic fingerprinting by geometric hashing (2022). https://doi.org/10.1007/s42979-022-01220-y

What Are "Good" Rhythms? Generating Rhythms Based on the Properties Set Out in *The Geometry of Musical Rhythm*

Paul Lascabettes[1(✉)] and Isabelle Bloch[2]

[1] Sorbonne Université, Ircam, CNRS, Ministère de la Culture, STMS, Paris, France
paul.lascabettes@ircam.fr
[2] Sorbonne Université, CNRS, LIP6, Paris, France
isabelle.bloch@sorbonne-universite.fr

Abstract. We propose in this article to generate good rhythms from geometric properties. This approach is based on the work by Toussaint, who investigated the properties that make a "good" rhythm good in his book *The Geometry of Musical Rhythm.* To do this, he analyzed the shapes of polygons corresponding to certain rhythms to derive geometric properties of a good rhythm. In this article, we propose to quantify these properties using original mathematical formulas. This scores each rhythm against several properties to measure how good a rhythm is, resulting in the generation of rhythms with k onsets and n pulses. When $k = 5$ and $n = 16$, we reveal that the *son* rhythm obtains the best score, as predicted by Toussaint at the end of his book. Applying the method for other values of k and n, we demonstrate that some of the rhythms with the highest scores are musically important rhythms, such as the *tresillo* rhythm with $k = 3$ and $n = 8$, the *fume-fume* rhythm with $k = 5$ and $n = 12$, the *samba* rhythm with $k = 7$ and $n = 16$ or the *Steve Reich's signature* rhythm with $k = 8$ and $n = 12$.

Keywords: Musical rhythm · Geometry of music · Generating music · Geometrical modeling of rhythms

1 Introduction

What Makes a "Good" Rhythm Good? is the central theme of Toussaint's book *The Geometry of Musical Rhythm* [14]. To address this question, he investigated the properties of six distinguished rhythms, each with $k = 5$ onsets and $n = 16$ pulses: *son*, *bossa-nova*, *gahu*, *rumba*, *shiko* and *soukous* rhythms. These "good" rhythms were chosen given their musical importance. They are represented in polygon notation in Fig. 1, which helps to understand the cyclical aspect of a rhythm. The originality of Toussaint's work is the geometric approach he uses

This work was partly supported by the chair of I. Bloch in Artificial Intelligence (Sorbonne Université and SCAI).

© The Author(s), under exclusive license to Springer Nature Switzerland AG 2024

T. Noll et al. (Eds.): MCM 2024, LNCS 14639, pp. 45–57, 2024.
https://doi.org/10.1007/978-3-031-60638-0_4

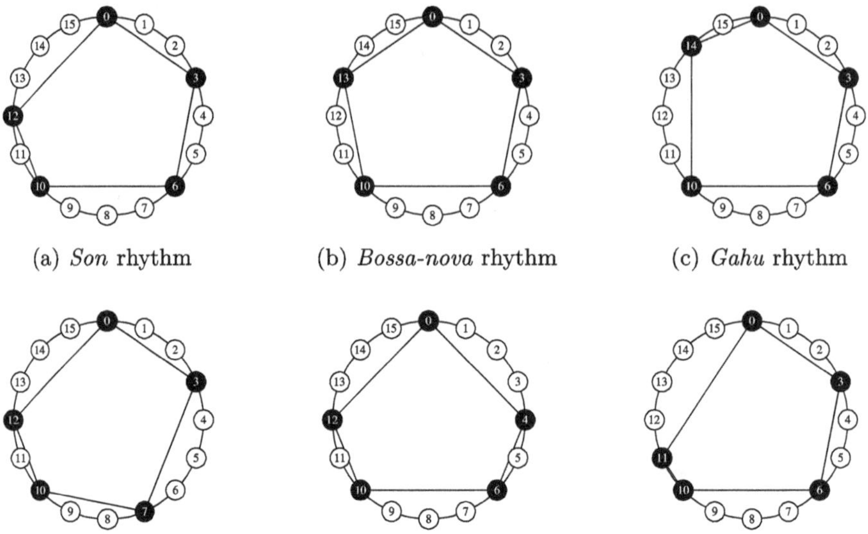

(a) *Son* rhythm (b) *Bossa-nova* rhythm (c) *Gahu* rhythm

(d) *Rumba* rhythm (e) *Shiko* rhythm (f) *Soukous* rhythm

Fig. 1. The six distinguished rhythms in polygon notation. The n pulses are indicated by numbers in the cyclic representation (from 1 to $n = 16$ here), and black points correspond to the k onsets ($k = 5$ here).

to describe the properties that make these rhythms good. In particular, only the shape of polygons associated with rhythms is analyzed, instead of using a more traditional representation such as a score. The properties of a "good" rhythm he proposed are therefore described using a mathematical, rather than musical, language. Among the six distinguished rhythms, he concluded that the *son* rhythm is the best one, as it satisfies the most properties [14]. But what about other rhythms that have $k = 5$ onsets and $n = 16$ pulses? Is there a rhythm as good or even better than the *son* rhythm? What are the "good" rhythms for other values of k onsets and n pulses? In this article, we propose mathematical models of the properties obtained by Toussaint to answer these questions and generate "good" rhythms with k onsets and n pulses.

We inherit from Toussaint's book the challenging term "good" rhythm, with the understanding that it underlines the theoretical interest of studying certain rhythmic properties in relation to the importance and pertinency of rhythms in their musical context. It is by no means an evaluation in an ethical sense. Rather, it points to the desire to have landmarks and measures of theoretical interest in the otherwise mindnumbing list of all combinatorially possible rhythms. In the following, we will omit the quotation marks for "good" with the understanding that the attribute stands for the value of fulfilment of rhythmic properties. This also includes the possibility for dissense among music theorists about the relevance of certain properties (see for example the criticism of some of Toussaint's featured properties [6]). Therefore, the purpose of this article is not to generate the best possible rhythms, which is a highly complex problem, but to extend Toussaint's work by providing non-binary mathematical formulas that measure each of these properties.

Previous works have already used the geometric properties of polygons associated with rhythms to generate good rhythms, such as balance [11], eveness [12] or interlocking reflection [13]. However, to our knowledge, there is no method that combines several properties with a non-binary score to generate good rhythms. To this end, in Sect. 2, we propose mathematical formulas for quantifying certain properties that characterize a good rhythm, in order to determine the level at which a rhythm with k onsets and n pulses satisfies each property, and to assign it a score aggregating all properties. Then, in Sect. 3, we apply this method to rhythms with $k = 5$ onsets and $n = 16$ pulses to generate good rhythms according to several properties. Finally, we apply the method to other values of k onsets and n pulses in Sect. 4 and demonstrate that some of the rhythms with the highest scores are musically important rhythms.

2 Characterizing the Properties of Good Rhythms

In this section, we present the properties we have considered in order to generate good rhythms. Most of these properties are from the book *The Geometry of Musical Rhythm* by Toussaint [14]. The main contribution of this article is to propose mathematical formulas to measure how a given rhythm satisfies each property. This quantification makes it possible to generalize the analysis to rhythms other than the six distinguished rhythms illustrated in Fig. 1. In particular, this enables to analyze a large database of rhythms, which would be impossible to do manually. Among the properties mentioned in [14], we have selected 12 that are particularly relevant for obtaining good rhythms. For each property, we normalize the value obtained by the mathematical formula with the *min-max normalization*[1] with the exception of the symmetry, shadow and fractal properties, to avoid overemphasizing these properties because we noticed that the min-max normalization gave too much importance to these properties (the values for rhythms with $k = 5$ onsets and $n = 16$ pulses lie in the interval $[0.6, 1]$, $[0.4, 1]$ and $[0.2, 1]$, respectively). We therefore sum the 12 values of the 12 properties to obtain a score between 0 and 12 for each rhythm (the sum was chosen for its compensation effect among the properties). By computing the score of all rhythms that have, for example, 5 onsets and 16 pulses, this allows to generate good rhythms with 5 onsets and 16 pulses based on geometric properties only.

In the remainder of this article, we define a rhythm R with k onsets and n pulses by $R = \{\overline{x_1}, \ldots, \overline{x_k}\}$ with $\overline{x_i} \in \mathbb{Z}_n$, where k and n are non-zero integers such that $k \leq n$. In this case, we set $x_i \in \mathbb{Z}$ the representative of $\overline{x_i} \in \mathbb{Z}_n$ such that $0 \leq x_i \leq n - 1$. For example, the *son* rhythm displayed in polygon notation in Fig. 1(a) is $\{\overline{0}, \overline{3}, \overline{6}, \overline{10}, \overline{12}\}$. In addition, we require in this article that $\overline{x_1} = \overline{0}$ in order to identify an onset as the beginning of the rhythm. We define

[1] The min-max normalization of x is defined by the formula $\frac{x-m}{M-m}$ where m and M are respectively the minimum and maximum values of the data.

$d_j = x_{j+1} - x_j$ the *duration* between two successive onsets x_j and x_{j+1} with $x_{k+1} = x_1 + n$, and the *interval* as the minimum duration between two onsets which are not necessarily successive. In the rest of this section, we present each of the 12 properties we have considered to generate good rhythms, explain the relevance of each property from a musical point of view and use existing or newly proposed mathematical formulas to quantify how much a rhythm satisfies each property in order to give it a score out of all 12 properties.

2.1 Eveness

The *eveness* of a rhythm characterizes how well the onsets are temporally spaced. If k divides n, then the polygon associated with the rhythm that maximizes eveness is a regular polygon. Otherwise it is the rhythm that best approximates the regular polygon. For example, with $k = 5$ and $n = 16$, the rhythm that maximizes eveness is the *bossa-nova* rhythm (or its rotations) represented in polygon notation in Fig. 1(b), which is close to a regular pentagon. Rhythms that maximize eveness given k onsets and n pulses are called *Euclidean* rhythms. There exists an algorithm for generating the Euclidean rhythms with k onsets and n pulses up to a rotation [12]. To quantify the eveness of a rhythm, Milne et al. have proposed the following formula [8], by identifying $\mathbb{Z}_n$ with vectors of norm 1 in the complex space:

$$\text{Eveness} = \frac{1}{k}\left|\sum_{j=1}^{k} e^{2i\pi\left(\frac{x_j}{n} - \frac{j}{k}\right)}\right|$$

This formula can be interpreted as the sum of the deviations between the polygon associated with the rhythm $\{e^{2i\pi\frac{x_j}{n}}\}_{0\leq j\leq k}$ and the regular polygon $\{e^{2i\pi\frac{j}{k}}\}_{0\leq j\leq k}$.

2.2 Balance

The *balance* of a rhythm measures how close the centroid of the associated polygon is to the center of the circle. This concept was introduced by Milne et al. and has also been generalized to musical scales [8,9]. The formula for quantifying the balance of a rhythm is as follows:

$$\text{Balance} = 1 - \frac{1}{k}\left|\sum_{j=1}^{k} e^{2i\pi\frac{x_j}{n}}\right|$$

Among rhythms with $k = 5$ onsets and $n = 16$ pulses, the rhythm that maximizes the balance is not the *bossa-nova* rhythm but the *son* rhythm (Fig. 1(a)), which demonstrates that eveness and balance are two different concepts.

2.3 Area

Another way of quantifying the dispersion of the onsets is to consider the *area* of the polygon associated with the rhythm. The larger the polygon area, the more dispersed the onsets, and vice versa. This idea has already been mentioned in [14], but rhythms with the largest area are barely addressed. We propose to use this property to characterize a good rhythm, which can be quantified as:

$$\text{Area} = \frac{1}{2}\left|\sum_{j=1}^{k} \sin\left(2\pi\frac{d_j}{n}\right)\right|$$

This formula can be interpreted as a sum of isosceles triangles at the center of the circle. Among rhythms with $k = 5$ onsets and $n = 16$ pulses, the *bossa-nova* rhythm is the one that maximizes area. It is important to note that this property is close to eveness and balance but is different.

2.4 Oddity

The *oddity* characterizes rhythms with an even number of pulses that do not have two onsets that are opposite to each other, i.e. two onsets that divide the rhythm into two sub-rhythms of equal length [2]. We can therefore quantify the oddity of a rhythm as its proportion of onsets that do not divide the rhythm in half, as follows for n even:

$$\text{Oddity} = 1 - \frac{1}{k}\left|R \cap (R + n/2)\right|,$$

where $R + n/2 = \{\overline{x_1} + \overline{n/2}, \ldots, \overline{x_k} + \overline{n/2}\}$ is the rhythm R shifted in time by half its length. This concept does not make sense for n odd, for which the oddity is not defined. Not all rhythms that maximize oddity are good. For example, the rhythm $\{\overline{0}, \overline{1}, \overline{2}, \overline{3}, \overline{4}\}$ maximizes oddity but is not musically relevant, which argues for the combination of different properties to generate good rhythms.

2.5 Closure

For rhythms with $n = 16$ pulses, Toussaint has defined the *closure* of a rhythm if the last onset is equal to 12, meaning that the last onset is located on the last strong beat [14]. Here, we propose to generalize this notion to give it more nuance. First, we propose to measure how an onset is a strong beat by computing its *orbit*, defined as: $\text{Orb}(\overline{x_i}) = \{\overline{x_i} \times \overline{y} \mid \overline{y} \in \mathbb{Z}_n\}$. For example, if $n = 16$, we have $\text{Orb}(\overline{4}) = \{\overline{0}, \overline{4}, \overline{8}, \overline{12}\}$ and $\text{Orb}(\overline{5}) = \mathbb{Z}_{16}$. With fewer elements in the orbit, the onset is a stronger beat. Therefore, we define the closure of a rhythm to quantify the level at which the last onset $\overline{x_k}$ of the rhythm is on a strong beat:

$$\text{Closure} = 1 - \frac{1}{n}\left|\text{Orb}\left(\overline{x_k}\right)\right|$$

This formula generalizes the idea introduced by Toussaint, but is useless if n is prime, as the strong and weak beats are more complex to determine.

2.6 Aperture

Inspired by the closure property, we propose here the *aperture* property. A rhythm satisfies this property if its second onset $\overline{x_2}$ is on a weak beat. We propose the following formula to quantify this property:

$$\text{Aperture} = \frac{1}{n} \left|\text{Orb}\left(\overline{x_2}\right)\right|$$

For example, if $n = 16$ and $\overline{x_2} = \overline{3}$, the aperture is maximal. The aperture creates tension while the closure makes a resolution. This new property partly explains why the *tresillo* rhythm $\{\overline{0}, \overline{3}, \overline{6}\}$ is one of the most important rhythms with $k = 3$ onsets and $n = 8$ pulses. In this case, the second onset is on a weak beat ($\overline{x_2} = \overline{3}$) which creates tension, while the last onset is not on a weak beat ($\overline{x_3} = \overline{6}$) which can be interpreted as a resolution.

2.7 Deepness

The *deepness* property characterizes rhythms where each interval between onsets occurs with a unique multiplicity, such that these multiplicities are the set $\{1, 2, \dots, k-1\}$ [5]. The formula that quantifies deepness is as follows:

$$\text{Deepness} = \frac{k(k-1)}{2} - \min_{0 \leq p \leq \lfloor n/2 \rfloor} \left(\sum_{j=1}^{\lfloor n/2 \rfloor} h_i - \max\left(0, j+1-p\right) \right),$$

where $H = \{h_1, \dots, h_{\lfloor n/2 \rfloor}\}$ is the ordered full interval-content histogram, and $\lfloor x \rfloor$ is the integer part of x. Rhythms that maximize the value of deepness are called *deep* rhythms. There are some links between Euclidean and deep rhythms. If a rhythm is Euclidean then it is deep, but the converse is false [5]. There is also an algorithm for generating deep rhythms [3]. Finally, some deep rhythms are also minimally even (unlike Euclidean rhythms, which are maximally even) which reveals that these concepts are related but also very different.

2.8 Entropy of Full Histogram

The *entropy of full histogram* ensures that interval multiplicities between onsets are similar [14]. This property can be quantified using the following formula:

$$\text{Entropy of Full Histogram} = -\sum_{j=1}^{\lfloor n/2 \rfloor} p_j \log_2\left(p_j\right),$$

where $P = \{p_1, \dots, p_{\lfloor n/2 \rfloor}\}$ is the normalized full interval-content histogram. Unlike the deepness property, this produces a rhythm with a flat histogram if zero-multiplicity intervals are not considered.

2.9 Entropy of Durations

The *entropy of durations* ensures that the durations between two successive onsets are similar to each other. This property is quantified with the following formula:

$$\text{Entropy of Durations} = -\sum_{j=1}^{k} d_j \log_2 (d_j)$$

We propose this property because it provides another method of obtaining a rhythm with temporally dispersed onsets.

2.10 Symmetry

One of the properties that characterizes a good rhythm is that the associated polygon has an axis of symmetry [14]. To quantify this property, we propose to compute the maximal proportion of symmetrical onsets according to the different axes with the following formula:

$$\text{Symmetry} = \max_{0 \leq j \leq n'} \frac{1}{k} \left| S_j(R) \cap R \right|,$$

where $S_j(R)$ is the symmetry of R with respect to the axis passing through the points $\{0, e^{i\pi \frac{j}{n}}\}$, i.e.: $S_j(R) = \{\overline{j} - \overline{x_i} \mid \overline{x_i} \in R\}$, and $n' = n - 1$ if n even or n if n odd.

2.11 Shadow

The *shadow* property characterizes rhythms that have a cyclic rhythmic contour that is the same as the cyclic rhythmic contour of the shadow of the rhythm [14]. We first introduce a function N of normalization such that $N(x) = 1$ if $x > 0$, -1 if $x < 0$ and 0 if $x = 0$, and we set $x_{k+j} = x_j + n$. The *cyclic rhythmic contour* CR of a rhythm R is defined by the cyclic contour of the rhythm durations: $CR = \{N(x_{i+2} - 2x_{i+1} + x_i)\}_{1 \leq i \leq k}$. Next, the *shadow* of a rhythm R is defined by the set $\{\frac{1}{2}(x_{i+1} - x_i)\}_{1 \leq i \leq k}$, and the cyclic rhythmic contour CS of the shadow of a rhythm is given by: $CS = \{N(x_{i+3} - x_{i+2} - x_{i+1} + x_i)\}_{1 \leq i \leq k}$. Consequently, the shadow property quantifies the similarity of CR and CS with respect to rotation, which is given by the following formula:

$$\text{Shadow} = \max_{j \in \{1,\ldots,k\}} \left| i \in \{1, \ldots, k\} : cr_i = cs_{i+j} \right|,$$

where $CR = \{cr_1, \ldots, cr_k\}$ is the cyclic rhythmic contour of the rhythm and $CS = \{cs_1, \ldots, cs_k\}$ is the cyclic rhythmic contour of the shadow of the rhythm. Among the six distinguished rhythms, only the *son* rhythm maximizes it.

2.12 Fractal

The *fractal* property characterizes the inclusion of a dilated R sub-rhythm in R. We propose the following formula to quantify this property:

$$\text{Fractal} = \frac{1}{k} \max_{R' \subseteq R/l, lR' \subseteq R} |R' \cup lR'|,$$

where $R/l = R \cap \{\overline{0}, \ldots, \overline{x_{\lfloor n/l \rfloor}}\}$ is the rhythm R until time $x_{\lfloor n/l \rfloor}$, and $lR' = \{\overline{lx_1}, \ldots, \overline{lx_k}\}$ is the rhythm dilated by l such that $2 \leq l \leq n$. For example, the rhythm $R = \{\overline{0}, \overline{5}, \overline{6}, \overline{10}, \overline{12}\}$ displayed in Fig. 2(e) in polygon notation maximizes the value of fractal because $R' = \{\overline{0}, \overline{5}, \overline{6}\}$ is included in R and $2R' = \{\overline{0}, \overline{10}, \overline{12}\}$ is also included in R and we have $\frac{1}{5}|R' \cup lR'| = 1$.

3 Generating Good Rhythms for $k = 5$ and $n = 16$

In this section, we present the results we obtained by applying the proposed method to rhythms with $k = 5$ onsets and $n = 16$ pulses. First, Table 1 provides the values of the six distinguished rhythms for all 12 properties described in Sect. 2.

Table 1. Values of the six distinguished rhythms for all 12 properties.

	Son	Bossa-nova	Gahu	Rumba	Shiko	Soukous
Eveness	0.99	1.0	0.97	0.98	0.97	0.94
Balance	1.0	0.94	0.85	0.91	0.92	0.92
Area	0.97	1.0	0.97	0.97	0.93	0.87
Oddity	1.0	1.0	0.5	1.0	0.5	0.5
Closure	0.86	0.0	0.57	0.86	0.86	0.0
Aperture	1.0	1.0	1.0	1.0	0.14	1.0
Deepness	0.33	1.0	0.0	0.33	1.0	0.0
Entropy H	0.97	1.0	0.97	0.97	0.94	0.87
Entropy D	0.37	0.0	0.81	0.56	0.0	0.81
Symmetry	1.0	1.0	0.6	0.8	1.0	0.8
Shadow	1.0	0.6	0.6	0.4	0.6	0.8
Fractal	0.8	0.6	0.6	0.6	0.6	0.6
Total	10.29	9.14	8.44	9.37	8.46	8.11

The last line of this table is the sum of the 12 values, and measures how good a rhythm is. The *son* rhythm is the one with the highest score among the six distinguished rhythms. However, what about other rhythms with $k = 5$ onsets and $n = 16$ pulses? We have applied the method to these rhythms such that the

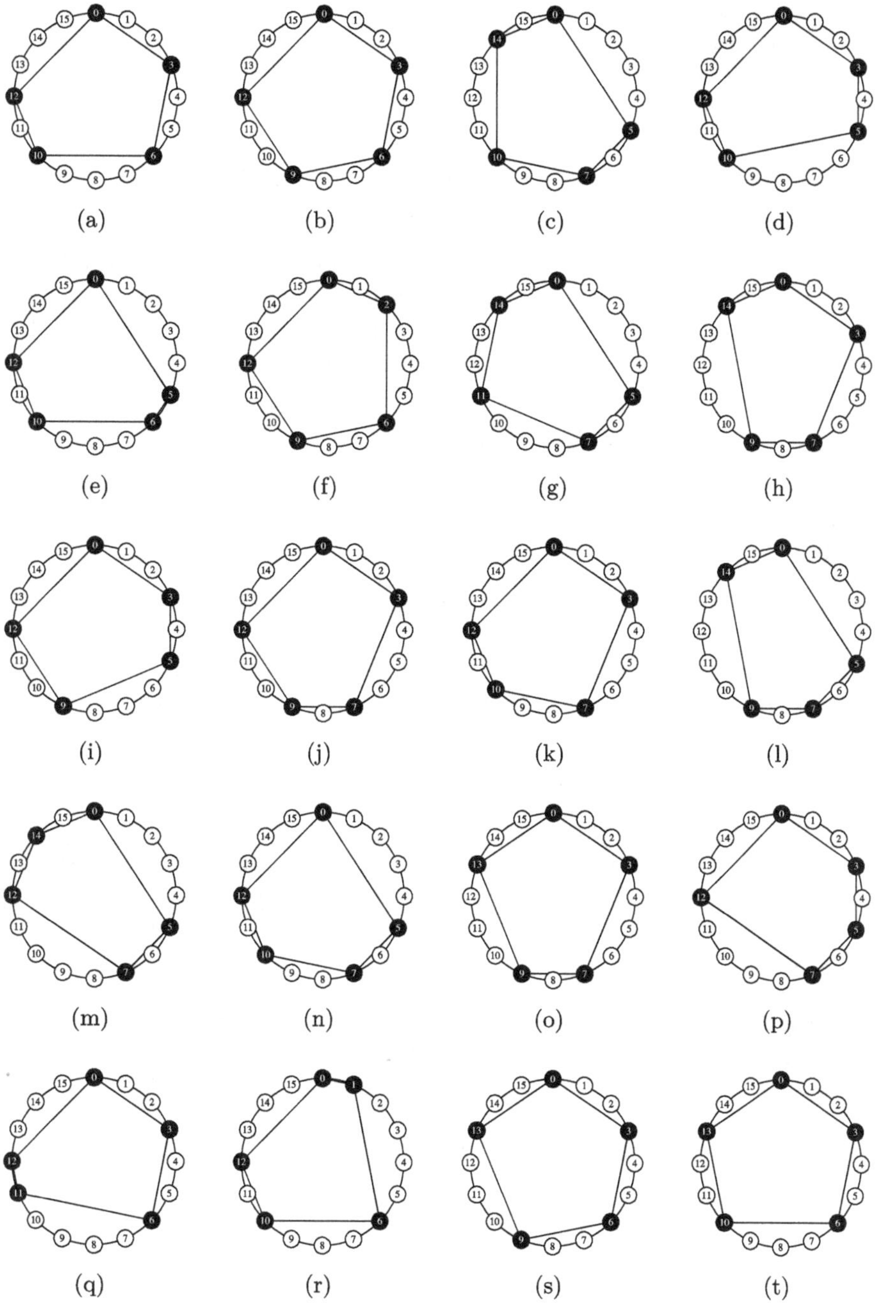

Fig. 2. Generation of the twenty rhythms with the best score for $k = 5$ and $n = 16$.

first onset is located at the origin ($\overline{x_1} = \overline{0}$). This generates many different good rhythms by selecting those with the best score. We have represented the twenty rhythms with the highest scores in Fig. 2 using polygon notation.

The *son* rhythm is the rhythm with the highest score, as predicted by Toussaint [14]. This can be explained by the fact that most of the properties come from his book, for which the *son* was already on top. Also, among these twenty rhythms, (k) is the *rumba* rhythm and (t) the *bossa-nova* rhythm. The other rhythms represented in Fig. 2 are rhythms that are close to the six distinguished rhythms, in the sense that they share common properties because they have a high score. It is also possible to generate "bad" rhythms by selecting rhythms with the worst score, as already done in [7]. For example, $\{\overline{0}, \overline{12}, \overline{13}, \overline{14}, \overline{15}\}$, $\{\overline{0}, \overline{2}, \overline{3}, \overline{4}, \overline{15}\}$ and $\{\overline{0}, \overline{2}, \overline{3}, \overline{14}, \overline{15}\}$ are the three rhythms with the worst scores.

4 Generating Good Rhythms for Other Values of k and n

In this section, we apply the method for other values of k onsets and n pulses and demonstrate that some rhythms with the best score are musically important. The three rhythms with the highest scores for onset and pulse values (k, n) equal to $(3, 8)$, $(5, 12)$, $(7, 16)$ and $(8, 12)$ are illustrated in Fig. 3 using polygon notation.

For $k = 3$ and $n = 8$, the rhythm with the highest score is the *tresillo* rhythm, shown on the left in Fig. 3(a), used in part in Latin and Afro-Cuban music. For $k = 5$ and $n = 12$, the rhythm with the highest score is the *fume-fume* rhythm, shown on the left in Fig. 3(b), frequently employed in West African music. For $k = 7$ and $n = 16$, the rhythm with the highest score is the *samba* rhythm necklace from Brazil, shown on the left in Fig. 3(c). Note that the actual *samba* rhythm is a rotation of it by 2 pulses: $\{\overline{0}, \overline{2}, \overline{5}, \overline{7}, \overline{9}, \overline{12}, \overline{14}\}$. For $k = 8$ and $n = 12$, the rhythm with the highest score is the *Steve Reich's signature* rhythm [15], shown on the left in Fig. 3(d), used in *Clapping Music* and in many of his compositions. However, with our method, the rhythm with the highest score is not always the most popular rhythm. For $k = 7$ and $n = 12$, one of the most popular rhythms is the *bembé* rhythm: $\{\overline{0}, \overline{2}, \overline{4}, \overline{5}, \overline{7}, \overline{9}, \overline{11}\}$. However, the *bembé* rhythm is not the one with the best score for $k = 7$ and $n = 12$ because it fails to satisfy the closure and aperture properties. The rhythm with the highest score is a rotation of the *bembé* rhythm by 1 pulse: $\{\overline{0}, \overline{1}, \overline{3}, \overline{5}, \overline{6}, \overline{8}, \overline{10}\}$.

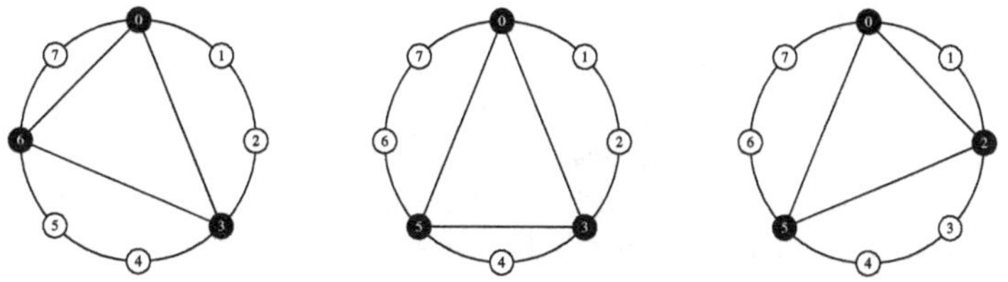

(a) $k = 3$ and $n = 8$

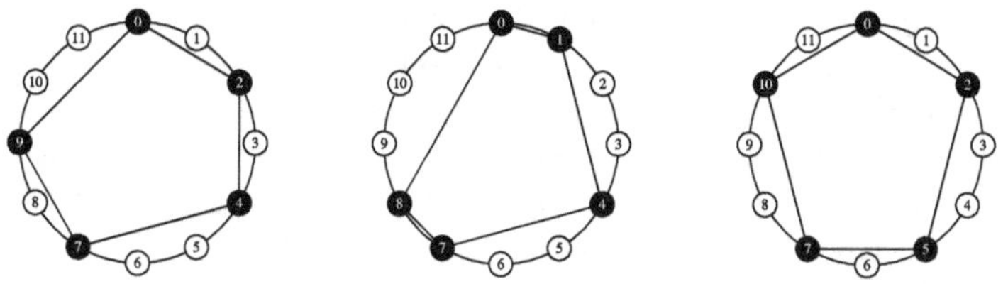

(b) $k = 5$ and $n = 12$

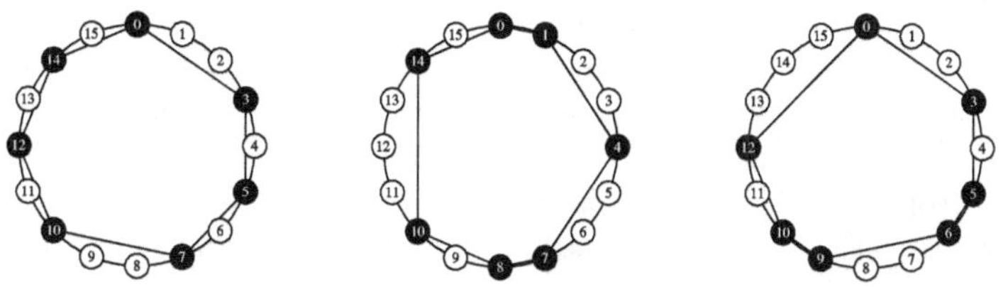

(c) $k = 7$ and $n = 16$

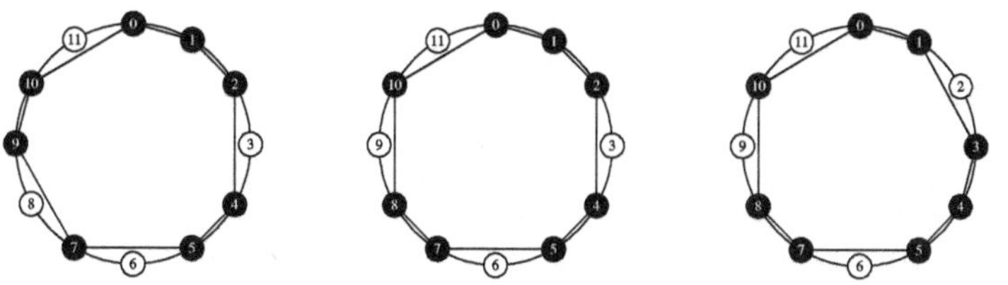

(d) $k = 8$ and $n = 12$

Fig. 3. Generation of the three rhythms with the best score (from left to right) for different values of k onsets and n pulses. In each case, the rhythm with the best score is a musically important rhythm: (a) the *tresillo* rhythm, (b) the *fume-fume* rhythm, (c) the *samba* rhythm and (d) the *Steve Reich's signature* rhythm.

5 Conclusion and Future Work

We have proposed in this paper a simple method for generating good rhythms based on a geometric approach. Despite the simplicity of this method, we have demonstrated that the results are the ones expected from Toussaint's findings, as some of the best generated rhythms are musically significant. Furthermore, defining scores, i.e. gradual satisfaction properties, instead of binary values allows, to some extent, to link qualitative concepts with concrete geometrical measures. In our future work, we will add other properties to the 12 we have chosen, and study their relevance. We also plan to go further in linking formal properties and musical meaning, for instance, high degree of deepness leads to good phase canons as done by Steve Reich [4]. In addition, perceptual analyses can be performed to determine whether the generated rhythms are indeed good and to determine whether certain properties are more important than others, as already done for eveness, balance or entropy [10]. We also believe that the case where n is odd or prime is worth studying, by adapting certain properties and analyzing the obtained results. Moreover, the proposed approach can be framed within the theory of *formal concept analysis*. Rhythms are objects O, satisfying a series of properties P, to some degree. This satisfaction relation can be seen as a fuzzy binary relation I, thanks to the gradual nature of our formulas, in $O \times P$, i.e. for $r \in O$ and $p \in P$, $I(r, p)$ is the degree to which rhythm r satisfies property p (score computed according to the relevant formula for p). From the formal context (O, P, I), fuzzy formal concepts can be derived, and organized in a lattice (see [1] for details on formal concept analysis). This structured way of representing rhythms and their properties could then be used for classification, or more generally as a support for the music-theoretical discourse about rhythms. Similar reasoning can be employed to generate musical scales by adapting certain geometric properties to the associated polygons, e.g. based on the work in [3]. Finally, we hope that these results, developed from a purely mathematical approach, can be applied in a musical context for composition and improvisation using the rhythms generated by this type of method.

References

1. Belohlavek, R.: Fuzzy Galois connections. Math. Log. Q. **45**(4), 497–504 (1999)
2. Chemillier, M., Truchet, C.: Computation of words satisfying the rhythmic oddity property (after Simha Arom's works). Inf. Process. Lett. **86**(5), 255–261 (2003)
3. Clough, J., Engebretsen, N., Kochavi, J.: Scales, sets, and interval cycles: a taxonomy. Music Theory Spectr. **21**(1), 74–104 (1999)
4. Cohn, R.: Transpositional combination of beat-class sets in Steve Reich's phase-shifting music. Perspect. New Music, 146–177 (1992)
5. Demaine, E.D., et al.: The distance geometry of music. Comput. Geom. **42**(5), 429–454 (2009)
6. Gómez-Martín, F.: A review of Godfried Toussaint's the geometry of musical rhythm. J. Math. Music **16**(2), 239–247 (2022)
7. Jafaripour, S., Nilforoushan, Z., Borna, K.: What makes a rhythm to be bad. J. Algorithms Comput. **52**(1), 65–82 (2020)
8. Milne, A.J., Bulger, D., Herff, S.A.: Exploring the space of perfectly balanced rhythms and scales. J. Math. Music **11**(2–3), 101–133 (2017)
9. Milne, A.J., Bulger, D., Herff, S.A., Sethares, W.A.: Perfect balance: a novel principle for the construction of musical scales and meters. In: Collins, T., Meredith, D., Volk, A. (eds.) Mathematics and Computation in Music. Lecture Notes in Computer Science(), vol. 9110, pp. 97–108. Springer, Cham (2015). https://doi.org/10.1007/978-3-319-20603-5_9

10. Milne, A.J., Herff, S.A.: The perceptual relevance of balance, evenness, and entropy in musical rhythms. Cognition **203**, 104233 (2020)
11. Milne, A.J., Herff, S.A., Bulger, D.W., Sethares, W.A., Dean, R.T.: XronoMorph: algorithmic generation of perfectly balanced and well-formed rhythms. In: NIME, pp. 388–393 (2016)
12. Toussaint, G.: The Euclidean algorithm generates traditional musical rhythms. In: BRIDGES: Mathematical Connections in Art, Music and Science, pp. 47–56 (2005)
13. Toussaint, G.: Generating "good" musical rhythms algorithmically. In: 8th International Conference on Arts and Humanities, Honolulu, Hawaii, pp. 774–791 (2010)
14. Toussaint, G.: The Geometry of Musical Rhythm: What Makes a "Good" Rhythm Good? CRC Press, Boca Raton (2019)
15. Yust, J.: Steve Reich's signature rhythm, and an introduction to rhythmic qualities. Music Theory Spectr. **43**(1), 74–90 (2021)

Euclidean Rhythm with Palindromic Rests

Paraj Mukherjee(✉)

Ramakrishna Mission Vidyamandira,
Belur Math, Howrah 711202, West Bengal, India
parajsphs@gmail.com

Abstract. Through a deep inspection of the rhythmic structures used in Indian Classical music, one can easily see how frequently Euclidean rhythms are used. There is a particular kind of Euclidean rhythm that is used very frequently: the rhythms in which the rests have a palindromic structure. In this paper, we characterize this family of rhythms. We look at the structural and geometric properties of such rhythms using tools that are very frequently used in live performances of Indian Classical music. Most of the properties have a certain combinatorial interest as well. The results we prove can be seen to be quite relevant in the context of Indian Classical music. We show how operations can be defined on such rhythms to construct a larger family of rhythms. Towards the end of the paper, we show applications of these rhythms in contemporary Indian Classical music.

Keywords: Euclidean rhythm · Indian Classical music · Erdös deep rhythm · palindromes

1 Introduction

Given n time intervals and $p < n$ onsets, we try to distribute the onsets as evenly as possible among the intervals. We may represent each rest by 0 and each onset by 1. The problem then becomes constructing a binary sequence of n bits with p 1's such that the 1's are spread as evenly as possible among the 0's [9]. The generated rhythm is denoted by $E(p, n)$. Let $k = n - p$ be the number of rests. To every corresponding $E(p, n)$, $V(p, k)$ denote the number formed by digits given by the number of rests of consecutive onsets. For example, for $E(5, 13) = 1001010010100$, the value of $V(5, 8) = 21212$. The $V(p, k)$ representation of any rhythm can be written in general. As all the Euclidean rhythms have this notion of *evenness*, it is perhaps very natural to ask what $E(p, n)$ would look like if the rests had a palindromic structure.[1] The problem of primary interest is when k and n are relatively prime numbers. The Euclidean rhythms can be generated by many methods, like some modification of Bresenham's line drawing algorithm using some matrix constructions [6]. In this paper, we use Bjorklund's algorithm

[1] Palindromes are the most symmetric numbers. Both the concepts of "evenness" and "symmetry", when put together, result in rhythms that are aesthetically "as uniform as possible".

© The Author(s), under exclusive license to Springer Nature Switzerland AG 2024
T. Noll et al. (Eds.): MCM 2024, LNCS 14639, pp. 58–69, 2024.
https://doi.org/10.1007/978-3-031-60638-0_5

[1]. Initially, we have a sequence of p 1's followed by k 0's. We can treat this sequence as n sequences of one bit each, and then distribute the 0's among 1's giving us some sequences of 2 bits. This counts as the first step. The process continues like this, distributing the remainders (that is, the sequences with the smallest number of bits) among sequences on the left side. The process ends when the remainder is exactly one, or we do not have any more zeroes to distribute. Then we concatenate the result. At the end of it, we obtain the left side forming the majority of the string and consisting of recursive, evenly spread onsets [2]. An example of the algorithm can be found in [9].

2 The Euclidean Rhythms with Palindromic Rests

We have p 1's and k 0's, $p < k$ and $gcd(p, k) = 1$. In that case, one remainder is left in the end. We characterize the values (p, k) for which $V(p, k)$ is a palindrome. We denote the set of all such rhythms by $\mathcal{T}$. The problem can be broken in two cases.

1. Case 1: $2p > k > p$; i.e., the number of remainders become less than the number of sequences on the left hand side right after 1 step
2. Case 2: $k > 2p, k > 3p, \ldots k > [\frac{k}{p}]p$; i.e., the number of remainders become less than the number of sequences on the left hand side after some finite steps greater than 1

We denote the phenomenon of the number of remainders becoming less than the number of sequences on the left hand side as a directional shift, i.e., the acting direction changes. In the second case, we have

$$1\underbrace{000....}_{[\frac{k}{p}]-1}1000....1000........\underbrace{0000....}_{k-([\frac{k}{p}]-1)p} \tag{1}$$

In this structure, $2p > k - ([\frac{k}{p}] - 1)p > p$ so the acting direction changes after 1 step. Therefore, every structure in the second case is isomorphic to one in the first case. Let P' be any structure in the second case, which is isomorphic to L' of the first case. For any (p, k), the structure itself follows Case 1 or is isomorphic to one following Case 1. Due to structural isomorphism, if $L' \in \mathcal{T}$, then $P' \in \mathcal{T}$. We call such a L' a base. On the other hand, we also see that if $P' \in \mathcal{T}$, then $L' \in \mathcal{T}$. From this one-to-one mapping, see that our problem gets reduced to Case 1 only, and solutions for Case 2 can be found accordingly. In the first case, the algorithm can go on for an arbitrary number of steps, depending on the values of k and p and the acting direction may again change arbitrarily many times after the second step.

Proposition 1. *$V(p, k)$ for some value of (p, k) for which the algorithm run 3 steps and does not change the acting direction after initially changing in the second step will be a palindrome*

Proof. The fact that (p, k) for which the condition holds will correspond to a solution can easily be seen by establishing a relation between k and p. After the first step, there are p 10's and $k - p$ 0's. After the second step, there are $k - p$ 100's and $2p - k$ 10's, and as the direction of the algorithm remains the same, $2k > 3p$. In the third step, there are $2p - k$ 10010's and $2k - 3p$ 100's. Since the algorithm terminates here, we have

$$2k - 3p = 1 \tag{2}$$

By using Bjorklund's algorithm, we can see that (p, k) except (1,2) and (3,5) satisfying (2) corresponds to a set of solutions for our problem. Hence,

$$\{(p, k) \mid 2k - 3p = 1; p \neq 1, 3\} \tag{3}$$

The first few values of (p, k) are $(5, 8), (7, 11), (9, 14), (11, 17), (13, 20)$, and so on. The structures of these base Euclidean rhythms can be easily found by running the algorithm itself. The smallest rhythm of this family which is found for $p = 5$, $k = 8$ looks like: $[\times \cdot \cdot \times \cdot \times \cdot \cdot \times \cdot \times \cdot \cdot \times]$ (in this representation, $\times$ denote the onsets and $\cdot$ denote the rests, the last onset in this representation corresponds to the first onset of the next cycle) with $V(p, k) = 21212$. One can see that $V(p, k)$ of the subsequent other base rhythms can be written by adding a "12" at the end of the $V(p, k)$ representation of previous base rhythm.

Let the set of these base rhythms be set A $V(p, k)$ representation of any element looks like

$$\underbrace{21212.....2}_{some\ finite\ times} \tag{4}$$

We now consider two subcases under Case 1.

2.1 SubCase 1

Let the acting direction change in the third step, i.e., $3p > 2k$. If the algorithm terminates after two more steps, the left part of the binary sequence starts as "1001010" but the right side ends as "10010" (which is the remainder unit). Hence, we have no solution. If the algorithm goes on for arbitrary finite steps, changing direction arbitrary many times, we get no solution. This is because the right side will always have "10" at the end, which does not make the rests have a palindromic structure.

2.2 SubCase 2

Let the algorithm not end after three steps. In this case, let the acting direction not change again in step 4. If the process terminates at the next step, we get no solution, as the left part of the binary sequence starts as "10010100" but the right side ends as "10010". After step 4, the process can go on for arbitrary finite steps, with the direction changing arbitrarily many times. In each possibility,

one can see that no solution can be found; every time either the end digits of $V(p,k)$ representation do not match, or even if they do match, the number is not a palindrome. Now let the acting direction change in step 4. If the process terminates in the next two steps, we have solutions!
Let the set of these base rhythms be set B. In that case, $V(p,k)$ representation looks like

$$\underbrace{2122212....2122212}_{some\ finite\ times} \tag{5}$$

One can see that $V(11,30) = 21222122212$, $V(15,41) = 212221222122212$, and so on. Observe that $V(p,k)$ of the subsequent other base rhythms can be written by adding a "2212" at the end of the $V(p,k)$ representation of the previous base rhythm. If the process does not terminate in the next two steps, one can look at the subsequent cases and see that no more solutions can be found.

We can now generate the whole set of solutions by considering Case 2. For example, from $E(5,13)$, we can generate $E(5,18)$, $E(5,23)$, $E(5,28)$, and so on. $E(5,18) = [\times\cdot\cdot\times\cdot\times\cdot\cdot\times\cdot\times\cdot\cdot\times\cdot\times\cdot\cdot\times]$. In general, when considering the base rhythms of set A, $V(p,k)$ generally looks like

$$\underbrace{(m+1)m(m+1)m....(m+1)]}_{some\ finite\ times},\ m \geq 1 \tag{6}$$

Let the set of these rhythms and the base rhythms of A be set τ_1. Similarly, by considering set B, $V(p,k)$ looks like

$$\underbrace{[(m+1)m(m+1)(m+1)(m+1)m....(m+1)(m+1)m(m+1)]}_{some\ finite\ times},\ m \geq 1 \tag{7}$$

Let the set of these rhythms and the base rhythms of B be set τ_2. From $V(p,k)$ we thus get the whole family $\mathcal{T}$ of Euclidean rhythms in which $V(p,k)$s are palindromes. $\mathcal{T} = \tau_1 \cup \tau_2$.

Due to slightly more simplicity in its structure, we study $V(p,k)$ instead of the rhythms themselves; all the results analogously apply for rhythms. The process by which this family of rhythms is generated holds little importance to any composer or musician; however, the sheer symmetry and overall geometry of distances do make these rhythms an attractive choice in any musical piece. The beauty or aesthetics of any rhythm depends on its mathematical structure, which in turn depends on the ability of one's mind to see these mathematical structures subconsciously in the musical piece (although there are many other factors beyond mathematical ones that shape musical beauty, general factors, namely, perceptual, cognitive, and cultural, determine the sense of aesthetics of a rhythm). We study a few properties of elements of $\mathcal{T}$, the proofs of most of the results discussed hereafter hold more importance than the results themselves from a musician's point of view. Many of these properties may be quite elementary, but the overall beauty of any rhythm depends on all of them.

Remark 1. It may so happen that the rhythm we get by Bjorklund's algorithm has no palindromic structure, but we can get one by considering some rotation of it. For example, $V(3,5) = 221$, which can be made a palindrome by a rotation. This rhythm, by definition, is also an Euclidean rhythm. The set of solutions can be expanded by considering these rotated rhythms as well.

3 Erdös-Deep

A rhythm $E(p,n)$ is caled an Erdös-deep rhythm if for every multiplicity in the set $\{1, 2, ..., p-1\}$ there exists a chordal distance with that multiplicity. The result that any element of $\mathcal{T}$ is Erdös deep follows from the result that any Euclidean rhythm where $gcd(p,n) = 1$ is Erdös-deep. Nevertheless, we demonstrate a proof that shows the beautiful geometry of chordal distances and multiplicities in elements of $\mathcal{T}$. The pattern of distances and the multiplicity pattern they follow are quite interesting from a musician's point of view.

Theorem 1. *Every element of $\mathcal{T}$ is an Erdös-deep rhythm.*

Proof. We illustrate the proof for set A, the proof for τ_1 follows similarly. Let $x \in A$ with p onsets. The set of multiplicities is : $\{1, 2, ..., \frac{p-1}{2}, \frac{p+1}{2}, ..., p-1\}$. Every distance δ is formed by adding a string of chordal distances, each of which cannot be in turn written as a sum of chordal distances; let it be represented by $(p_1 p_2 ... p_n)$. The distance 6, formed by (33) is unique and occurs only 1 time. Through some simple combinatorial counting of δ by writing the $V(p,k)$ of x, one can see that the distance 2, i.e., (2) occurs $\frac{p-1}{2}$ times, and the distance 3, i.e., (3) occurs $\frac{p+1}{2}$ times. Continuing this counting, we can see that the distance 5, i.e., (23) occurs $p-1$ times, distance 7, i.e., (232) occurs $\frac{p-3}{2}$ times, distance 10, i.e., (2323) occurs $p-2$ times. In general, distance (2323...23) with number of 2's = number of 3's = q (let) satisfying $p - q > \frac{p+1}{2}$ occurs $p - q$ times and distance (2323...2) with number of 2's = m, number of 3's = $m - 1$ satisfying $\frac{p-2m-1}{2} > 1$ occurs $\frac{p-2m-1}{2}$ times. Hence, we exhaust the whole set of multiplicities in this process. We can do a similar analysis of the relation between distances and multiplicities for elements of B and extend that to τ_2. This completes the proof.

4 Almost Winograd-Deep

As introduced by Winograd in a 1966 class project report [10], we call a rhythm $Winograd-deep$ if every chordal distance $1, 2, ..., [\frac{n}{2}]$ has a unique multiplicity. Now, several distances do not occur in elements of $\mathcal{T}$, for example, distances like 1, 4, 9. So, no element of $\mathcal{T}$ is $Winograd-deep$. We define a rhythm $almost\ Winograd-deep$ if the chordal distances in the set $\{1, 2, ...[\frac{n}{2}]\}$ have a unique multiplicity; the occurence of every chordal distance is not necessary.

Theorem 2. *Every element of $\mathcal{T}$ is almost Winograd-deep.*

Proof. Let $x \in \tau_1$. We have proved in Sect. 1 that x is Erdös deep. The available set of multiplicities is: $\{1, 2, ..., p-1\}$. Let $D_1, D_2, ...D_{p-1}$ be the distances we used in Sect. 3. Some of these are chordal distances while some of them are not. If D_j's are chordal distances, we can see easily that the chordal distance associated with any multiplicity is unique. We name it set E. Let D_i's be not chordal, then we have $n - D_i < [\frac{n}{2}]$ so $n - D_i$'s become chordal. The multiplicity of D_i = multiplicity of $n - D_i = l$ (let), then $n - D_i$ is the only chordal distance associated with l. We name the set of all $n - D_i$'s as F. We have, $E \cap F = \phi$, and as $E \cup F$ is the set of all chordal distances occurring in x, all the distances occurring in $\{1, 2, ...[\frac{n}{2}]\}$ have unique multiplicities. The proof for any element of τ_2 follows similarly.

In the process, we have found another set of distances (which is the set of all possible chordal distances in x) that can be used to prove that x is Erdös-deep. Let G = set of all chordal distances of x, H = set of all multiplicities, and we consider the natural mapping $f : G \longrightarrow H$ where g_1 is mapped to h_1 if g_1 occurs h_1 times. As x is almost-Winograd deep, f is one-one. Also, since f is Erdös-deep, f is surjective. Therefore, f is a bijective mapping. We sum all this up and state the following theorem, which is the heart of the last two sections.

Theorem 3. *For every $x \in \mathcal{T}$ there exists a bijective map between the set of all chordal distances and the set of all multiplicities.*

5 Homometric Rhythms

Homometric rhythms are those with the same histogram or multiset of distances. For any rhythm, all the rotated structures are trivially homometric to one another. It is easy to prove that these are the only homometric rhythms for any member of $\mathcal{T}$.

Theorem 4. *For every member x of $\mathcal{T}$ the rotated rhythms are the only ones homometric to x.*

If we speak about isospectral vertices, each rhythm is homometric to itself, and by the symmetry of all the rhythms along the vertical axis, each pair of the symmetric vertices is therefore isospectral to each other. For example if we mark the vertices of $E(5, 13)$ in its geometric representation by $\{A_0, A_1, ..., A_{12}\}$, vertices $(A_3, A_{10}), (A_5, A_8)$ are pairwise isospectral. Now if we consider the rotated rhythms, we can take the isospectral vertices accordingly in the same way.

For any $x \in A$, all the possible rhythms formed by rotations are evenly spaced. Total number of combinations possible by arranging digits of $V(p, k)$ is equal to the structures where chordal distance 2 occurs in the pattern (11), plus the structures where chordal distance 2 occurs in the pattern (2) (total number of such structures $= p$, it is the same as the number of ways $\frac{p-1}{2}$ 1's can be distributed among p places so that two 1's are not side by side). All these rhythms are the only homometric rhythms of x that have an onset at the first pulse. Also, the $V(p,k)$ representation of each of these structures can be made a palindrome by giving a particular rotation! This matches the fact that these are homometric to x, we can say that an isomorphism exists between each of them. All of the p rhythms can be obtained from x by suitable transformations. We select a string of consecutive digits of $V(p,k)$ starting from left digit. For each such selection, a symmetric selection from the right digit can also be made. Observe that by swapping any string (reversing the digits of the string in $V(p,k)$) we get one homometric rhythm of x. Strings with 2 as the extreme right digit (when taking selections starting from the left digit) give us the same $V(p,k)$, so we do not consider them. In $E(p,n)$, there are p beats, total number of 1's in $V(p,k) = \frac{p-1}{2}$, total number of strings $= \frac{p-1}{2} \times 2 = p-1$, swapping each of which gives us one rhythm homometric to x. One can also observe this phenomenon by looking at its geometric representation. For example, in $E(5,13)$, $A_3 \rightarrow A_2$, $A_{10} \rightarrow A_{11}$, $A_5 \rightarrow A_6$ and $A_{10} \rightarrow A_{11}$, $A_8 \rightarrow A_7$ and $A_3 \rightarrow A_2$ swappings give us all the homometric rhythms that have an onset at the first pulse. Like this, such swappings can be studied for more complicated rhythms.

These rhythms can also be looked upon as a group action of S_n on the elements of $\mathcal{T}$. Let f be the group action. Then $A_3 \rightarrow A_2$ swapping can be looked upon as a group action of the cycle $S = (3\ 4)$ on the rhythm. f is defined as

$$\{A_j : j \in \{1,2,...,13\}\} \rightarrow \{A_{S(j)} : j \in \{1,2,...,13\}\}. \tag{8}$$

6 Constructing New Rhythms

In the previous section, we saw how any pair of isospectral vertices look for any rhythm of $\mathcal{T}$ and any of its homometric pair. An operation which involves isospectral vertices is illustrated below.

Definition 1 (Pumping [7]). *Let P_1, P_2 be a homometric pair corresponding $E(p,n)$ with isospectral vertices p_x, p_y respectively. $S = \{+s_i, -s_i : i \in \{1,..,r\}\}$. We define a (m,r,S) pumping of P_1, P_2 $(m \geq 1, r \geq 0)$ as a new pair of rhythms with number of time intervals $= n'$ and number of onsets $= k'$ with $n' = mn$ and $k' = k + 2r$. These are formed by replacing p_x in P_1 with $\{p_x, p_{x+s_i}, p_{x-s_i} : i \in \{1,...,r\}\}$ and p_y in P_2 with $\{p_y, p_{y+s_i}, p_{y-s_i} : i \in \{1,...,r\}\}$.*

It is quite apparent that any pair of rhythms formed by any pumping will be homometric to one another, and one can be obtained by giving the other one a suitable rotation. For the sake of constructing new rhythms, we can also apply different pumping to the two homometric rhythms; they would not remain homometric anymore. We can also apply pumping to a separate pair of isospectral points and take the union of all points to form a new rhythm.

Pumping two rhythms is a way to systematically create new rhythms from $\mathcal{T}$. Like this, we can use various other operations and algorithms to construct new rhythms. Two very simple operations are addition and addition by changing the components. In addition, for rhythms $P_1(V(p,k_1))$ and $P_2(V(p,k_2))$, $V(p,k)$

representation of $P_1 + P_2$ is defined as $V(p, k_1 + k_2)$ and the rests are distributed by component addition. If two rhythms have different numbers of onsets, one cannot define the usual component addition. This is where we can perform addition by changing the components. In that case, the components get adjusted sideways before doing the addition. One can see that multiple such adjustments are possible for any two rhythms. For example, $V(p, k)$ representation of $E(7, 11) + E(5, 18)$ can be written as 2555552 by arranging the components in a symmetric way about each other and performing usual addition. Some other operations are: concatenation of two rhythms, taking complement of a rhythm, breaking the rhythms (the breaking is done such that there is a beat in the first rhythm formed at the end) and introducing a set of onsets. We can also consider Euclidean rhythms with a different number of onsets and the same number of time intervals as an element of $\mathcal{T}$ and use them accordingly in some compositions.
We can extend some of the ideas. After pumping a rhythm and itself and taking union of both the rhythms, we can divide the figure vertically (resulting in dividing the rhythm symmetrically), and then consider a rhythm homometric to one half; the corresponding rhythm homometric to the other half will be symmetric to the first one. We can then take the union of the points and form another rhythm; further properties of such rhythms can be studied. This can be done by first studying what the homometric rhythms of any arbitrary rhythm look like. One can explore this by taking the necessary ideas from Patterson's paper in crystallography [8].

7 Arithmetic Progressions

The (n, p, k) values of the set of solutions $\mathcal{T}$ can all be written in a matrix form. It is a symmetric matrix, and we observe that arithmetic progressions (APs) can be formed by taking elements from any particular row (or column). One can think of characterizing all APs formed by taking elements from the matrix.

Theorem 5. *Let $\{a_n\}$ be a sequence of arbitrary Euclidean rhythms such that $\{V(p_n, k_n)\}$ forms an arithmetic progression satisfying either of the following conditions:*

1. *there exist some number of rests between the last onset and the first onset (of the next cycle) in a_1*
2. *the common difference of $\{V(p_n, k_n)\}$ is not a multiple of 10*

Then $\{a_n\}$ contains infinitely many Euclidean rhythms with palindromic rests [5].

Proof. It can be easily proved that we cannot find any Euclidean rhythm with palindromic rests if both conditions do not hold for any sequence $\{a_n\}$. Let either of them hold. Let d be the common difference of $\{V(p_n, k_n)\}$, $10|d$ (i.e., the second condition does not hold, and the first condition holds). Then by the first condition, $10 \not| V(p_1, k_1)$. Let $V(p_1, k_1) = [v_q v_{q-1} ... v_1 v_0]$; we define $V'(p_1, k_1) = [v_0 v_1 ... v_{q-1} v_q]$. Let $d = 2^m 5^n$, then for any $l = max\{q, m, n\} + r$ $(r \in \mathbb{Z})$ the elements $V(p_1, k_1) + 10^l V'(p_1, k_1) \in \{V(p_n, k_n)\}$ and are of the form

$$[v_0 v_1 ... v_{q-1} v_q 0....0 v_q v_{q-1} ... v_1 v_0] \tag{9}$$

which in turn are either of the form $[(m+1)m(m+1)m....(m+1)]$ or $[(m+1)m(m+1)(m+1)(m+1)m(m+1)....(m+1)m(m+1)]$ and thus the corresponding rhythms are Euclidean rhythms with palindromic rests in $\{a_n\}$. Now let $10 \not| d$, $d = 2^m 5^n c$. We can show that there exists $e = max\{q, m, n\} + r$ $(r \in \mathbb{Z})$ such that $10^e \equiv 1 \ (mod\ c)$. If there exist some number of rests between the last onset and the first onset (of the next cycle) in a_1, there exist $s \in \{0, 1, ..., c-1\}$ such that $V'(p_1, k_1).10^{e-q} + s \equiv 0 \ (mod\ c)$. Consider

$$x = V'(p_1, k_1).10^{es+e-q} + 10^{es} + 10^{(s-1)e} + .. + 10^e + V(p_1, k_1)$$

We have,

$$\begin{aligned} x - V(p_1, k_1) &= V'(p_1, k_1).10^{es+e-q} + 10^{es} + 10^{(s-1)e} + .. + 10^e \\ &\equiv V'(p_1, k_1).10^{se}.10^{e-q} + \underbrace{1 + 1 + ... + 1}_{s} \ (mod\ c) \end{aligned}$$

Thus,

$$x - V(p_1, k_1) \equiv V'(p_1, k_1).10^{e-q} + s \equiv 0 \ (mod\ c)$$

So, $c | x - V(p_1, k_1)$. We already know that $2^m, 5^n | x - V(p_1, k_1)$; putting them together, we get $d | x - V(p_1, k_1)$ i.e., $x \in \{V(p_n, k_n)\}$. Also,

$$x = [v_0 v_1 ... v_{q-1} v_q \underbrace{0..0}_{e-q-1} 10..010..010..010..01 \underbrace{0..0}_{e-q-1} v_q v_{q-1} ... v_1 v_0] \tag{10}$$

so x is a palindrome. Also, infinitely many such x are present in $\{V(p_n, k_n)\}$ since $10^{ze} \equiv 1 \ (mod\ c)$ for any $z \in \mathbb{N}$. Note that if there are no rests between the last onset and first onset (of the next cycle) in a_1, then we consider a_2 and carry on the same steps. Hence infinitely many $V(p_i, k_i)$s are palindromes, and the corresponding rhythms are Euclidean rhythms with palindromic rests.

We can try to generalize this result slightly to any arbitrary AP of rhythms. If $\{a_n\}$ is a sequence of arbitrary rhythms such that $\{V(p_n, k_n)\}$ forms a AP satisfying either of the two conditions, then $\{a_n\}$ contains infinitely many rhythms with palindromic rests. These rhythms may not be Euclidean, but we can think of them as a series of operations on Euclidean rhythms. For example, if we consider using only the first two operations defined in Sect. 6, and if they are in τ_1, we can write them as a series of operations on Euclidean rhythms, or some rotation of an Euclidean rhythm, or some broken-down Euclidean rhythms and a residue rhythm (a rhythm that is neither of the three). But this cannot be done in all cases; we need some extra conditions. A further discussion on this is beyond the scope of the current manuscript.

8 Applications in Indian Classical Music

We present some applications of Euclidean rhythms with palindromic rests in Indian Classical music.
The rhythms that are performed in Indian Classical Music are mostly cyclic in nature. Each *taal* (a regular rhythmic pattern of equally placed beats of any composition) can be differentiated by the number of beats occuring in each of the loops. Each cycle is in turn sub-divided into segments/divisions of varying measurements. For example, *teen taal* is the *taal* of 16 beats following the division 4/4/4/4. *Jhaap taal* is the taal of 10 beats, following the division 2/3/2/3 and *dhamar taal* is the taal of 14 beats, following the the division 5/2/3/4. The first beat of each cycle is called the *sam*. It is the most emphatic of strokes since it commences the cycle. It provides the centre for the exposition and all compositions are usually set so that their emphasis too is on this beat.

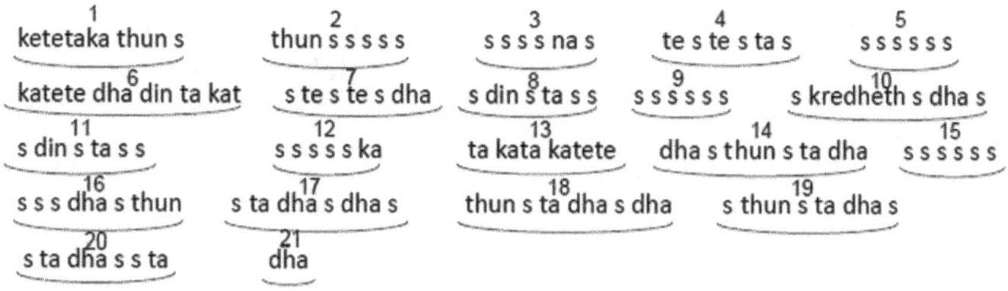

Fig. 1. *tehai* in *teen taal*

Figure 1 is one part of a *tehai*. A *tehai* is repeating a rhythmic piece identically three times with equal interval in between the repetitions in a particular *taal* such that it ends at the *sam*. The phrases written are made of *bols* (syllables used to define the rhythmic patterns). The *s*'s represent the rests, and every phrase written inside a bracket corresponds to a particular beat, which is given by the number written above the bracket. This *tehai* follows *tisra jaati* (3 micro beats in one beat) and is in *teen taal*. The length of each rest is $\frac{1}{6}$. Euclidean rhythm with $V(p,k) = 43434$ is used here. There is a gap of 1 length between each part of the *tehai*, so the length of the *tehai* $= 21 \times 3 + 2 = 65 = 64 + 1$. This shows that the *tehai* comes back to *sam* after 4 cycles (loops). Although the pattern of rests is easy to observe, we see that a set of onsets is present between these rests rather than a single onset, which should not be the case! However, theoretically, each of these sets of onsets gets reduced to a single onset at some very high tempo, so all of these are Euclidean rhythms at some *laya* (the speed or tempo at the which the rhythm is played). This is a very beautiful concept, and it shows the importance of tempo in any musical composition. Figure 2 is one part of a *tehai* following *chatusra jaati* (4 micro beats in one beat) and is in *jhaap taal*. The length of each rest is $\frac{1}{4}$. Euclidean rhythm with $V(p,k) = 2121212$ is used here. There are no rests between each part

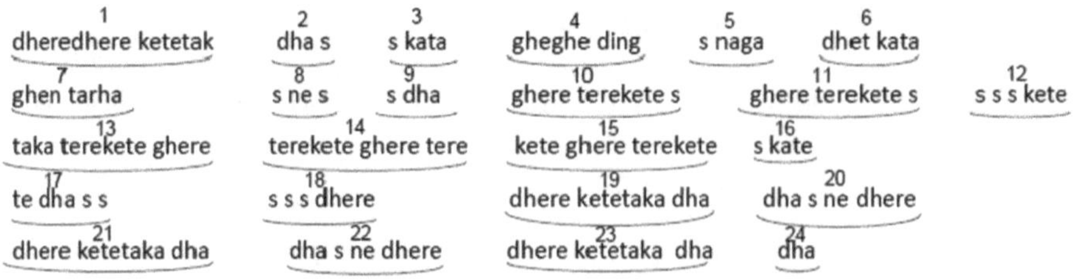

Fig. 2. *tehai* in *jhaap taal*

of the *tehai*, *dha* of the 24th beat takes $\frac{1}{2}$ length of the beat, total length of *tehai* $= 23.5 \times 3 = 70 + \frac{1}{2}$ and so it comes to the *sam* after 7 cycles. Figure 3 is one part of a *tehai* following *chatusra jaati* and is in *dhamar taal*. Rhythm with the $V(p, k)$ representation 53535, which can in turn be generated by the addition operation we defined earlier, is used here. The operation is between the Euclidean rhythms with $V(p, k) = 21212$ and 32323. There are no rests between each part of the *tehai*, the *dha* of the 24th beat takes $\frac{1}{2}$ length of the beat. Therefore, total length of the *tehai* $= 23.5 \times 3 = 70.5 = 70 + \frac{1}{2}$, so it comes to *sam* after 5 cycles.

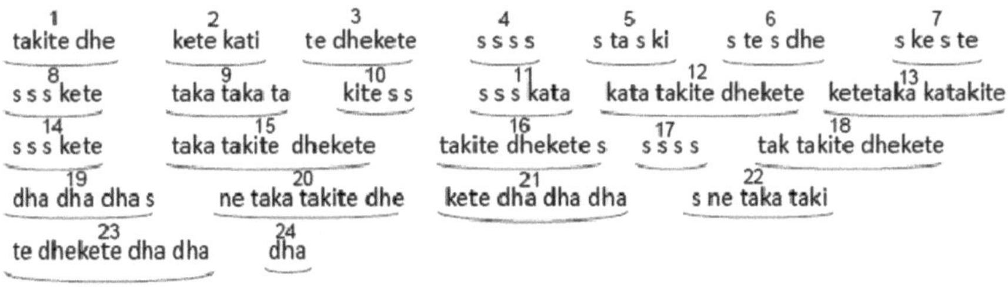

Fig. 3. *tehai* in *dhamar taal*

9 Concluding Remarks

In this paper, we have restricted ourselves to analysing these rhythms using tools that are widely used in live performances of Indian Classical music, which are mostly mathematically simple in nature. One can think of analysing them using various other advanced mathematical methods. However, while analysing rhythms, there is always a restriction to the amount of mathematical abstractness one uses in analysing them. Any such analysis should always have some reflection on the practical demonstration of such rhythms. In a sense, one can say that many areas of theoretical research on rhythms are in some way restricted by how mathematically advanced musical compositions are; one needs to make advancements in the types of practises of music in order to pave the way for new theoretical interpretations of rhythms. The compositions in Sect. 8 show how widely these rhythms are used in Indian Classical music.

Based on our analysis and intuition, we believe that any finite-length AP is totally contained in a particular row (or column) (Sect. 7). However, at present, we are unable to provide a definitive proof for this claim. We can analyse the rhythms further by adopting tools used by Toussaint in [4]. Since every rhythm can be reinterpreted as a scale, the Euclidean rhythms with palindromic rests can be further analysed by drawing ideas from Clough and Douthett's work [3].

Acknowledgement. I would like to thank Shibam Mondal and Ankit Saha for their ideas and help throughout. I would also like to express deep gratitude to Professor Jean Paul Allouche for his suggestions and the anonymous referees for their valuable feedback. I would like to thank my father, Sunando Mukherjee, who introduced me to music and who also helped me compose the compositions in Sect. 8. I would also like to thank my mother Ranjita Mukherjee and my tabla *Guruji* Subrata Manna, for providing me with the finest musical exposure since my childhood.

References

1. Bjorklund, E.: A metric for measuring the evenness of timing system rep-rate patterns. SNS ASD Tech Note, SNS-NOTECNTRL-100 (2003)
2. Bjorklund, E.: The theory of rep-rate pattern generation in the SNS timing system. SNS ASD Tech Note, SNS-NOTE-CNTRL-99 (2003)
3. Clough, J., Douthett, J., Krantz, R.: Maximally even sets: a discovery in mathematical music theory is found to apply in physics. In: Bridges: Mathematical Connections in Art, Music, and Science, pp. 193-200. Central Plain Book Manufacturing Winfield (2000)
4. Gómez-Martín, F., Taslakian, P., Toussaint, G.: Structural properties of Euclidean rhythms. J. Math. Music **3**(1), 1–14 (2009)
5. Harminc, M., Sotdk, R.: Palindromic numbers in arithmetic progressions. Fibonacci Q. **36**, 259–261 (1998)
6. Morrill, T.: On the Euclidean algorithm: rhythm without recursion. Bull. Aust. Math. Soc. **107**(3), 361–367 (2023)
7. O'Rourke, J., Taslakian, P., Toussaint, G.: A pumping lemma for Homometric rhythms (2008)
8. Patterson, A.L.: Ambiguities in the x-ray analysis of crystal structures. Phys. Rev. **65**(5–6), 195 (1944)
9. Toussaint, G.: The Euclidean algorithm generates traditional musical rhythms. In: Renaissance Banff: Mathematics, Music, Art, Culture, pp. 47-56 (2005)
10. Winograd, T.: An analysis of the properties of 'deep' scales in a t-tone system. Unpublished, May 17 (1966)

Categorical and Algebraic Approaches to Music

Finding Homometric Multiplets

Franck Jedrzejewski(✉)

CEA Paris Saclay University, Saclay, France
franckjed@gmail.com

Abstract. In the article *On Extended Z-triples* (1981), David Lewin proved that Z-sets introduced by Allen Forte, which are neither transposed nor inverted forms and which have the same interval vector are not necessarily pairs. In a $N = 16$ degrees equal temperament, he exhibited triples of pcsets which are Z-associated. In this paper, we give a general formula for all homometric triples of pentachords, and some new methods to construct homometric multiplet.

Z-related sets were introduced in 1977 by Allen Forte in *The Structure of Atonal Music* [2]. They were already known to crystallographers who had studied them from the 1940s, such as A.L. Patterson [15]. More recently, the algebraist J. Rosenblatt generalized the problem to arbitrary group algebras and related it to the phase retrievial problem [16,17]. In biology, the phase retrievial problem of finding a set from a multiset difference is known as the *partial digest problem.* In music theory, it has been extensively studied by [3,4,11–13,18]. The enumeration of the Z-related sets in equal temperaments of up to thirty-one notes per octave has been done in [9,19].

1 The Z Relation

We assume that the reader is familiar with the Z-relation, otherwise he might consult [2,6,7]. For convenience, we recall the main definitions and results. Homometric sets (or sets in Z-relation) are sets of numbers whose distances between all the elements of each set are the same. The *distance set* for a set $X = \{x_0, ..., x_{n-1}\}$, denoted by ΔX, is a multiset defined by:

$$\Delta X = \{x_j - x_i \bmod N, 0 \leq i \leq j \leq n-1\}.$$

Theorem 1. *Two sets A and B of $\mathbb{Z}_N$ are Z-related (or homometric) if they have the same interval content:*

$$(A \ \mathcal{Z}_N \ B) \Leftrightarrow \mathrm{ic}(A) = \mathrm{ic}(B)$$

The two sets share the same interval vector. Since transposition or invertion do not change the interval content, we get a lot of trivially Z-related sets. To avoid these trivial cases, we consider set classes up to transposition and inversion.

© The Author(s), under exclusive license to Springer Nature Switzerland AG 2024
T. Noll et al. (Eds.): MCM 2024, LNCS 14639, pp. 73–84, 2024.
https://doi.org/10.1007/978-3-031-60638-0_6

Let $A \subset \mathbb{Z}_N$, the *mask polynomial* is the polynomial defined by

$$A(x) = \sum_{a \in A} x^a \in \mathbb{Z}[x]/(x^N - 1)$$

The *reflection* of A is the polynomial

$$A^*(x) = \sum_{b \in I(A)} x^b$$

whose exponents are the inversion $I(A)$ of A mod N, where I denotes the inversion $I(x) = -x \mod N$.

The *Patterson function* of the set $A \subset \mathbb{Z}_N$ is the autocorrelation function given by the convolution product

$$P(x) = A * A^*(x) = A(x)A(x^{-1}) \mod (x^N - 1)$$

The mask polynomial of the distance set ΔA is the Patterson function,

$$\Delta A(x) = A(x)A(x^{-1})$$

Theorem 2. *The Patterson function's coefficients are the components of the interval function.*

$$A * A^*(x) = \sum_{c_i \in \text{ifunc}(A,A)} c_i x^i = \sum_{n=0}^{N-1} \text{iv}(A)(n) x^n$$

Theorem 3. *Two sets A and B of $\mathbb{Z}_N$ are homometric if and only if they have the same Patterson function.*

Rosenblatt [17] has introduced the notion of *spectral unit*, which allows us to show the homometry of two sets.

Definition 1. *A unit $S(x)$ of the algebra group $\mathbb{Q}[\mathbb{Z}_N]$ is called a* spectral unit *if $S(x)S(x^{-1}) = 1 \mod (x^N - 1)$.*

Theorem 4. *Two sets A and B are homometric if and only if there exists a spectral unit $S \in \mathbb{Q}[\mathbb{Z}_N]$ such that*

$$S(x)B(x) = A(x) \mod (x^N - 1)$$

Example 1. Let $N = 8$, $A = \{0, 3, 4, 5\}$ and $B = \{0, 1, 3, 4\}$. The polynomial $S(x) = (1 + x^2 - x^4 + x^6)/2$ verifies $S(x)S(x^{-1}) \equiv 1$. Moreover, $S(x)B(x) = xA(x)$. Thus B and $A' = T_1(A) = \{1, 2, 3, 6\}$ are homometric. Since $A(-1) = B(-1) = 0$, the solution is not unique. The polynomial

$$S(x) = \frac{1}{4}(-3 + x + x^2 + x^3 + x^4 + x^5 + x^6 + x^7)$$

is a spectral unit such that $S(x)B(x) = x^5 A(x)$. The sets B and $T_5(A)$ are homometric, and therefore, A and B are homometric.

Rosenblatt [17] used these results to find all non-trivially homometric sets of 4 elements.

Theorem 5. *If A and B are two non trivially homometric sets of 4 elements, then A and B are one of the following types: for $N = 4(r+s)$,*

$$A = \left\{0, \frac{N}{2}, \frac{N}{2} + s, \frac{N}{2} + s + r\right\} \text{ and } B = \left\{0, \frac{N}{2} - r, \frac{N}{2}, \frac{N}{2} + s\right\}$$

and if $N = 13n$ is a multiple of 13,

$$A = \{0, n, 4n, 6n\} \text{ and } B = \{0, 2n, 3n, 7n\}$$

O'Rourke, Taslakian and Toussaint introduced the notion *isospectral vertices*. Then, they prove a result known as the *pumping lemma* (see [14]).

Definition 2. *Two vertices $a \in A$ and $b \in B$ are isospectral if they have the same histogram of distances to all other vertices in their respective sets.*

2 Homometric Triples

In the classical system of 12 tones, A. Forte has shown that there is only one homometric pair of 4 notes ({0,1,4,6}, {0,1,3,7}), three pairs of 5 notes ({0, 1, 2, 4, 7}, {0, 1, 3, 5, 6}), ({0, 1, 2, 5, 8}, {0, 1, 3, 8, 9}), ({0, 1, 2, 5, 9}, {0, 1, 3, 4, 8}), and 15 pairs of 6 notes, such as 6-Z29 and 6Z-50 ({0, 1, 3, 6, 7, 10}, {0, 1, 3, 6, 8, 9}).

A systematic study of the first values of N shows that there is only one 4-notes homometric pair in $\mathbb{Z}_8$ ({0, 1, 2, 5}, {0, 1, 3, 4}), and three 5-notes homometric pairs in $\mathbb{Z}_{10}$. In the $\mathbb{Z}_{13}$ universe, there is a unique 4-notes pair ({0, 1, 3, 9}, {0, 1, 4, 6}) and only two 5-notes pairs ({0, 1, 2, 3, 6, 10}, {0, 1, 2, 4, 5, 8}) and ({0, 1, 2, 4, 7, 9}, {0, 1, 3, 5, 7, 8}).

From $N \geq 16$ elements, David Lewin noticed that multiplets appear. But there are no multiplets for some values of N, such as $N = 17$ and $N = 19$. Note that multiplets can exist even if N is a prime number. For instance, there exists 11 triples of 9 elements for $N = 23$, such as ({0, 1, 2, 3, 4, 6, 8, 11, 20}, {0, 1, 2, 4, 5, 7, 9, 10, 21}, {0, 1, 2, 4, 5, 16, 18, 19, 21}).

In his paper, *On extended Z-Triples* [10], David Lewin gave the three triples of Z-associated hexachords mod 16, and the two triples of Z-associated heptachords. There also exist three quadruples of Z-associated octachords, which exhaustively complete the list of homometric multiplets mod 16, namely:

$$\{0,1,2,3,4,7,9,12\}, \{0,1,2,3,5,8,9,14\}, \{0,1,2,3,5,10,12,13\}, \{0,1,2,4,5,7,9,10\} \bmod 16$$

$$\{0,1,2,3,5,8,11,12\}, \{0,1,2,3,6,8,9,13\}, \{0,1,2,4,5,10,11,13\}, \{0,1,2,5,6,8,9,11\} \bmod 16$$

$$\{0,1,2,4,5,8,10,12\}, \{0,1,2,4,6,8,9,12\}, \{0,1,2,4,6,9,10,14\}, \{0,1,2,4,8,10,12,13\} \bmod 16$$

After $N = 16$, the next space that has homometric multiplets is the universe of third-tones ($N = 18$). This space was used by 20th century micro-interval pioneers Julian Carrillo, Ivan Wyschnegradsky and Alois Hába, and later by composer Maurice Ohana who wrote for third-tones zithers. Composers Iannis Xenakis and Jean-Etienne Marie have also explored the universe of third-tones, as well as composers who have studied twelfth-tones such as Erza Sims and Franz Richter Herf. Some composers have systematically composed in N-EDO (equal division of octave) for all values of N, as one can hear in the *39 dissonant etudes* of the Australian based composer Warren Burt. The particularity of this temperament ($N = 18$) is that it only has a triple of 5 notes. This is a particular case of the following result, which gives all the homometric triples of the pentachords.

Theorem 6. *All non trivially homometric triples of pentachords in the N-tone universe (up to translations and inversions) are given by the following sets:*

$$A_1 = \left\{0, a, \frac{N}{6} + a, \frac{N}{3} + a, \frac{N}{2}\right\}$$
$$A_2 = \left\{0, a, \frac{N}{6} + a, \frac{N}{2}, \frac{5N}{6} + a\right\}$$
$$A_3 = \left\{0, a, \frac{N}{2}, \frac{2N}{3} + a, \frac{5N}{6} + a\right\}$$

where $N \equiv 0 \bmod 6$ and a is some positive integer.

Proof. To show that the three sets are homometric, we give three different proofs. The first is to calculate the difference table of each set and check that they are equal. The three sets will then have the same interval content.

$$\begin{aligned} &A_1 - A_1 = A_2 - A_2 = A_3 - A_3 = \\ &\{0, a, N/6 - a, N/6, N/6 + a, N/3 - a, N/3, N/3 + a, N/2 - a, N/2, N/2 + a, \\ &\quad -N/3 - a, -N/3, -N/3 + a, -N/6 - a, -N/6, -N/6 + a, -a\} \end{aligned}$$

Another way to show this result is to consider the spectral unit

$$S(x) = \left(1 - x^{N/3} + x^{N/2} + x^{5N/6}\right)/2$$

which verifies the relations:

$$S(x)S(x^{-1}) = 1, \quad A_1(x)S(x) = A_2(x), \quad A_2(x)S(x) = A_3(x)$$

A third way to show this result is to note that the points $\{\frac{N}{3} + a\} \in A_1$ and $\{\frac{5N}{6} + a\} \in A_2$ are isospectral, as well as $\{\frac{N}{6} + a\} \in A_2$ and $\{\frac{2N}{3} + a\} \in A_3$. That's why we can exchange them without changing the internal distances.

Finally, to show that there are no other triples, it is enough to show that the spectral unit is unique. Since $A_1(x)$ never vanishes on the multiplicative group in $\mathbb{C}$ of the Nth roots of unit, then $S(x)A_1(x) = A_2(x) \bmod (x^N - 1)$ has a unique solution $S(x)$.

Example 2. This theorem gives all the triples of pentachords, and concerns $6n$-tones universes. For third-tones ($N = 18$), the triple is obtained for $a = 1$: $A_1 = \{0, 1, 4, 7, 9\}$, $A_2 = \{0, 1, 4, 9, 16\}$ and $A_3 = \{0, 1, 9, 13, 16\}$. The points $7 \in A_1$ and $16 \in A_2$ are isospectral, as well as the points $4 \in A_2$ and $13 \in A_3$. For quatertones ($N = 24, a = 1$), we get a second triple $A_1 = \{0, 1, 5, 9, 12\}$, $A_2 = \{0, 1, 5, 12, 21\}$ and $A_3 = \{0, 1, 12, 17, 21\}$. For fifth-tones ($N = 30$), one has two triples, for $a = 1$ and $a = 2$. In the sixth-tones universe, one has three homometric triples of pentachords (for $a = 1, 2, 3$), etc.

The following results give three triples of hexachords. Their proof consists of calculating the sets of differences or to exhibit a spectral unit.

Theorem 7. *The sets*

$$A_1 = \left\{0, a, \frac{N}{9} + a, \frac{2N}{9} + a, \frac{N}{3}, \frac{2N}{3}\right\}$$
$$A_2 = \left\{0, a, \frac{N}{9} + a, \frac{N}{3}, \frac{2N}{3}, \frac{8N}{9} + a\right\}$$
$$A_3 = \left\{0, a, \frac{N}{3}, \frac{2N}{3}, \frac{7N}{9} + a, \frac{8N}{9} + a\right\}$$

form a homometric triple for $N \equiv 0 \bmod 9, N \geq 27$ *and* a *is some positive integer.*

Proof. The proof of the homometry of the three sets is based on the spectral unit
$$S(x) = \left(1 - x^{2N/9} + x^{N/3} - x^{5N/9} + x^{2N/3} + 2x^{8N/9}\right)/3$$
which allows the passage from one set to another $A_1(x)S(x) = A_2(x)$.

If we apply the transformation in which each element of $A_i = (x_0, x_1, .., x_5)$ is preserved except the point x_{4-i} which is transformed into $x_{4-i} + N/3$, we obtain the following result.

Theorem 8. *The sets*

$$B_1 = \left\{0, a, \frac{N}{9} + a, \frac{N}{3}, \frac{5N}{9} + a, \frac{2N}{3}\right\}$$
$$B_2 = \left\{0, a, \frac{N}{3}, \frac{4N}{9} + a, \frac{2N}{3}, \frac{8N}{9} + a\right\}$$
$$B_3 = \left\{0, \frac{N}{3}, \frac{N}{3} + a, \frac{2N}{3}, \frac{7N}{9} + a, \frac{8N}{9} + a\right\}$$

form a homometric triple for $N \equiv 0 \bmod 9, N \geq 27$ *and* a *is some positive integer.*

Proof. The same spectral unit as in the Theorem 7 shows that the sets are homometric.

If we apply the transformation in which each element of $A_i = (x_0, x_1, .., x_5)$ is preserved, except the points x_2 of A_1, x_1 of A_2 and x_5 of A_3 which are transformed into $x \to x + N/3$, we get the following result.

Theorem 9. *The sets*

$$C_1 = \left\{0, a, \frac{2N}{9} + a, \frac{N}{3}, \frac{4N}{9} + a, \frac{2N}{3}\right\}$$
$$C_2 = \left\{0, \frac{N}{9} + a, \frac{N}{3}, \frac{N}{3} + a, \frac{2N}{3}, \frac{8N}{9} + a\right\}$$
$$C_3 = \left\{0, a, \frac{2N}{9} + a, \frac{N}{3}, \frac{2N}{3}, \frac{7N}{9} + a\right\}$$

form a homometric triple for $N \equiv 0 \bmod 9, N \geq 27$ and a is some positive integer.

Proof. The same spectral unit as in the Theorem 7 shows that the sets are homometric.

For heptachords and for multiples of 10, the following triple is homometric.

Theorem 10. *The sets of heptachords*

$$A_1 = \left\{0, a, \frac{N}{10} + a, \frac{N}{5} + a, \frac{3N}{10} + a, \frac{2N}{5} + a, \frac{N}{2}\right\}$$
$$A_2 = \left\{0, a, \frac{N}{10} + a, \frac{N}{5} + a, \frac{3N}{10} + a, \frac{N}{2}, \frac{9N}{10} + a\right\}$$
$$A_3 = \left\{0, a, \frac{N}{10} + a, \frac{N}{5} + a, \frac{N}{2}, \frac{4N}{5} + a, \frac{9N}{10} + a\right\}$$

are homometric for $N \equiv 0 \bmod 10, N \geq 20$ and $a = 1, 2, ..., N/10 - 1$ is a positive integer.

Proof. The spectral unit

$$S(x) = \left(1 - x^{2N/5} + x^{N/2} + x^{9N/10}\right)/2$$

shows that the sets are homometric.

For octachords and for multiples of 12, the following triple is homometric.

Theorem 11. *The sets of the following triple for octachords*

$$A_1 = \left\{0, a, \frac{N}{12} + a, \frac{N}{6} + a, \frac{N}{4} + a, \frac{N}{3} + a, \frac{5N}{12} + a, \frac{N}{2}\right\}$$
$$A_2 = \left\{0, a, \frac{N}{12} + a, \frac{N}{6} + a, \frac{N}{4} + a, \frac{N}{3} + a, \frac{N}{2}, \frac{11N}{12} + a\right\}$$
$$A_3 = \left\{0, a, \frac{N}{12} + a, \frac{N}{6} + a, \frac{N}{4} + a, \frac{N}{2}, \frac{5N}{6} + a, \frac{11N}{12} + a\right\}$$

are homometric for $N \equiv 0 \bmod 12, N \geq 24$ *and* $a = 1, 2, ..., N/12 - 1$ *is a positive integer.*

Proof. The proof is based on the spectral unit

$$S(x) = \left(1 - x^{5N/12} + x^{N/2} + x^{11N/12}\right)/2$$

3 Homometric Multiplets

The theorems that we have just established for triples become generalized and form multiplets of more than 3 sets. Theorem 7 is generalized in the following way.

Theorem 12. *The sets of* $k = (r + 3)$ *notes*

$$A_1 = \left\{0, a, \frac{N}{3r} + a, \frac{2N}{3r} + a, \ldots, \frac{(r-1)N}{3r} + a, \frac{N}{3}, \frac{2N}{3}\right\}$$
$$A_2 = \left\{0, a, \frac{N}{3r} + a, \frac{2N}{3r} + a, \ldots, \frac{(r-2)N}{3r} + a, \frac{N}{3}, \frac{2N}{3}, \frac{(3r-1)N}{3r} + a\right\}$$
$$\ldots$$
$$A_r = \left\{0, a, \frac{N}{3}, \frac{2N}{3}, \frac{(2r+1)N}{3r} + a, \ldots, \frac{(3r-1)N}{3r} + a\right\}$$

are homometric for $N = 9r, r \geq 3$ *and* $a = 1, 2, ..., r - 2$ *is a positive integer.*

Proof. The spectral unit

$$S(x) = \left(1 + x^{2N/3} + x^{N/3} + 2x^{(3r-1)N/3r} - x^{(r-1)N/3r} - x^{(2r-1)N/3r}\right)/3$$

shows that the sets are homometric.

The generalization of Theorems 10 and 11 gives:

Theorem 13. *The sets of* $k = (r + 2)$ *notes*

$$A_1 = \left\{0, a, \frac{N}{2r} + a, \frac{N}{r} + a, \frac{3N}{2r} + a, \ldots, \frac{(r-1)N}{2r} + a, \frac{N}{2}\right\}$$
$$A_2 = \left\{0, a, \frac{N}{2r} + a, \frac{N}{r} + a, \frac{3N}{2r} + a, \ldots, \frac{(r-2)N}{2r} + a, \frac{N}{2}, \frac{(2r-1)N}{2r} + a\right\}$$
$$A_3 = \left\{0, a, \frac{N}{2r} + a, \frac{N}{r} + a, \ldots, \frac{(r-3)N}{2r} + a, \frac{N}{2}, \frac{(2r-2)N}{2r} + a, \frac{(2r-1)N}{2r} + a\right\}$$
$$\ldots$$
$$A_{r-2} = \left\{0, a, \frac{N}{2r} + a, \frac{N}{r} + a, \frac{N}{2}, \frac{(r-1)N}{r} + a, \ldots, \frac{(2r-1)N}{2r} + a\right\}$$

are homometric for $N = 4r, r \geq 5$ *and* $a = 1, 2, ..., r - 2$ *is a positive integer.*

Proof. The proof is based on the spectral unit

$$S(x) = \left(1 - x^{(r-1)N/2r} + x^{N/2} + x^{(2r-1)N/2r}\right)/2$$

The quarter-tone universe is one of the most popular spaces. It has been used by very many composers. Its homometric structure is very rich. However, there are no homometric t-tuples for $t = 7$, 9, 10 and 11, and the theorems that we have established do not give all multiplets. For instance, the 24-EDO has a sextuplet of 7 notes

$$\{0,1,2,6,9,12,17\}, \{0,1,3,7,8,13,16\}, \{0,1,3,8,9,12,19\},$$
$$\{0,1,3,8,15,16,21\}, \{0,1,3,9,12,19,20\}, \{0,1,4,6,9,17,18\} \bmod 24$$

an octuplet of 9 notes

$$\{0,1,2,4,6,9,12,16,17\}, \{0,1,2,4,6,9,14,17,18\}, \{0,1,2,4,8,9,12,14,17\},$$
$$\{0,1,2,4,9,10,14,17,22\}, \{0,1,2,4,9,14,16,17,20\}, \{0,1,2,6,9,10,12,14,17\},$$
$$\{0,1,3,5,7,8,13,16,17\}, \{0,1,3,5,8,9,13,15,16\} \bmod 24$$

and three very strange dodecuplets of 12 elements:

$$\{0,1,2,3,4,5,8,12,13,15,18,20\}, \{0,1,2,3,4,6,7,11,14,16,18,19\},$$
$$\{0,1,2,3,4,7,9,11,12,14,15,19\}, \{0,1,2,3,4,7,12,14,16,17,19,20\},$$
$$\{0,1,2,3,5,6,7,10,13,15,17,18\}, \{0,1,2,3,5,6,8,13,14,15,18,22\},$$
$$\{0,1,2,3,5,6,8,13,16,17,18,20\}, \{0,1,2,3,5,6,10,12,13,14,19,22\},$$
$$\{0,1,2,3,5,6,12,13,14,17,20,22\}, \{0,1,2,3,5,8,13,15,16,17,19,20\},$$
$$\{0,1,2,3,6,8,10,11,13,14,15,18\}, \{0,1,2,4,5,12,13,14,16,18,19,21\} \bmod 24$$

$$\{0,1,2,3,4,7,8,10,13,15,18,19\}, \{0,1,2,3,4,7,10,13,15,18,19,20\},$$
$$\{0,1,2,3,5,6,9,11,12,14,18,19\}, \{0,1,2,3,5,6,11,12,14,18,19,21\},$$
$$\{0,1,2,3,5,8,9,11,12,14,19,20\}, \{0,1,2,3,5,8,11,12,14,19,20,21\},$$
$$\{0,1,2,3,6,7,8,10,13,15,16,19\}, \{0,1,2,3,6,7,9,11,12,14,17,18\},$$
$$\{0,1,2,3,6,8,9,10,13,15,16,21\}, \{0,1,2,3,7,8,9,11,12,14,17,20\},$$
$$\{0,1,2,4,5,7,8,10,13,17,18,19\}, \{0,1,2,6,7,9,11,12,14,15,17,18\} \bmod 24$$

$$\{0,1,2,3,4,7,8,12,14,16,17,19\}, \{0,1,2,3,4,7,9,11,14,15,16,19\},$$
$$\{0,1,2,3,4,8,10,11,13,15,16,20\}, \{0,1,2,3,5,6,10,11,13,15,17,18\},$$
$$\{0,1,2,3,5,6,10,13,15,17,18,19\}, \{0,1,2,3,5,7,8,13,15,16,17,20\},$$
$$\{0,1,2,3,5,8,9,13,14,15,17,22\}, \{0,1,2,3,5,8,10,12,13,14,17,18\},$$
$$\{0,1,2,3,5,9,10,11,14,15,17,22\}, \{0,1,2,3,5,9,10,12,13,14,19,21\},$$
$$\{0,1,2,3,7,9,11,12,14,15,16,19\}, \{0,1,2,4,5,6,9,11,14,16,17,18\} \bmod 24$$

Another equal temperament that is widely used is the 31-note temperament. It comes from the division of the tone into 5 parts (and the semitone into 3 parts) since $5 \times 5 + 2 \times 3 = 31$. This division was proposed by Nicolo Vicentino in *L'antica musica ridotta alla moderna prattica* (1555) and commented on by many authors including Francisco Salinas who, in his *De musica* (1577) proposed to build the famous *archicembalo*. But it is Christiaan Huygens who published the first table of this temperament and of which we republish the complete works on music and sound [1]. The 31-tet was taken up in the twentieth century by the Dutch physicist Adriaan Fokker who built a special organ for this system and promoted this music. Nowadays, Ere Lievonen and the Scala ensemble, founded by the Huygens-Fokker Foundation to experience microtonality give concerts all over the world. Some pieces have become famous such as *Solar Neon* of Ned McGowan premiered in 2010 or *Fantasia for Fokker Organ 4 hands* of Sebastian Dumitrescu. Precursor works had been composed by supporters of Harry Partch and of the just intonation, such as Dean Drummond who uses 31-just-tuning in his zoomozophone, and composes the amazing *Cloud Garden* in 1974. In this temperament, there are many triplets and quadruplets of 9 notes or more, but there is only one hexachord quintuplet (see Fig. 1) : {0, 1, 3, 8, 12, 18}, {0, 1, 3, 10, 14, 26}, {0, 1, 4, 6, 13, 21}, {0, 1, 4, 10, 12, 17}, {0, 1, 8, 11, 13, 17} mod 31.

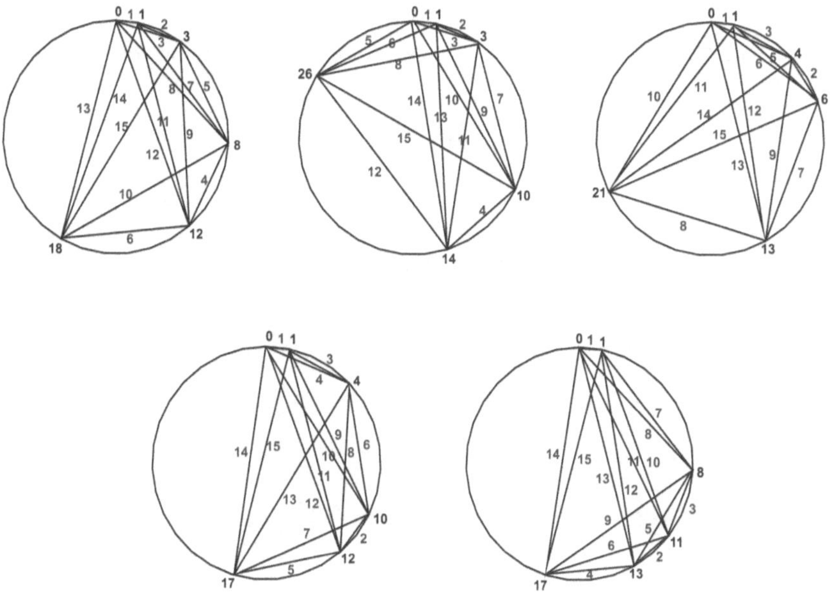

Fig. 1. The unique quadruplet of nonachords for $N = 31$.

4 Musical Applications

Tom Johnson tackled homometries in 2013 and since has continued to deal with it. Although he does not use micro-interval resources, he works with $\mathbb{Z}_N$ rhythms. And as a composer, has written extensively on music [5,7] and recently published in a collection all of his texts and explanatory notes [8]. *Intervals, for flexible instrumentation* (2013) connects in different ways the 48 transforms of homometric sets $\{0, 1, 3, 7\}$ and $\{0, 1, 4, 6\}$. For instance, the connection of homometric sets by tritons depicted on Fig. 2, leads to harmonic progressions (Fig. 3).

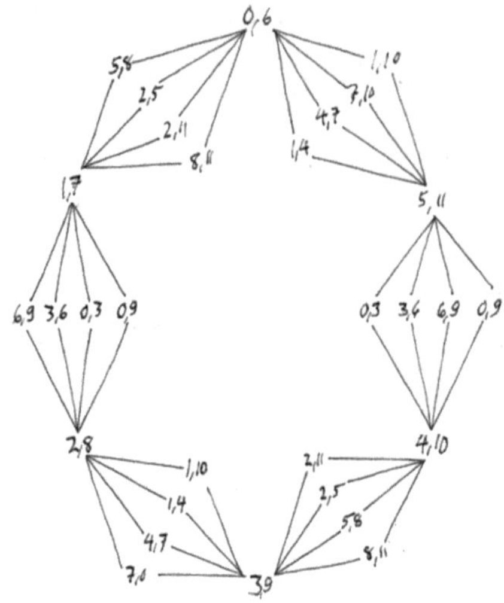

Fig. 2. Tom Johnson, Connections of homometric sets by tritons.

Fig. 3. Tom Johnson, The harmonic progressions.

Fig. 4. Tom Johnson, Slight Variations.

The *Slight Variations* (2015) for piano, composed two years later, are built on the Z-related hexachords $\{0, 1, 3, 4, 5, 7\}$ and $\{0, 2, 3, 4, 5, 8\}$. Since each formation can be read, say Tom, either clockwise or counter-clockwise in twelve different positions, they permit a total of forty-eight variations (see Fig. 4)

More recent is the piece composed for two wood-blocks *Homometry 108* (2019), based on 108 homometric chords and three permutations. The drawing Fig. 5 shows these three pairs in a special way. Four points in each circle are black, and two points are white. To make the five-note chords one simply

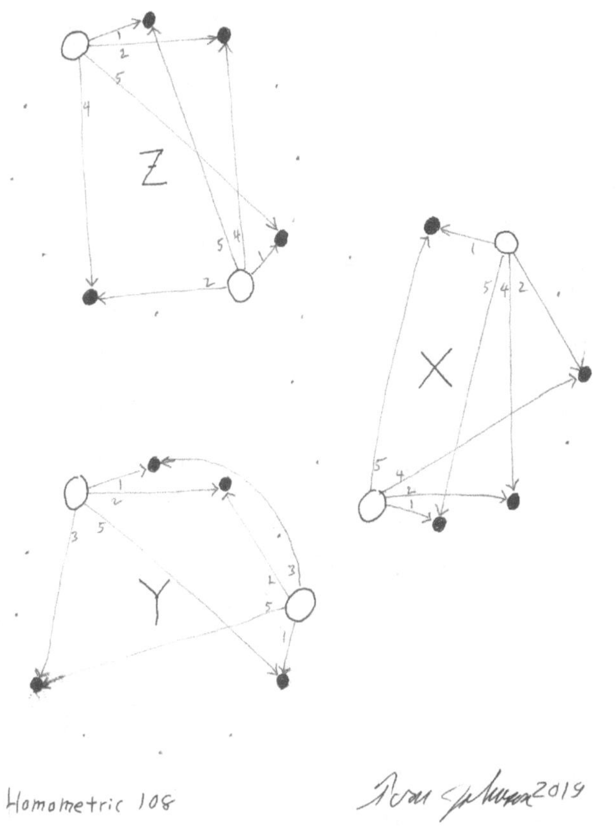

Fig. 5. Tom Johnson, Homometry 108.

adds one of the white points to the four black points. The two white points are isospectral, meaning that they each have the same distances with the other four notes, as indicated by the arrows. Thus in all three systems the two five-note pairs have the same number of distances and are homometric.

References

1. Caddeo, R., Hascher, X., Jedrzejewski, F., Papadopoulos, A.: Christiaan Huygens: É crits sur la musique et le son, (in French). Hermann, Paris (2020)
2. Forte, A.: The Structure of Atonal Music, 2nd edn. Yale University Press, New Haven (1977)
3. Genuys, G., Popoff, A.: Homometry in the dihedral groups: lifting sets from $\mathbb{Z}_n$ to D_n. In: Agustin-Aquino, O., Lluis-Puebla, E., Montiel, M. (eds.) Mathematics and Computation in Music. Lecture Notes in Computer Science(), vol. 10527, pp. 38-49. Springer, Cham (2017). https://doi.org/10.1007/978-3-319-71827-9_4
4. Goyette, J.S.: The Z-Relation in Theory and Practice, Ph.D. thesis, University of Rochester, New York (2012)

5. Johnson, T., Jedrzejewski, F.: The structure of Z-related sets. In: Yust, J., Wild, J., Burgoyne, J.A. (eds.) Mathematics and Computation in Music. Lecture Notes in Computer Science(), vol. 7937, pp. 128–137. Springer, Berlin (2013). https://doi.org/10.1007/978-3-642-39357-0_10
6. Jedrzejewski, F.: A Compendium of Musical Mathematics. World Scientific, Singapore (2024)
7. Johnson, T., Jedrzejewski, F.: Looking at Numbers. Birkhauser, Basel (2013)
8. Johnson, T.: Finding Music. Writings 1961-2018. MusikTexte, Köln (2019)
9. Lemke, P., Skiena, S.S., Smith, W.D.: Reconstructing sets from interpoint distances. In: Aronov, B., Basu, S., Pach, J., Sharir, M. (eds.) Discrete and Computational Geometry. Algorithms and Combinatorics, vol. 25, pp. 597–631. Springer, Berlin (2003). https://doi.org/10.1007/978-3-642-55566-4_27
10. Lewin, D.: On extended Z-triples. Theory Pract. **7**, 38–39 (1981)
11. Lewin, D.: Generalized Musical Intervals and Transformations. Yale University Press, New Haven (1987)
12. Mandereau, J., et al.: Z-relation and homometry in musical distributions. J. Math. Music **5**(2), 83–98 (2011)
13. Mandereau, J., et al.: Discrete phase retrieval in musical structures. J. Math. Music **5**(2), 99–116 (2011)
14. O'Rourke, J., Taslakian, P., Toussaint, G.: A pumping lemma for homometric rhythms. In: Proceedings of the 20th Canadian Conference on Computational Geometry, pp. 121–123 (2008)
15. Patterson, A.L.: Ambiguities in the x-ray analysis of crystal structures. Phys. Rev. **65**(5–6), 195–201 (1944)
16. Rosenblatt, J., Seymour, P.D.: The structure of homometric sets. Siam J. Alg. Discrete Methods **3**(3), 343–350 (1982)
17. Rosenblatt, J.: Phase retrieval. Commun. Math. Phys. **95**, 317–343 (1984)
18. Soderberg, S.: Z-related sets as dual inversions. J. Music Theory **39**(1), 77–100 (1995)
19. Wild, J.: Enumerating set-classes and Z-related tuplets in equal temperaments of up to thirty-one notes per octave. Unpublished Graduate Seminar Paper, McGill University (1996)

The Sandwich-Lemma: The Recursive Structure of Super-Syntonic and Super-Diatonic Automorphisms

Thomas Noll[1(✉)], David Clampitt[2], and Mariana Montiel[3]

[1] Departament de Creació i Teoria Musical, Escola Superior de Música de Catalunya, Barcelona, Spain
tnoll@esmuc.cat, thomas.mamuth@gmail.com
[2] School of Music, Ohio State University, Columbus, USA
clampitt.4@osu.edu
[3] Department of Mathematics and Statistics, Georgia State University, Atlanta, USA
mmontiel@gsu.edu

Abstract. In the context of the transformational modelling of Pairwise Well-Formed (PWWF) modes, we study transformations σ_f and τ_g on four letters a, b, c, d, which we call *super-syntonic* and *super-diatonic* morphisms, respectively. They are constructed from underlying *syntonic* and *diatonic* morphisms f and g on two letters a, b. From f and g, which are instances of Sturmian morphisms, one also constructs the Well-Formed (WF) syntonic and diatonic projections of the PWWF mode in question. From a super-syntonic morphism σ_f one obtains the PWWF authentic triadic mode $\pi_{d\to a}(\sigma_f(c)|\sigma_f(b)||\sigma_f(a))$ and the PWWF plagal triadic mode $\pi_{d\to a}(\sigma_f(d)||\sigma_f(c)|\sigma_f(b))$. In Sect. 2 we complete an earlier investigation by proving that the *kaleidoscope maps* σ and τ are monoid homomorphisms, i.e. $\sigma_{f_1 f_2} = \sigma_{f_1}\sigma_{f_2}$ and $\tau_{g_1 g_2} = \tau_{g_1}\tau_{g_2}$. In Sect. 3 we show that the recursive "sandwich" construction of syntonic and diatonic morphisms can be lifted to the level of super-syntonic and super-diatonic morphisms. This result enables us to show, that super-syntonic and super-diatonic morphisms, which are constructed in this way, are automorphisms of the free group F_4 with four generators.

Keywords: Pairwise Well-Formed Scales · Kaleidoscope Transformations · Diatonic and Syntonic Modes · Sturmian Morphisms · Algebraic Combinatorics on Words

1 Introduction: F_4-Automorphisms and PWWF Modes

The present paper is a contribution to the transformational modelling of (non-singular) Pairwise Well-Formed modes along the lines of [12–14]. In particular it fulfils a promise, which Noll and Clampitt announced in [14]. The concept of a *Pairwise Well-Formed Scale* (PWWF scale) has been introduced and developed in [3], and summarized in [4]. The underlying concept of a *(non-degenerate) Well-Formed Scale* (WF scale) applies to *generated scales* and highlights the structural

© The Author(s), under exclusive license to Springer Nature Switzerland AG 2024
T. Noll et al. (Eds.): MCM 2024, LNCS 14639, pp. 85–96, 2024.
https://doi.org/10.1007/978-3-031-60638-0_7

role of the generating interval in connection with a circular binary step interval pattern. PWWF scales are not generated, and their circular step interval patterns are made of three different step intervals. It would be misleading, though, to think of a PWWF scale as a "weaker" type of scale. Its ternary step interval pattern is, in fact, a combination of two WF step interval patterns and represents a more complex type of scale. In particular there is an alternation of two of the step intervals, which are of the same multiplicity.[1] Under the modal point of view the circular step interval patterns are replaced by the step interval fillings of a division of the octave ambit. In the WF case one distinguishes authentic and plagal divisions of the octave into generator and co-generator. In the PWWF case one considers a ternary division of the octave in the form of a generic triad.

The basic idea behind the transformational approach is to regard the acts of step interval fillings as transformations. And in the WF case there is a genuine mathe-musical realisation of this idea in terms of automorphisms of the free group F_2 with two generators. The refined processes of zooming into smaller and smaller step intervals corresponds to the concatenation of these automorphisms. An attempt towards an analogous transformational approach to the PWWF case with *kaleidoscope transformations* acting on PWWF words in [12] succeeds with respect to the aspect of filling the generic triad, but suffers from the violation of closure of these transformations under concatenation. Therefore we intend to get to the bottom of a new idea from [14], namely to bring a fourth letter into play. We study certain transformations σ_f on four letters, which we call *super-syntonic* and *super-diatonic* morphisms, respectively. They are constructed from underlying *syntonic* and *diatonic* morphisms on two letters, corresponding to the WF projections of the PWWF mode in question. From a super-syntonic morphism σ_f one obtains the authentic triadic mode $\pi_{d\to a}(\sigma_f(c)|\sigma_f(b)||\sigma_f(a))$ and the plagal triadic mode $\pi_{d\to a}(\sigma_f(d)||\sigma_f(c)|\sigma_f(b))$. Thereby $\pi_{d\to a} : \{a,b,c,d\}^* \to \{a,b,c\}^*$ denotes the projection of the letter d to the letter a, leaving the other letters a, b, c unchanged.

We show that the recursive "sandwich" construction of syntonic and diatonic morphisms from [12] can be lifted to the level of super-syntonic and super-diatonic morphisms. The main goal of this paper is then to show, that these super-syntonic and super-diatonic morphisms are automorphisms of the free group F_4. The difficulty thereby is the invertibility. If one applies the *Graham construction* for super-syntonic morphisms to the inverses of syntonic morphisms, one does not obtain the inverses of the corresponding super-syntonic morphisms. Therefore, we express all these morphisms as concatenations of Nielsen transformations [11], which generate the automorphism group $Aut(F_4)$. Besides letter permutations and letter inversions these are the *prepend* and *append* substitutions $P_{x,y}$ and A_{xy} for letters $x \neq y \in \{a,b,c,d\}$:

$$P_{xy}(x) = yx \text{ and } P_{xy}(z) = z \text{ for } z \neq x, \quad A_{xy}(x) = xy \text{ and } A_{xy}(z) = z \text{ for } z \neq x.$$

[1] There is one singular type PWWF scale with seven notes whose step interval pattern is *abacaba*, where all three step intervals are of different multiplicities, which is exemplified by the "Hungarian minor" scale. It is not in the scope of this article and we skip the attribute non-singular.

2 The Graham Maps and the Kaleidoscope Symmetry

In this section we build substitutions on words on the four letters $\{a, b, c, d\}$ from syntonic and diatonic morphisms, which are substitutions on words on the two letters $\{a, b\}$. Noll and Clampitt [12] characterise the implied redundancy with the metaphor of a kaleidoscope. Later we will understand that these super-syntonic and super-diatonic substitutions are F_4-automorphisms. But for now we have to treat them as monoid morphisms of the word monoid $\{a, b, c, d\}^*$.

2.1 Syntonic and Diatonic Morphisms

In the case of the free group F_2 with two generators a and b, Nielsen's prepend- and append-substitutions are also known as *Sturmian morphisms* $G = P_{ba}, \tilde{G} = A_{ba}, D = P_{ab}, \tilde{D} = A_{ab}$. The letter exchange is denoted by E. Together they generate the *Sturmian monoid* $\mathcal{S} = \left\langle G, \tilde{G}, D, \tilde{D}, E \right\rangle \subset Aut(F_2)$. The submonoid of standard morphisms is $\mathcal{S}_{st} = \langle G, D, E \rangle$.[2] We define syntonic and diatonic morphisms in terms of their commutative images in $GL_2(\mathbb{N}) = \{\begin{pmatrix} a & b \\ c & d \end{pmatrix} \,|\, ad - bc = \pm 1, a, b, c, d \in \mathbb{N}\}$. We start with the submonoids of all matrices with non-negative entries within the congruence subgroups $\Gamma_0(2)$ and $\Gamma^0(2)$, i.e. $\Sigma = \{\begin{pmatrix} a & b \\ c & d \end{pmatrix} \in SL_2(\mathbb{N}) \,|\, c \equiv 0 \, mod \, 2\}$, and $\Delta = \{\begin{pmatrix} a & b \\ c & d \end{pmatrix} \in SL_2(\mathbb{N}) \,|\, b \equiv 0 \, mod \, 2\}$. Further we extend them by the matrices $\begin{pmatrix} 1 & 1 \\ 2 & 1 \end{pmatrix}$ and $\begin{pmatrix} 1 & 2 \\ 1 & 1 \end{pmatrix}$ of determinant -1 in $GL_2(\mathbb{N})$, respectively, and define

$$\overline{\Sigma} := \left\langle \Sigma, \begin{pmatrix} 1 & 1 \\ 2 & 1 \end{pmatrix} \right\rangle, \quad \overline{\Delta} := \left\langle \Delta, \begin{pmatrix} 1 & 2 \\ 1 & 1 \end{pmatrix} \right\rangle.$$

Definition 2.1. *A* syntonic morphism *is a Sturmian morphism* $f \in \mathcal{S}$, *whose incidence matrix* M_f *is in* $\overline{\Sigma}$. *A* diatonic morphism *is a Sturmian morphism* $g \in \mathcal{S}$, *whose incidence matrix* M_g *is in* $\overline{\Delta}$.

We denote the monoids of syntonic and diatonic morphisms with the symbols:

$$\overline{Syn} := \{f \in \mathcal{S} \,|\, M_f \in \overline{\Sigma}\}, \quad \overline{Dia} := \{g \in \mathcal{S} \,|\, M_g \in \overline{\Delta}\}$$

Syntonic morphisms f have the property that the multiplicity $|f(a)|_b$ of the letter b in $f(a)$ is even and its multiplicity $|f(b)|_b$ in $f(b)$ is odd. This is the basis for the Graham construction on the word $f(bba)$, which has an even multiplicity $2|f(b)]_b + |f(a)]_b$ of the letter b. Diatonic morphisms g have the property that the multiplicity $|g(b)|_a$ of the letter a in $g(b)$ is even and its multiplicity $|g(a)|_a$ in $g(a)$ is odd, which is the basis for an analogous Graham construction.

[2] For a subset X of monoid morphisms (either of $\{a, b\}^*$ or $\{a, b, c, d\}^*$) let $\langle X \rangle$ denote the monoid generated by X.

2.2 The Graham Construction

Starting from a word over the two-letter alphabet $\{a, b\}$, the idea is basically to replace every other instance of the letter b by an instance of c (this is the Graham construction for PWWF words in 3 letters a, b, c (see [3] for the music-theoretical context and [7] for an earlier combinatorial application) and to replace every power of a by the corresponding power of d, whenever it follows an instance of b which is being replaced by c (this is our extension of the Graham construction to 4 letters). We first define the syntonic Graham map γ and its Kaleidoscope-opposite γ^*:

Definition 2.2. *We consider the map* $\gamma : \{a, b\}^* \to \{a, b, c, d\}^*$ *where for any word* $w \in \{a, b\}^*$

$$\gamma(w)_k := \begin{cases} c \text{ if } w_k = b \text{ and } |w_1 \dots w_k|_b \text{ is odd,} \\ d \text{ if } w_k = a \text{ and } |w_1 \dots w_k|_b \text{ is odd,} \\ b \text{ if } w_k = b \text{ and } |w_1 \dots w_k|_b \text{ is even,} \\ a \text{ if } w_k = a \text{ and } |w_1 \dots w_k|_b \text{ is even.} \end{cases}$$

γ *shall be called the* syntonic Graham map. *Further, let* $\gamma^* = E_{(bc)(ad)} \circ \gamma$ *denote the associated* kaleidoscope opposite. *Thereby* $E_{(bc)(ad)}$ *denotes the letter permutation* $b \leftrightarrow c, a \leftrightarrow d$.

Definition 2.3. *Consider a syntonic morphism* $f \in \overline{Syn}$. *Then we associate with* f *the following super-syntonic morphism* $\sigma_f : \{a, b, c, d\}^* \to \{a, b, c, d\}^*$:

$$\sigma_f(a) := \gamma(f(a)), \sigma_f(b) := \gamma^*(f(b)), \sigma_f(c) := \gamma(f(b)), \sigma_f(d) := \gamma^*(f(a)).$$

The image of the monoid $\overline{Syn}$ *under* σ *is denoted as* $\overline{SuperSyn}$.

Analogously, we apply the same procedure to diatonic morphisms with sensitivity to the parity of the multiplicities of the letter a:

Definition 2.4. *We consider the map* $\delta : \{a, b\}^* \to \{a, b, c, d\}^*$ *where for any word* $w \in \{a, b\}^*$

$$\delta(w)_k := \begin{cases} a \text{ if } w_k = a \text{ and } |w_1 \dots w_k|_a \text{ is odd,} \\ b \text{ if } w_k = b \text{ and } |w_1 \dots w_k|_a \text{ is odd,} \\ c \text{ if } w_k = b \text{ and } |w_1 \dots w_k|_a \text{ is even,} \\ d \text{ if } w_k = a \text{ and } |w_1 \dots w_k|_a \text{ is even.} \end{cases}$$

δ *shall be called the* diatonic Graham map. *Let further* $\delta^* = E_{(bc)(ad)} \circ \delta$ *denote the associated* kaleidoscope opposite.

Definition 2.5. *Consider a diatonic morphism* $g \in \overline{Dia}$. *Then we associate with* g *the following super-diatonic morphism* $\tau_g : \{a, b, c, d\}^* \to \{a, b, c, d\}^*$:

$$\tau_g(a) := \delta(g(a)), \tau_g(b) := \delta^*(g(b)), \tau_g(c) := \delta(g(b)), \tau_g(d) := \delta^*(g(a)).$$

The image of the monoid $\overline{Dia}$ *under* τ *is denoted as* $\overline{SuperDia}$.

The maps $\sigma : \overline{Syn} \to \overline{SuperSyn}$ and $\tau : \overline{Dia} \to \overline{SuperDia}$ shall be called the (syntonic and diatonic) kaleidoscope homomorphisms, respectively. In Subsect. 2.3 we show that the homomorphy property is fulfilled.

2.3 The Kaleidoscope Maps are Monoid Homomorphisms

We verify that the super-syntonic image of a composed syntonic morphism σ_{fg} coincides with the composition of the associated super-syntonic images, i.e. $\sigma_f\sigma_g$.

Proposition 2.6. *Consider a factorisation $w = uv \in \{a,b\}^*$ of a two-letter-word in two factors u and v. Then the Graham images $\gamma(w), \gamma^*(w)$, $\gamma(u), \gamma^*(u)$, and $\gamma(v), \gamma^*(v)$ are interrelated as follows:*

$$\gamma(w) = \begin{cases} \gamma(u)\gamma(v) \text{ when } |u|_b \text{ is even,} \\ \gamma(u)\gamma^*(v) \text{ when } |u|_b \text{ is odd.} \end{cases}$$
$$\gamma^*(w) = \begin{cases} \gamma^*(u)\gamma^*(v) \text{ when } |u|_b \text{ is even,} \\ \gamma^*(u)\gamma(v) \text{ when } |u|_b \text{ is odd.} \end{cases}$$

Proof. Let $|u|_b = n$ and $|v|_b = m$, and we write the factors u and v explicitly as

$$\begin{aligned} u &= (a^{s_1}ba^{r_1})(a^{s_2}ba^{r_2})...(a^{s_n}ba^{r_n}) \quad \text{for some } r_1,...,r_n,s_1,...,s_n \geq 0 \\ v &= (a^{t_1}ba^{u_1})(a^{t_2}ba^{u_2})(a^{t_3}ba^{u_3})\ldots \quad \text{for some } t_1,...,t_m,u_1,...,u_m \geq 0 \end{aligned}.$$

It turns out that the splitting of the a-powers into the neighboring factors, such as of $a^{r_1+s_2}$ into a^{r_1} and a^{s_2}, can be arbitrarily chosen.

We apply γ and γ^* to u and v, respectively:

$$\begin{aligned} \gamma(u) &= (a^{s_1}cd^{r_1})(d^{s_2}ba^{r_2})\ldots \begin{cases} (d^{s_n}ba^{r_n}) \text{ when } n \text{ is even,} \\ (a^{s_n}cd^{r_n}) \text{ when } n \text{ is odd.} \end{cases} \\ \gamma^*(u) &= (d^{s_1}ba^{r_1})(a^{s_2}cd^{r_2})\ldots \begin{cases} (a^{s_n}cd^{r_n}) \text{ when } n \text{ is even,} \\ (d^{s_n}ba^{r_n}) \text{ when } n \text{ is odd.} \end{cases} \\ \gamma(v) &= (a^{t_1}cd^{u_1})(d^{t_2}ba^{u_2})(a^{t_3}cd^{u_3})\ldots \\ \gamma^*(u) &= (d^{t_1}ba^{u_1})(a^{t_2}cd^{u_2})(d^{t_3}ba^{u_3}) \end{aligned}$$

First suppose, that n is even. Then we obtain:

$$\begin{aligned} \gamma(w) &= (a^{s_1}cd^{r_1})(d^{s_2}ba^{r_2})\ldots(d^{s_n}ba^{r_n})(a^{t_1}cd^{u_1})(d^{t_2}ba^{u_2})\ldots = \gamma(u)\gamma(v) \\ \gamma^*(w) &= (d^{s_1}ba^{r_1})(a^{s_2}cd^{r_2})\ldots(a^{s_n}cd^{r_n})(d^{t_1}ba^{u_1})(a^{t_2}cd^{u_2})\ldots = \gamma^*(u)\gamma^*(v) \end{aligned}$$

Now suppose, that n is odd. Then we analogously obtain:

$$\begin{aligned} \gamma(w) &= (a^{s_1}cd^{r_1})(d^{s_2}ba^{r_2})\ldots(a^{s_n}cd^{r_n})(d^{t_1}ba^{u_1})(a^{t_2}cd^{u_2})\ldots = \gamma(u)\gamma^*(v) \\ \gamma^*(w) &= (d^{s_1}ba^{r_1})(a^{s_2}cd^{r_2})\ldots(d^{s_n}ba^{r_n})(a^{t_1}cd^{u_1})(d^{t_2}ba^{u_2})\ldots = \gamma^*(u)\gamma(v) \end{aligned}$$ □

Lemma 2.7. *The parity $Mod(|w|_b, 2)$ of the multiplicity of the letter b in any word $w \in \{a,b\}^*$ remains invariant under the application of any syntonic morphism f, i.e. $Mod(|w|_b, 2) = Mod(|f(w)|_b, 2)$.*

Proof. Consider a syntonic morphism f with its incidence matrix

$$M_f = \begin{pmatrix} |f(a)|_a & |f(b)|_a \\ |f(a)|_b & |f(b)|_b \end{pmatrix} = \begin{pmatrix} r & s \\ 2u & v \end{pmatrix} \in \overline{\Sigma}$$

and a word $w \in \{a,b\}^*$. Then we obtain $|f(w)|_b = 2u|w|_a + v|w|_b$, hence $Mod(|f(w)|_b, 2) = Mod(v|w|_b, 2) = Mod(|w|_b, 2)$ because v is odd.

By virtue of Lemma 2.7 we may immediately formulate the following generalization of Proposition 2.6.

Corollary 2.8. *Consider a factorisation* $w = uv \in \{a,b\}^*$ *of a two-letter-word in two factors* u *and* v *and consider a syntonic morphism* f*. Then the Graham images* $\gamma(f(w)), \gamma^*(f(w))$, $\gamma(f(u)), \gamma^*(f(u))$, *and* $\gamma(f(v)), \gamma^*(f(v))$ *are interrelated as follows:*

$$\gamma(f(w)) = \begin{cases} \gamma(f(u))\gamma(f(v)) \text{ when } |u|_b \text{ is even,} \\ \gamma(f(u))\gamma^*(f(v)) \text{ when } |u|_b \text{ is odd.} \end{cases}$$
$$\gamma^*(f(w)) = \begin{cases} \gamma^*(f(u))\gamma^*(f(v)) \text{ when } |u|_b \text{ is even,} \\ \gamma^*(f(u))\gamma(f(v)) \text{ when } |u|_b \text{ is odd.} \end{cases}$$

Proposition 2.9. *For any syntonic morphism* f *the following two commutation relations hold:*

$$\gamma \circ f = \sigma_f \circ \gamma, \text{ and } \gamma^* \circ f = \sigma_f \circ \gamma^*$$

Proof. Consider any word $w \in \{a,b\}^*$. We have to show that $\gamma(f(w)) = \sigma_f(\gamma(w))$. We do this by induction over the length k of the word w. Looking at the word $w = a^{\pm 1}$ of length 1 we have $|w_b| = 0$ is even, hence $\sigma_f(\gamma(w)) = \sigma_f(a) = \gamma(f(a)) = \gamma(f(w))$. Analogously, for $w = b$ of length 1 we have $|w_b| = 1$ is odd, and hence $\sigma_f(\gamma(w)) = \sigma_f(\gamma(b)) = \sigma_f(c) = \gamma(f(b)) = \gamma(f(w))$. Now we consider a word w of length k and write $w = uw_k$ and suppose that $\gamma(f(u)) = \sigma_f(\gamma(u))$. We distinguish four cases (applying proposition 2.6 and corollary 2.8):

1. $w_k = a$ and $|u|_b$ is even:

$$\begin{aligned} \sigma_f(\gamma(ua)) = \sigma_f(\gamma(u)\gamma(a)) &= \sigma_f(\gamma(u))\sigma_f(a) = \gamma(f(u))\gamma(f(a)) \\ = \gamma(f(u)f(a)) &= \gamma(f(ua)) \qquad = \gamma(f(w)). \end{aligned}$$

2. $w_k = a$ and $|u|_b$ is odd:

$$\begin{aligned} \sigma_f(\gamma(ua)) = \sigma_f(\gamma(u)\gamma^*(a)) &= \sigma_f(\gamma(u))\sigma_f(d) = \gamma(f(u))\gamma^*(f(a)) \\ = \gamma(f(u)f(a)) &= \gamma(f(ua)) \qquad = \gamma(f(w)). \end{aligned}$$

3. $w_k = b$ and $|u|_b$ is even:

$$\begin{aligned} \sigma_f(\gamma(ub)) = \sigma_f(\gamma(u)\gamma^*(b)) &= \sigma_f(\gamma(u))\sigma_f(c) = \gamma(f(u))\gamma^*(f(b)) \\ = \gamma(f(u)f(b)) &= \gamma(f(ub)) \qquad = \gamma(f(w)). \end{aligned}$$

4. $w_k = b$ and $|u|_b$ is odd:

$$\begin{aligned} \sigma_f(\gamma(ub)) = \sigma_f(\gamma(u)\gamma(b)) &= \sigma_f(\gamma(u))\sigma_f(b) = \gamma(f(u))\gamma(f(b)) \\ = \gamma(f(u)f(b)) &= \gamma(f(ub)) \qquad = \gamma(f(w)). \end{aligned}$$

□

Corollary 2.10. *For any given syntonic morphisms* $f, g \in \overline{Syn}$ *we have* $\sigma_{fg} = \sigma_f \sigma_g$

Proof. We apply the commutation relations from proposition 2.9 to the words $g(a)$ and $g(b)$, respectively and obtain:

$$\sigma_{fg}(a) = \gamma(f(g(a))) = \sigma_f(\gamma(g(a))) = \sigma_f(\sigma_g(a)),$$
$$\sigma_{fg}(b) = \gamma^*(f(g(b))) = \sigma_f(\gamma^*(g(b))) = \sigma_f(\sigma_g(b)). \qquad \square$$

An analogous result holds for the composition of diatonic morphisms and their super-diatonic images under τ:

Proposition 2.11. *For any given diatonic morphisms $f, g \in \overline{Dia}$ we have $\tau_{fg} = \tau_f \tau_g$.*

The proof is analogous to the syntonic case.

3 Syn-Diatonic Recursion and the Sandwich-Lemma

In this section we extend our findings on the basis of generators on three levels, namely firstly for syntonic and diatonic matrices, secondly for syntonic and diatonic morphisms and thirdly for super-syntonic and super-diatonic morphisms.

3.1 Recursive Generation of Syntonic and Diatonic Morphisms

The monoid $SL_2(\mathbb{N})$ is freely generated by the two matrices $R = \begin{pmatrix} 1 & 1 \\ 0 & 1 \end{pmatrix}$ and $L = \begin{pmatrix} 1 & 0 \\ 1 & 1 \end{pmatrix}$. And while the congruence groups $\Gamma_0(2)$ and $\Gamma^0(2)$ are finitely generated, it turns out that the monoids $\Sigma = \Gamma_0(2) \cap SL_2(\mathbb{N})$ and $\Delta = \Gamma^0(2) \cap SL_2(\mathbb{N})$ of those congruence matrices with non-negative entries are not. The two matrices R and LL generate a proper submonoid $\Sigma_0 = \langle R, LL \rangle \subset \Sigma$ and the two matrices RR and L generate a proper submonoid $\Delta_0 = \langle RR, L \rangle \subset \Delta$. Any matrix $X \in \Sigma \backslash \Sigma_0$, which is not a product of two Σ_0-elements is necessarily of the "sandwich-form" $X = LRYRL$ for some $Y \in \Delta$. To see this, let us first suppose that X is not a product of Σ_0-elements, then there must be a single instance of L which must be followed by an instance of R. But as LR itself is not syntonic, it cannot be followed only be instances of LL and R. At some stage there must be a single instance of L, preceded by an instance of R (simply because LRL alone is not syntonic). The same kind of argument can be then recursively applied to general Σ-elements (see Proposition 4.3 in [12]). Likewise, any matrix $Y \in \Delta \backslash \Delta_0$, which is not a product of two Δ_0-elements is necessarily of the form $Y = RLXLR$ for some $X \in \Sigma$. Successively we obtain larger and larger submonoids $\Sigma_1 = \langle \Sigma_0, LR\Delta_0 RL \rangle$ and $\Delta_1 = \langle \Delta_0, LR\Sigma_0 RL \rangle$, $\Sigma_2 = \langle \Sigma_1, LR\Delta_1 RL \rangle$ and $\Delta_2 = \langle \Delta_1, LR\Sigma_1 RL \rangle$, etc. In the limit we obtain $\Sigma = \bigcup_{n=0}^{\infty} \Sigma_n$ and $\Delta = \bigcup_{n=0}^{\infty} \Delta_n$. Including $U = \begin{pmatrix} 1 & 1 \\ 2 & 1 \end{pmatrix}$ and $V = \begin{pmatrix} 1 & 2 \\ 1 & 1 \end{pmatrix}$ at all stages of this recursion gives us also access to syntonic and diatonic matrices of determinant -1: $\overline{\Sigma}_n := \langle \Sigma_n, U \rangle$, $\overline{\Delta}_n := \langle \Delta_n, U \rangle$, and obtain again $\overline{\Sigma} := \langle \Sigma, U \rangle$, $\overline{\Delta} := \langle \Delta, V \rangle$.

The monoid $\mathcal{S}_{st}$ of Standard Morphisms is in bijection to the monoid $GL_2(\mathbb{N})$ via commutative images. In particular we have $M_G = R$, $M_D = L$, $M_E = J$ with $J = \begin{pmatrix} 0 & 1 \\ 1 & 0 \end{pmatrix}$. Furthermore we have $M_{GDE} = RLJ = U$ and $M_{DGE} = LRJ = V$. Thus we can transfer the same recursive construction to standard syntonic morphisms Syn_{st} and standard diatonic morphisms Dia_{st}.

$$
\begin{array}{llll}
Syn^0_{st} & := \langle G, DD, GDE \rangle, & Dia^0_{st} & := \langle D, GG, DGE \rangle \\
Syn^1_{st} & := \langle Syn^0_{st}, DG \cdot Dia^0_{st} \cdot GD \rangle, & Dia^1_{st} & := \langle Dia_{st}, GD \cdot Syn^0_{st} \cdot DG \rangle, \\
 & \vdots & & \vdots \\
Syn^{n+1}_{st} & := \langle Syn^n_{st}, DG \cdot Dia^n_{st} \cdot GD \rangle, & Dia^{n+1}_{st} & := \langle Dia^n_{st}, GD \cdot Syn^n_{st} \cdot DG \rangle, \\
Syn_{st} & := \bigcup_{n=0}^{\infty} Syn^n_{st} & Dia_{st} & := \bigcup_{n=0}^{\infty} Dia^n_{st}
\end{array}
$$

The same procedure can be now applied in order to define larger submonoids $Syn \subset \overline{Syn}$ and $Dia \subset \overline{Dia}$:

$$
\begin{aligned}
Syn^0 &:= \left\langle G, \tilde{G}, DD, \tilde{D}D, \tilde{D}\tilde{D}, DGE, D\tilde{G}E, \tilde{D}GE, \tilde{D}\tilde{G}E \right\rangle \\
Dia^0 &:= \left\langle D, \tilde{D}, GG, \tilde{G}G, \tilde{G}\tilde{G}, GDE, G\tilde{D}E, \tilde{G}DE, \tilde{G}\tilde{D}E \right\rangle
\end{aligned}
$$

$$
\begin{array}{llll}
Syn^1 & := \langle Syn^0, DG \cdot Dia^0 \cdot GD \rangle, & Dia^1 & := \langle Dia_{st}, GD \cdot Syn_0 \cdot DG \rangle, \\
 & \vdots & & \vdots \\
Syn^{n+1} & := \langle Syn^n, DG \cdot Dia^n \cdot GD \rangle, & Dia^{n+1} & := \langle Dia^n, GD \cdot Syn^n \cdot DG \rangle, \\
Syn & := \bigcup_{n=0}^{\infty} Syn^n & Dia & := \bigcup_{n=0}^{\infty} Dia^n
\end{array}
$$

3.2 The Sandwich-Lemma

In this subsection we show that all super-syntonic morphisms $f \in Syn$ are F_4-automorphisms. The proof follows the recursive construction of Syn as described in the previous subsection. And the main challenge in this recursion is the mutual encapsulation of diatonic and syntonic morphisms in the two "sandwiches" $DGgGD$ and $GDfGD$, respectively. The following result provides the Nielsen representations for the "sandwich-halves", respectively.

Proposition 3.1 (Sandwich-Lemma). *For $g \in Dia$ and $f \in Syn$ we have:*

$$\sigma_{DGgGD} = \rho\tau_g\tilde{\rho} \text{ and } \tau_{GDfDG} = \tilde{\rho}\sigma_f\rho, \text{ where}$$

$$\rho = P_{dc}P_{ab}P_{bd}P_{ca} = (ba, cdb, bac, cd), \tilde{\rho} = P_{ba}P_{cd}P_{ac}P_{db} = (dca, ab, dc, abd).$$

Proof. We prove it for the case, where the "insides" are (super-)diatonic (g and τ_g) and the asserted "outside" equation relates super-syntonic morphisms: $\sigma_{DGgGD} = \rho\tau_g\tilde{\rho}$. The proof for the mirrored equation is completely analogous. The equation actually entails equalities for the images of the four letters, namely $\sigma_{DGgGD}(x) = \rho\tau_g\tilde{\rho}(x)$ for $x = a, b, c, d$. By virtue of the kaleidoscope relations for σ_{DGgGD} on the left side and all three factors $\rho, \tau_g, \tilde{\rho}$ on the right side, it is sufficient to do the proof for the images of the letters $x = a$ and $x = c$.

Furthermore we may observe, that for both sides the images of c are prefixes of the images of a. Both Graham maps (syntonic and diatonic) have the property that equal prefixes of arguments become equal prefixes of the corresponding images. On the left side $\sigma_{DGgGD}(c) = \gamma(DGg(ab))$ is therefore a prefix of $\sigma_{DGgGD}(a) = \gamma(DGg(aba))$ and on the right side is $\rho(\tau_g(\tilde{\rho}(c))) = \rho(\tau_g(dc))$ a prefix of $\rho(\tau_g(\tilde{\rho}(a))) = \rho(\tau_g(dca))$. As the word lengths $|\gamma(DGg(ab))|$ and $|\rho(\tau_g(dc))|$ also coincide, we can conclude that the validity of the Sandwhich-Lemma for the images under a imply its validity for the images under c. Writing out the morphisms DG, GD, ρ and $\tilde{\rho}$ we have to prove the following equation:

$$\gamma((ba, bab)(g(a)g(b)g(a)) = (ba, cdb, bac, cd)(\delta^*(g(a))\delta(g(b))\delta(g(a)))$$

In the next step we use the symmetry in $g(a)g(b)g(a)$ and $\delta^*(g(a))\delta(g(b))\delta(g(a))$ to further reduce the equation to be proven. On the left side, where $g(b)$ is enclosed between two copies of $g(a)$, we have to apply the syntonic Graham map γ, which is sensitive to the parity of the letter b-frequency. The parity of $|(ba, bab)(g(b))|_b = |g(b)|_a + 2|g(b)|_b$ is even, because $|g(b))|_a$ is even. The parity of $|(ba, bab)(g(a))|_b = |g(a)|_a + 2|g(a)|_b$ is odd, because $|g(a)|_a$ is odd. Taken together these two facts imply, that the $g(a)$-induced prefix and the $g(a)$-induced suffix of length $|(ba, bab)(g(a))| = 2|g(a)|_a + 3|g(a)|_b$ in $\gamma((ba, bab)(g(a)g(b)g(a))$ are always Kaleidoscope-opposites of each other. In the equation, to be proven, this is confirmed by the fact, that $\delta^*(g(a))$ and $\delta(g(a))$ are Kaleidoscope opposites and that this property is preserved under the application of the Kaleidoscope-symmetric automorphism $\rho = (ba, cdb, bac, cd)$. It remains to be shown that

$$\gamma((ba, bab)(g(a)g(b))) = (ba, cdb, bac, cd)(\delta^*(g(a))\delta(g(b)))$$

Now we proceed inductively along the prefixes of $g(a)g(b)$. First we write $g(a) = uxv$ with prefix u, letter x and some suffix v. And we write $\tilde{u} = DG(u)$. And under the assumption that $\gamma(\tilde{u}) = \rho(\delta^*(u))$, we show that $\gamma(\tilde{u}DG(x)) = \rho(\delta^*(ux))$. The parity of $|\tilde{u}|_b = |(ba, bab)(u)|_b = |u|_a + 2|u|_b$ coincides with the parity of $|u|_a$. If $|u|_a$ is even, which includes the case where x is the first letter, then $|\tilde{u}|_b$ is even and the decisive first letter in the image of $DG(x)$ after $\tilde{u}$ is b, both for $x = a$ and $x = b$, and thus we find:

$$\begin{aligned}\gamma((ba, bab)(ua)) &= \gamma(\tilde{u}ba) &= \gamma(\tilde{u})cd &= \rho(\delta^*(u))cd &= \rho(\delta^*(u))\rho(d) = \rho(\delta^*(ua))\\ \gamma((ba, bab)(ub)) &= \gamma(\tilde{u}bab) &= \gamma(\tilde{u})cdb &= \rho(\delta^*(u))cdb &= \rho(\delta^*(u))\rho(b) = \rho(\delta^*(ub))\end{aligned}$$

Analogously, we find for odd $|u|_a$:

$$\begin{aligned}\gamma((ba, bab)(ua)) &= \gamma(\tilde{u}ba) &= \gamma(\tilde{u})ba &= \rho(\delta^*(u))ba &= \rho(\delta^*(u))\rho(a) = \rho(\delta^*(ua))\\ \gamma((ba, bab)(ub)) &= \gamma(\tilde{u}bab) &= \gamma(\tilde{u})bac &= \rho(\delta^*(u))bac &= \rho(\delta^*(u))\rho(c) = \rho(\delta^*(ub))\end{aligned}$$

When we proceed with the proof to include $g(b)$ as well, we have to take into consideration that γ on the left side is applied to $DG(g(a)g(b)) = DGg(ab)$ as one word, while δ is applied to $g(b)$ separately. Therefore it is important to note that $|g(a)|_a$ is odd. And hence, for prefixes u of $g(b)$, the parity of $|g(a)u|_a$ is opposite to the parity of $|u|_a$ alone. And while the calculation on the left side proceeds as before, we now obtain the Kaleidoscope opposite $\rho(\delta(g(b)))$ of $\rho(\delta^*(g(b)))$ on the right side. □

Corollary 3.2. *Super-syntonic morphisms σ_f for $f \in Syn$ and super-diatonic morphisms τ_g for $g \in Dia$ are automorphisms of the free group F_4.*

Proof. In accordance with the homomorphy of σ and τ, with the sandwich lemma, and in the light of the recursive definition of the monoids Syn and Dia it is sufficient to have Nielsen representations for the 18 generators (two times 9) for the syntonic and diatonic morphisms in the initial monoids Syn^0 and Dia^0, i.e. for

$$\{G, \tilde{G}, DD, \tilde{D}D, \tilde{D}\tilde{D}, DGE, D\tilde{G}E, \tilde{D}GE, \tilde{D}\tilde{G}E\}$$
$$\{D, \tilde{D}, GG, \tilde{G}G, \tilde{G}\tilde{G}, GDE, G\tilde{D}E, \tilde{G}DE, \tilde{G}\tilde{D}E\}$$

We have listed them in Table 1.

Table 1. Nielsen representations for elementary super-syntonic morphisms and super-diatonic morphisms

$f \in Syn$ $(f(a), f(b))$	σ_f $\sigma_f(a,b,c,d)$	$g \in Dia$ $(g(a), g(b))$	τ_g $\tau_g(a,b,c,d)$
G (a, ab)	$P_{bd}P_{ca}$ (a, db, ac, d)	D (ba, b)	$P_{ac}P_{db}$ (ca, b, c, bd)
$\tilde{G}$ (a, ba)	$A_{ba}A_{cd}$ (a, ba, cd, d)	$\tilde{D}$ (ab, b)	$A_{ab}A_{dc}$ (ab, b, c, dc)
DD (bba, b)	$P_{ab}P_{ac}P_{dc}P_{db}$ (cba, b, c, bcd)	GG (a, aab)	$P_{ba}P_{bd}P_{cd}P_{ca}$ (a, dab, adc, d)
$D\tilde{D}$ (bab, b)	$A_{db}A_{ac}P_{ab}P_{dc}E_{(ad)}$ (cdb, b, c, bac)	$G\tilde{G}$ (a, aba)	$P_{cd}P_{ba}A_{bd}A_{ca}E_{(bc)}$ (a, dca, abd, d)
$\tilde{D}\tilde{D}$ (abb, b)	$A_{ac}A_{ab}A_{db}A_{dc}$ (acb, b, c, dbc)	$\tilde{G}\tilde{G}$ (a, baa)	$A_{bd}A_{ca}A_{cd}A_{ba}$ (a, bda, cad, d)
DGE (bab, ba)	$P_{ab}P_{dc}P_{ca}P_{bd}E_{(ab)(cd)}$ (cdb, ba, cd, bac)	GDE (ab, aba)	$P_{cd}P_{ba}P_{db}P_{ac}E_{(ab)(cd)}$ (ab, dca, abd, dc)
$D\tilde{G}E$ (bba, ba)	$P_{ab}P_{dc}A_{bd}A_{ca}E_{(acdb)}$ (cba, ba, cd, bcd)	$\tilde{G}DE$ (ba, baa)	$A_{bd}A_{ca}P_{ab}P_{dc}E_{(acdb)}$ (ca, bda, cad, bd)
$\tilde{D}GE$ (abb, ab)	$A_{db}A_{ac}P_{ba}P_{cd}E_{(abdc)}$ (acb, db, ac, dbc)	$G\tilde{D}E$ (ab, aab)	$P_{ba}P_{cd}A_{db}A_{ac}E_{(abdc)}$ (ab, dab, adc, dc)
$\tilde{D}\tilde{G}E$ (bab, ba)	$A_{ac}A_{db}A_{ba}A_{cd}E_{(ac)(bd)}$ (cdb, db, ac, bac)	$\tilde{G}\tilde{D}E$ (ba, aba)	$A_{ca}A_{bd}A_{ab}A_{dc}E_{(ac)(bd)}$ (ca, dca, abd, bd)

A detailed verification requires 18 calculations, which the reader can carry out easily. We illustrate this here for the example of the *Zarlino* morphism $f = DGE$ (left half of the 6th row in Table 1): The syntonic morphism $DGE(a, b) = DG(b, a) = D(ab, a) = (bab, ba)$ entails the super-syntonic morphism

$$\sigma_{EGE} = (\gamma(bab), \gamma^*(ba), \gamma(ba), \gamma^*(bab)) = (cdb, ba, cd, bac).$$

We verify the asserted Nielsen representation $P_{ab}P_{dc}P_{ca}P_{bd}E_{(ab)(cd)}$ for σ_{DGE}, consisting of 4 prepend-substitutions and the letter permutation $E_{(ab)(cd)}$:

$$\begin{aligned} P_{ab}P_{dc}P_{ca}P_{bd}E_{(ab)(cd)}(a,b,c,d) &= P_{ab}P_{dc}P_{ca}P_{bd}(b,a,d,c) \\ &= P_{ab}P_{dc}P_{ca}(db,a,d,c) \\ &= P_{ab}P_{dc}(db,a,d,ac) \\ &= P_{ab}(cdb,a,cd,ac) \\ &= (cdb,ba,cd,bac) \end{aligned}$$

4 Outlook

There is a desire to know whether all super-syntonic and super-diatonic morphisms are F_4 automorphisms. But there remains a gap which is manifest in the difference sets $\overline{Syn}\backslash Syn$, $\overline{Dia}\backslash Dia$ and has to do with the 11 conjugates of $DGGD$ (analogously for $DGGD$). For all these conjugates $DGGD$, $DGG\tilde{D}$, $DG\tilde{G}D$, $DG\tilde{G}\tilde{D}$, ..., $\tilde{D}\tilde{G}\tilde{G}\tilde{D}$ one finds Nielsen representations, but only four of them (in each case) have a corresponding sandwich lemma. They are shown in Table 2.

Table 2. "Sandwichable" super-syntonic and super-diatonic morphisms.

$f \in Syn$	σ_f
$(f(a), f(b))$	$\sigma_f(a,b,c,d)$
$DGGD$	$(P_{dc}P_{ab}P_{bd}P_{ca})(P_{ba}P_{cd}P_{ac}P_{db})$
$(bababba, babab)$	$(cdbacba, bacdb, cdbac, bacdbcd)$
$DG\tilde{G}\tilde{D}$	$(P_{dc}P_{ab}P_{bd}P_{ca})(A_{bd}A_{ca}A_{ab}A_{dc})$
$(bababba, babba)$	$(cdbacba, bacba, cdbcd, bacdbcd)$
$\tilde{D}\tilde{G}GD$	$(A_{ac}A_{db}A_{cd}A_{ba})(P_{ba}P_{cd}P_{db}P_{ac})$
$(abbabab, abbab)$	$(acbacdb, dbcdb, acbac, dbcdbac)$
$\tilde{D}\tilde{G}\tilde{G}\tilde{D}$	$(A_{ac}A_{db}A_{ba}A_{cd})(A_{ca}A_{bd}A_{dc}A_{ab})$
$(abbabab, babab)$	$(acbacdb, bacdb, cdbac, dbcdbac)$
$g \in Dia$	τ_g
$(g(a), g(b))$	$\tau_g(a,b,c,d)$
$GDDG$	$(P_{ba}P_{cd}P_{ac}P_{db})(P_{dc}P_{ab}P_{bd}P_{ca})$
$(ababa, ababaab)$	$(abdca, dcabdab, abdcadc, dcabd)$
$\tilde{G}\tilde{D}DG$	$(A_{bd}A_{ca}A_{ab}A_{dc})(P_{dc}P_{ab}P_{bd}P_{ca})E_{(ad)(bc)}$
$(baaba, baababa)$	$(cadca, bdabdca, cadcabd, bdabd)$
$GD\tilde{D}\tilde{G}$	$(P_{ba}P_{cd}P_{db}P_{ac})(A_{ac}A_{db}A_{cd}A_{ba})E_{(ad)(bc)}$
$(abaab, ababaab)$	$(abdab, dcabdab, abdcadc, dcadc)$
$\tilde{G}\tilde{D}\tilde{D}\tilde{G}$	$(A_{ca}A_{bd}A_{dc}A_{ab})(A_{ac}A_{db}A_{ba}A_{cd})$
$(baaba, babaaba)$	$(abdca, bdabdca, cadcabd, dcabd)$

What we know is, that any syntonic morphism f has a standard conjugate f' for which $\sigma_{f'}$ is a F_4-automorphism. We conjecture that every σ_f has a Nielsen representation whose number of prepend or append substitutions is twice the number of Gs, $\tilde{G}$s, Ds or $\tilde{D}$s in the representation of f. This should also hold for the diatonic case. If this can be proven, there is a next challenge on the agenda, namely to investigate the generated groups $\langle\langle\overline{SuperSyn}\rangle\rangle$ and $\langle\langle\overline{SuperDia}\rangle\rangle$ and to explore also their inverted elements and composites musically.

References

1. Bulgakova, D.V., Buzhinsky, N., Goncharov, Y.O.: On balanced and abelian properties of circular words over a ternary alphabet ArXiv:2012:15818, WORDS 2021 (2021)
2. Chuman, Y.: Generators and relations of $\Gamma\_0(N)$. Kyoto J. Math. **13**(2), 381–390 (1973)
3. Clampitt, D.: Pairwise Well-Formed Scales: Structural and Transformational Properties. Ph.D. diss, SUNY at Buffalo (1997)
4. Clampitt, D.: Mathematical and musical properties of pairwise well-formed scales. In: Klouche, T., Noll, T. (eds.) MCM 2007. CCIS, vol. 37, pp. 464–468. Springer, Heidelberg (2009). https://doi.org/10.1007/978-3-642-04579-0_46
5. Clampitt, D., Noll, T.: Modes, the height-width duality, and Handschin's tone character. Music Theory Online **17**(1) (2011). http://www.mtosmt.org/issues/mto.11.17.1/mto.11.17.1.clampitt_and_noll.html
6. De Luca, A.: Sturmian words: structure, combinatorics, and their arithmetics. Theor. Comput. Sci. **183**, 45–83 (1997)
7. Graham, R.L.: Covering the positive integers by disjoint sets of the form $\{[n\alpha+\beta] : n = 1, 2, \ldots\}$. J. Comb. Theory (A) **15**, 354–358 (1973)
8. Kassel, C., Reutenauer, C.: Sturmian morphisms, the braid group B_4, Christoffel words and bases of F_2 (2005). https://doi.org/10.48550/arXiv.math/0507219
9. Koblitz, N.: Introduction to Elliptic Curves and Modular Forms. Springer, New York (1984)
10. Lothaire, M.: Algebraic Combinatorics on Words. Cambridge University Press, Cambridge (2002)
11. Nielsen, J.: Die Isomorphismengruppe der freien Gruppen. Math. Ann. **33**, 169–209 (1924)
12. Noll, T., Clampitt, D.: Kaleidoscope substitutions and pairwise well-formed modes: Major-Minor duality transformationally revisited. J. Math. Music **12**(3) (2018)
13. Noll, T., Clampitt, D.: Exploring the syntonic side of major-minor tonality. In: Montiel, M., Gomez-Martin, F., Agustín-Aquino, O.A. (eds.) MCM 2019. LNCS (LNAI), vol. 11502, pp. 125–136. Springer, Cham (2019). https://doi.org/10.1007/978-3-030-21392-3_10
14. Noll, T., Clampitt, D.: Transformations for pairwise well-formed modes. In: Montiel, M., Agustín-Aquino, O.A., Gómez, F., Kastine, J., Lluis-Puebla, E., Milam, B. (eds.) MCM 2022. LNCS, vol. 13267, pp. 140–152. Springer, Cham (2022). https://doi.org/10.1007/978-3-031-07015-0_12
15. Paquin, G., Reutenauer, C.: On the superimposition of Christoffel words. Theoret. Comput. Sci. **412**, 402–418 (2011)

Hidden Categories: A New Perspective on Lewin's Generalized Interval Systems and Klumpenhouwer Networks

Alexandre Popoff[1(✉)] and Moreno Andreatta[2,3]

[1] Paris, France
al.popoff@free.fr

[2] CNRS/Institute for Advanced Mathematical Research, ITI CREAA, University of Strasbourg, Strasbourg, France
andreatta@math.unistra.fr

[3] IRCAM, Paris, France
andreatta@ircam.fr

Abstract. In this work we provide a categorical formalization of several constructions found in transformational music theory. We first revisit David Lewin's construction of a Generalized Interval System (GIS) to show that even a subset of the GIS conditions already implies a sequence of functors between categories. When all the conditions in Lewin's definition are fullfilled, this sequence involves the category of elements $\int_{\mathbf{G}} S$ for the group action $S\colon \mathbf{G} \to \mathbf{Sets}$ implied by the GIS structure. By focusing on the role played by categories of elements in such a context, we reformulate previous definitions of transformational networks in a **Cat**-based diagrammatic perspective, and present a new definition of categorical transformational networks, or *CT-Nets*, in more general musical categories. We show how such an approach provides a bridge between algebraic, geometrical, and graph-theoretical approaches in transformational music analysis. We end with a discussion on the new perspectives opened by such a formalization of transformational theory, in particular with respect to **Rel**-based transformational networks which occur in well-known music-theoretical constructions such as Douthett's and Steinbach's Cube Dance.

Keywords: Transformational Music Theory · Generalized Interval System · David Lewin · Category Theory · Transformational Networks · CT-Nets · Cube Dance

1 Introduction

Transformational theory represents a challenging topic in contemporary "mathemusical" research. It not only constitutes a turning point in the field of music analysis but also naturally leads to fundamental questions about the object-based

A. Popoff—Independent Researcher.

© The Author(s), under exclusive license to Springer Nature Switzerland AG 2024

T. Noll et al. (Eds.): MCM 2024, LNCS 14639, pp. 97–110, 2024.
https://doi.org/10.1007/978-3-031-60638-0_8

vs. operation-based duality in the formalization of musical structures and processes. Starting from the original group-based contributions by David Lewin in the field of transformational music theory and analysis, the role of category theory has been quickly recognized [13–15,17,24] as a useful foundation in order to build extended generalizations. Following previous work on the categorical formalization of Lewin's transformational theory and transformational networks in music analysis, this paper has two main goals. In a first part, we show how a restricted form of Lewin's definition of Generalized Interval Systems implicitly defines diagrams of categories. In case all the conditions of Lewin's definition are fullfilled, these diagrams involve the notion of categories of elements. We take this observation as the starting point for further generalization and discuss the importance of such categories for transformational music theory and for graph-based geometrical approach. This shift in focus allows us, in a second part, to revisit Klumpenhouwer networks and to provide a new definition of categorical transformational networks ("CT-Nets"), as well as that of morphisms between them.

2 Revisiting Lewin's Generalized Interval Systems from a Categorical Perspective

In this section, we will show that a Generalized Interval System (GIS) of Lewin [10] inherently defines a category, which coincides with the category of elements for the group action the GIS defines due to the additional conditions imposed by Lewin.

2.1 Lewin's Generalized Interval Systems

We recall here Lewin's definition of a Generalized Interval System (GIS) for transformational music analysis.

Definition 1 (Lewin, 1987). *A Generalized Interval System (GIS) is a triple* $(X, IVLS, int)$ *where*

- X *is a set of musical elements,*
- $IVLS$ *is a group (the group of intervals for the GIS), and*
- *int is a function* $X \times X \to IVLS$

such that

1. *for all* x, y, *and* z *in* X, $int(x, y) * int(y, z) = int(x, z)$, *and*
2. *for all* $x \in X$ *and* $g \in IVLS$, *there exists a unique* $y \in X$ *such that* $int(x, y) = g$.

We know from Vuza [28, p. 270] and Kolman [9, p. 157] that the data of a GIS is equivalent to the data of a simply transitive (right) group action of $IVLS$ on the set X, or equivalently, of a representable functor $\mathbf{IVLS}^{op} \to \mathbf{Sets}$[1]Note that

[1] By an abuse of notation, for a given group G, we will notate throughout this paper its corresponding single-object category as $\mathbf{G}$. All functors are assumed to be covariant.

Lewin [10, pp 157–158] also discusses how a GIS can be obtained from a simply transitive left action of a group $STRANS$ on a set, by constructing a group $IVLS$ isomorphic to $STRANS^{\mathrm{op}}$. This result was extended to an equivalence of categories between the category of GISs and the category of left simply transitive actions by Fiore [5, p. 10]. Note that if we replace condition (1) by $\mathrm{int}(y,z) * \mathrm{int}(x,y) = \mathrm{int}(x,z)$, we obtain the data of a simply transitive left group action of $IVLS$ on X.

The condition (2) is necessary to construct the group action of $IVLS$ on X from the data of the GIS. Indeed, we can observe that it enforces the surjectivity of the interval function int. Without condition (2), the function $\mathrm{int}\colon X \times X \to IVLS$ may not be surjective, and there would then be elements of $IVLS$ for which we cannot define their action on X. However we show in the following proposition that condition (1) alone in Lewin's definition inherently defines the data of a category $\mathbf{C}$ along with a functor to the group associated to the function int. Note that we use the alternative form of condition (1) to agree with the usual composition of morphisms in a category.

Proposition 1. *We consider a triple (X, G, int) where*

- *X is a set of musical elements,*
- *G is a group, and*
- *int is a function $X \times X \to G$ such that for all x, y, and z in X, $int(y,z) * int(x,y) = int(x,z)$.*

This defines the data of a category $\mathbf{C}$ along with a functor $\mathbf{C} \to \mathbf{G}$.

Proof. We construct $\mathbf{C}$ as the category having X as its set of objects, such that for all $(x,y) \in X^2$, $\mathrm{Hom}(x,y) = \{(x,y)\}$. This is called the *indiscrete category on X*. Composition of morphisms is straightforward, with $(y,z)(x,y) = (x,z)$. By the properties of the function int, we can immediately define a functor from $\mathbf{C}$ to $\mathbf{G}$. □

We now recall the definition of the category of elements (also called the Grothendieck construction) for a functor $S\colon \mathbf{C} \to \mathbf{Sets}$.

Definition 2. *Let $S\colon \mathbf{C} \to \mathbf{Sets}$ be a functor from a category $\mathbf{C}$ to $\mathbf{Sets}$. The category of elements $\int_{\mathbf{C}} S$ is defined as the category having*

- *objects of the form (X, x) with X an object of $\mathbf{C}$ and x an element of the set $S(X)$, and*
- *morphisms between objects (X, x) and (Y, y) of the form (x, f, y) with f being a morphism of $\mathbf{C}$ such that $y = S(f)(x)$.*

There is a canonical projection functor $\pi_S\colon \int_{\mathbf{C}} S \to \mathbf{C}$ sending each object (X, x) to X.

If both condition (1) and (2) are used in Proposition 1, we then obtain the following immediate corollary.

Corollary 1. *Let (X, G, int) be a triple as defined in Proposition 1, such that for all $x \in X$ and $g \in G$, there exists a unique $y \in X$ such that $int(x, y) = g$. Then, the category* **C** *defined in Proposition 1 is the category of elements $\int_{\mathbf{G}} S$ for the simply transitive group action $S\colon \mathbf{G} \rightarrow$* **Sets** *derived from (X, G, int)*[2].

In the following subsection, we explore how these results provide insights into the relationship between transformational music theory and category theory.

2.2 From Lewin's Generalized Interval Systems to Categories, and to Graphs

As we have seen above, the condition (2) enforces the existence of a simply transitive group action of G on X, i.e. a functor $S\colon \mathbf{G} \rightarrow \mathbf{Sets}$ mapping the single object of $\mathbf{G}$ to X and which is representable. This gives rise to a sequence of functors of the form $\int_{\mathbf{G}} S \rightarrow \mathbf{G} \rightarrow \mathbf{Sets}$. Recent advances in transformational music theory have made use of various groups acting on sets of musical elements. Such groups do not always act in a simply transitive manner: this is the case for example of the T/I group acting on pitch classes, or Hook's UTT group [6] acting on major and minor triads. Despite the absence of a corresponding GIS structure, they nevertheless are useful and widely used tools for transformational analysis [12].

In this regard, dropping condition (2) altogether encourages us to shift our focus to categories rather than group actions, and to consider the functor $\mathbf{C} \rightarrow \mathbf{G}$ implied by Proposition 1 as a key notion. We will show in this section how we can generalize this observation by going beyond the definition of the interval function int.

As a first observation, it can be noted that for any group G acting on a set X we can always form the corresponding category of elements $\int_{\mathbf{G}} S$. While properly defining an interval function int might be complicated if the group does not act in a simply transitive way, the definition of $\int_{\mathbf{G}} S$ and the associated functor $\int_{\mathbf{G}} S \rightarrow \mathbf{G}$ already contains the necessary properties of composability and associativity.

Recent research has emphasized the significance of extending beyond Lewin's original group-based approach, by, for instance, exploring the use of groupoids or categories in general [21,23]. In this view, the group $\mathbf{G}$ and its action on a set is replaced with a general functor $S\colon \mathbf{C} \rightarrow \mathbf{Sets}$, and we can thus consider the corresponding category of elements $\int_{\mathbf{C}} S$ together with the canonical projection functor $\int_{\mathbf{C}} S \rightarrow \mathbf{C}$.

This generalization can be extended further by recognizing that the definition of the category of elements expands beyond functors to **Sets**. In particular, this notion can be defined for functors $S\colon \mathbf{C} \rightarrow \mathbf{Rel}$, in which **Rel** is the category of finite sets and binary relations between them [27]. Such functors have been shown to occur in transformational music analysis, for example as in the algebraic

[2] In the particular case of a group action, the category of elements $\int_{\mathbf{G}} S$ is also called the *action groupoid*. As discussed in the next section, this paper considers more general cases and we thus retain "category of elements" as a common unifying term.

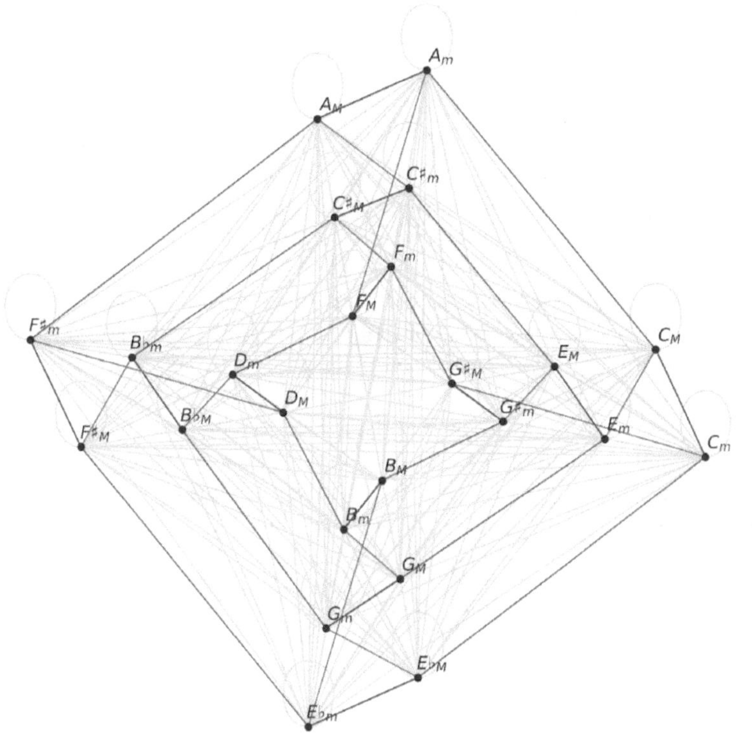

Fig. 1. The quiver corresponding to the category of elements for the simply transitive action of the PLR group on the set of major and minor triads. Edges (corresponding to invertibles morphisms) are represented with undirected edges colored in gray for clarity, except for morphisms projecting on R (red), P (blue), and L (green) in the PLR group. This subquiver corresponds to the "chicken-wire torus" [4]. Best viewed in color. (Color figure online)

formalization of the "Cube Dance" graph [19,22]. One advantage of using the category of elements $\int_{\mathbf{C}} S$ is that it "forgets" the target category of the functor S, whether it is **Rel** or **Sets**, which allows one to consider them on an equal footing.

In passing, it can also be observed that the category of elements is a discrete analogue to the fundamental groupoid of a topological space: objects of the fundamental groupoid are points of the space, and 1-morphisms are paths between these points. This reconciles with Lewin's view of transformations as ways to pass from one object to another[3].

Finally, it can be noted that the notion of category of elements establishes a bridge between algebraic transformational approaches and geometrical

[3] During the redaction of this manuscript, the authors have been made aware by Dmitri Tymoczko of his current work on groupoids. Category of elements are encountered as a common point between his approach, rooted in geometrical and topological considerations, and ours, which stems from algebraic ones.

graph-based approaches which have been studied extensively [2–4,18,25,26]. Indeed, it is well known that there is a forgetful functor from **Cat** to **Quiv**, the category of quivers, i.e. directed multigraphs ("multidigraphs"). The forgetful functor returns the underlying multidigraph of a category and forgets all information about composition of arrows. Certain well-known musical graphs can be obtained from the application of this forgetful functor. For example, the so-called "chicken-wire torus" [4] is a sub-multidigraph of the multidigraph obtained by application of the forgetful functor $\mathbf{Cat} \to \mathbf{Quiv}$ on the category of elements $\int_{PLR} S$ for the usual action S of the PLR group on the set of major and minor triads. This is illustrated in Fig. 1, in which the multidigraph obtained from the category of elements is depicted in gray (since the action is simply transitive, the corresponding quiver is a complete one), with the "chicken-wire torus" highlighted in color.

This colored multidigraph is a special case of the category-based general case, in which we consider the canonical projection functor $\int_{\mathbf{C}} S \to \mathbf{C}$ for a general functor $S\colon \mathbf{C} \to \mathbf{Sets}$. In this context, applying the forgetful functor $\mathbf{Cat} \to \mathbf{Quiv}$ induces additional structure, as shown in the construction below.

Construction 1. *Given a functor* $\mathbf{U} \to \mathbf{V}$, *the forgetful functor* $\mathbf{Cat} \to \mathbf{Quiv}$ *induces a coloring on nodes and edges of the multidigraph underlying* $\mathbf{U}$. *The nodes are colored in the set of nodes of the multidigraph corresponding to* $\mathbf{V}$, *while the edges are colored in the set of edges of same multidigraph.*

In the above example, the category of elements $\int_{\mathbf{PLR}} S$ and the associated functor $\int_{\mathbf{PLR}} S \to \mathbf{PLR}$ for the usual action S of the PLR group on the set of major and minor triads gives an edge-colored multidigraph in which each edge is colored (labelled) by an element of the PLR group. Note that we have colored the edges in Fig. 1 in gray merely for clarity. The nodes have only one color since we are considering a single-object category.

3 A New Definition of Transformational Networks

By following our shift of focus on categories of elements, we consider in this section a new definition of transformational networks which relates musical objects and transformations between them in a consistent diagram. Transformational networks first appeared in Lewin's work [10], and so-called *Klumpenhouwer networks* were developed by Klumpenhouwer [7,8,11] to specifically study pitch classes and their transformations by the T/I group. In recent years, transformational networks have been revisited with new definitions from a categorical perspective [16,20,21,24]. Our starting point is the definition of [21] which we recall below.

Definition 3. *Let* $\mathbf{C}$ *be a category, and* S *a functor from* $\mathbf{C}$ *to the category* **Sets** *of (small) sets. Let* Δ *be a small category and* R *a functor from* Δ *to* **Sets** *with non-empy values. A PK-net of form* R *and of support* S *is a 4-tuple* (R, S, F, ϕ), *in which* F *is a functor from* Δ *to* $\mathbf{C}$, *and* ϕ *is a natural transformation from* R *to* SF.

As per the discussion above, our objective is to shift the focus from **Sets**-based functors to the corresponding categories of elements, which we do by leveraging the following theorem.

Theorem 2. *Assume we have a PK-net (R, S, F, ϕ) of form R and of support S. Then there exists a functor H_F between the category of elements $\int_\Delta R$ and $\int_{\mathbf{C}} S$ such that the top square in the following diagram commute.*

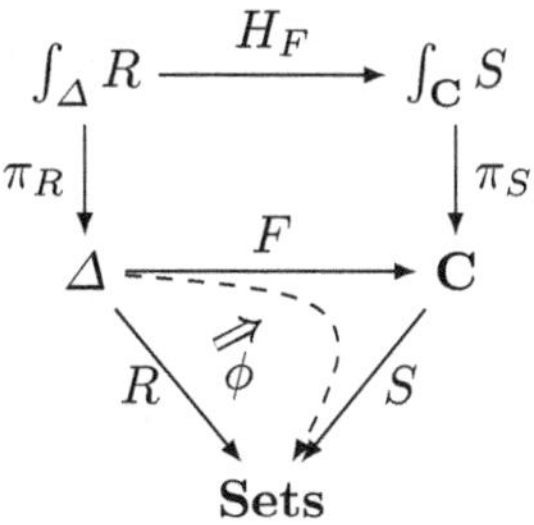

Proof. We construct the functor H_F as follows.

- For any object (X, x) of $\int_\Delta R$, the functor H_F sends it to $(F(X), \phi_X(x))$.
- For any morphism (x, f, y) between two objects (X, x) and (Y, y) of $\int_\Delta R$, the functor H_F sends this morphism to $(\phi_X(x), F(f), \phi_Y(y))$.

The construction in the first item is consistent, since by the definition of the natural transformation ϕ, $\phi_X(x)$ is indeed an element of $SF(X)$. Similarly, the construction in the second item is also consistent, since for any (x, y) such that $y = R(f)(x)$, we have $\phi_Y(y) = SF(f)(\phi_X(x))$. We have $H_F((y, f', z)(x, f, y)) = H_F((y, f', z))H_F((x, f, y))$ by horizontal composition of the natural transformation. □

The definition of PK-nets allows us to define networks in which nodes are sets, and not necessarily singletons. In this diagram, the role of the functor $\int_\Delta R \to \Delta$ is to remember which objects of $\int_\Delta R$ belong to the same set. The fact that the top square is commutative ensures that objects corresponding to elements of the same set are transformed by the same transformations.

While this possibility gives more generality to transformational networks, the majority of actual transformational analyses in the musicological literature uses networks with only one musical object per node, i.e. singletons. In this case, $\int_\Delta R$ is the same category as Δ, and since the data of the content of sets is no longer relevant, this makes the functor $\int_\Delta R \to \Delta$ somewhat unnecessary. We will thus now consider the specific case of singleton-based networks and drop this functor altogether. We can then rename $\int_\Delta R$ as Δ to indicate its new role as a category defining the skeleton of the transformational network to be considered.

We therefore obtain a sequence of functors of the form $\Delta \xrightarrow{M} \int_{\mathbf{C}} S \xrightarrow{\pi_S} \mathbf{C}$ which encodes transformational networks with a single musical element on each node. The nodes and arrows of the network are given by the objects and morphisms of the category Δ. The functor M maps the objects of Δ to the objects of $\int_{\mathbf{C}} S$ which, as we have seen above, corresponds to musical elements. In other words, the functor M labels the nodes of the network with musical elements. In addition, it maps the arrows of the network to morphisms of $\int_{\mathbf{C}} S$, which are then given a label in $\mathbf{C}$ (for example, group elements if $\mathbf{C}$ is a group) by the functor $\pi_S \colon \int_{\mathbf{C}} S \to \mathbf{C}$. Thus, the functor $\pi_S \circ M$ labels the arrows of Δ with morphisms (transformations) of $\mathbf{C}$.

Example 1. The transformational network represented below

$$\begin{array}{ccc} B & \xrightarrow{T_2} & C_\sharp \\ {\scriptstyle I_3}\downarrow & & \downarrow{\scriptstyle I_9} \\ E & \xrightarrow[T_4]{} & G_\sharp \end{array}$$

can be conceived as a sequence of functors $\Delta \xrightarrow{M} \int_{\mathbf{T/I}} S \xrightarrow{\pi_S} \mathbf{T/I}$, where $S \colon \mathbf{T/I} \to \mathbf{Sets}$ is the usual action of the T/I group on pitch-classes, Δ is the commutative square $\begin{array}{ccc} X & \xrightarrow{f} & Y \\ {\scriptstyle h}\downarrow & & \downarrow{\scriptstyle g} \\ Z & \xrightarrow[k]{} & W \end{array}$, and where

$$M(X) = B, M(Y) = C_\sharp, M(Z) = E, M(W) = G_\sharp, \text{and}$$

$$\pi_S \circ M(f) = T_2, \pi_S \circ M(g) = I_9, \pi_S \circ M(h) = I_3, \pi_S \circ M(k) = T_4.$$

In a final generalization, we can observe that, since we do not consider **Sets**-based (or **Rel**-based) functors anymore, we can consider all categories in our definition of a transformational network, and not just category of elements. We thus obtain the following definition.

Definition 4. *A categorical transformational network (CT-Net) is a sequence of functors of the form* $\Delta \xrightarrow{M} \mathbf{C}_{el} \xrightarrow{\pi} \mathbf{C}_T$*, where*

- Δ *is a category representing the skeleton of the network,*
- $\mathbf{C}_{el}$ *is a category whose objects represent musical objects of interest, with morphisms between them, and*
- $\mathbf{C}_T$ *is a category whose objects represent classes of musical objects, and musical transformations between them.*

The nodes of the network are labelled by the images of the objects of Δ *by* M*, and the edges of the network are labelled by the images of the morphisms of* Δ *by* $\pi \circ M$*.*

Note that we do not *a priori* impose conditions on the categories $\mathbf{C}_{\mathrm{el}}$ and $\mathbf{C}_T$, nor on the functor π. In fact, $\mathbf{C}_T$ may well be $\mathbf{C}_{\mathrm{el}}$ itself, in which case transformations are individualized for each musical object.

Since our definition of transformational networks is entirely categorical, all constructions in **Cat** (pullbacks, pushforwards, equalizers, etc.) can in fact be used to construct new CT-Nets, as shown in the following example.

Example 2. Consider the two networks represented below,

$$\begin{array}{ccc} C\sharp & \xrightarrow{I_5} & E \\ {\scriptstyle I_2}\downarrow & \swarrow{\scriptstyle T_7} & \\ E & & \end{array} \qquad\qquad \begin{array}{ccc} C\sharp & \xrightarrow{I_5} & E \\ {\scriptstyle T_5}\downarrow & \swarrow{\scriptstyle I_{10}} & \\ F\sharp & & \end{array}$$

viewed as in the previous example as sequences of functors $\Delta \overset{M}{\underset{M'}{\rightrightarrows}} \int_{\mathbf{T/I}} S \longrightarrow \mathbf{T/I}$. Taking the equalizer on the functors M and M' returns a new sequence of functors

$$\Delta' \longrightarrow \Delta \overset{M}{\underset{M'}{\rightrightarrows}} \int_{\mathbf{T/I}} S \longrightarrow \mathbf{T/I}$$

and thus a new CT-Net representing the network each have in common, i.e. $C\sharp \xrightarrow{I_5} E$. Since such categorical constructions can readily be implemented computationally, and more easily than the previously exposed transformational network frameworks, this opens new perspectives for computational score analysis. Furthermore, as mentioned above, the passage from **Cat** to **Quiv** is functorial which allows to translate these constructions into the corresponding multidigraphs (and morphisms between them).

4 Morphisms of CT-Nets

In this section, we revisit the notions of morphisms of transformational networks and explore their implications for diagrams involving categories of elements. We will build upon the morphisms of PK-Nets defined in previous work, and in this view we will revert to category of elements for functors $\mathbf{C} \to \mathbf{Sets}$ before giving the final definition in the general case.

We first consider the notion of *complete homographies* introduced in the formalization of PK-Nets [20]. We recall their definition below.

Definition 5. *Let* (R, S, F, ϕ) *be a PK-Net, and assume we have a functor* $S' \colon \mathbf{C}' \to \mathbf{Sets}$. *A complete homography is a pair* (N, ν) *with*

- *$N\colon \mathbf{C} \to \mathbf{C}'$, and*
- *$\nu\colon SF \to S'FN$ such that $\nu = \tilde{\nu}F$, where $\tilde{\nu}$ is a natural transformation from S to $S'N$*

Applying (N, ν) on the PK-net (R, S, F, ϕ) transforms it into the new PK-Net $(R, S', FN, \nu \circ \phi)$.

When $\mathbf{C}$ and $\mathbf{C}'$ are groups, with simply transitive group actions, this is known as a morphism of GIS [5]. For a given complete homography, we can apply Theorem 2 on the diagram formed by S and $S'N$. This defines a functor H_N from $\int_{\mathbf{C}} S$ to $\int_{\mathbf{C}'} S'$. Starting from a transformational network $\Delta \xrightarrow{M} \int_{\mathbf{C}} S \xrightarrow{\pi_S} \mathbf{C}$, we thus obtain a new transformational network, as shown in the diagram below.

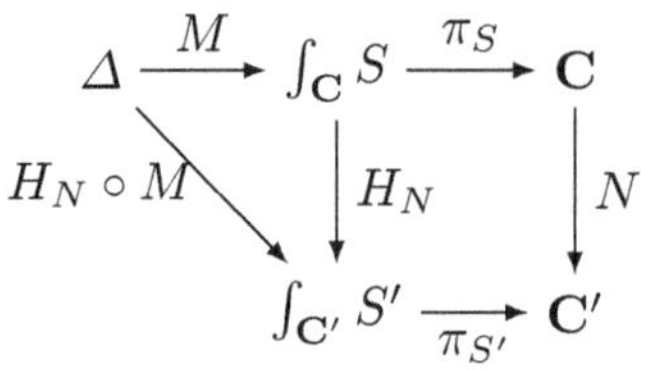

Secondly, we consider *local transformations* [23], which are natural transformations $\hat{\nu}\colon F \to F'$, extended by S such that $\phi' = (S\hat{\nu})\phi$. In such a case, it can readily be seen that the natural transformation $\hat{\nu}\colon F \to F'$ induces a natural transformation $\eta_{\hat{\nu}}\colon H_F \to H_{F'}$ between the corresponding functors $H_F\colon \int_\Delta R \to \int_{\mathbf{C}} S$ and $H_{F'}\colon \int_\Delta R \to \int_{\mathbf{C}} S$. We then obtain the following diagram.

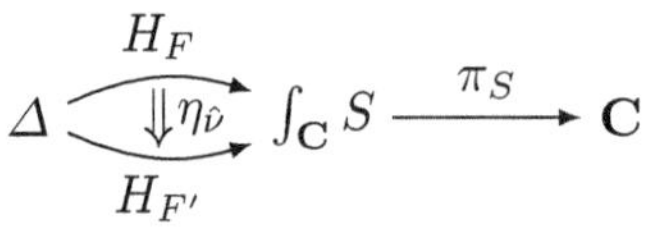

By combining these two notions, and by generalizing it through the possibility of changing the category Δ, we propose a unified notion of morphism of CT-Nets as defined below.

Definition 6. *A morphism between two categorical transformational networks $\Delta \xrightarrow{M} \mathbf{C}_{el} \xrightarrow{\pi} \mathbf{C}_T$ and $\Delta' \xrightarrow{M'} \mathbf{C'}_{el} \xrightarrow{\pi'} \mathbf{C'}_T$ is defined as a 4-tuple (I, N, H_N, ν) such that the following diagram commutes.*

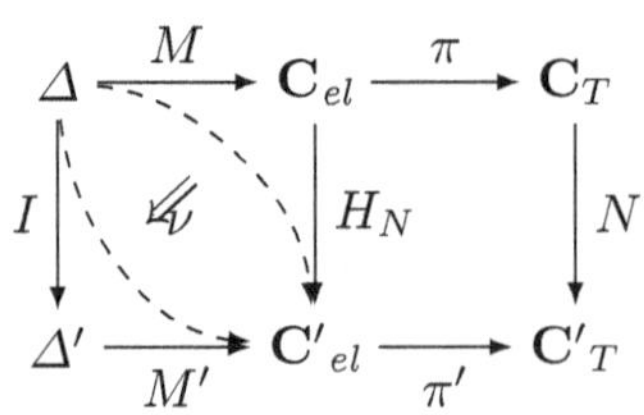

Morphisms of CT-Nets are composable, via the appropriate stacking of the corresponding diagrams.

In the above definition of a morphism of CT-Nets, the rightmost square represents a global transformation of the musical elements and the corresponding transformations between them, independently of the specific network Δ. As detailed above, if $\mathbf{C}_{\text{el}}$ is the category of elements for a given group action, this corresponds to a morphism of a group action. The leftmost square represents a local transformation of the network nodes: the components of the natural transformation ν are morphisms in $\mathbf{C'}_{\text{el}}$, which can be given labels via π' in $\mathbf{C'}_T$. We give below an example of such a combination of global and local transformation.

Example 3. The two transformational networks shown below are $\langle T_2 \rangle$-isographic in the sense of Klumpenhouwer.

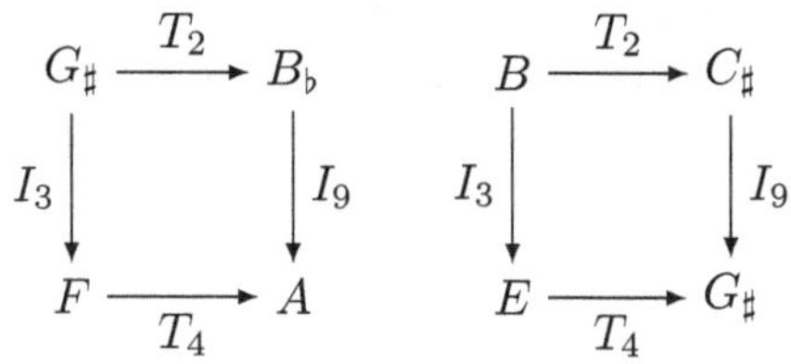

However, no morphism of the T/I group action can transform the first into the second. This can nevertheless be achieved with a combination of a global transformation with a local one, as shown below. We refer the reader to [20] for more details about global and local transformations of networks.

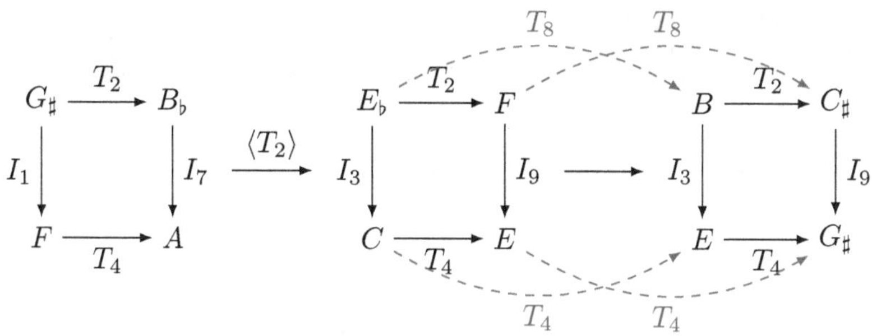

5 Conclusions

In this work, we have revisited Lewin's original framework of Generalized Interval Systems from a categorical perspective, by stressing the importance of diagrams $\mathbf{C}_{\text{el}} \to \mathbf{C}_T$, in which $\mathbf{C}_{\text{el}}$ is a category of musical objects which maps onto $\mathbf{C}_T$, a category of musical transformations. We have shown that Lewin's definition of GIS implictly contains the definition of such a diagram, and that the condition for simple transitivity turns $\mathbf{C}_{\text{el}}$ into the category of elements $\int_{\mathbf{G}} S$ for the corresponding group action $S\colon \mathbf{G} \to \mathbf{Sets}$. This consideration extends beyond group and group actions however, and category of elements $\int_{\mathbf{C}} S$ for a given functor $S\colon \mathbf{C} \to \mathbf{Sets}$ or even $S\colon \mathbf{C} \to \mathbf{Rel}$ may be considered. Ultimately, one may consider diagrams $\mathbf{C}_{\mathrm{el}} \to \mathbf{C}_T$ in which $\mathbf{C}_{\mathrm{el}}$ is not necessarily a category of elements.

Drawing from previous work on the formalization of transformational networks, we have proposed a new categorical definition of transformational networks, hence called CT-Nets, as a sequence of functors $\Delta \to \mathbf{C}_{\mathrm{el}} \to \mathbf{C}_T$, in which Δ represents the skeleton of the network. The definition operates directly in **Cat**, and thus unifies the consideration of **Sets**- and **Rel**-based functors at once. The definition of morphisms of CT-Nets results from previous considerations of global and local morphisms, the former corresponding to global transformations of the musical action, independently of the network being considered, the latter corresponding to individual node transformations. As diagrams, CT-Nets can undergo the machinery of all constructions in **Cat** (pullbacks, pushforwards, equalizers, etc.) to give new networks. In addition, the forgetful functor $\mathbf{Cat} \to \mathbf{Quiv}$ readily gives the corresponding multidigraphs (and morphisms between them), bridging algebraic and geometrical graph-based methods in music theory.

This new categorical formalization of transformational networks also opens further research perspectives. For example, while categories of elements can be constructed for functors $\mathbf{C} \to \mathbf{Rel}$, Theorem 2 cannot be applied to relational PK-nets as introduced in [22] if the natural transformation ϕ is a binary relation and not simply a function. In such a case, the machinery of profunctors [1] may be used between the corresponding category of elements, with new implications regarding the musicological meaning of transformations between musical objects.

Acknowledgments. We thank Andrée Ehresmann for many fruitful discussions on this topic. We also thank the reviewers for their detailed and critical comments on the first version of this paper. We acknowledge the support by the Interdisciplinary Thematic Institute CREAA, as part of the ITI 2021–2028 program of the Université de Strasbourg, the CNRS, and the Inserm (funded by IdEx Unistra ANR-10-IDEX-0002, and by SFRI-STRAT'US ANR-20-SFRI-0012 under the French Investments for the Future Program).

References

1. Bénabou, J.: Les distributeurs. Université Catholique de Louvain, Institut de Mathématique Pure et Appliquée, Technical report (1973)
2. Cohn, R.: Weitzmann's regions, my cycles, and Douthett's dancing cubes. Music Theory Spectr. **22**(1), 89–103 (2000). http://www.jstor.org/stable/745854
3. Cohn, R.: Audacious Euphony: Chromaticism and the Triad's Second Nature. Oxford University Press, Oxford (2012)
4. Douthett, J., Steinbach, P.: Parsimonious graphs: a study in parsimony, contextual transformations, and modes of limited transposition. J. Music Theory **42**(2), 241–263 (1998). http://www.jstor.org/stable/843877
5. Fiore, T.M., Noll, T., Satyendra, R.: Morphisms of generalized interval systems and *PR*-groups. J. Math. Music **7**(1), 3–27 (2013). https://doi.org/10.1080/17459737.2013.785724
6. Hook, J.: Uniform triadic transformations. J. Music Theory **46**(1/2), 57–126 (2002). http://www.jstor.org/stable/4147678

7. Klumpenhouwer, H.: A Generalized Model of Voice-Leading for Atonal Music. Ph.D. thesis, Harvard University (1991)
8. Klumpenhouwer, H.: The inner and outer automorphisms of pitch-class inversion and transposition: some implications for analysis with Klumpenhouwer networks. Intégral **12**, 81–93 (1998). http://www.jstor.org/stable/40213985
9. Kolman, O.: Transfer principles for generalized interval systems. Perspect. New Music **42**(1), 150–189 (2004)
10. Lewin, D.: Generalized Music Intervals and Transformations. Yale University Press, New Haven (1987)
11. Lewin, D.: Klumpenhouwer networks and some isographies that involve them. Music Theory Spectr. **12**(1), 83–120 (1990)
12. Lewin, D.: Musical Form and Transformation. Oxford University Press, Oxford (2007)
13. Mazzola, G.: Gruppen und Kategorien in der Musik. Heldermann (1985)
14. Mazzola, G.: Geometrie der Töne : Elemente der mathematischen Musiktheorie. Birkhäuser (1990)
15. Mazzola, G., et al.: The Topos of Music: Geometric Logic of Concepts, Theory, and Performance. Birkhäuser Verlag, Basel (2002). https://doi.org/10.1007/978-3-0348-8141-8
16. Mazzola, G., Andreatta, M.: From a categorical point of view: K-Nets as limit denotators. Perspect. New Music **44**(2), 88–113 (2006). http://www.jstor.org/stable/25164629
17. Mazzola, G., Andreatta, M.: Diagrams, gestures and formulae in music. J. Math. Music **1**(3), 199–200 (2007). https://doi.org/10.1080/17459730701774269
18. Piovan, L.A.: A Tonnetz model for pentachords. J. Math. Music **7**(1), 29–53 (2013). https://doi.org/10.1080/17459737.2013.769637
19. Popoff, A.: On the use of relational presheaves in transformational music theory. J. Math. Music **16**(1), 51–79 (2022). https://doi.org/10.1080/17459737.2020.1825845
20. Popoff, A., Agon, C., Andreatta, M., Ehresmann, A.: From K-Nets to PK-Nets: a categorical approach. Perspect. New Music **54**(2), 5–63 (2016). https://doi.org/10.7757/persnewmusi.54.2.0005
21. Popoff, A., Andreatta, M., Ehresmann, A.: A categorical generalization of klumpenhouwer networks. In: Collins, T., Meredith, D., Volk, A. (eds.) MCM 2015. LNCS (LNAI), vol. 9110, pp. 303–314. Springer, Cham (2015). https://doi.org/10.1007/978-3-319-20603-5_31
22. Popoff, A., Andreatta, M., Ehresmann, A.: Relational poly-Klumpenhouwer networks for transformational and voice-leading analysis. J. Math. Music **12**(1), 35–55 (2018). https://doi.org/10.1080/17459737.2017.1406011
23. Popoff, A., Andreatta, M., Ehresmann, A.: Groupoids and wreath products of musical transformations: a categorical approach from poly-klumpenhouwer networks. In: Montiel, M., Gomez-Martin, F., Agustín-Aquino, O.A. (eds.) MCM 2019. LNCS (LNAI), vol. 11502, pp. 33–45. Springer, Cham (2019). https://doi.org/10.1007/978-3-030-21392-3_3
24. Rahn, J.: Cool tools: Polysemic and non-commutative nets, subchain decompositions and cross-projecting pre-orders, object-graphs, chain-hom-sets and chain-label-hom-sets, forgetful functors, free categories of a net, and ghosts. J. Math. Music **1**(1), 7–22 (2007)
25. Reenan, S., Bass, R.: Types and applications of $p_{3,0}$ seventh-chord transformations in late nineteenth-century music. Music Theory Online **22**(2) (2016)
26. Rockwell, J.: Birdcage flights: a perspective on inter-cardinality voice leading. Music Theory Online **15**(5) (2009)

27. Sobociński, P.: Relational presheaves as labelled transition systems. In: Pattinson, D., Schröder, L. (eds.) CMCS 2012. LNCS, vol. 7399, pp. 40–50. Springer, Heidelberg (2012). https://doi.org/10.1007/978-3-642-32784-1_3
28. Vuza, D.T.: Some mathematical aspects of David Lewin's book generalized musical intervals and transformations. Perspect. New Music **26**(1), 258–287 (1988)

Voice and Math: The Art of Singing in Light of Mathematical Music Theory

Juliana Spector[1], Maria Mannone[2,3,4](✉), Mariana Montiel[5], and Francisco Gómez[6]

[1] Universidad Politécnica de Madrid, Madrid, Spain
jspectorsoprano@gmail.com
[2] Institute of Physics and Astronomy, Universität Potsdam, Potsdam, Germany
maria.mannone@uni-potsdam.de
[3] MIRMI, Technische Universität München, Munich, Germany
[4] DSMN - ECLT, Università Ca' Foscari di Venezia, Venice, Italy
[5] Department of Mathematics and Statistics, Georgia State University, Atlanta, USA
mmontiel@gsu.edu
[6] Universidad Politécnica de Madrid, Madrid, Spain
francisco.gomez@upm.es

Abstract. *Elements of a Vocal Gesture Theory*, previously developed as an abstract approximation, have served as the basis for the present work in which the theoretical aspect of this development is refined to accommodate concrete realities that the singer confronts. In particular, the connection between musical thinking (symbolic gestures) and performance (concrete gestures) can be described by a mathematical approach. On the one hand, singers have to imagine the vocal effect they wish to obtain, and, on the other hand, they must carry out concrete physical inner movements of their larynx, vocal folds, etc., to achieve the desired effect. Mathematical Gesture Theory can provide a framework to help understand how such *imaginary* and *embodied movements* relate one to another, as well as how they are connected in the practice of singing. Additionally, voice pedagogy, in general, is tremendously subjective and, as in all creative processes, what works for one singer does not necessarily function for another. We hope to offer a very broad and flexible framework in which the essence of the relation between symbolic and physical gestures is captured and allows for multiple concrete models. We also offer some very basic experimental evidence as proof of concept of this emerging Vocal Gesture Theory, considering an application to Italian *belcanto*, with the analysis of the head voice and the chest voice in a Verdi's aria.

Keywords: gesture theory · voice physiology · *belcanto*

1 Introduction

The use of the term *gesture* in the present study coincides with the theoretical development presented in [9]. In the cited chapter the elements of a "future vocal

© The Author(s), under exclusive license to Springer Nature Switzerland AG 2024
T. Noll et al. (Eds.): MCM 2024, LNCS 14639, pp. 111–122, 2024.
https://doi.org/10.1007/978-3-031-60638-0_9

gesture theory" were developed. The present work builds upon that foundation, while treating the concrete physiological processes with more precision. To create a flexible framework that can capture the relationship between *symbolic gestures* (what the singer imagines is occurring and how it can be controlled), and *physical gestures* (what actually happens), we have made use of Mathematical Gesture Theory [11,12], in which Category Theory has played a fundamental role since its genesis. This framework must also consider that, while there is no fixed and general rule that will work for all singers, there is a range of possibilities that can certainly be systematized and introduced as part of voice pedagogy. Indeed, the search to understand what happens in the process of singing, and how to control the only instrument that is visually hidden from the performer and the public, has a long history [11,13,15,16]. Remember, for example, the famous vocal pedagogue Manuel García (1805–1906), who is considered the father of the laryngoscope, another of many cases in which music stimulates scientific discovery. In principle, the vocal theory of gestures can be applied to a variety of musical genres and styles. However, as has already been done in [9], we focus on *Italian belcanto*, given its importance for Western music and the worldwide fascination and interest it provokes. As an additional reason, we mention that *the practice of Opera singing in Italy* has recently (December 2023) been included among the immaterial treasures protected by UNESCO.[1]

It is relevant to mention that, as often occurs in interdisciplinary endeavors, one of the authors was studying the use/lack of use of the chest voice in the soprano and its relation to microphone technology, with a focus on voice pedagogy. The researcher, without knowing about vocal gesture theory nor the modeling of imaginary and physical gestures, emphasized the attempt by vocal instructors to use metaphors and images to affect the physical process and the outcome. Another of the authors saw a complementarity between the two research endeavors, that is, on the one hand, the formulation of a mathematical vocal gesture a theory and, on the other hand, the search for theory that can explain the physical and mental (symbolic) processes involved in singing (especially in view of being able to carry out MRI studies in the near future). That is how this research and its present aims arose.

The work will be developed according to the following scheme: In Sect. 2, we will describe the vocal anatomy of some organs involved in singing, in particular, those necessary for chest voice in belcanto singing. Basic physiological aspects of the singing process will then be described, as these elements are essential if we are to define parameters and create spaces and categories for imaginary and physical voice gestures. We will also explore ways of relating the aspects introduced, and touch on different theories in this respect. In Sect. 3, starting from the anatomy, we will formalize the vocal gestures by exploiting the formalism of vocal gesture theory. As in the creation of any type of model, once the parts have been defined the key is the way they are related. Therefore, we will set the stage we will set the stage and establish the framework to enhance and provide depth to the information that we will obtain through future MRI studies.

[1] https://ich.unesco.org/en/RL/the-practice-of-opera-singing-in-italy-01980.

In Sect. 4, we will exemplify the idea of vocal hypergestures through an experiment that captures the use of the chest voice in the soprano. By superimposing on the spectrogram the laryngeal parameters as defined and modeled via the mathematical description of vocal gestures, a particular diagrammatic image of the framework is presented. In Sect. 5, we will discuss our findings and perspectives toward future work. Section 6 will briefly present conclusions derived from the presented work. We hope to overcome, as much as possible, both subjective and/or rigid approaches.

2 The Anatomy of Chest Voice in *belcanto* Singing

> "Science understands too little of singing, and the singer too little of science." –Lilli Lehman [7]

There is not enough vocabulary to capture the complex vocal phenomena inside the singer. Voice professionals often resort to hand gestures akin to conductors and pianists when they require musical expressivity from their singing. Unlike the piano, in which one can visualize the hand movements of the player, the singer only relies on sensory guidelines. Thus, hand gestures emerged over time to compensate for the absence of visual references. Students-singers and well-rounded professionals cup their hands over their heads, stretch their arms, hold their bellies, point up or down, bend their knees, etc. In addition, instructors also evoke imagery to convey the desirable vocal effects. The following statements derive from a personal experience of one of the authors as a voice student: "Imagine you're drinking water," "Pretend you're gagging when you sing a high note," "Sing through your nose," "Sing this on one leg," "Imagine you're evacuating your bowels." Such as teaching approach fails to correlate singing with those sensations. The vocal gestures referred to in this article relate to muscular functions disassociated from any external gesture. By knowing how the voice apparatus works, the singer may connect with the "internal eye" of the body. The knowledge of vocal physiology

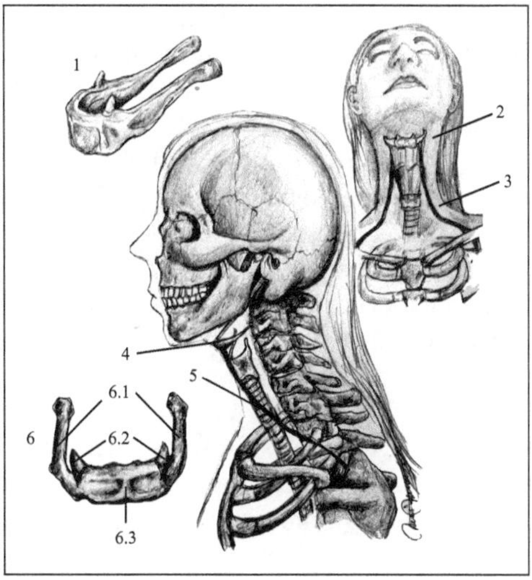

Fig. 1. Anatomy of the hyoid bone and the omohyoid muscle. 1. hyoid bone, lateral view; 2. hyoid bone; 3. omohyoid muscle; 4. hyoid bone 5. omohyoid muscle; 6. hyoid bone, frontal view, detail; 6.1. greater horns. Drawings by M. Mannone.

operates alongside the singer's sensory experience. The larynx moves down when we drink water; thus, we can assume that singing requires the same laryngeal direction to achieve effortless beauty. Singing is a muscular enterprise engaging several groups of muscles in the back, chest, ribs, neck, and head. Each singer develops a unique set of resources in response to their repertoire. Artistic singing demands strong muscles to regulate the breath pressure under adducted vocal folds to produce a consistent tone. Singers acquire the power and endurance by regularly expanding the muscles of the chest, back, and ribs, which keeps the diaphragm's position low and the lungs full of air. The sensation of muscular tension against the chest appears because of the low larynx, activating a resonant foundation in the entire vocal diapason. The larynx moves down with the assistance of the hyoid bone/omohyoid muscle mechanism (Fig. 1). The hyoid bone is uniquely placed below the lower jaw atop the thyroid cartilage. At this location, the hyoid bone supports the floor of the tongue, which allows the oropharyngeal passage to form distinct vocal sounds. The double omohyoid muscle is attached to the hyoid bone and to the scapulae bones. The tongue forms and shapes vowels and consonants. It is the "steering wheel" of the singing mechanism. If the interested reader goes to [7], it can be seen that the singer's tongue changes shape as he exemplifies the different singing styles.

As it will be described in more detail in the Discussions, we have preferred to start with simple elements when moving from physiology to our mathematical formalism (Sect. 3), and to little by little build up the whole structure which will eventually includes the tongue. Opera singers maintain the base of the tongue elevated to form a narrow oropharyngeal space as it facilitates the *falsetto* tones and access to higher pitches. The muscular fibers contract through inhalations, lowering the hyoid bone and positioning the larynx in a desirable place for singing. Time and experience cultivate the voluntary use of the muscles that control the larynx. The singer controls the omohyoid muscle through breathing, which keeps the back expanded throughout the musical phrase. This action prompts the hyoid bone to set the throat in a lower position, thus forming an elongated passage in the oropharyngeal area. The expanse created by the opening of the oropharyngeal area allows full resonant tones to radiate from the singing voice. The vocal outcome generated by the coordination between the low larynx and the oropharyngeal area is also known as the core voice, or chest voice. Inside the larynx, the glottal space closes (adduction) and the vocal folds vibrate to create pitch. The singer controls the air pressure that drives pitch change and masters the vocal technique through the knowledge of physiology and the role it plays in singing. One conclusion comes to mind: breath control development is critical to conquering the muscular movements that activate proper singing. The conscious management of the air pressure and the forms of the oropharyngeal passage will produce good quality tones. Technological advances enable the visibility of the internal functions of our bodies, benefiting the mastery of the singing technique while allowing the will of the singer to create effortless tones. As the singer observes how the vocal organs behave during singing, the vocal artistry can be transmitted through the teaching. More details can be found in [2–6].

3 Mathematical Modeling

After our analysis of hyoid bone and omohyoid muscle anatomy and motion, we can move toward their mathematical formalization. The intermediate, fundamental step is the schematization of their essential structure and movements, necessary for breathing and thus also for singing. Let us start with inhalation, whose basic movements are sketched in Fig. 2. We notice the downward motion of the hyoid bone, the downward contraction of the omohyoid muscle, and the consequent expansion of the lung volume, also corresponding to the increase of inter-scapula distance.

The inverse movements are required for the exhalation: upward relaxation of the omohyoid muscle, upward motion of the hyoid bone, restriction of lungs, and decrease of inter-scapula distance. The inhalation movement corresponds to the lowering of diaphragm, not represented in the sketch for reasons of space and to focus on the omohyoid, and vice versa for the exhalation. The described movements "live" in the 3-dimensional space+time, to which human bodies belong to.

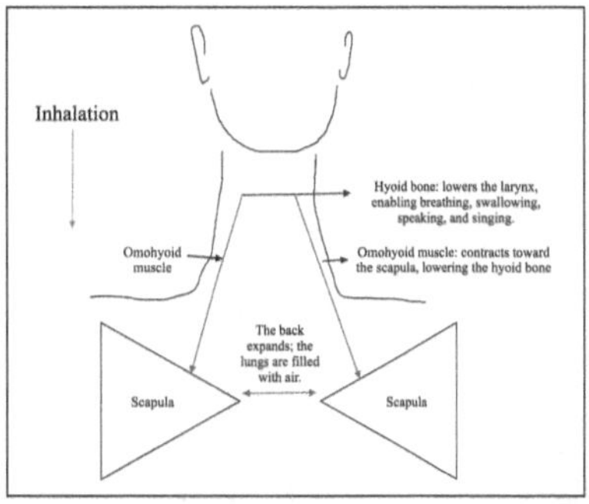

Fig. 2. Starting point of the mathematical model: essential scheme of the motion of the hyoid bone, omohyoid muscle, and lung expansion. This model suggests our formalization. Sketch by J. Spector.

However, let us now perform the first step toward abstraction, and consider a Cartesian monoidal category. That is, a category C with a terminal object and binary products. Then we define a monoidal category structure on C as the category-theoretic product. We remember that, in any category admitting a product of two objects and a terminal object, the operation $(X, Y) \mapsto X \otimes Y$ gives a symmetric monoidal structure, with the terminal object as the unit. Now we consider a space of "states," a space for the hyoid bone position, a space for the omohyoid muscle position, and a space for the lung volume:

$$\text{(hyoid bone position)} \otimes \text{(omohyoid muscle position)} \otimes \text{(lung volume)} \qquad (1)$$

where we use $\otimes$, the tensor product, to denote the separation of these (now conceptual) spaces. We are using the tensor product before having defined our spaces. In future investigations (see also Sect. 6) we intend to consider different interpretations of the tensor product. For now we stick to the direct product. The formalization is here quite general. We are moving from physiological depiction, to its schematization, to the progressive formalization. The states "up" and "down" can be schematized as 1 and 0, respectively, in analogy with the excited versus ground level of a 2-level atom in physics. Thus, the up-state position of the hyoid bone can be represented with 1, and its low-state with 0. Its downward movement occurring in the inhalation step, as shown in Fig. 2, can thus be schematized as a *gesture* from the state 1 to the state 0: $1 \rightarrow 0$. For the omohoyid muscle, Fig. 2 represents a contraction, that is, a movement from a "relaxed state" 0 to a "tense state" 1, that is, $0 \rightarrow 1$. Finally, for the lungs, we have a movement from a "restricted state" 0 to an "expanded state" 1, that is, again $0 \rightarrow 1$. The overall movement of inhalation can thus be described as:

$$inhalation = (1 \rightarrow 0) \otimes (0 \rightarrow 1) \otimes (0 \rightarrow 1). \tag{2}$$

Correspondingly, the inverse motion, the exhalation, can be described as:

$$exhalation = (0 \rightarrow 1) \otimes (1 \rightarrow 0) \otimes (1 \rightarrow 0). \tag{3}$$

With a more compact notation, we can characterize the *state breathe-in* with $[0, 1, 1]$ and the *state breathe-out* with $[1, 0, 0]$, where we introduce a notation partially inspired by quantum states. Thus, the *gestures* will be the paths between these two states:

$$inhalation : [100] \rightarrow [011], \quad exhalation : [011] \rightarrow [100]. \tag{4}$$

Here, we considered the downward motion of the hyoid bone from 1 to 0, while we indicated the movement of the omohyoid muscle from 0 to 1 to give an idea of the switch from a *relaxed* to a *contracted, tense* state. The actions of inhalation and exhalation can be considered as transitions between states. Thus, we can envisage a continuum of states (r, s, t), such as the state $(0.5, 0.5, 0.5)$ which could be well understood as being half-way of an inhalation. With this simple consideration, we can recover a multi-value logic [10]. We notice that the symbol "$\rightarrow$" is either embedded inside the single spaces, or considered as an overall transition between the overall states. We are thus gently introducing the concept of *mathematical gesture*, *hypergesture*, and *nested gesture*, being inspired by the mathematical theory of musical gestures and its developments. In fact, the gesture $1 \rightarrow 0$, as a path connecting two points, can be schematized as $\cdot \rightarrow \cdot$, and it constitutes the most basic example of gesture. In the mathematical theory of musical gestures [9], the structure (called *digraph*, a type of oriented graph) of this kind of gestures, connecting two points in an abstract and unspecified space, is denoted with the symbol $\uparrow$. In the same theory, a *musical gesture* is in general described as a mapping between a digraph Δ to a curve x in a topological space $\mathbf{X}$. In [9], the set of all gestures of the digraph Δ to the topological space (seen as a digraph) is described as $Hom(\Delta, \mathbf{X})$, that is, Hom-Set in the category of digraphs. Here, we use the following notation to indicate a single gesture:

$$\Delta \rightarrow x \tag{5}$$

Δ is called *skeleton* of the gesture, and the points inside the topological space $\mathbf{X}$ and the curves connecting them, constituting x, are the *body* of the gesture. With a slight abuse of notation, we can include the detail of the movement already in the arrow symbol. Thus, the inhalation becomes:

$$(\downarrow\rightarrow \mathbf{h}) \otimes (\uparrow\rightarrow \mathbf{o}) \otimes (\uparrow\rightarrow \mathbf{l}), \tag{6}$$

where **h**, **o**, **l** denote the spaces of movements up/down, contracted/relaxed of the hyoid bone, omohyoid muscle, and lungs, respectively. Concerning physiology, the hyoid bone lowers, the omohyoid muscle contracts (from the relaxed state to the tense state, thus $\uparrow$), and the lungs expand. We can group the digraph-information of "local" gestures inside $(\uparrow \otimes \uparrow \otimes \uparrow)$, and the curve-information of local gestures inside $(\mathbf{h} \otimes \mathbf{o} \otimes \mathbf{l})$. Thus, (6) becomes a gesture under the identification of the two sides of (7):

$$(\uparrow\rightarrow \mathbf{h}) \otimes (\uparrow\rightarrow \mathbf{o}) \otimes (\uparrow\rightarrow \mathbf{l}) = (\uparrow \otimes \uparrow \otimes \uparrow) \rightarrow (\mathbf{h} \otimes \mathbf{o} \ \otimes \mathbf{l}). \tag{7}$$

While our tensor product is a direct product, and we are treating movements of the considered body parts as gestures inside independent spaces, here we know that there are physiological and anatomical constraints: for instance, it is not possible to breathe restricting the space between the scapula, connected with the lungs' volume. If we add phonation, we can consider another space, where 1 denotes "yes" and 0 denotes "no." According to the scope of the above-mentioned study [9], more spaces could be added. For instance, three spaces characterize the sound of the human voice: phonation, myoelasticity, and turbulence, whose values at each time constitute a triple in a 3-dimensional space [10].

In [9], the author also considered a parametrization with respect to timbre. In fact, each combination of vocal parameters gives a different output in terms of spectral characterization. For instance, given a paramater $a \in A$, where A is a space of parameters, a gesture q having digraph $\uparrow$ and body in X lives in the space $(\uparrow\rightarrow_A X)$. Assigned a specific value $a \in A$, we obtain $q(a) : [0,1] \rightarrow \mathbf{X}$ as a specific gesture in the spectral space. Similarly, one can obtain a parametrization with respect to time: a time gesture is thus defined as $(\Delta \rightarrow T)$. A simultaneous parametrization with respect to spectrum and time can be formalized as a gesture v such that $(\Delta \rightarrow A \otimes T) \xrightarrow{\sim} (\Delta \rightarrow A) \otimes (\Delta \rightarrow T)$. Let us make a little note about the concept of *time*. We can talk of a "thought gesture," that is, a gesture which is happening only along the *imaginary axis* of time, when there is, for instance, a motion in the spectral space, to which there is no corresponding variation of the (real) physical time. That is, a singer can think of modifying their timbre, without performing any actual change, or even without singing any note, but only thinking of it. In our examples, the parametrization could be constituted by suitable choices of hyoid bone, omohyoid muscle, and scapula distance (or lungs' volume). However, the notation we proposed, with the definition of two basic states, and gestures as the transitions between them (read: superposed states with different coefficients at each time), appears as more compact and flexible with respect to the notation proposed in [9]. In fact, that notation can result excessively heavy for practical purposes. To recover in our model a separate description of time, we can consider the transition between 0 and 1 in the space of *complex time*, where 0 stands for *imaginary time only*, and 1 for *real time only*. A gesture from the imaginary to the real time, as also described in [9], can correspond to the progressive realization of a vocal performing movement, from the mind of the singer to the physiological reality of the human body and physical reality of the sound. In the mathematical theory of musical gestures, the concept of gesture has been extended to the *hypergesture*. If a gesture is considered as a point, the hypergesture transforming a gesture into another one can be seen as a *path* between them [8]. Deviating from an earlier notation for hypergestures in [8], given two gestures

$$g_1 = (\Delta \rightarrow x),\ g_2 = (\Gamma \rightarrow y), \tag{8}$$

the path between them becomes:

$$h_1 = g_1 \rightarrow g_2 = (\Delta \rightarrow x) \dashrightarrow (\Gamma \rightarrow y), \tag{9}$$

where the middle arrow $\dashrightarrow$ can be seen as one arrow of a digraph of the hypergesture, whose two nodes are the gestures.

With a slight abuse of definition already proposed in [8], we are now considering a (hyper)gesture not as the mapping between the skeleton and the body, but as the body itself connecting two points in a topological space. If we see each gesture as a point, the (higher-level) curve connecting them is another gesture, that is, a hypergesture. By the same argument, we can define nested gestures. As it is proved in Theorem 1 in [8], gestures' and hypergestures' structures can be defined recursively. We end this section mentioning that the mathematical theory of musical gestures has been extended to the comparison between gestures on different musical instruments, and even between music and the visual arts [8]. Other developments concern homology of mathematical gestures and applications of the gesture theory to music analysis [1]. Getting back to the formalization of human singing, we can see breathing as a sequence of complementary hypergestures, and singing itself as a hypergesture. Singing a whole opera requires a highly-nested structures of arias, connecting passages, single phrases, and ultimately, breaths, where each breath, with inhalation and exhalation, can be related with the preparation and beat of the conductor's baton, and the same difference with the Greek *thesis* and *arsis*.

4 Experiment

The example that we propose here refers to the gesture definition from the auditory output. We can infer the composition of movements of the vocal organs by analyzing the sound. However, we will get the proper information on vocal organs' gestures through the MRI analysis. Thus, as a proof of concept of our formalization of the use of the chest voice for *belcanto* singing, we analyze a passage from the opera *Don Carlo* by Giuseppe Verdi, in particular, a fragment from the aria by Princess Eboli, performed with the use of the chest voice by a soprano (not a mezzo-soprano, as has been usual in the past 20 years or so). The corresponding spectrogram and score are shown in Fig. 3. The whole passage can be described in terms of hypergestures. Let us focus on the first part of the example (Fig. 4). The main involved gestures are inhaling, exhaling, and inhaling. During exhaling, five notes are sung (four pitches, as the last one is repeated); thus, the skeleton of the corresponding gesture Ph in the phonation

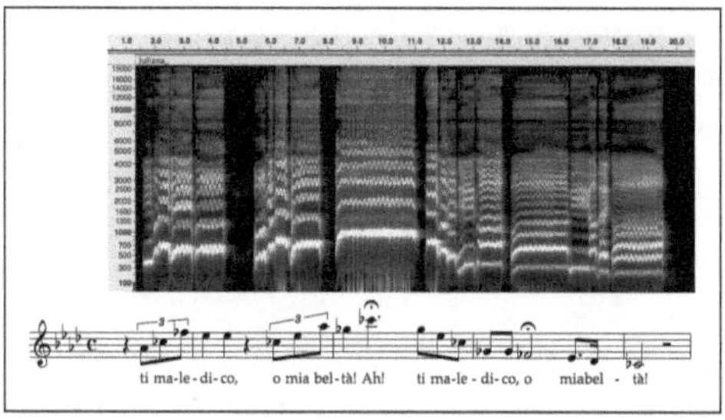

Fig. 3. A passage from Princess Eboli's aria *O Don Fatale* in *Don Carlo* by G. Verdi. The lyrics *Ti maledico, o mia beltà!* can be translated as "I curse you, my beauty!". The passage is performed with chest and head voice by J. Spector. The spectrogram is obtained with the software Audacity. The horizontal axis indicates time measured in seconds, and the vertical axis indicates pitch height measured in Hertz.

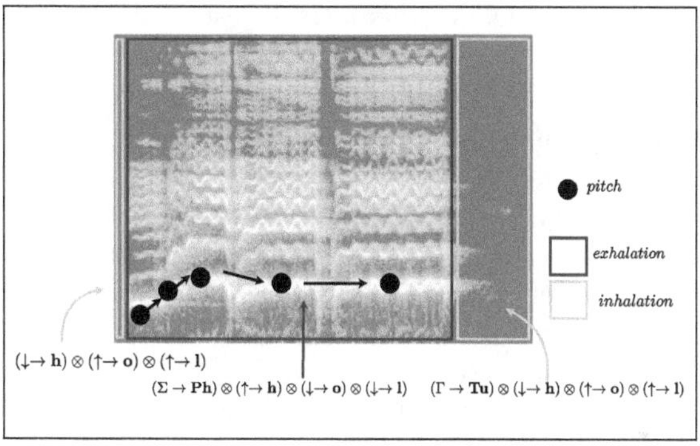

Fig. 4. Gestural analysis from a detail of Fig. 3. The last two pitches are the same, so the last pitch gesture can be considered as a loop.

space is $\Sigma = \cdot \to \cdot \to \cdot \to \overset{\circlearrowleft}{\cdot}$, and the sequence can be constituted by the following steps: (1) inhalation and silence: $(\downarrow\to \mathbf{h}) \otimes (\uparrow\to \mathbf{o}) \otimes (\uparrow\to \mathbf{l})$; (2) exhalation and singing: $(\Sigma \to \mathbf{Ph}) \otimes (\uparrow\to \mathbf{h}) \otimes (\downarrow\to \mathbf{o}) \otimes (\downarrow\to \mathbf{l})$; (3) inhalation and turbulence: $(\Gamma \to \mathbf{Tu}) \otimes (\downarrow\to \mathbf{h}) \otimes (\uparrow\to \mathbf{o}) \otimes (\uparrow\to \mathbf{l})$. In the last step, we introduced a gesture in the space of turbulence **Tu**, because the spectrogram shows some noise or residual air vibration, probably in correspondence of the singer's inhalation, with a simple skeleton $\Gamma = \cdot \to \cdot$. In our analysis, we are neglecting syllables pronunciation and diction, but the interested reader can add this layer to the gestural analysis as an exercise. Figures 3 and 4 show the output of vocal gestures in terms of produced sound. Further research will compare these representations with the information from the magnetic resonance imaging (MRI), for a precise correspondence with the inner movements of hyoid bone, omohyoid muscle, and lungs over time. Eventually, we will also incorporate brain imaging, to be able to see what happens at the "imaginary" (symbolic) level. In [14] there is evidence that the "will" of the singer can have a marked effect on the vocal gestures, to a sometimes thought impossible extent, something that needs to be understood, studied, and systematized.

5 Discussion

The need to create a general mathematical model for the physical gestures that occur in singing arose in the context of its practice and pedagogy. The gestures that control the sound and the music making (with all that this means: pitch, dynamics, etc.) are much less visible in the voice than in other instruments. However, there is a physiological reality that is common to all singers and, as was seen in Sect. 2, it is possible to extract the physical aspects and begin to model what occurs so as to be able to understand the whole gamut of differences that exist in the vocal product, and to have a more objective way to control what is possible and feasible. Mathematical gesture theory was constructed, precisely, for a related kind of challenge (Sect. 3). Once the formalism has been established, it is necessary to employ some proof of concept to see if the model, with its elaborated language, can reasonably capture the reality. In the present work, this experiment is explained in Sect. 4, and consists of a spectrogram that gives us some information about the soprano register when the core voice is consciously used and when it is not. We employ that information to test the vocal gesture-theory model as it stands now, and if it is a practical tool when explaining the differences we see. However, in the near future we plan to carry out MRI studies to give a much better idea of how the model adjusts to the reality of gestures. As it is explained in [9, 1319–1321], "for every choice of parameter we get a spectrum... the gesture that transforms a spectrum into another one is a spectral hypergesture." As each gesture can be seen as a curve, the geometrical interpretation of a hypergesture is a surface formed by the morphisms between them as in Equation (9) of Sect. 2. Once we are able to carry out and process the MRI studies, we will be able to actually observe the parameters defined in Sect. 2, as to the movement of the hyoid bone, omohyoid muscle plus the lung volume. Already according to [14], the knowledge of vocal organs and the precise will of the singer can help better shape the vocal gestures to obtain the desired vocal output. For the purposes of the present work, we created a spectrogram which has the usual pitch and volume parameters for a version of a brief excerpt of the aria *O Don Fatale* from Verdi's opera *Don Carlo*, using the core voice and the head voice. We then superimposed the parameters defined in the section on mathematical modeling, as the information from the score and the spectrogram allowed us to deduce the gestures and to express them with the mathematical formalism. As further developments of vocal gesture theory, we can develop the quantum-inspired definition of states of Eq. (4), and in general, thinking of a multi-value logic to describe vocal states, and suitable operators to define the transitions between them [10]. In further research, we can join the formalism of the mathematical theory of gestures with the quantum vocal theory of sound [10], whose first experiments in quantum-computing strategies have been successfully applied to vocal analysis, for instance, to the separation between phonation and turbulence or myoelastic effects, thanks to the corresponding operations of measure. A connection between these emerging approaches can lead to a deeper understanding of the vocal phenomena. In our analysis, we mostly focused on hyoid bone and omohyoid muscle. A more complete analysis should also include the use of tongue for sound production and diction, fundamental for singing. A 3D model could be defined, where every state of the tongue is a 3D-manifold and each movement of the tongue as cobordism (i.e., a 4D-manifold with time added as the 4th dimension). This would bring us in the Category 4Cob of cobordisms, which has a non-cartesian tensor-product in form of the disjoint union, which is very appropriate for parallel processes. These cobordisms will then play the role of the topological spaces for the bodies of the gestures. As a digraph, we may select n measure-points capable to grasp the essence of gestures. The digraph Δ would be a disjoint union of n arrows, as the ones in Fig. 4, mapped on the union of the timelines of the measured points. Similarly, we can enrich the model with other parts of the vocal apparatus. In this way, we can also model the production of phonemes. The classic singing exercises of performing a vocalise on the same vowel, or changing the vowel while keeping the same pitch, would thus be described with tensor products!

6 Conclusions

The present work has drawn upon Mathematical Gesture Theory as applied to the art of singing. This particular application is discussed theoretically, in terms of abstract spaces and parameters, and contextualized in the domain of vocal physiology. The research presented in this study opens the way toward forthcoming studies in which the model presented will consider new parameters, and its validity and efficiency can be tested with a more sophisticated technology, such as MRI experiments. This paper is part of a larger work dedicated the role of the chest voice, fundamentally—but not exclusively—in sopranos dedicated to Belcanto. However, the ongoing development of a vocal gesture theory can have a far-reaching impact in terms of performance and pedagogy in all types of genres, as the theory itself is transversal in this sense.

Author contributions. J. Spector: Sect. 2, M. Mannone: Sects. 3, 4, 6, M. Montiel: Abstract, Sects. 1, 5. M. Montiel and F. Gómez revised the manuscript.

References

1. Arias-Valero, J.S., Lluis-Puebla, E.: Explicit presentations of topological categories of gestures. J. Math. Music **17**(2), 213–243 (2023)
2. Cumming, K.: Wiring the Singer. Classical Singer and the Microphone. Doctoral dissertation. University of Newcastle (2019)
3. Hahn, R.: On Singers and Singing. Lectures and an Essay. Translated by Leopold Simoneau. O.C. Editions Gallimard (1957). Translation by Amadeus Press (1990)

4. Hoss, C.K.: Classical and Musical Theater Vocal Pedagogy for Female Collegiate Singers: An Observation and Study. Liberty University School of Music (2023)
5. Mitchell, H.F., Kenny, D.T.: The impact of 'open throat' technique on vibrato rate, extent and onset in classical singing. Logoped. Phoniatr. Vocol. **29**(4), 171–182 (2004)
6. Lauri-Volpi, G.: Parallel Voices. Translated and annotated by Daniele D. Godor. Bongiovanni Editore, Bologna, Italy (2022)
7. Lehman, L.: How to Sing. Macmillan, New York (1902)
8. Mannone, M.: Knots, Music, and DNA. J. Creat. Music Syst. **2**(2) (2018)
9. Mazzola, G., et al.: Elements of a future vocal gesture theory. In: The Topos of Music III: Gestures. CMS, pp. 1313–1331. Springer, Cham (2017). https://doi.org/10.1007/978-3-319-64481-3_31
10. Mannone, M., Rocchesso, D.: Quanta in sound, the sound of quanta: a voice-informed quantum theoretical perspective on sound. In: Miranda, E.R. (ed.) Quantum Computing in the Arts and Humanities. Springer, Cham (2022). https://doi.org/10.1007/978-3-030-95538-0_6
11. Mannone, M., Mazzola, G.: Hypergestures in complex time: creative performance between symbolic and physical reality. In: Collins, T., Meredith, D., Volk, A. (eds.) MCM 2015. LNCS (LNAI), vol. 9110, pp. 137–148. Springer, Cham (2015). https://doi.org/10.1007/978-3-319-20603-5_14
12. Mazzola, G., Andreatta, M.: Diagrams, gestures and formulae in music. J. Math. Music **1**(1), 23–46 (2007)
13. Netter, M.D., Frank, H.: Atlas of Human Anatomy. Novartis, Basel (1997)
14. Schutte, H.K., Stark, J.A., Miller, D.G.: Change in singing voice production, objectively measured. J. Voice **17**(4), 495–501 (2003)
15. Reid, C.: Bel Canto. Principles and Practices. Joseph Patelson Music House, New York (1971)
16. Tommasini, A.: Wearing a Wire at the Opera, Secretly, Of Course. The New York Times, 28 June 2013

Structural and Transformational Relations Between Z-Related Hexachords

Qiuwan Zhao(✉)

Eastman School of Music, Rochester, NY 14604, USA
qzhao17@u.rochester.edu

Abstract. Z-related pitch-class sets (pc-sets) have captured particular interest in set theory in music due to their similarity in intervallic structures alongside obscurity in transformational relations. This paper explores the structural and transformational relationships between Z-related hexachords in $\mathbb{Z}_{12}$. We demonstrate that common subsets, precisely those with prime forms [06] and [016], play a crucial role in defining the structure of complementary Z-related hexachords. We present a categorization of Z-related hexachords based on the discrete Fourier transform (DFT). The categorization aligns with the relative positions of the [06] subsets in the hexachord. Moreover, we introduce the method of *dyad expansion* for interpreting the transformational relationships between Z-related hexachords.

Keywords: Interval content · Set theory · Transformational theory · Z-relation

1 Introduction

The comprehensive study of pc-sets and set classes, initiated by Allen Forte in 1973 [1], laid the groundwork for understanding Z-related pc-sets in $\mathbb{Z}_{12}$. Forte's work identified the Z-related pc-sets and provided the first list of set classes in $\mathbb{Z}_{12}$. A pc-set in $\mathbb{Z}_{12}$ is an unordered collection of pitch classes denoted by integers from 0 to 11, and pc-sets belong to the same set class if they can be related through transposition or inversion. Subsequent studies by Rahn [10] and Lewin [5] developed the mathematical framework for set theory and transformational theory in music. Building upon these concepts, Morris categorized pc-sets by *set groups* and uncovered numerous crucial features of Z-related pc-sets [6,7]. Lewin [3] and Quinn [9] applied the DFT to the analysis of musical structures. More recent studies on Z-related pc-sets include works by Goyette [2] and Nuño [8].

This paper focuses on the Z-related hexachords in $\mathbb{Z}_{12}$—hexachords that share the same interval content (IC) but cannot be related by transformation or inversion. Section 2 investigates the common subsets (the [06] and [016] subsets) between Z-related hexachords and how the subsets are crucial for determining

© The Author(s), under exclusive license to Springer Nature Switzerland AG 2024
T. Noll et al. (Eds.): MCM 2024, LNCS 14639, pp. 123–134, 2024.
https://doi.org/10.1007/978-3-031-60638-0_10

complementary Z-related hexachords[1]. Section 3 applies the DFT to Z-related hexachords and establishes four categories regarding *the degrees of imbalance* of Fourier balances introduced by Quinn [9], among which Z-related hexachords will belong to at least one. Section 4 introduces the method of *dyad expansion* as a transformation between Z-related hexachords and provides a visualization of the Z-relation as the reconfiguration of cells in the IFUNC table proposed by Lewin [5]. Section 5 provides a conclusion of the paper and several future research possibilities.

2 The [06] and [016] Subsets in Z-Related Hexachords

Section 2 explores the common subsets (the [06] and [016] subsets) of Z-related hexachords and how the subsets are determinants for complementary Z-related hexachords. We will introduce some definitions and preliminaries in Sect. 2.1 and then focus on Z-related hexachords with two [06] subsets in Sect. 2.2 and Z-related hexachords with one [06] subset in Sect. 2.3.

2.1 Definitions and Preliminaries

Definition 1 (Z-relation). *In music, two homometric pc-sets that are not transposition or inversion-related are called "Z-related." Two pc-sets X, Y are homometric if $IC(X) = IC(Y)$, or equivalently, $IFUNC(X, X) = IFUNC(Y, Y)$.*

The interval content (IC) of a pc-set X in $\mathbb{Z}_{12}$ can be represented by the interval vector of the form $(i_1, i_2, i_3, i_4, i_5, i_6)$, where i_j is the cardinality of $\{\{u, v\} | u, v \in X; u - v \mod 12 = j\}$.

The interval function (IFUNC) from pc-set X to pc-set Y in $\mathbb{Z}_{12}$ can be represented by a vector of the form $(p_0, p_1, \ldots, p_{11})$, where p_q is the cardinality of $\{(x, y) | x \in X, y \in Y, y - x \mod 12 = q\}$. The i_j in $IC(X)$ is equal to p_j in $IFUNC(X, X)$, for $j = 1, \ldots, 6$.

In $\mathbb{Z}_{12}$, 15 pairs of hexachordal set classes are Z-related ($15 \cdot 2 = 30$ set classes in total). A full list is included in Table 1, in which the names[2] of two Z-related hexachordal set classes are listed in columns 1 and 3, their prime forms are listed in columns 2 and 4, and their shared interval content is listed in column 5.

As of notation, we would use "Z-hexachord" to denote a pc-set that belongs to a hexachordal set class involved in Z-relation (one of the 30 set classes) and use "Z-related pair" to denote a pc-set that is Z-related with the original pc-set. For example, we would call {0,1,2,3,5,6} a Z-hexachord and {0,1,2,3,4,7} a Z-related pair of {0,1,2,3,5,6}, and vice versa.

We will now introduce some theorems and propositions related to the Z-hexachords. The mathematical expressions in the following discussions will be in $\mathbb{Z}_{12}$.

[1] In this paper, we will use prime forms to denote subsets. For example, a [06] subset is a subset with prime form [06], which can be any transposition or inversion of the pc-set {0,6}.

[2] We use the set class names from Forte's 1973 work [1] to denote the set classes.

Theorem 1. *Following the Hexachord Theorem, which states that two complementary hexachords have the same interval content, any Z-hexachord and its complement are Z-related.*

Proposition 1. *Every Z-hexachord contains either one or two [06] subsets.*

Proof. If a hexachord X does not contain any [06] subset, then for any pitch class $a \in X$, $a + 6$ would be in the complement of X. Hence, X and its complement can be related by a transposition of 6 semitones and are not Z-related. If a hexachord X contains three [06] subsets, X and its complement can be related through transposition or inversion, which can be proved by cases. Therefore, in either scenario, X would not be a Z-hexachord by Theorem 1. A hexachord cannot contain more than three [06] subsets due to the cardinality limit. An example of a Z-hexachord with one [06] subset is {0,1,3,4,5,7}. An example of a Z-hexachord with two [06] subsets is {0,1,2,4,6,7}. □

Proposition 2. *Every Z-hexachord contains at least one [016] subset.*

Proof. By Proposition 1, since any Z-hexachord X contains at least one [06] subset, let pc-set $\{a, a + 6\} \subset X$. Suppose that X does not contain any [016] subset, then $a+11$, $a+1$, $a+5$, and $a+7$ cannot be in X since any of them would create a [016] subset with $\{a, a+6\}$. To make X a hexachord, we need to choose four pitch classes among the six pitch classes left, which are $\{a + 2, a + 3, a + 4, a + 8, a + 9, a + 10\}$. The only choice that would avoid creating a [016] subset is $\{a + 2, a + 4, a + 8, a + 10\}$. However, the resulting $X = \{a, a + 2, a + 4, a + 6, a + 8, a + 10\}$ and its complement are transposition-related and not Z-related, which forms a contradiction. □

As proved in Proposition 1, Z-hexachords contain either one or two [06] subsets. We will discuss the cases of Z-hexachords with two [06] subsets in Sect. 2.2 and the cases of Z-hexachords with one [06] subset in Sect. 2.3.

2.2 Z-Related Hexachords with Two [06] Subsets

The following discussions will demonstrate that a pair of complementary Z-hexachords with two [06] subsets can be uniquely determined by knowing all the [06] and [016] subsets of one of the two Z-hexachords. All mathematical expressions will be in $\mathbb{Z}_{12}$.

Proposition 3. *Any Z-hexachord with two [06] subsets contains at least two [016] subsets.*

Proof. For any Z-hexachord with two [06] subsets, there are three cases: the two [06] subsets are (interval) 01 apart, 02 apart, or 03 apart. Pc-sets $\{a, a+6\}$ and $\{a+i, a+i+6\}$ are two [06] subsets that are $0i$ apart, where $i \in \mathbb{Z}_{12}$. One can check by placing two [06] subsets on a clock-like circle that if the two [06] subsets are 01 apart, there are already two [016] subsets. Moreover, it is impossible to construct a Z-hexachord with one [016] subset and two [06] subsets 02 apart or to construct a hexachord with one [016] subset and two [06] subsets 03 apart. Since any Z-hexachord with two [06] subsets contains at least one [016] subset (Proposition 2) but cannot contain exactly one [016] subset, it must contain at least two [016] subsets. □

Proposition 4. *Let X be a Z-hexachord with two [06] subsets. Suppose X and its complement X' both have two 06 subsets that are 01 apart. Then, X can be uniquely determined by the pitch classes of all its [016] subsets, and X' can also be uniquely determined by the pitch classes of all its [016] subsets.*

Proof. Let X and X' be complementary Z-related hexachords with two [06] subsets that are 01 apart. For X and X' to be Z-hexachords, the [06] subsets in X and X' would need to be $X_1 = \{a, a+6, a+1, a+7\}$ and $X_1' = \{a+3, a+4, a+9, a+10\}$, respectively. We would need to choose two pitch classes among $\{a+2, a+5, a+8, a+11\}$ to make X a hexachord. Any pitch class chosen would create a [016] subset with X_1. Since the two [06] subsets in X are 01 apart, the pitch classes of all the [016] subsets include the pitch classes of the two [06] subsets. Therefore, X could be uniquely determined by all its [016] subsets. Since X and X' are mutual complements, X' can also be uniquely determined by all its [016] subsets. □

Proposition 5. *For any Z-hexachord X with two [06] subsets that are 02 apart, the Z-hexachord can be partitioned into two disjoint [016] subsets X_1 and X_2 and can be uniquely determined by the pitch classes of all its [016] subsets, since $X = X_1 \sqcup X_2$.*

Proof. Let X be a Z-hexachord with two [06] subsets 02 apart, denoted by $X_1 = \{a, a+6, a+2, a+8\}$. Suppose for contradiction that X does not have two disjoint [016] subsets. Then, $a+11$, $a+3$, $a+5$, and $a+9$ cannot be in X. Since if any one of them is in X, only one [016] subset would be created with X_1, and by Proposition 3, the remaining one pitch class of the X would need to create at least one [016] subset with the five pitch classes already chosen, which would inevitably form two disjoint [016] subsets in X. Therefore, we would need to choose two pitch classes in $\{a+1, a+4, a+7, a+10\}$. However, any choice would result in an X not qualifying as a Z-hexachord. □

Proposition 6. *Any Z-hexachord that has two [06] subsets 03 apart can be uniquely determined by the pitch classes of all its [06] and [016] subsets.*

Proof. Let X be a Z-hexachord with two [06] subsets 03 apart. The pitch classes of the two [06] subsets constitute four pitch classes in X. The knowledge of the two [016] subsets would give information about the other two pitch classes in X since the two [06] subsets are 03 apart. □

Proposition 7. *A pair of complementary Z-hexachords with two [06] subsets can be uniquely determined by knowing all the [06] and [016] subsets of one of the two Z-hexachords.*

Proof. For two complementary Z-hexachords X and X' with two [06] subsets, either the two [06] subsets in X and the two [06] subsets in X' are both 01 apart, or at least one pair of [06] subsets are 02 or 03 apart. In either case, by Propositions 4, 5, and 6, at least one of X and X' can be uniquely determined by the pitch classes of all its [06] and [016] subsets, and the other one can be uniquely determined by taking the complement. □

In Fig. 1, the green line segments in each subfigure suggest the [06] and [016] subsets of the Z-hexachords that belong to the set classes notated in the first line below the subfigure, and the blue line segments suggest the [06] and [016] subsets in the complementary Z-hexachords, which belong to the set classes in the second line. The green line segments connect six nodes on the circle, symbolizing the six pitch classes in a Z-hexachord. Therefore, the knowledge of all the [06] and [016] subsets of any Z-hexachord from set classes in the first line below each subfigure could uniquely determine the Z-hexachord and its complementary Z-related pair.

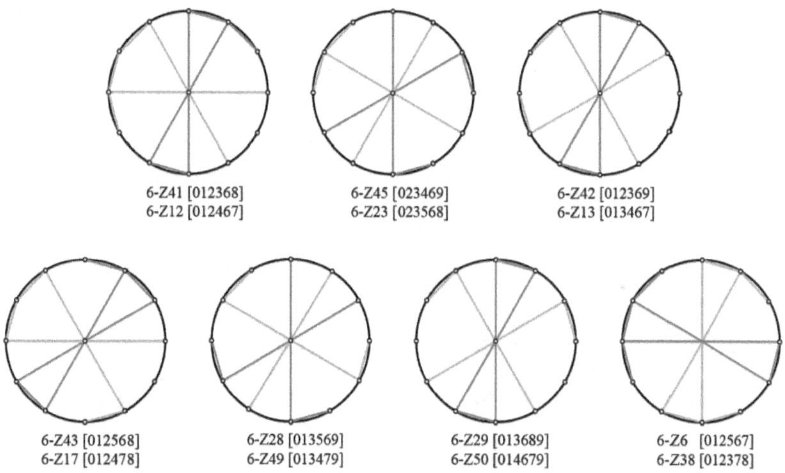

Fig. 1. The [06] and [016] subsets in complementary Z-hexachords with two [06] subsets.

2.3 Z-Related Hexachords with One [06] Subset

There are eight pairs of Z-related hexachordal set classes with one [06] subset, which can be divided into three categories:

1. The [06] subset in a Z-hexachord and the [06] subset in its complement are 03 apart: 6-Z37 and 6-Z4, 6-Z48 and 6-Z26.
2. The [06] subset in a Z-hexachord and the [06] subset in its complement are 02 apart: 6-Z36 and 6-Z3, 6-Z47 and 6-Z25, 6-Z11 and 6-Z40, 6-Z44 and 6-Z19.
3. The [06] subset in a Z-hexachord and the [06] subset in its complement are 01 apart: 6-Z39 and 6-Z10, 6-Z46 and 6-Z24.

While the conclusion in Sect. 2.2 cannot be directly applied to Z-hexachords with one [06] subset, if the [06] subset in a Z-hexachord and the [06] subset in its complement are 03 apart, then a unique pair of complementary Z-related hexachords can be determined with the knowledge of all the [06] and [016] subsets of the two Z-hexachords. In Fig. 2, the green line segments are related to Z-hexachords from the set classes in the first line under each subfigure, and the blue line segments are related to their complementary pairs, which belong to the set classes in the second line. The solid line segments connect all the [06] and [016] subsets in the two complementary Z-hexachords. The dotted line segments connect the nodes (which symbolize pitch classes) that are inferred to be in one Z-hexachord but not its complement to avoid additional [016] subsets.

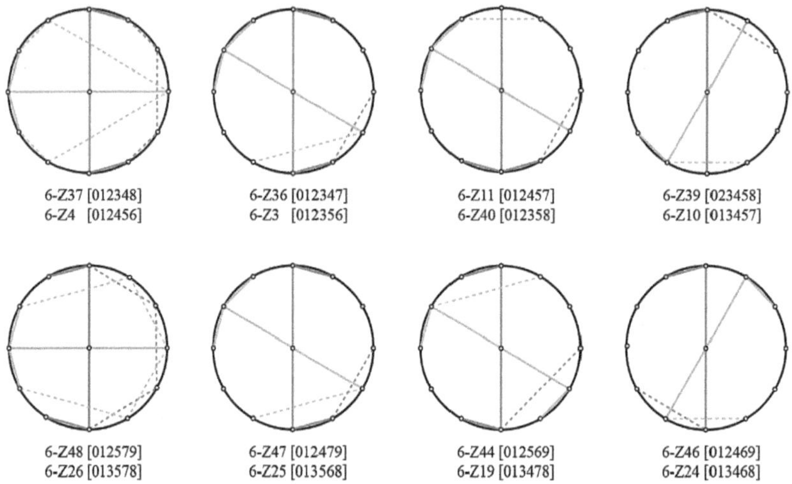

Fig. 2. The [06] and [016] subsets in complementary Z-hexachords with one [06] subset.

3 Z-Related Hexachords and the DFT

3.1 Overview of the DFT on Pitch-Class Sets

From Lewin's first paper in 1959 [3] to later research by Quinn in 2004 [9], *the magnitude of the DFT coefficients* became known for reflecting the *shape* of musical structures, including pc-sets.

Definition 2. *Let X be a pc-set in $\mathbb{Z}_n$. The* Discrete Fourier Transform (DFT) *of X is the transform of its characteristic function $\mathbf{1}_X = (x_0, x_1, \ldots, x_{n-1})$, where $x_j = 1$ if pitch class j is in X, and $x_j = 0$ otherwise. The DFT of X is defined as the vector $\mathcal{F}_X = \hat{\mathbf{1}}_\mathbf{X} = (\hat{x}_0, \hat{x}_1, \ldots, \hat{x}_{n-1})$, where*

$$\hat{x}_j = \sum_{x \in X} e^{-2\pi i \cdot j \cdot x/n}.$$

Here, e is the base of the natural logarithm, i is the imaginary unit, and $j, x \in \mathbb{Z}_n$.

Theorem 2 (Lewin's Lemma). *For a pc-set X in $\mathbb{Z}_{12}$, the DFT of $IFUNC(X, X)$ (a vector of length 12 defined in Definition 1) is of the form* $\widehat{IFUNC(X, X)} = (\hat{p}_0, \hat{p}_1, \ldots, \hat{p}_{11})$, *where*

$$\hat{p}_j = \sum_{k=0}^{11} p_k \cdot e^{-2\pi i \cdot j \cdot k/12}.$$

Here, $j \in \mathbb{Z}_{12}$. By Lewin's Lemma,

$$\forall j \in \mathbb{Z}_{12}, \quad \widehat{IFUNC(X, X)}(j) = |\mathcal{F}_X(j)|^2.$$

By Theorem 2, the magnitude of the DFT coefficients stores the information about the IC of a pc-set by *inverse Fourier transform*. Therefore, we could define *homometric pc-sets* in terms of the DFT.

Definition 3. *Let X, Y be two pc-sets in $\mathbb{Z}_n$,*

$$\begin{aligned} X, Y \text{ homometric} &\iff IFUNC(X, X) = IFUNC(Y, Y) \\ &\iff IC(X) = IC(Y) \\ &\iff |\mathcal{F}_X| = |\mathcal{F}_Y|. \end{aligned}$$

Z-related pc-sets are homometric, with the corresponding DFT coefficients of each pc-set having the same magnitude.

Lewin introduced five Fourier properties (FOURPROPs) which a pc-set may or may not possess for interpreting the Fourier transform in his 1959 [3] and 2001 [4] articles, as summarized below:

Let X be a pc-set.

- X has FOURPROP(1) if X can be expressed as a disjoint union of tritone sets and/or augmented-triad sets. (exceptional property)
- X has FOURPROP(2) if, for any (0167)-set K, X has the same number of notes in common with $T_3(K)$, as it has in common with K. (tritone property)
- X has FOURPROP(3) if, for any augmented-triad set A, X has the same number of notes in common with $T_6(A)$, as it has in common with A. (augmented-triad property)
- X has FOURPROP(4) if X has the same number of notes in common with each of the three diminished-seventh-chord sets. (diminished-seventh-chord property)
- X has FOURPROP(6) if X has the same number of notes in one whole-tone set as in the other. (whole-tone-scale property)

For any pc-set X in $\mathbb{Z}_{12}$, if the magnitude of the nth DFT coefficient $|\mathcal{F}_X(n)| = 0$, where $n = 1, 2, 3, 4,$ or 6, then X possesses FOURPROP(n).

In 2004, Quinn introduced Fourier balances, which provide graphical representations of Lewin's five Fourier properties plus FOURPROP(5) [9]. Subfigures (a) to (f) in Fig. 3 correspond to Fourier properties (1) to (6), respectively. The Fourier balances not only tell if a set is "balanced on" or possesses the corresponding Fourier properties but also describe the *degree of imbalance* of a set on such properties. The degree of imbalance can be calculated in three ways, with the unit of measurement 'Lewin,' abbreviated as 'Lw':

(1) The degree of imbalance can be calculated through *vector summation*, where each pitch class is considered a vector from the middle of the circle to the particular plate where it is contained.
(2) The degree of imbalance on Fourier balance n can be determined by multiplying the pc-set by n and considering it on Fourier balance 1.
(3) For any pc-set X in $\mathbb{Z}_{12}$, if the magnitude of the nth DFT coefficient $|\mathcal{F}_X(n)| \neq 0$, where $n = 1, 2, 3, 4, 5,$ or 6, then $|\mathcal{F}_X(n)|$ is equal to the *degree of imbalance* of X on Fourier balance n.

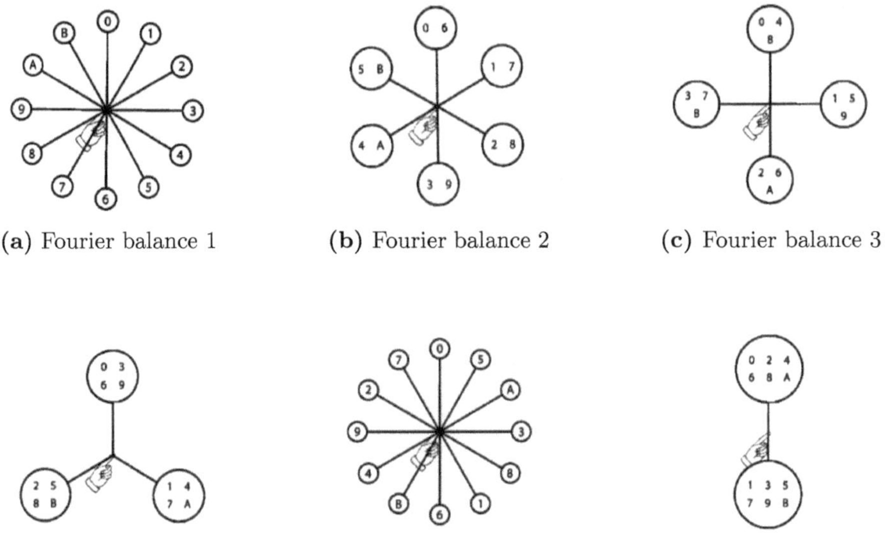

(**a**) Fourier balance 1 (**b**) Fourier balance 2 (**c**) Fourier balance 3

(**d**) Fourier balance 4 (**e**) Fourier balance 5 (**f**) Fourier balance 6

Fig. 3. The Fourier balances (Quinn 2004, 78–85).

The reason to include Fourier balance 5 is that although all and only pc-sets with FOURPROP(1) will possess FOURPROP(5), which makes FOURPROP(5) seem unnecessary, the degrees of imbalances of Fourier balances 1 and 5 may not be the same for pc-sets that do not possess FOURPROP(1) and FOURPROP(5).

3.2 Application of the DFT to Z-Related Hexachords

By calculating the magnitudes of the DFT coefficients for all the Z-related set classes of cardinality 6 in $\mathbb{Z}_{12}$, we could discover that every set class falls into at least one of the following four categories regarding the degrees of imbalance of Fourier balances:

(1) The degrees of imbalance on Fourier balances 1 and 5 are the same.
(2) The degrees of imbalance on Fourier balances 2 and 4 are the same.
(3) The degrees of imbalance on Fourier balances 3 and 6 are the same.
(4) The degrees of imbalance on all Fourier balances (1–6) are distinct.

In Table 1, columns 1–4 list the names and prime forms of Z-related set classes of cardinality 6. Column 5 lists the shared interval content of the Z-related set classes. Column 6 lists the magnitude of $\hat{x}_0, \hat{x}_1, \ldots, \hat{x}_6$ in the DFT vector $\mathcal{F}_X = (\hat{x}_0, \hat{x}_1, \ldots, \hat{x}_{11})$ rounded to two decimal places. $|\hat{x}_i| = |\hat{x}_{12-i}|$, for $i = 7, \ldots, 11$. The last column shows the categories to which the Z-related set classes belong.

Table 1. The categorizations of Z-related hexachords based on the magnitude of the DFT coefficients.

Name	PF	Name	PF	IC	|DFT|	Category
6-Z6	[012567]	6-Z38	[012378]	(421242)	(6,1.41,3.46,1.41,0,1.41,0)	1
6-Z11	[012457]	6-Z40	[012358]	(333231)	(6,2.24,1.73,1.41,1.73,2.24,0)	1,2
6-Z44	[012569]	6-Z19	[013478]	(313431)	(6,1,1.73,3.16,1.73,1,0)	1,2
6-Z36	[012347]	6-Z3	[012356]	(433221)	(6,2.91,1.73,1.41,1.73,1.24,0)	2
6-Z37	[012348]	6-Z4	[012456]	(432321)	(6,2.73,2,2,0,0.73,2)	3
6-Z39	[023458]	6-Z10	[013457]	(333321)	(6,2.39,1,2,1.73,1.51,2)	3
6-Z41	[012368]	6-Z12	[012467]	(332232)	(6,1.73,2.65,0,1.73,1.73,2)	1
6-Z42	[012369]	6-Z13	[013467]	(324222)	(6,1.93,1.73,1.41,3,0.52,0)	4
6-Z43	[012568]	6-Z17	[012478]	(322332)	(6,1,2.65,2,1.73,1,2)	1,3
6-Z45	[023469]	6-Z23	[023568]	(234222)	(6,1.73,1,0,3,1.73,2)	1
6-Z28	[013569]	6-Z49	[013479]	(224322)	(6,1,1,2,3,1,2)	1,3
6-Z47	[012479]	6-Z25	[013568]	(233241)	(6,1.24,1.73,1.41,1.73,2.91,0)	2
6-Z48	[012579]	6-Z26	[013578]	(232341)	(6,0.73,2,2,0,2.73,2)	3
6-Z46	[012469]	6-Z24	[013468]	(233331)	(6,1.51,1,2,1.73,2.39,2)	3
6-Z29	[013689]	6-Z50	[014679]	(224232)	(6,0.52,1.73,1.41,3,1.93,0)	4

There are some similarities between Z-hexachords in $\mathbb{Z}_{12}$ that belong to the same category:

- By the second way of computing the degree of imbalance in Sect. 3.1, a Z-hexachord X belongs to Category 1 if and only if the pc-set $M = \{f(x) = 5x + b \mod 12 | x \in X\}$ belongs to the same set class as X or is one of X's Z-related pairs.
- For any Z-hexachord X that belongs to Category 2, X can be partitioned into disjoint subsets of the form:

$$(X = ([03] \cup [03] \cup [01])) \wedge (X = ([024] \cup [026]) \vee ([048] \cup [026]))$$

- For any Z-hexachord X that belongs to Category 3, X can be partitioned into disjoint subsets of the form:

$$(X = ([06] \cup [06] \cup [04]) \vee ([06] \cup [02] \cup [04]))$$
$$\wedge\, (X = ([01] \cup [05] \cup [04]) \vee ([01] \cup [05] \cup [02]) \vee ([01] \cup [03] \cup [02]))$$

Here, all the subsets are represented by prime forms.

Proposition 8. *Let X and Y be two Z-hexachords. If the relative placement of the [06] subsets in X and its complement and the relative placement of the [06] subsets in Y and its complement are the same, then the four set classes to which X, the complement of X, Y, and the complement of Y belong share at least one common category regarding the degrees of imbalance of Fourier balances.*

For example, pc-set $X = \{0, 1, 2, 4, 6, 9\}$ belongs to 6-Z46, and pc-set $Y = \{0, 2, 3, 4, 5, 8\}$ belongs to 6-Z39. As shown in Fig. 2, the two [06] subsets in X and its complement are interval 01 apart, so are the two [06] subsets in Y and its complement. The set classes that X, the complement of X, Y, and the complement of Y belong to are 6-Z46, 6-Z24, 6-Z39, and 6-Z10 respectively, which all belong to Category 3 in terms of the degrees of imbalance of Fourier balances.

4 Z-Related Hexachords and Dyad Expansion

4.1 Procedure of Dyad Expansion

All Z-related hexachords can be related through the method of *dyad expansion.*

Definition 4. *Let $A = \{a_1, a_2\}$ be a dyad of pitch classes in $\mathbb{Z}_{12}$, which is an ordered pair of two pitch classes. The result of applying dyad expansion of magnitude d to A is the dyad of the form $\{a_1 - d, a_2 + d\}$, where $d, a_1 - d, a_2 + d \in \mathbb{Z}_{12}$.*

Since the prime form of a Z-hexachord has either 2 or 4 elements different from the prime form of its Z-related pair, if we apply transformation or inversion to let the two Z-related hexachords compact to the same side[3] and let 0 in the prime form correspond to the same pitch class, then the resulting pc-sets would vary by either one or two pair(s) of dyads.

If the resulting pc-sets after the preliminary transposition or inversion differ by a pair of dyads, the two elements in each dyad can be viewed as inversionally related, and the axes of inversion would be the same for both dyads. Therefore, we could *expand* one dyad to the other. If the resulting pc-sets differ by two pairs of dyads, the two dyads in each pair are also expansion-related.

For example, Z-hexachord {11,0,1,2,4,5} can be transformed to its Z-related pair {0,1,2,3,4,7} by first transposing to {0,1,2,3,5,6} and expanding the dyad {5,6} by magnitude 1 to {4,7}.

[3] For pc-set $X = \{X_1, X_2, \ldots, X_n\}$, where $X_i > X_{i-1}, \forall i = 2, \ldots, n$, if $X_{n-1} - X_1 < X_n - X_2$, the collection is compacted to the left; otherwise, it is compacted to the right. If $X_{n-1} - X_1 = X_n - X_2$, then if $X_{n-2} - X_1 < X_n - X_3$, the collection is compacted to the left; otherwise, it is compacted to the left, and so on.

4.2 Visual Representation of the Z-Relation

Let X and X' be any Z-related hexachords in $\mathbb{Z}_{12}$. After the preliminary transposition or inversion mentioned in Sect. 4.1, we could partition X into two subsets: the expanded subset S and the remainder set $X-S$. Similarly, we could partition X' into S' and $X'-S'$. Since X and X' are Z-related, and $X = S \sqcup (X-S)$, $X' = S' \sqcup (X'-S')$, by the definition of IFUNC introduced in Definition 1,

$$\begin{aligned} &IFUNC(X,X)\\ =&IFUNC(S,X)+IFUNC(X-S,S)+IFUNC(X-S,X-S)\\ =&IFUNC(S',X')+IFUNC(X'-S',S')+IFUNC(X'-S',X'-S')\\ =&IFUNC(X',X'). \end{aligned}$$

Since the remainder sets $X-S = X'-S'$,

$$IFUNC(S,X)+IFUNC(X-S,S) = IFUNC(S',X')+IFUNC(X'-S',S') \quad (1)$$

We can visually display Eq. 1 by placing a Z-hexachord and itself on an IFUNC table. We list the remainder set from the top left corner, followed by the expanded subset. Figure 4 provides the framework of the IFUNC table for Z-hexachord X, with the expanded subset S as a dyad. The differences between the IFUNC tables of two Z-related hexachords of this form will be elements in cells No. 1–18.

Let elements in cells No. 1–9 of the X to X IFUNC table be $x_1,\ldots,x_9$ respectively. Then, all elements in cells No.1–18 of the X to X and X' to X' IFUNC tables can be expressed in terms of $x_1,\ldots,x_9$, and d, the magnitude of dyad expansion. By Eq. 1, the set of all elements in cells No. 1–18 of the X to X IFUNC table and the set of all elements in cells No. 1–18 of the X' to X' IFUNC table are bijectively related. Therefore, one possible interpretation of the Z-relation between two Z-related hexachords is that intervals in cells No. 1–18 of the IFUNC table are "reordered."

	$(X-S)_1$	$(X-S)_2$	$(X-S)_3$	$(X-S)_4$	S_1	S_2
$(X-S)_1$	0				No.10	No.14
$(X-S)_2$		0			No.11	No.15
$(X-S)_3$			0		No.12	No.16
$(X-S)_4$				0	No.13	No.17
S_1	No.1	No.2	No.3	No.4	0	No.18
S_2	No.5	No.6	No.7	No.8	No.9	0

Fig. 4. The framework of IFUNC tables for Z-related hexachords whose expanded subset is a dyad.

5 Conclusion

The paper explores the structural and transformational relations between Z-related hexachords in $\mathbb{Z}_{12}$. We investigate common subsets among the Z-related hexachords and derive that [06] and [016] subsets are structurally determinant for complementary Z-related hexachords with two [06] subsets. The application of the DFT allows for categorizations based on the degrees of imbalance of Fourier balances, which aligns with categorizations based on the relative positions of the [06] subsets in the pc-set. Moreover, the dyad expansion method presents a novel interpretation of the Z-relation, which can be visually displayed as the reconfiguration of cells within IFUNC tables.

Several avenues are possible for future work on Z-related pc-sets. While this study is limited to Z-related hexachords in $\mathbb{Z}_{12}$, similar transformational methods may apply to sets of other cardinalities and set spaces. Furthermore, the common partitions of Z-related hexachords derived in Sect. 3.2 can be more rigorously proved. For practical applications, employing the structural characteristics and transformations identified in this paper in actual musical analysis and composition would be a valuable extension of the research.

Acknowledgements. The author would like to thank Henry Klumpenhouwer for valuable inspiration and the anonymous reviewers for their constructive feedback.

References

1. Forte, A.: The Structure of Atonal Music. Yale University Press, New Haven (1973)
2. Goyette, J.: The Z-relation in theory and practice. Ph.D. thesis, University of Rochester (2012)
3. Lewin, D.: Intervallic relations between two collections of notes. J. Music Theory **3**(2), 298–301 (1959)
4. Lewin, D.: Special cases of the interval function between pitch-class sets x and y. J. Music Theory **45**(1), 1–29 (2001)
5. Lewin, D.: Generalized Musical Intervals and Transformations. Oxford University Press, New York (2007)
6. Morris, R.: Set groups, complementation, and mappings among pitch-class sets. J. Music Theory **26**(1), 101–144 (1982)
7. Morris, R.: Pitch-class complementation and its generalizations. J. Music Theory **34**(2), 175–245 (1990)
8. Nuño, L.: A detailed list and a periodic table of set classes. J. Math. Music **15**(3), 267–287 (2021)
9. Quinn, I.: A unified theory of chord quality in equal temperaments. Ph.D. thesis, University of Rochester (2004)
10. Rahn, J.: Basic Atonal Theory. Schirmer Books, New York (1980)

Quantum Music

Quantum Memory and Mathematical Gestures: Two Perspectives on Verdi and Wagner

Alberto Avitabile[1], Omar Costa Hamido[2], and Maria Mannone[3,4,5](✉)

[1] Department of Mathematics, La Sapienza University, Rome, Italy
albe.avit@gmail.com
[2] Centre for Interdisciplinary Studies (CEIS20), University of Coimbra, Coimbra, Portugal
ocostaha@uci.edu
[3] Institute of Physics and Astronomy, Universität Potsdam, Potsdam, Germany
maria.mannone@uni-potsdam.de
[4] MIRMI, Technische Universität München, Munich, Germany
maria.mannone@tum.de
[5] DSMN and ECLT, Ca' Foscari University of Venice, Venice, Italy
maria.mannone@unive.it

Abstract. Concepts and methods from theoretical physics and abstract mathematics can inspire new techniques to analyze the structure of musical compositions, seen as the unfolding of phenomena over time. Here, we consider an adaptation to music of a criterion to measure memory in quantum states (degree of non-Markovianity) and elements from the mathematical theory of musical gestures. We analyze salient features of two opera pieces, from Verdi's *La forza del destino* and Wagner's *Tristan und Isolde*, commenting the obtained results and the limits of these techniques.

Keywords: non-Markovianity · category theory · opera · orchestra

1 Introduction

What makes a piece of music interesting and effective in conveying movement and expression is deeply linked to the composer ability to create and resolve—or *not* resolve—expectations of the listener. This depends on how music is perceived and experienced. The ability to recognize and predict patterns is rooted in a deep and ancient area of our brain, where our ability to recognize melodic, harmonic, rhythmic, timbric, and dynamical patterns, and to make predictions based on scheme recognition, resides. These elements are experienced in neighboring moments in time, almost superposed in our *memory*, and then compared and classified by adequate criteria of similarity. In this framework, gesture theory [14] can constitute a formal tool to make statements about diagrammatic thinking and organization, ultimately formalizing the notion of *gestural similarity* through an extended notion of homotopy of gestures [12]. Pattern-similarity

© The Author(s), under exclusive license to Springer Nature Switzerland AG 2024
T. Noll et al. (Eds.): MCM 2024, LNCS 14639, pp. 137–146, 2024.
https://doi.org/10.1007/978-3-031-60638-0_11

recognition is related with *memory*. Musical memory can be analyzed borrowing concepts and tools from physics, and in particular, quantum mechanics. Nowadays, quantum computing and quantum mechanics are more and more used for music analysis and music-making [3,6,15–17]. The measure of memory of quantum states across time [18], i.e., their degree of non-Markovianity [5,19], has produced a criterion to measure the degree of memory in musical compositions [10,13]. These can be seen as phenomena unfolding through time in the space of sounds, and thus they can be investigated with tools from physics.

In this article, we want to compare the quantitative results obtained with this method with the intuition of the listener, formalized through gesture theory and the smooth language of category theory.

Can these two techniques, the quantum-based measure of memory, and the gesture theory, provide comparable results? Is the degree of memory similar to the similarities retrieved via a diagram-based analysis of the salient points of the scores? We address these questions analyzing two opera pieces composed by G. Verdi and R. Wagner, respectively. In Sect. 2 we present the adopted methods in more details; in Sect. 3 our results, and in Sect. 4 our discussions and hints for future research.

2 Methods

2.1 Musical Non-Markovianity

Here, we summarize the key ideas of the considered methods. Let us start with non-Markovianity. In theoretical physics, the *memory* of a quantum state can be quantified considering its specificity, that is, its *distinguishability* with respect to another state. If the distance between the states diminishes across time, the memory is lost. Vice versa, if it is kept or increases [5], the dynamics is non-Markovian [19]. The rate of variation σ between two states, described by density matrices ρ_1, ρ_2, is:

$$\sigma = \sigma(\rho_1, \rho_2) = \frac{d}{dt} D(\rho_1, \rho_2) = \frac{d}{dt} \left(\frac{1}{2} tr|\rho_1 - \rho_2| \right), \tag{1}$$

where $|\rho_1 - \rho_2| = \frac{1}{2} tr \left[\sqrt{(\rho_1 - \rho_2)^T (\rho_1 - \rho_2)} \right]$. The degree of memory $\mathcal{N}$ is obtained integrating the variation rate over time. However, in music there is memory $\mathcal{M}$ between two sequences if they are similar: the smaller the changes, the higher the thematic memory, thus conceptually $\mathcal{M} = 1 - \mathcal{N}$. Defining opportune density matrices ρ_i, ρ_j with respect to a pair of parameters for two finite musical sequences i and j, for the sake of simplicity we directly denote their variation as the trace of their difference,

$$\dot{d}_1 = \sigma_{ij} = \sigma(\rho_i, \rho_j) = \frac{1}{2} tr(\rho_j - \rho_i) \tag{2}$$

or as

$$\dot{d}_2 = \frac{1}{2} tr \left[\sqrt{(\rho_j - \rho_i)^T (\rho_j - \rho_i)} \right]. \tag{3}$$

The *musical non-Markovianity degree* $\mathcal{M}_C$, comprised between 0 (min value) and 1 (max value) with respect to the considered parameters, is defined as [13]:

$$\mathcal{M}_C = \frac{1}{1 + r\sum_i (if\,\sigma_i > 0)\sigma_i}, \tag{4}$$

where $r = n^+/n^T$ is the fraction of positive rates (the number of positive σ) over the total rates (the total number of computed σ).

As an example of correspondence between thematic memory and degree of non-Markovianity, we can consider two different musical sequences [13], exemplifying two "extremes" of the spectrum of musical memory. In Fig. 1a, we can see a musical sequence composed by two identical bars. The degree of onset-pitch non-Markovianity $\mathcal{M}_C$ computed as the variation rate between the first and the second sequence is equal to 1, that is, the highest possible value of memory. In Fig. 1b, we can observe a musical sequence with two completely different bars. In this case, the degree of onset-pitch $\mathcal{M}_C$ is equal to 0.3, closer to the minimum possible value, 0.

(a)

(b)

Fig. 1. Two "extreme" examples of different degrees of musical memory from [13]: a short sequence constituted by two identical bars, with $\mathcal{M}_C = 1$ (a), and another sequence with two completely different bars, with $\mathcal{M}_C = 0.3$ (b).

The matrices ρ_i contain statistical information on the distribution of features in a musical sequence, with respect to a pair of parameters. To obtain them, we first compute the normalized distance between each note, and the mean value (of the first and the second parameter) of the sequence. We then count the number of notes inside each interval of the first (and then of the second) parameter, in the order [0, 0.25], [0.25, 0.5], [0.5, 0.75], [0.75, 1]. With this information, we obtain a matrix, whose rows contain the intervals of the first parameter, and whose columns contain the intervals of the second parameter. Then, we divide all elements by the total number of notes in the sequence, getting 1 as the sum of all the matrix entries.

As an example of computation, we present the onset-pitch matrices corresponding to the sequences of Fig. 1. Equation (5) presents the density matrix computed from each one of the bars of the sequence in Fig. 1a:

$$\rho_1 = \rho_2 = \begin{pmatrix} 0.2 & 0.2 & 0.2 & 0.2 \\ 0 & 0 & 0 & 0 \\ 0 & 0 & 0 & 0 \\ 0 & 0 & 0.2 & 0 \end{pmatrix}, \tag{5}$$

and Eq. (6) shows the matrices corresponding to the first and second bar, respectively, of the example in Fig. 1b:

$$\rho_3 = \begin{pmatrix} 0.12 & 0.25 & 0.12 & 0.12 \\ 0.12 & 0 & 0 & 0.12 \\ 0 & 0 & 0 & 0 \\ 0.125 & 0 & 0 & 0.12 \end{pmatrix}, \quad \rho_4 = \begin{pmatrix} 0 & 0 & 0 & 0 \\ 0 & 0 & 0 & 0 \\ 0 & 0 & 0 & 0 \\ 1 & 0 & 0 & 0 \end{pmatrix}. \tag{6}$$

The computation of these matrices, given a MIDI sequence or a real-time MIDI controller, is automatized. The code[1] to compute with MAX/MSP a set of matrices starting from a midi file is available at [7].

2.2 Mathematical Theory of Musical Gestures

A *musical gesture* can be intuitively seen as a path connecting points in space and time, as the movements of a dancer between a position and another [11]. In musical performance, gestures are necessary to *unfold* over time the symbolic content of a score, to obtain sound and shape its physical parameters of pitch, loudness, and timbre. A mathematical definition of musical gestures is thus necessary. In [14], musical gestures have been defined for the first time in mathematical terms, as a mapping from an oriented graph (a digraph), called *skeleton*, and a curve in a topological space, called *body*. The topological space can be not only the space-time where the performers moves, but also a space of parameters. For instance, in the framework of score analysis, we can think of orchestral performers' gestures producing music [12], in turn constituted by gestures in the spaces of melody, rhythm, and harmony. For the sake of simplicity, in this work we are more focused on melody.

The distinction between skeleta and bodies is also a distinction between discrete and continuous gestures. The movements of the dancers between consecutive steps and poses actually constitute the body of the gesture. Formally, we consider a topological space X (ideally a subspace of some parameter space $\mathbb{R}^n$ modeling the corresponding perceptual dynamics), and a fixed digraph Γ. One can define a topological digraph $\overrightarrow{X}$ canonically associated to X, whose vertices

[1] Original code in C by MM; version for MAX/MSP by OCH. The original idea of a musical non-Markovianity degree is due to the physicist Giuseppe Compagno.

are points of X and whose arrows are continuous paths in X that connect two given points. We define a Γ-gesture in X as a morphism of digraphs:

$$g : \Gamma \to \overrightarrow{X}. \tag{7}$$

Thus, a (Γ-shaped) gesture is the concrete realization, in some appropriate space X, of an abstract diagram scheme Γ. The set $\widehat{\mathcal{D}}(\Gamma, \overrightarrow{X})$ of all Γ-gestures in X carries a natural structure of topological space in itself, making it possible to consider *gestures in a space of gestures.*[2] This process could be iterated, but 2-gestures are just enough to introduce a homotopy language for gestures, that generalizes traditional homotopy theory of curves, in a natural and expected way.

The advantage of working with gestures rather than with just diagrams or curves is that, on the one hand, the topological realization of a digraph leads to more flexible (i.e., continuous) relations among musical elements that better reflect the compositional process of motivic variation inside a given piece, impossible to grasp in the discrete world of graphs. On the other hand, gestures of different shapes are to be intended as a generalization of continuous curves in space, which we recover as $\uparrow$-gestures.

Here, we focus on gestures in the pitch-time space $X = \mathbb{R}^2$, in which the musical score (and its diagram analysis) is embedded in a gestural way, and where we recognize that several different motivic variations can be expressed in terms of local homotopy of gestures, in a similar fashion as that suggested in [2].

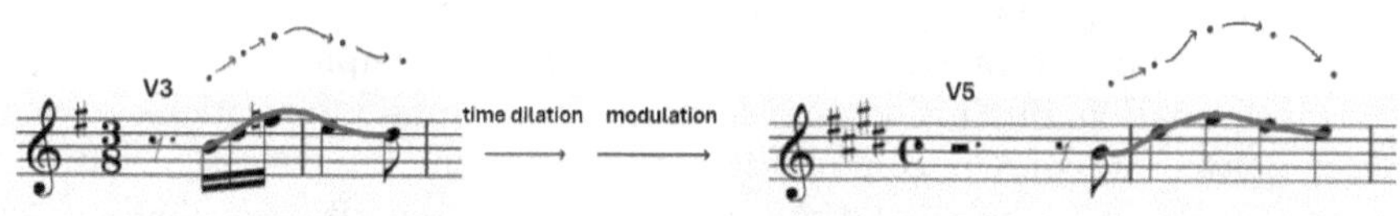

Fig. 2. Before a time dilation, the first three notes spelling a B-diminished (B°) are modulated to an E-major triad (E_{maj}) in second inversion. Then the stepwise descending motion is imitated. (Color figure online)

Of course, the standard topology on $\mathbb{R}^2$ can't differentiate between gestures that are—or aren't—similar, so the question of which topology could provide a notion of continuity able to grasp musically relevant features of (dis)-similarity arises. This question, related to heuristic arguments concerning the notion of *continuous motivic variation*, is for now left open. We do not delve here into these details, and we limit our focus to the following gestural-similarity relations, inspired by the examples given in [2]: time translation-dilation, pitch translation-dilation, both chromatic and diatonic, melodic extensions, melodic contractions, neighboring notes and appoggiaturas. We can use these ideas to analyze musical

[2] Traditionally referred to as *hypergestures*, or 2-gestures.

sequences and compare them in terms of progressive transformations. We can indicate each note with a point, and the intervals as arrows. Changing the shape of the arrows, being inspired by neumatic notation in Gregorian chant, we can schematically indicate the differences between intervals; see the above-diagrams in Fig. 2, with two examples from Verdi's *La forza del destino*. In the same figure, a blue line is superposed to the actual melody on the pentagram. The above-diagrams are oriented according to the flowing of time. The first sequence in Fig. 2 can be transformed into the second one through a time dilation followed by a modulation.

3 Results

3.1 Non-Markovianity Degree for Verdi and Wagner

We analyze the degree of memory concerning pitch and onset (frequency-start matrices) of the *Prelude* from *Tristan und Isolde* by R. Wagner and the *Ouverture* from *La forza del destino* by G. Verdi. We divide both pieces into four main sequences: bars 1–32, 33–62, 63–88, 89–122 for Wagner and sections A-B, C-D-E-F, G-H-I-J-L-M, N-O-P for Verdi, see Table 1. Negative rates between two consecutive musical segments, or between two macro-sequences, denote a decreasing similarity between them. Only the positive rates contribute to the memory degree $\mathcal{M}_C$. Wagner's piece is characterized by a progressive transformation of the motivic material, with blended elements and less separated sequences (*infinite melody*). The information is more clear observing the pairwise rates, with the comparison between consecutive sequences ($\sigma_{1,2}$, $\sigma_{2,3}$, $\sigma_{3,4}$) for Wagner. The rate of variation measured with $\dot{d}_2$ emphasizes the progressive transformation, diminishing memory. In Verdi's music, repetitions are more similar between them. Concerning style, we notice the progressive variation of elements in Wagner's piece, and the more consistent repetition of thematic elements in Verdi's piece. We also notice that, for both pieces, the rate $\sigma_{4,1}$, that is, the comparison between pitch-onset matrices of the fourth (last) macro-sequence and first macro-sequence is the highest, probably denoting the coherence between initial and final elements in both compositions. We do not have to forget that the considered musical compositions are ouvertures of operas, with the precise role of presenting to the listener the main themes that will be heard in the following hours, and the style to connect them and relate one element with the other. Finally, in Table 2 we present the rates of variation between the main themes of both pieces (Fig. 3), connecting memory with gestural analysis.

3.2 Gestural Analysis of Verdi and Wagner

Here we propose two groups of salient gesture classes obtained with the methods described above. Classes inside the same piece, and between the two pieces, can in turn be compared by arrows heuristically describing motivic variations obtained through sections and extensions. Figure 3 shows our choice of themes, labeled

Table 1. Macro-sequences variation rates and memory degree.

		variation rates				memory $\mathcal{M}_C$
		$\sigma_{1,2}$	$\sigma_{2,3}$	$\sigma_{3,4}$	$\sigma_{4,1}$	
Verdi	$\dot{d}_1$	0.2	−0.2	−0.14	0.04	**0.89**
	$\dot{d}_2$	0.05	5E−04	0.04	0.08	0.86
Wagner	$\dot{d}_1$	−0.17	0.06	−0.08	0.19	**0.97**
	$\dot{d}_2$	0.08	0.02	0.05	0.11	0.79

Table 2. Variation rates (as trace distances) between the themes in Fig. 3. V_i refers to Verdi's themes, and W_i to Wagner's ones.

	themes' variation rates								memory
	σ_{V_1,V_2}	σ_{V_2,V_3}	σ_{V_3,V_4}	σ_{V_4,V_5}	σ_{V_5,V_6}	σ_{V_6,V_7}	σ_{V_7,V_8}		$\mathcal{M}_C^*$
Verdi	0.06	0.13	0.12	0.11	0.12	0.14	0.19		0.53
	σ_{W_1,W_2}	σ_{W_2,W_3}	σ_{W_3,W_4}	σ_{W_4,W_5}	σ_{W_5,W_6}	σ_{W_6,W_7}	σ_{W_7,W_8}		$\mathcal{M}_C^*$
Wagner	0.10	0.13	0.18	0.11	0.22	0.12	0.06		0.52
	σ_{V_1,W_1}	σ_{V_2,W_2}	σ_{V_3,W_3}	σ_{V_4,W_4}	σ_{V_5,W_5}	σ_{V_6,W_6}	σ_{V_7,W_7}	σ_{V_8,W_8}	
Verdi/Wagner	0.05	0.117	0.07	0.25	0.08	0.16	0.26	0.08	

with V_i, W_i, $i = 1,...,8$ (V for Verdi and W for Wagner). At the "atomic" level of melodic cells, small transformations lead one element into the other, also considering cross-comparisons between the two pieces. This confirms the small variation rates of Table 2. Also, we notice that homotopic transformations can relate different materials. What is needed, and will be investigated in further analysis, is a *degree of similarity*, as mentioned in [12], requiring experimental validation and perceptual analysis, that is, extra-mathematical references.

4 Discussion and Conclusions

We analyzed the structure and key elements of two whole pieces, by Verdi and Wagner, respectively, in light of an analytic technique derived from category theory and gesture theory, and of a memory measurement based on quantum mechanics, comparing their results. The first is a formal tool that requires extra-mathematical elements for the definition of a similarity degree; the second is essentially a statistical method which loses the detail of information. These methods can be considered as complementary, as they catch different aspects. In general, each analytic method highlights some aspect and neglects other ones. Comparing different analytical techniques allows one to gain different perspectives on works of art. Further research can include measurements of timbre variations and thus orchestration differences, along with a mobile window, allowing a more fine-grained analysis of memory.

This article is devoted to the compared presentation of two analytical techniques, the first based on quantum-defined musical memory, and the second

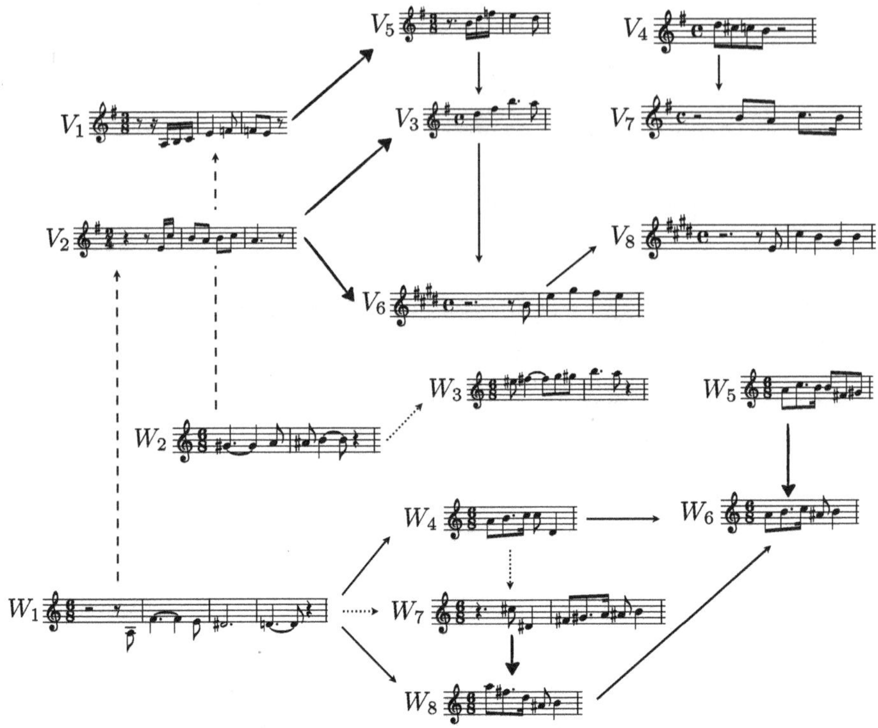

Fig. 3. A diagram synthesis of the relations: full-line arrows represent direct homotopy equivalence of gestures, thick-line arrows represent contractions, dotted arrows represent extensions, dashed arrows represent cross-comparisons. V_i $(i = 1, .., 8)$ stands for the selection of Verdi's sequences, and W_i $(i = 1, .., 8)$ for the selection of Wagner's ones, see Table 2 for the memory degrees of their comparison.

based on mathematical gesture theory. While we mostly focused here on pitches and onsets, the presented techniques can be applied also to the other musical parameters, such as loudness, duration, and timbre. The melodic aspect is the mostly investigated in the literature. A comparison with already existing and established analytical techniques for melodies is out of the scope of this work. However, we mention here three well-known approaches, and the key differences with the discussed techniques. For instance, Steve Larson [8] analyzed musical structures as the result of shaping forces. This metaphorical approach, with an emphasis on tonal relationships, was later mathematically formalized by Peter beim Graben and Reinhard Blutner [4]. This approach is probably more focused on the "reasons" behind specific structures, while the non-Markovian and gesture analysis is inherently descriptive. David Tempereley proposed a probabilistic model of melody perception [20]. Being quantum-based, the probabilities are intrinsically present in the non-Markovian approach that is, however, not based on perception. Listeners' perception can be compared to the degree of musical memory as part of a different and more comprehensive analysis. Markovian approaches, where memory is restricted, have been used in the framework of generative music theory [9]. Further research can also involve the comparison between the non-Markovian degree of musical sequences and the information from the Fourier transform of their symbolic content, following the approach proposed by Emmanuel Amiot [1].

What this study ultimately intends to highlight is the auxiliary role of the formal tools of mathematics and physics to investigate particular aspects of orchestral masterpieces.

Acknowledgments. The authors thank Prof. Domenico Fiorenza, supervisor (jointly with MM) of the MSc thesis in Mathematics by AA, for his suggestions on algebraic geometry. The work by OCH is supported by the European Project IIMPAQCT, under grant agreement Nr. 101109258 (DOI 10.3030/101109258).

Funded by the European Union

Author contributions. Conceptualization and writing: AA, OCH, and MM; gestures: AA; non-Markovianity: MM and OCH; revision: MM.

References

1. Amiot, E.: Music Through Fourier Space. Springer, Cham (2016). https://doi.org/10.1007/978-3-319-45581-5
2. Arias-Valero, J.S., Lluis-Puebla, E.: Explicit presentations of topological categories of gestures. J. Math. Music **17**(2), 213–243 (2023). https://doi.org/10.1080/17459737.2022.2111612
3. beim Graben, P.: Gauge symmetries of musical and visual forces. J. Math. Arts **17**(3–4), 347–382 (2023). https://doi.org/10.1080/17513472.2023.2281895
4. beim Graben, P., Blutner, R.: Quantum approaches to music cognition. J. Math. Psychol. **91**, 38–50 (2019). https://doi.org/10.1016/j.jmp.2019.03.002
5. Breuer, H.P., Petruccione, F.: Theory of Open Quantum Systems. Oxford University Press, New York (2002)
6. Fugiel, B.: Quantum-like melody perception. J. Math. Music **17**(2), 319–331 (2022)
7. Hamido, O.C., Mannone, M.: Musical_non-Markovianity (v1.0.0). Zenodo (2023). https://doi.org/10.5281/zenodo.8152691
8. Larson, S.: Musical Forces: Motion, Metaphor, and Meaning in Music. Indiana University Press (2012)
9. Lerdahl, F., Jackendoff, R.: Generative Theory of Tonal Music. MIT Press, Cambridge (1983)
10. Mannone, M., Hamido, O.C.: Quantum memory of musical compositions. In: Miranda, E., Jansen, K. (eds.) 2nd International Symposium on Quantum Computing and Musical Creativity (ISQCMC), Berlin (2023). https://doi.org/10.5281/zenodo.10206714

11. Mannone, M., Turchet, L.: Shall we (math and) dance? In: Montiel, M., Gomez-Martin, F., Agustín-Aquino, O.A. (eds.) MCM 2019. LNCS, vol. 11502, pp. 84–97. Springer, Cham (2019). https://doi.org/10.1007/978-3-030-21392-3_7
12. Mannone, M.: Introduction to gestural similarity in music. J. Math. Music **12**(2), 63–87 (2018)
13. Mannone, M., Compagno, G.: Characterisation of the degree of musical non-Markovianity. J. Creative Musical Syst. **6**(1), 1–21 (2022)
14. Mazzola, G., Andreatta, M.: Diagrams, gestures and formulae in music. J. Math. Music **1**(1), 23–46 (2007)
15. Miranda, E.R. (ed.): Quantum Computer Music. Springer, Cham (2022). https://doi.org/10.1007/978-3-031-13909-3
16. Putz, V., Svozil, K.: Quantum music. Soft. Comput. **21**, 1467–1471 (2017)
17. Rocchesso, D., Mannone, M.: A quantum vocal theory of sound. Quantum Inf. Process. **19**(292), 1–28 (2020). https://doi.org/10.1007/s11128-020-02772-9
18. Preskill, J.: Lecture notes for physics 219: quantum computation. California Institute of Technology (2004)
19. Rivas, Á., Huelga, S.F., Plenio, M.B.: Quantum non-Markovianity. Rep. Prog. Phys. **77**(9), 094001 (2014)
20. Tempereley, D.: A probabilistic model of melody perception. Cogn. Sci. **32**, 418–444 (2008)

Intro to Quantum Harmony: Chords in Superposition

Christopher Dobrian[1](✉) and Omar Costa Hamido[2]

[1] University of California, Irvine, CA 92697-2775, USA
dobrian@uci.edu
[2] Centre for Interdisciplinary Studies (CEIS20), University of Coimbra, Coimbra, Portugal
ocostaha@uci.edu

Abstract. Correlations between quantum theory and music theory—specifically between principles of quantum computing and musical harmony—can lead to new understandings and new methodologies for music theorists and composers. The quantum principle of *superposition* is shown to be closely related to different interpretations of musical meaning. Superposition is implemented directly in the authors' simulations of quantum computing, as applied in the decision-making processes of computer-generated music composition.

Keywords: Harmony · Quantum Theory · QAC · Superposition

1 Quantum Theory and Music Theory

1.1 Superposition

The purpose of this article is to draw correlations between quantum theory and music theory, specifically between principles of quantum computing and musical harmony. We believe that these correlations may be interesting and useful to music theorists and composers. We will demonstrate how these relationships can form new metaphors for understanding musical harmony, and we will provide some possible concrete implementations in experimental arenas such as computer-generated composition.

The most notable correspondence is derived from the quantum principle of *superposition*.[1] In quantum computing, a fundamental unit known as the *qubit* (**qu**antum **b**inary dig**it**) does not have an established state—0 or 1—as a bit does in conventional computing. Rather, the qubit is in an unmeasurable state which can be thought of as the superposition (linear combination) of two *basis states* (vectors) known as $|0\rangle$ and $|1\rangle$ ("ket 0" and "ket 1"). In practice, one might think of a qubit as having a probability of being 0 and a probability of being 1, with the sum of those two probabilities being 1, but those probability values are only knowable by measuring (observing) the qubit, which has the effect of collapsing it to a classical representation of 0 or 1.

This quality of being unknowable until observed is known to most people through the thought experiment of Schrödinger's cat. In that hypothetical, a cat is placed for one

[1] C.f., [1, 2].

© The Author(s), under exclusive license to Springer Nature Switzerland AG 2024
T. Noll et al. (Eds.): MCM 2024, LNCS 14639, pp. 147–158, 2024.
https://doi.org/10.1007/978-3-031-60638-0_12

hour in a box, in which is also placed a contraption that may or may not release poisonous gas into the box, depending on whether or not a single radioactive atom decays in that hour. During that hour, the alive-or-dead state of the cat is in superposition, unknowable until the moment of observation, when the box is opened. The thought experiment demonstrates that 1) a microcosmic happenstance on the order of a single atom (viz. a single qubit) can have a larger-scale effect, 2) the classical binary (alive or dead, 1 or 0) is reified only by observation, and 3) until that moment of resolution, the situation is ambiguous (except to the cat).

Music theorists and composers deal with a metaphorically comparable sort of ambiguity all the time. The meaning[2] of a given musical chord, in terms of its tonal function in its current context, frequently may have more than one interpretation. Music analysts strive either to remove that ambiguity by identifying the "true" function of the chord, or to draw some conclusions about the composer's process based on the way the composer treats the ambiguity.

The role of a composer is different from that of a theorist. One important distinction is between a theorist's analytical viewpoint, looking at existing music and trying to discern why and how it works, and a composer's creative viewpoint, looking at musical potential and trying to decide where to go. In the case of the analyst, the music pre-exists as a fixed entity, and the progress from one chord to another is a known and unalterable fact. In the case of a composer, by contrast, the musical "present" contains infinite potential to proceed anywhere, because the composer has free will. Infinite free will, however, is terrifying and unmanageable for most composers, which is why they self-impose constraints that limit their possibilities[3].

1.2 Chord Ambiguity and Equal Temperament

Composers exploit the inherent ambiguity of chords as a means to pivot to a more or less unexpected subsequent chord. This is a well-known theoretical concept, one which has enabled composers for centuries to achieve musical variety and modulate from one tonal center to another. Indeed, in jazz and in classical music of the last 150 years, the listener may be suspended in a state of constant ambiguity as to where the music is headed. Constant chromaticism, constant modulation, and atonality all utilize tonal ambiguity, and are all made possible by the inherent neutrality of twelve-tone equal temperament.

Twelve-tone equal temperament is the tuning system that is generally accepted as the norm in Western music, and that is assumed by every Western harmony text, be it classical, jazz, atonal, etc. Yet, tonal harmonic theory is derived from the diatonic scale, which grew out of just intonation and vastly predates equal temperament. Therefore, there is an inherent tension between the uniform intervals of the chromatic scale in equal temperament and the just intonation that is implicit in the diatonic scale and its chromatic

[2] C.f., [3].

[3] "I experience a sort of terror when, at the moment of setting to work and finding myself before the infinitude of possibilities that present themselves, I have the feeling that everything is permissible to me... Will I then have to lose myself in this abyss of freedom?... I have no use for a theoretic freedom. Let me have something finite, definite—matter that... Presents itself to me together with limitations. I must in turn impose mine upon it."—Igor Stravinsky [4], pp. 63–64.

alterations. This is common knowledge to most theorists, yet it is considered important in some contexts and conveniently ignored in others, a paradox made possible by the use of twelve-tone equal temperament.

For example, the fundamental tonal concept of the "circle of fifths" depends on the enharmonic equivalency inherent in equal temperament. A stack of twelve perfect fifths only creates a circle because, at some point on the circle, a pitch is respelled as its enharmonic equivalent, such as F#/Gb (Fig. 1).[4]

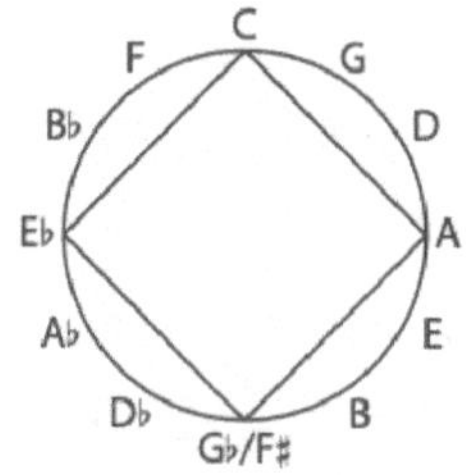

Fig. 1. A diminished seventh chord depicted on the circle of fifths.

Classical composers, theorists, and analysts quite reasonably insist on correct spelling of pitches consistent with the underlying tonal theory. For example, the third of a G# major triad is spelled B# rather than C. Theory textbooks and analyses distinguish between the interval of a diminished fourth and a major third, or between a minor seventh and an augmented sixth, in terms of their harmonic role, while in the same text also touting the interchangeability of enharmonic spellings for purposes of tonal ambiguity and modulation. The theory insists that these enharmonically spelled intervals, which are in fact identical in equal temperament, are also paradoxically different because of their tonal function, but that that difference can magically evaporate through mental respelling. Indeed, that "magic" is what enables much of the interesting and beautiful chromaticism of romantic music and jazz.

2 Superposition and the Diminished Seventh Chord

The most obvious example of this magic is the diminished seventh chord. The utility of the diminished seventh chord as a means of misdirection or modulation relies on enharmonic respelling, especially the enharmonic identity between the minor third and the augmented second. A diminished seventh chord is a stack of three minor thirds, causing the interval of an augmented second from the seventh of the chord going up to the root (Fig. 2).

The sound of the augmented second is identical to that of the minor third, so the diminished seventh chord effectively divides the circle of fifths equally into four minor thirds (Fig. 1). Any diminished seventh chord can be spelled in four different ways, by

[4] In equal temperament, the interval of a perfect fifth is tempered by -1.955 cents ($^1/_{12}$ of the Pythagorean comma) relative to a just fifth, such that after ascending twelve perfect fifths from C, B# is exactly the same as C.

Fig. 2. The intervals of the diminished seventh chords.

considering any of its four pitches to be the root of the chord (Fig. 3). By implication, each note of the chord can therefore potentially function in any of four ways[5], as either the root, third, fifth, or seventh of the chord. Similarly, the two tritones present in the diminished seventh chord are each ambiguous in their function due to these enharmonic reinterpretations, potentially functioning as the interval of either an augmented fourth or a diminished fifth.

Fig. 3. Respellings of the same set of four pitch classes.

The essential function of the diminished seventh chord is as a leading tone chord in minor tonalities, or as an altered (borrowed) leading tone chord in major keys. In both instances, one chromatic alteration is needed, because the diminished seventh chord doesn't occur naturally in either the major scale or the natural minor scale. The root of the chord functions as a leading tone, wanting to resolve upward by semitone, and the diminished seventh of the chord inherently pulls downward by semitone. So, since any of the four pitches of the chord may function as the root, any time that a diminished seventh chord occurs, it has the potential to proceed to (at least!) eight different chords, four major chords and four minor chords. This ambiguity and versatility led to the diminished seventh chord's most common usage, as a secondary leading tone seventh chord (an implied secondary dominant), used either as a fleeting "tonicization" of a non-tonic degree of the scale without actually leaving the home key, or as an important force in modulating to a new key (Fig. 4).

Fig. 4. Secondary diminished seventh chords as passing chords and as a modulating device.

[5] We will omit, for now, "wrong" spellings, such as B# Eb F# Bbb.

The important thing to note for our discussion of this chord's relationship to quantum theory is that, for the listener, the chord's function—its directionality and its tonal meaning—is highly ambiguous until the moment of its resolution. One can say that the diminished seventh chord is a complex quantum state, and its resolution is the measurement of that quantum state. For a composer or an improviser, the chord is in a state of multiplicity (viz. superposition) until a decision is made to choose one of the multiple possible resolutions. For the listener, the chord may seem at first to have a primary "most likely" interpretation, but the chord's true function is evident only in retrospect, based on the chord that succeeds it. This is not to suggest that listeners are literally performing realtime functional analysis of the chords they hear, but that the ambiguity of the diminished seventh chord, and its multiple enharmonic representations, permit the composer or improviser to conceptually reinterpret (effectively, respell) the chord to have a different root, to lead to a different, less expected chord. The resolution, which may be unexpected, makes musical sense in light of that conceptual respelling.

In the example below (Fig. 5), starting in the key of C, the second chord is heard initially as a C#°7 chord leading to ii. Once we land on the third chord, however, we retrospectively reinterpret (respell) the soprano note of the second chord from Bb to A#. From the third chord, B major, the bass note of the fourth chord only really makes sense as an E# leading to F#. However, on the downbeat of the second measure it falls to E, revealing retrospectively that the E# may actually have been functioning as an F (or alternatively, that the A major chord may be some sort of deceptive cadence in F# minor). From there, in the second chord of the second measure, the tenor note can be heard either as B# leading us to C# or as C falling to B. The correct interpretation is revealed to us only retrospectively when we arrive on the E major chord.

Fig. 5. Modulation via respelled secondary diminished seventh chords.

Because of the diminished seventh chord's inherent ambiguity, when it is used consecutively or in sequence, or is reinterpreted in succession and points to far-related tonal centers, the listener tends to lose a sense of tonal gravity, and becomes less confident in the directionality of the music. This ability for the chord to create a sense of wandering, and to go almost anywhere, led composer and theorist Arnold Schoenberg to refer to it as a vagrant (*vagierende*) chord [5].

The foregoing discussion is known territory to music theorists, yet the relevancy to quantum computing has rarely[6] been noted. We have demonstrated that this particular

[6] C.f., [6–8].

chord shares the key characteristics of Schrödinger's cat: 1) a microcosmic occurrence (in this case, which note of the diminished seventh chord is considered the root) has a larger-scale effect (the progression to the subsequent chord and possibly to a far-removed key), 2) the true nature of the chord is reified only by observation of its resolution, and 3) until that moment of resolution, the situation is ambiguous; the chord is in superposition.

3 Practical Application: Quantum Decision Making

So much for the metaphorical correlations between harmonic ambiguity and quantum superposition. But what does this imply in terms of its practical applications? We will direct our attention first to the potential use of quantum circuits in computer-generated composition. The probabilistic nature of quantum computing makes it useful for conceptualizing and implementing computer decision making in tonal chord progressions. In this initial discussion, we continue using the well-studied diminished seventh chord as an archetypical example of an ambiguous chord, and suggest an algorithm and circuit that could be resolved as a compositional decision to proceed to one chord or another.

There are three distinct pitch class (PC) sets that form diminished seventh chords: $\{0, 3, 6, 9\}$, $\{1, 4, 7, 10\}$, and $\{2, 5, 8, 11\}$. Any transposition of any of these results in a version of one of them. So, if a composer or improviser decides to use a diminished seventh chord, there are effectively only three PC sets to choose from. In practice, such a choice is not made arbitrarily. It is influenced by considerations of voice leading, short- or long-term harmonic targets (goals for resolution), and the degree of ambiguity desired. Some choices might be more stylistically appropriate or common than others, and the choice might well even be influenced by a related decision made elsewhere in the piece. Given the chord's vagrant ability to resolve to so many possible subsequent chords, however, it's literally impossible for a composer/improviser to make a "bad" choice, since a "good" subsequent chord exists in every case. So, our hypothetical composer/improviser (i.e., our computer program) need only choose one of the three options.

Once a PC set has been chosen, whether or not that choice was made with a particular root note in mind, any of the four pitch classes can be treated as the root (leading tone), which determines the chord's conceptual spelling and thus the likely root of its successor chord. For example, the PC set $\{0, 3, 6, 9\}$ can be spelled CEbGbBbb (or B#D#F#A), D#F#AC, F#ACEb, or ACEbGb (or even GXB#D#F#), and thus can imply resolution to PC 1, 4, 7, or 10. Again, such choices are not made arbitrarily in actual practice. But the fact remains that, once the diminished seventh chord has been stated or implied, any of four pitch classes may be considered its root. The relative likelihood(s) of the subsequent chord(s) for the listener will again depend on melodic motion, a sense of the current key, stylistic traits of the music, and comparable occurrences elsewhere in the piece. Yet, at the local chord-to-chord level, each note may be the root with equal probability. So here, too, the composer/improviser, having chosen a pitch class set, can then confidently assign any of those four pitch classes to be the chord root.

The leading tone diminished seventh chord, when not being used as a secondary or vagrant chord, can occur in a minor key by raising the subtonic note of the scale to be the leading tone, or it can occur in a major key by lowering the submediant note of the scale. Thus, a leading tone diminished seventh chord progresses quite readily to either a major or minor triad a semitone higher. This gives the computer composer one additional choice, in this case a binary one, to resolve to either a major or minor triad.

The preceding description outlines one very simple algorithm for chord progression, based on the ambiguity inherent in a particular chord type, and rooted in the ambiguities of chord spelling enabled by twelve-tone equal temperament. One chooses one of three pitch class sets, one then ascribes the role of chord root to one of the four pitch classes in that set, and one then chooses one of two chord qualities to which to resolve. This simple decision-making process provides twenty-four unique two-chord progressions ($3 \times 4 \times 2 = 24$), proceeding from any starting chord. The choices may be made probabilistically, either using equal probabilities with no particular consideration of directionality or key center, or using weighted probabilities to make some results more likely than others. The procedure can be carried out independently of considerations of voice leading, which may be determined by a separate algorithm.[7] In our example here, we'll use equal probability distribution.

The conventional way to choose one of n possibilities, each with equal likelihood, is to generate a random number within a range that's divisible into n equal regions. E.g. (in the C programming language):

```
int n = rand()%24;
```

where values 0 to 23 represent the 24 possible two-chord progressions. However, to demonstrate the applicability of quantum computing to this problem, we will use a method that requires only a very small number of binary (0-or-1) decisions. In quantum computing terms, we need only to measure one or two qubits for each decision.

A choice between two possibilities, such as whether the final chord should be major or minor, is made by measuring a single qubit.[8] A choice among four possible alternatives, such as which note of the pitch class set will be considered the root of the chord, requires two qubits (Fig. 6).

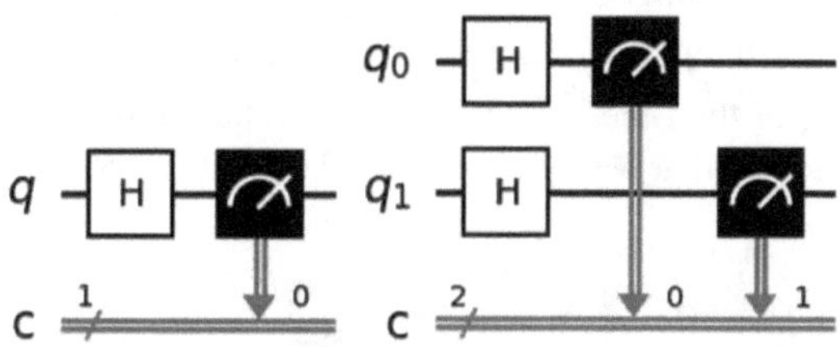

Fig. 6. One qubit q and two qubits q_0 and q_1, with Hadamard gates H, measured as classical bits c.

[7] This is another deliberate oversimplification for the sake of this simple demonstration. In practice, a human composer assuredly considers voice-leading concurrently with harmonic progression, and chord choices may depend upon melodic considerations. In a future article we will demonstrate computer decision making based on voice-leading and melodic implications.

[8] This is achieved by use of a Hadamard gate (notated as H In a circuit diagram)—a quantum logical gate that puts the qubit in a state of superposition. Upon measurement, the state is collapsed into a classical state, either 0 or 1.

In classical computing, a binary choice can be made by isolating a single bit of a random number; a four-way choice requires two bits, and so on for all powers of 2. For a choice among three options, such as deciding which pitch class set to use for the choice of diminished seventh chord, conventional computing does not provide a perfect way to make an equal-probability three-way choice using equal-probability binary choices, because 3 and 2 are mutually prime numbers. The most efficient substitute is to make a four-way choice with two random bits, and consider one of the four results to be a null choice that requires redoing the process [9]. This method theoretically has no upward bound, but in practice will require an average of $2.\overline{6}$ binary choices. This is where quantum computing provides a different method.

In quantum computing, the basis states of $|0\rangle$ and $|1\rangle$ can have unequal probabilities of being measured. This can be transformed by rotating the state vector on the Bloch sphere representation[9] of the qubit (Fig. 7).

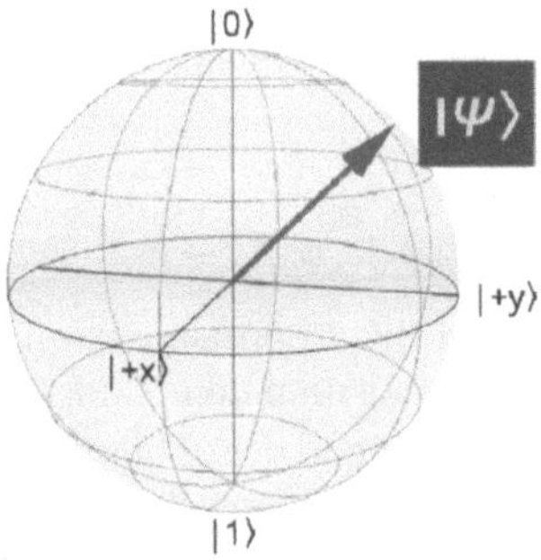

Fig. 7. Bloch sphere representation displaying an arbitrary state vector $|\Psi\rangle$.

In our specific case, rotating the qubit around the x axis in the amount of $arccos(^1/_3)$ ($\cong$1.23) will give it a probability distribution of $^2/_3\,|0\rangle$ and $^1/_3\,|1\rangle$. When that qubit, q_0, is measured as 1, which happens $^1/_3$ of the time, we can return that choice immediately as 01; otherwise, $^2/_3$ of the time, we measure a second qubit, q_1, which has an equal probability distribution so that we return either 00 or 10 (Fig. 8).

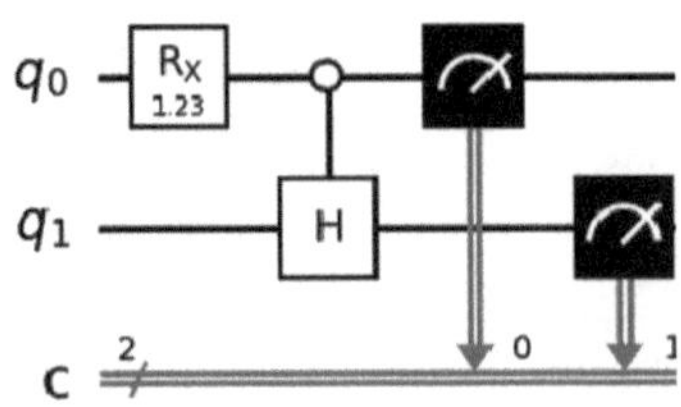

Fig. 8. Three-way choice with equal probability.

9 The "Bloch sphere" is an abstract 3-dimensional representation of the computational space provided by a single qubit. C.f., [10].

For implementing these experiments, we have been using the Quantum-computing Assisted Composition package (The QAC Toolkit) [11] in the Max programming environment. In that package, the **och.microqiskit** object provides a subset of the IBM open-source Quantum Information Science Kit (Qiskit) framework for quantum computing simulation. Specifying a circuit with the characteristics described above results in the three options 00, 01, and 10 being produced with equal probability (Fig. 9).

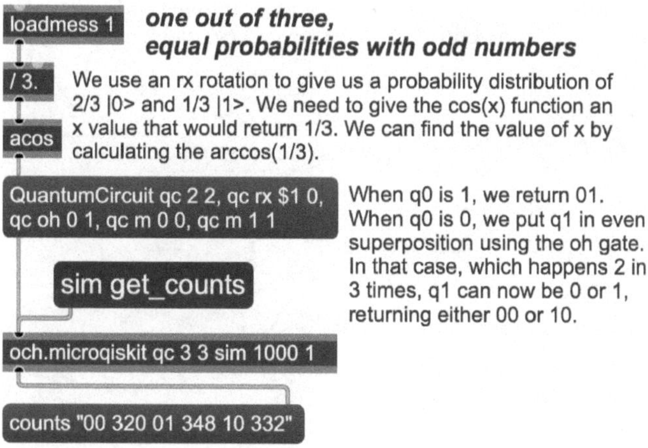

Fig. 9. Three-way choice with equal probability, using rotation by *arccos(1/3)* (Rx) and anti-controlled (open) Hadamard gate (H).

Equipped with these quantum-computing-inspired methods for handling ambiguity, we can generate compositional choices of diminished seventh chords and their resolutions. We'll provide here two examples of a series of such choices made with this method, in conjunction with some very simple rules for voice leading (favoring conjunct motion, etc.). Starting with a C major triad, we use the computer to choose a subsequent diminished seventh chord, to which we attribute the status of chord root to one the four pitch classes (or, equivalently, in our program we attribute the role of root, third, fifth, or seventh to the lowest voice, which then determines the role that must be fulfilled by the other voices), and then resolve that chord to either a major or minor triad whose root is a semitone above. In these examples, every decision is made locally, which is to say without including consideration of any other factors such as larger harmonic context, long term goals, melodic contours, etc. The result is simply an alternating succession of diminished seventh chords and their consequent triad resolutions.

The first example (Fig. 10) shows a computer-composed progression consisting alternately of triads and diminished seventh chords. (Such frequent and consistent use of this alternating pattern would surely grow tiresome if carried on for too long, but it serves well here to demonstrate the idea and the methodology.) First, the program selected the PC set {0, 3, 6, 9} using the above-described three-way decision method. It then attributed the role of seventh of the chord to the bass. The bass voice was already on a note of the selected PC set, so by staying where it is it optimally fulfills the criteria for conjunct voice leading, making C the seventh of the chord, thus making D# the chord root. This decision having been made, the program chose the most conjunct available voice leading to arrive at all the other needed pitches of the diminished seventh chord. It then decided that it should resolve to an E minor chord (selecting that over an E major chord), and resolved the voices "correctly": the leading tone resolves upward by semitone, the seventh of the chord resolves downward by semitone, the fifth of the chord by preference resolves downward by step to the third of the next chord, and the third of the chord in this case opted for the nearest neighbor as its tone of resolution. From that point, the entire process was repeated twice more to produce the four-measure example. Note that in this instance the entire progression can be analyzed as staying in C major.

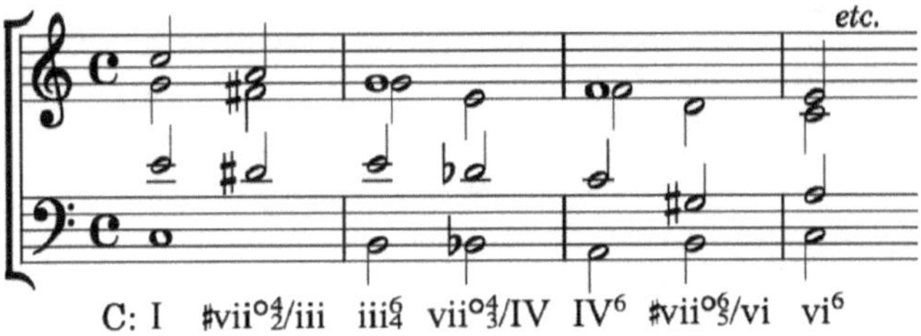

Fig. 10. Example of computer decision making for chord progression.

Another run of the program using the same methodology takes us immediately to a very different destination (Fig. 11). In this case, the program selected the PC set {1, 4, 7, 10}, and it gave the bass the role of the third of the chord. The most conjunct voice leading possible for the bass is to move from C to C# (from PC 0 to PC 1). For the listener, this will most logically imply #vii°7/ii, but because the computer has decided that C# is to be considered the third of the chord, the actual root of the chord is A#, leading to B minor (or B major). Thus, already on the downbeat of measure 2, we land on a chord that does not belong to C major, indicating that we may be in the process of modulating. Subsequent computer decisions take us even further afield, to the far-related key of F# major.

Fig. 11. The same method produces different results.

4 Future Directions

For this experiment and this article, we have focused upon the diminished seventh chord as the most studied example of chord ambiguity. However, there is no dearth of other chord types that lend themselves well to this interpretation of being in a state of superposition. For example, the major-minor seventh chord type (a.k.a. the dominant seventh chord), which differs from the diminished seventh chord only by a single semitone, similarly has many possible interpretations and usages. It can be used as a dominant seventh chord resolving either authentically or deceptively in either a major or minor key, as a German augmented sixth chord in either major or minor, as a subtonic seventh chord in a minor key, as a subtonic "backdoor" dominant chord common in jazz and popular music either in major or in minor, as a tritone substitution for the dominant seventh chord (bII7), as the result of double chromatic passing tones toward another dominant seventh chord, and so on.

Indeed, ambiguity of chord function is a crucial part of classical harmonic theory. Any major or minor triad belongs to more than one key, and therefore can be used to pivot from one key to another, a technique well known in classical music. The ambiguous harmonic function of such a pivot chord is revealed only upon its resolution to an unexpected chord or to a chord that does not belong to the initial key. Similarly, any diad can potentially imply membership in a variety of chords consisting of three or more pitches. (For example, the interval of a minor third can be the root and third of a minor triad, or the third and fifth of a major triad, or the fifth and seventh of a dominant seventh chord, or any two pitches of a diminished seventh chord, etc.) At the extreme limit, one might even say that every single pitch exists in a state of superposition, because a composer or improviser has free will to proceed from that pitch to anywhere at all. The notion that all musical sounds are in a state of multiplicity until they are "measured" by their resolution may be such a truism as to be almost useless, but it definitely alludes to the relationship between music theory and quantum theory, and can be used as a compositional tool, and as a mechanism for computer-generated music.

This idea of quantum compositional decision making has been extended to other chord types, into the realm of voice leading decisions and melodic composition, and non-standard theoretical models of harmony. We are undertaking an in-depth exploration of modeling the musical phenomenon of multiplicity of meaning and ambiguity as *superposition*. As a result of that correlation, we are evaluating the role of other key quantum concepts such as *entanglement* for future experimentation and more sophisticated computer-generated composition, creating new tools and methods for using quantum theory in a practical way. The role of equal temperament—equal divisions of the octave—as a catalyst for the functional ambiguity of pitch intervals implies future research in other equal-tempered divisions of the octave. Perhaps most importantly from the standpoint of creating new musical possibilities, the use of purely numerical music theory, dispensing with traditional note names, will enable this experimentation to go well beyond classical and jazz theory into other chord constructions and other harmonic orders.

Audio and software examples relevant to this article are available online at: https://harmony.quantumland.art/mcm24

Acknowledgments. The work by Omar Costa Hamido is supported by the European Project IIMPAQCT, under grant agreement Nr. 101109258.

Funded by
the European Union

Disclosure of Interests. The authors have no competing interests to declare that are relevant to the content of this article.

References

1. Bernhardt, C.: Quantum Computing for Everyone. The MIT Press, Cambridge (2019)
2. Larson, S.: Musical Forces: Motion, Metaphor, and Meaning in Music. Indiana University Press, Bloomington (2012)
3. Meyer, L.: Meaning in music and information theory. J. Aesthet. Art Critic.Aesthet. Art Critic. **15**(4), 412–424 (1957)
4. Stravinsky, I.: Poetics of Music in the Form of Six Lessons. Harvard University Press, Cambridge (1942)
5. Schoenberg, A.: Theory of Harmony (1911). Roy E. Carter (transl.). University of California Press, Los Angeles (1983)
6. Blutner, R.: Modelling tonal attraction: tonal hierarchies, interval cycles, and quantum probabilities. Soft. Comput.Comput. **21**, 1401–1419 (2017)
7. Souma, S.: Exploring the application of gate-type quantum computational algorithm for music creation and performance. In: Miranda, E.R. (ed.) Quantum Computer Music, pp. 83–103. Springer, Cham (2022). https://doi.org/10.1007/978-3-031-13909-3_5
8. Fugiel, B.: Quantum-like melody perception. J. Math. Music **17**(2), 319–331 (2023)
9. Knuth, D.E., Yao, A.C.: The complexity of nonuniform random number generation. In: Traub, J.F. (ed.) Algorithms and Complexity: New Directions and Recent Results. Academic Press, New York (1976)
10. Nielsen, M.A., Chuang, I.L.: Quantum Computation and Quantum Information. Cambridge University Press, Cambridge (2011)
11. Hamido, O.C.: QAC: quantum-computing aided composition. In: Miranda, E.R. (ed.) Quantum Computer Music, pp. 159–195. Springer, Cham (2022). https://doi.org/10.1007/978-3-031-13909-3_8

Quantum Tonality: A Mathemusical Playground

Peter beim Graben[1] and Thomas Noll[2](✉)

[1] Bernstein Center for Computational Neuroscience, Humboldt University of Berlin, Berlin, Germany
peter.beimgraben@b-tu.de

[2] Departament de Creació i Teoria Musical, Escola Superior de Música de Catalunya, Barcelona, Spain
tnoll@esmuc.cat, thomas.mamuth@gmail.com

Abstract. We adopt some basic ideas on quantum-theoretical modeling of tonal attraction and develop them further in an alternative direction. Fitting Gaussian Mixture Models (GMM) to the Krumhansl-Kessler (KK) probe tone profiles for static attraction opens the possibility to investigate the underlying wave function as the stationary ground state of an anharmonic quantum oscillator with a *schematic Hamiltonian* involving a perturbation potential. We numerically verify the fulfilment of the associated stationary Schrödinger equation and also inspect its excited states as a solution basis for the corresponding time-dependent Schrödinger equation. With their help we calculate the temporal evolution of any initial state. As an example, we study the dynamics of transpositions of the stationary KK wave function across Regener's line of fifths. This offers potential models for dynamic tonal attraction and also for the behavior of deflected key profiles within the Hamiltonian dynamics of a given tonality. Finally, we sketch further directions of investigation within this mathemusical experimental playground.

Keywords: Quantum Theory · Music Theory · Pitch Class Profiles · Krumhansl-Kessler Probetone Profiles · Gaussian Mixture Models · Quantum Harmonic Oscillator · Schematic Hamiltonian

1 Introduction

In the present paper we experiment with a specific concept of *tonality* that enters music theory from studies in cognition on the one hand and computer aided analysis on the other. The psychological Krumhansl-Kessler (KK) probe tone profiles [10] for the major and minor keys describe *static tonal attraction* of the various pitch classes in a tonal context and thereby reflect traces of enculturation within the Western musical mentality [9,18]. Related profiles also result from statistical analyses of pitch class distributions in musical corpora and several authors have discussed connections between both kinds of data, e.g. [2,9]. In a more refined way, pitch class profiles are also used to represent

© The Author(s), under exclusive license to Springer Nature Switzerland AG 2024
T. Noll et al. (Eds.): MCM 2024, LNCS 14639, pp. 159–172, 2024.
https://doi.org/10.1007/978-3-031-60638-0_13

the (fuzzy) pitch class content of local score segments and the investigation of their Discrete Fourier Transforms (DFT) provides valuable analytical insights through the tonal interpretation of the phases of prominent Fourier coefficients (see [1,7,20]). The KK profiles have also been interpreted as manifestations of pitch class hierarchies. For example, in his book *Tonal Pitch Space*, Fred Lerdahl [11] adopted the idea of embedded pitch hierarchies from Diana Deutsch und John Feroe [4], and analyzed tonal chord progressions in terms of such hierarchies. Moreover, melodic probe tone profiles and pitch class transition probabilities play an important role for *dynamic tonal attraction* [9,19]. These remarks indicate that pitch class profiles enter music theory in connection with various approaches to the investigation of tonality. Upon this wider background it is desirable to understand the implications of a more recent approach to the mathematical interpretation of the KK profiles, proposed by Reinhard Blutner and Peter beim Graben in [5,6] using ideas from quantum theory.

Starting point of the present study is the observation in [14, Fig. 1] that the deformed quantum cosine model $\psi : \mathbb{R}/2\pi\mathbb{Z} \to \mathbb{C}$ over the continuous circle of fifths [6] can be equivalently described by a Gaussian wave function $\psi : \mathbb{R} \to \mathbb{C}$ over Regener's line of fifths $\mathbb{R}$ as configuration space [15], retaining the probe tone positions along the circle of fifths up to enharmonic equivalence, such that the KK profiles emerge as mixed quantum states, i.e. as weighted sums

$$p(x) = |\psi(x)|^2 = \alpha \cdot \rho(x-a) + \beta \cdot \rho(x-b) + \gamma \cdot \rho(x-c) \tag{1}$$

of three transposed instances of a Gaussian probability density function $\rho(x)$, where $a, b, c \in \mathbb{R}$ refer to the positions of the three components of the tonic triad. This finding leads to two important consequences. (*i*) it links the quantum music approach to statistical machine learning and data science [16], and (*ii*) it allows the application of the quantum theory of harmonic and anharmonic oscillators [12,17].

2 Quantum Wave Functions for Gaussian Mixture Models

This section summarizes the results from a longer companion article by the authors of the present submission [8]. First we recall that a Gaussian wave packet

$$\psi(x) = \sqrt[4]{\frac{E_0}{\pi}} \mathrm{e}^{-\frac{E_0 x^2}{2}} \tag{2}$$

with energy

$$E_0 = \frac{1}{2\sigma^2} \tag{3}$$

and variance σ^2 solves the stationary Schrödinger equation of the harmonic oscillator

$$-\psi''(x) + E_0^2 x^2 \psi(x) = E_0 \psi(x) \tag{4}$$

for the ground state energy $E = E_0$ [17].

Next, we consider a generalized Gaussian mixture model (GMM) [Eq. (1)] of the form

$$p(x) = \sum_{k=1}^{N} \alpha_k \, \rho(x - a_k) \tag{5}$$

where

$$\rho(x) = |\psi(x)|^2 = \sqrt{\frac{E_0}{\pi}} \mathrm{e}^{-E_0 x^2} = \frac{1}{\sqrt{2\pi\sigma^2}} \mathrm{e}^{-\frac{1}{2}\left(\frac{x}{\sigma}\right)^2} \tag{6}$$

is a normal distribution density with variance $\sigma^2 \in \mathbb{R}$, whose mean 0 is shifted to a tone position $a_k \in \mathbb{R}$ in each corresponding summand for the given tonality, altogether constituting a tonal context $\{a_1, \ldots, a_N\} \subset \mathbb{R}$ [6]. The coefficients $\alpha_k \in \mathbb{R}$ are the respective weights in the convex linear combination (5) with $\sum_{k=1}^{N} \alpha_k = 1$. Such GMM play an important role in the domain of statistical machine learning [16]. Here, they serve as a mathematical model for the schematic traces of musical enculturation [9,18].

In the companion article [8], we show that there exists a perturbation $W : \mathbb{R} \to \mathbb{R}$ of the harmonic oscillator potential such that the ground state solution $\eta_0(x)$ (for the same energy eigenvalue $D_0 = E_0$) of the anharmonic stationary Schrödinger equation [12]

$$-\eta_n''(x) + \left[E_0^2 x^2 + W(x)\right] \eta_n(x) = D_n \eta_n(x) \tag{7}$$

has the probability density function $|\eta_0(x)|^2 = p(x)$ of the GMM (5). Without proof we mention here, that the perturbation potential is given by the formula

$$W(x) = E_0^2 \frac{\sum_{kl} [(a_k - 3a_l)x + a_l(2a_l - a_k)] \alpha_k \alpha_l \, \rho(x - a_k) \rho(x - a_l)}{\sum_{kl} \alpha_k \alpha_l \, \rho(x - a_k) \, \rho(x - a_l)} \tag{8}$$

where $\rho(x)$ and the parameters E_0, α_k and a_k are the defining ingredients of the GMM (c.f. Eqs. (3, 5) and (6)). The differential operator

$$H = -\frac{\partial^2}{\partial x^2} + E_0^2 x^2 + W(x) \tag{9}$$

is referred to as the *schematic Hamiltonian* associated with the GMM (5) reflecting a statistically acquired tonal schema [9,18].

3 Fitting Gaussian Mixture Models for Static Tonal Attraction

In the present paper we explore these results for the case of two concrete examples, namely a GMM (1) for the KK probe tone profiles in C major and C minor [10]. Table 1 lists both data sets (major and minor) in a normalized form:

Table 1. Normalized KK probe tone profiles for C-major (top) and C-minor (bottom).

$D\flat$	$A\flat$	$E\flat$	$B\flat$	F	C	G	D	A	E	B	$F\sharp$
$-\frac{5}{6}\pi$	$-\frac{2}{3}\pi$	$-\frac{1}{2}\pi$	$-\frac{1}{3}\pi$	$-\frac{1}{6}\pi$	0	$\frac{1}{6}\pi$	$\frac{1}{3}\pi$	$\frac{1}{2}\pi$	$\frac{2}{3}\pi$	$\frac{5}{6}\pi$	π
0.	0.0203	0.0127	0.0064	0.2365	0.524	0.3764	0.159	0.1819	0.2734	0.0827	0.0369
0.0184	0.1893	0.3733	0.1052	0.1301	0.4982	0.2905	0.1236	0.1288	0.0079	0.0447	0.

The note names $G\flat, D\flat, A\flat, E\flat, B\flat, F, C, G, D, A, E, B, F\sharp$ in the line-of-fifth order (with $G\flat$ added to the left side with the same values as for $F\sharp$) occupy the equidistant positions $x \in \{-\pi, -\frac{5}{6}\pi, \ldots, -\frac{1}{6}\pi, 0, \frac{1}{6}\pi, \ldots, \frac{5}{6}\pi, \pi\}$, centered around the note C at the origin $x = 0$ from the original quantum model over the circle of fifths. We write $\mathrm{pos}(X) \in \mathbb{R}$ for the position of the note X at the real line of fifths [15]. The original values are scaled such that the norms of the linear interpolations $p(x)$ yield unity. We fit several GMM onto the KK data in analogy with the approach in [6], i.e. we choose GMM of the form (1) with different constraints upon the parameters.

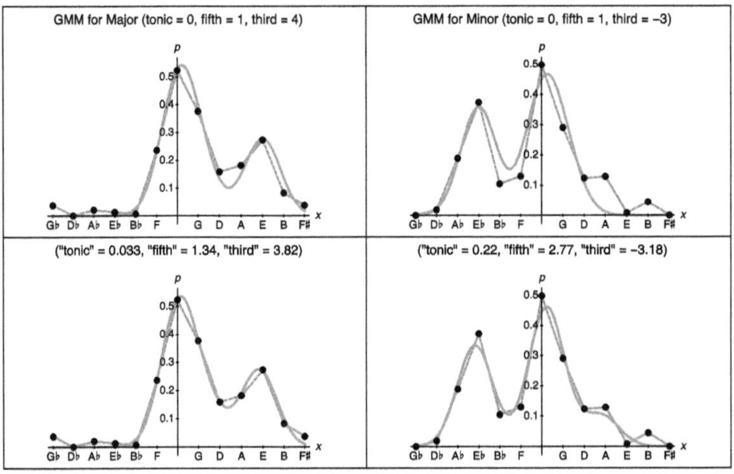

Fig. 1. Gaussian Mixture Models for the KK probe tone profiles for major (left) and minor (right). Top: means are chosen as triadic positions $a = 0 = \frac{\pi}{6}\mathrm{pos}(\text{tonic})$, $b = \frac{\pi}{6} = \frac{\pi}{6}\mathrm{pos}(\text{fifth})$, and $c = \frac{\pi}{6}\mathrm{pos}(\text{third})$ ($= \frac{2\pi}{3}$ or $-\frac{\pi}{2}$, respectively) and the parameters $\alpha = 0.487468$, $\beta = 0.203587$ (major) and $\alpha = 0.405881$, $\beta = 0.194719$ (minor) are fitted, respectively. The correlation between the KK data and the corresponding values of the GMM are 0.989 (for major) and 0.935 (for minor). Bottom: means a, b, c are fitted to the data as well. The correlations are 0.995 (for major) and 0.975 (for minor).

The choice of the circle of fifths interval $[-\pi, \pi]$ motivates the inheritance of the ground state energy

$$E_0 = \frac{36 \ln 2}{\pi^2} \tag{10}$$

from the *Lerdahl interpolation* in [3].[1] In order to fit the most general GMM to the KK data we have five parameters: a, b, c and α, β (γ results from the convexity constraint $\alpha + \beta + \gamma = 1$). The locations of the three mean values a, b, c are musically constrained through the assumption of the formative role of the tonic triads, i.e. $a = \text{pos}(C), b = \text{pos}(G), c = \text{pos}(E)$ for the C major context and $a = \text{pos}(C), b = \text{pos}(G), c = \text{pos}(E_\flat)$ for the C minor context. Beyond this it is also interesting to compare these choices with "musically ignorant" fits to see how far they deviate from the musically motivated ones.[2]

The results, displayed in Fig. 1, indicate that in the case of the major key profile there is no significant difference between the two fits. In the case of the minor key profile, however, there are noteworthy differences in connection with the raised scale degrees $\sharp\hat{6}$ and $\sharp\hat{7}$ which deserve a separate investigation. For the scope of the present paper we focus on the case of the major key.[3]

4 Wave Functions of Static Tonal Attraction as Schematic Stationary Ground States

We proceed with the C major GMM (1) for tone positions $a = 0, b = \frac{\pi}{6}, c = \frac{2\pi}{3}$ from the top left cell of Fig. 1, with $\alpha = 0.487468$, $\beta = 0.203587$ and $\gamma = 0.3994$ defining our candidate for the associated schematic ground state wave function as $\eta_0(x) = \sqrt{p(x)}$.

Our next goal is to verify that $\eta_0(x)$ is indeed the stationary ground state of the schematic Hamiltonian (9) by solving the eigenvalue problem of the operator H. For this purpose, it is useful to take advantage of the Hermite basis of the harmonic oscillator, described by (11).

$$\varphi_n(x) = \sqrt[4]{\frac{E_0}{\pi}} \frac{1}{\sqrt{2^n n!}} h_n(\sqrt{E_0} x) \mathrm{e}^{-\frac{E_0 x^2}{2}}, \tag{11}$$

where $h_n(x)$ denotes the Hermite polynomial of order n. These Hermite states $\varphi_n(x)$ are solutions of the unperturbed Schrödinger equation (4) for the eigenvalues $E_n = 2E_0 \left(n + \frac{1}{2}\right)$, i.e.

$$-\varphi_n''(x) + E_0^2 x^2 \varphi_n(x) = E_n \varphi_n(x). \tag{12}$$

---

[1] Note that we are able to confirm Lerdahl's speculation about a Hamiltonian principle of musical least effort [11] in the present framework, as the ground state energy E_0 can be obtained from a corresponding variational principle.

[2] The authors are acutely aware of the danger of overfitting, in particular as the shape of p evokes a connection between the elephant of Antoine de Saint-Exupéry's *The Little Prince* and *John von Neumann's elephant*: "With four parameters I can fit an elephant, and with five I can make him wiggle his trunk." [13].

[3] Fitting with a higher energy parameter $E = 3.53406$ one can even achieve a correlation of 0.996338, unfortunately, the match for the raised leading tone $\sharp\hat{7}$ is not yet convincing.

When we apply the schematic Hamilton operator (9) to a superposition state

$$\psi(x) = \sum_n P_n \varphi_n(x) \tag{13}$$

with coefficients $P_n = \langle \psi, \varphi_n \rangle = \int_{-\infty}^{\infty} \psi(x)\varphi_n(x)\,\mathrm{d}x$, we obtain

$$H\psi(x) = \sum_n P_n \left[-\frac{\partial^2}{\partial x^2} + E_0^2 x^2 + W(x) \right] \varphi_n(x) = \sum_n P_n [E_n + W(x)] \varphi_n(x).$$

The coefficients P_n for eigenstates of the schematic Hamilton operator form eigenvectors of the infinite matrix $M = (e_{mn} + w_{mn})_{m,n \in \mathbb{N}}$ where

$$e_{mn} = E_n \delta_{mn} \quad \text{and} \quad w_{mn} = \int_{-\infty}^{\infty} \varphi_m(x) W(x) \varphi_n(x)\,\mathrm{d}x \tag{14}$$

(for details see the corresponding companion paper [8]). The potentials for the harmonic oscillator and its perturbation $W(x)$ in our concrete example (see (8)) are shown in Fig. 2:

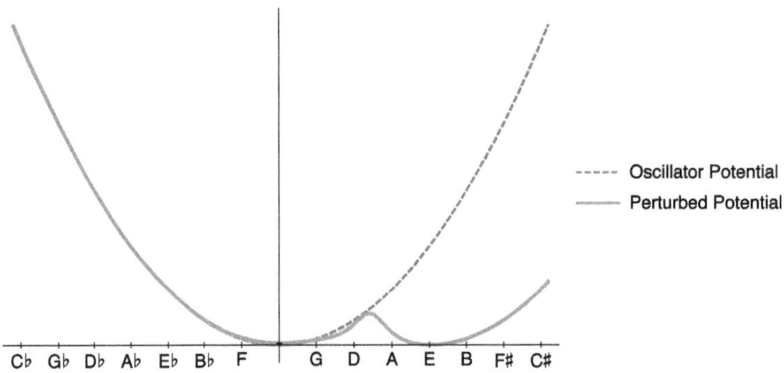

Fig. 2. The unperturbed potential of the harmonic oscillator with ground energy E_0 and perturbed anharmonic potential for the C major GMM $p(x)$.

For a numeric validation of the schematic Schrödinger equation, we calculate eigenvectors and eigenvalues of M in growing (but finite) number of dimensions by truncating the infinite matrix after N rows and columns: $M_N = (e_{mn} + w_{mn})_{m,n \leq N}$.

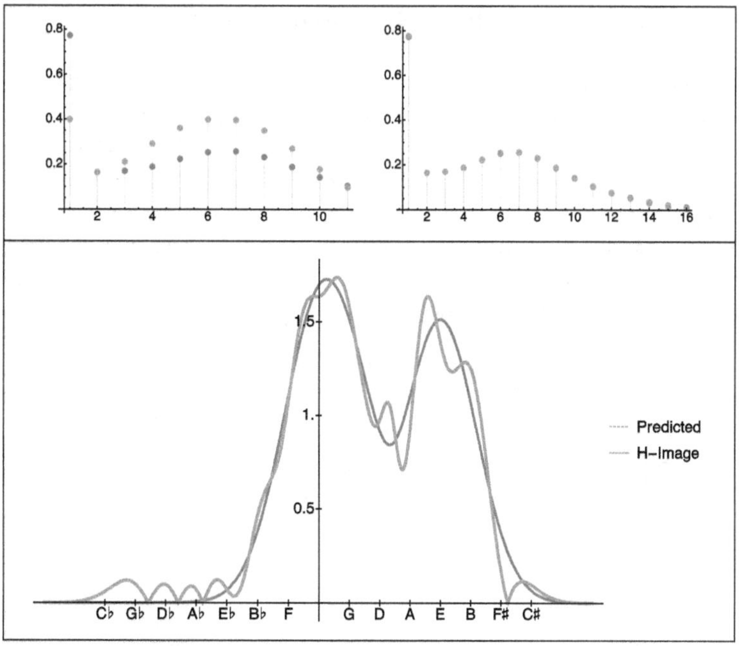

Fig. 3. Top: Comparison of the first N coordinates P_n from $\eta_0(x) = \sum_n Q_n \varphi_n(x)$ with the coefficients of the "smallest" eigenvectors $v_N = (Q_0, \ldots, Q_N)$ of the truncated matrices M_N for $N = 10$ (left) and $N = 15$ (right). Bottom: Comparison of the stretched wave function $E_0\eta_0(x)$ with the action of the (truncated) Hamilton operator H_{15} on the superposition $\sum_{n=0}^{15} Q_n\varphi_n(x)$.

The upper segment of Fig. 3 shows list plots of the first 11 (top left) and the first 16 (top right) coordinates P_n of $\eta_0(x)$ with respect to the Hermite basis (12) and compares them with the coordinates of the "smallest" eigenvectors of M_{10} and M_{15}, respectively (i.e. the eigenvectors with the corresponding smallest eigenvalues). These eigenvalues (2.617... and 2.531...) are close to the Lerdahl ground state energy $E_0 = 2.528\ldots$ (10), while one can still observe a clear deviation between the coefficients in the case of 11 dimensions (top left side of Fig. 3) the deviation already gets smaller in the case of 16 dimensions (top right side of Fig. 3). The plot in the lower segment of Fig. 3 illustrates the fulfillment of the Schrödinger equation. The dashed curve shows the wave function $E_0\eta_0(x)$, i.e. our predicted ground state stretched by the predicted energy eigenvalue E_0, while the continuous curve shows the image of the smallest eigenstate under the action of H_N, which is the approximation of the schematic Hamilton operator in terms of the truncated matrix M_N. For growing N the little oscillations of the Hermite components become smaller and smaller and their superposition approaches the predicted wave function.

5 Excited States of the Schematic Hamiltonian

In this section we explore an immediate consequence of the fact that we are in the possession of a Hamilton operator. Apart from the ground state we have an infinite sequence of higher energy eigenvalues and an infinite basis of corresponding excited stationary states, which form an orthonormal basis for the space of square-integrable complex functions over the line of fifths $\mathbb{R}$. In order to explore them, we may inspect their approximations from the sequence of eigenvectors of the truncated matrices M_N (Fig. 5). Figure 4 suggests that—for growing N—the eigenvalues of M_N converge in quadratic mean the eigenvalues of H from above, too.

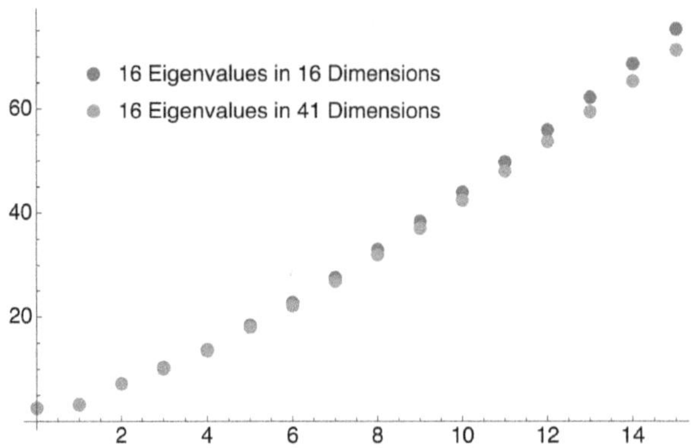

Fig. 4. The two list plots show the first 16 eigenvalues of M_{15} and M_{40}, respectively

Mathematically, we will now use the excited states in order to calculate the time-development of given initial states under the schematic Hamiltonian according to the time-dependent Schrödinger equation.

$$H\Psi(x,t) = \mathrm{i}\frac{\partial}{\partial t}\Psi(x,t). \tag{15}$$

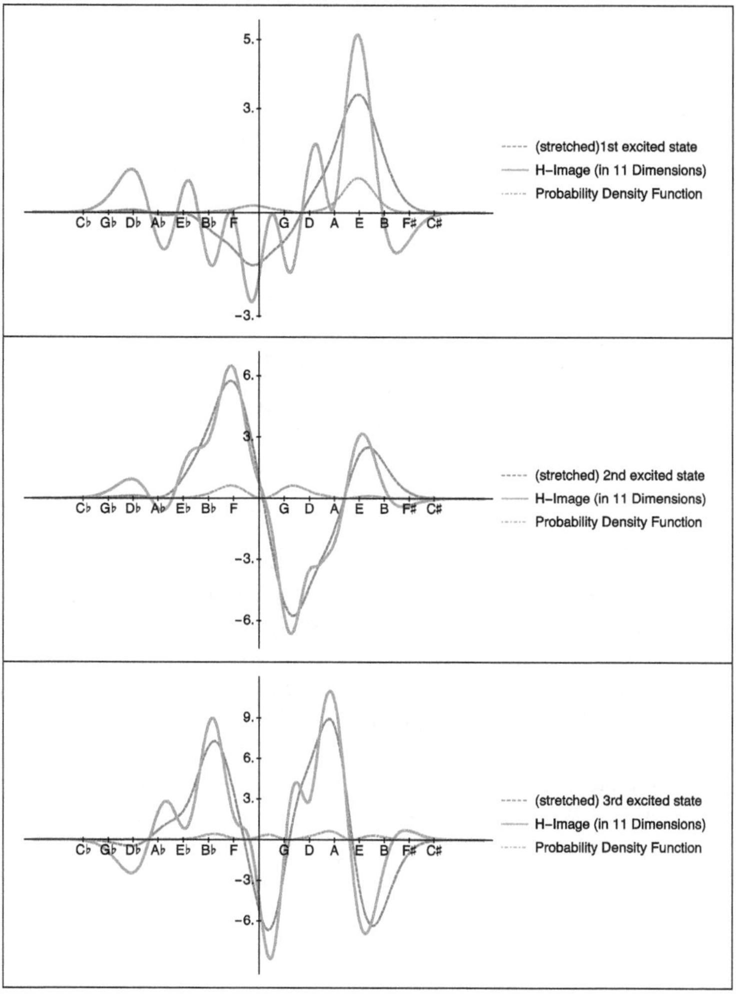

Fig. 5. Approximation of the 1st (top), 2nd (middle) and 3rd (bottom) excited states of the schematic Hamilton operator H by means of the 2nd, 3rd and 4th smallest eigenvalues and associated eigenvectors $v_k = (Q_{k,0}, \ldots, Q_{k,10})$ of the truncated matrices M_{10}. The dashed curves show the superpositions $E_k \sum_{n=0}^{N} Q_{nk}\varphi_n(x)$ $(k = 1, 2, 3)$, which are stretched by the eigenvalues $\tilde{D}_1 = 3.41579$, $\tilde{D}_2 = 7.29815$ and $\tilde{D}_3 = 11.1759$, respectively. The continuous curves show the actions of the (truncated) Hamilton operator H_{10} on the superposition $\eta_k(x) = \sum_{n=0}^{10} Q_{nk}\varphi_n(x)$. The dot-dashed curves show the associated probability density functions $|\eta_k(x)|^2$.

6 Time Development of Deflected Key Profiles

The excited states of the schematic Hamiltonian H, denoted $\eta_0, \eta_1, \eta_2, \ldots$, form an orthonormal basis of the space of complex square-integrable functions on $\mathbb{R}$. The stationary ground state is the KK profile η_0 of static tonal attraction.

Suppose we have a representation of a wave function

$$\psi(x) = \sum_k R_k \eta_k(x), \tag{16}$$

whose coefficients are obtained by the orthogonal projections

$$R_k = \langle \eta_k, \psi \rangle = \int_{-\infty}^{\infty} \eta_k(x)\psi(x)\,\mathrm{d}x. \tag{17}$$

Then, the solution of the time-dependent Schrödinger equation (15) is universally given by

$$\Psi(x,t) = \sum_k R_k \mathrm{e}^{-\mathrm{i}D_k t}\eta_k(x) \tag{18}$$

with the eigen energies D_k of the schematic Hamiltonian (9).

Assuming that $\eta_0(x)$ represents the ground state of a given tonality (here C major), it seems reasonable to associate other (major) tonalities with transpositions $\eta_0(x-a)$ of $\eta_0(x)$. However, regarding the underlying dynamics of the C major tonality such transpositions are not autonomous representations of other tonalities at all. While $\eta_0(x)$ is stationary, $\eta_0(x-a)$ for $a \neq 0$ is not. The coherent states of the (unperturbed) quantum harmonic oscillator may serve as prime examples for our situation. They are the quantum-theoretical analogues for a deflected pendulum. We will therefore baptise the wave functions $\eta_0(x-a)$ (in the role as initial states) as *deflected key profiles*. In order to explore their time developments we need to calculate the orthogonal projections

$$R_k = \int_{-\infty}^{\infty} \eta_k(x)\eta_0(x-a)\,\mathrm{d}x. \tag{19}$$

Figure 6 gives a simultaneous flipbook-like trace (of the absolute values) of two time developments, namely of the fifth deflected wave functions $\eta_0(x-\frac{\pi}{6})$ in sharp-ward and $\eta_0(x+\frac{\pi}{6})$ in flat-ward direction. Both wave functions show the tendency to eventually arrive at the location of the stationary ground state. But in the details they behave quite differently and they never seem to match the precise shape again. From the numeric computation in 30 dimensions alone we cannot tell whether the deviations arise completely from the finite-dimensional approximation. A rough resemblance with the well-understood behavior of the coherent states of the harmonic oscillator cannot be dismissed out of hand.

7 Sonification Experiments

In parallel to our goal to approach theoretical and empirical questions about tonality through the interpretation of the schematic Hamiltonian and time developments of various prepared initial states, there is also a desire to establish a

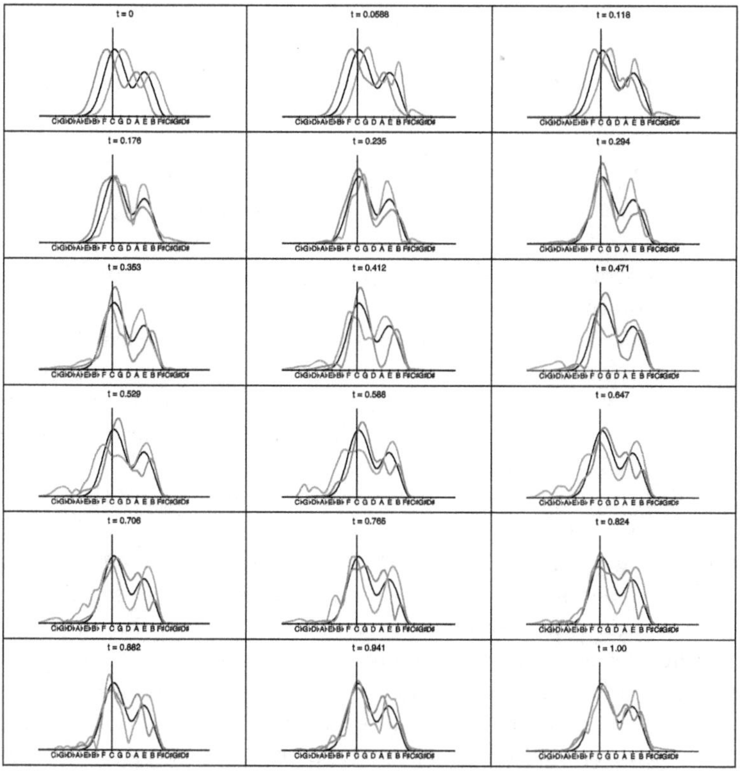

Fig. 6. Time Development of the deflected wave functions $|\eta_0(x - \frac{\pi}{6})|$ of one fifth-unit to the sharp side and functions $|\eta_0(x + \frac{\pi}{6})|$ of one fifth-unit to the flat side.

more direct hands-on connection to a musical level of reality in an experimental attitude. In this section we briefly sketch our ideas for a *gradus ad Parnassum* towards sonifications of time developments of wave functions. And while also envisaging inspirations from the role of quantum measurements in physics, we will start from a purely "meta-physical" point of view by designing an audible medium for representations of wave functions as faithful as possible. On the basic level we build a *wave-player*—still pretending to be ignorant of our concept of *quantum tonality*. And on the successive levels we attempt to introduce step by step some musical ideas to exemplify aspects of tonality.

First Species. With respect to the magnitude of the wave function at a given tone position the idea suggests itself to use the volume parameter. For a faithful sonification of the complex phase we intend to realise the illusion of travelling on a circle of sounds. And apart from the aesthetics of building something from scratch there is a good practical reason to realize this with the help of the *Operator* synthesizer in *Ableton Live*. This software instrument generalizes the concept of FM-synthesis and consists of four flexibly configurable oscillator modules which can be embedded into each other. A circle of sounds can be obtained, for example, by tuning the modulation indices of three different oscillators acting on a basic oscillator in a coordinated way. In order to apply this to our deflected tonality wave function, we create a single track for each tone position along the line of fifths and each "tonality state" at time t is only represented by the resulting sound at this moment t. With the availability of a multi-channel audio system each of the single tone-position-tracks may travel in space and the phase may additionally be represented by the angle of the travelling sound source.

Second Species. While still relying on the sonification of each state of a time-development by the sound parameters of that particular moment t in time, we may extract a local tonal stratification (tonic, chord, scale, chromatic) and express the transitions within in these layers as separate processes with different parameter choices from the first species.

Third Species. In tonal music, tonality unfolds over time and the notes acquire individual qualities (scale degrees and modal character within each of the layers of tonal stratification), which then change subtly (or dramatically) through processes of tonal modulation. Depending on the transitions of notes along the stratification layers in the second species we obtain a segmentation of the time development into different allocation configurations. Our third species entails two challenges: (i) to achieve an exemplification of tonality for each allocation configuration in terms of discrete metric-rhythmical patterns, where individual notes express their tonal meanings as "stable notes", "passing notes", "suspensions" and (ii) to achieve an integration of these compositional processes on the macro-level with continuous changes of the wave functions. The combination of these two challenges points towards some kind of *minimal music*. A third challenge is the question of a temporal expression of phase, but which comes with possible contradictions similar to those happening in time travels in science fiction literature.

Fourth Species. The three species above are dedicated to the sonification of the time development of a single wave function, such as the deflected wave functions shown in Fig. 6. It can likewise be applied to the time development of other prepared initial states, e.g. to those of the coherent states from the quantum harmonic oscillator which are plausible candidates as models for single notes in a tonal context. Here comes the idea of quantum measurement into play as a means for musical interaction. In this scenario the sonifications of time developments in species 1, 2, 3 could be thought of inaccessible virtually infinite processes. The activity of playing them consists of their preparation in terms of initial states, the shaping of time flows and the execution of measurements, leading to projections of the wave function to an eigenstate of the measured observable.

8 Conclusion

Our novel quantum approach to tonal attraction combines impulses from statistical learning theory,—which has also been applied to the enculturation of acquired tonal schemata [9,18] in terms of Gaussian Mixture Models—with well-established findings from the quantum theory of harmonic and anharmonic oscillators. We are able to attribute to any GMM fit of the musical major and minor modes its associated schematic Hamilton operator. We choose the attribute *schematic*, as our approach is driven by the idea to relocate the "seat of tonality" in the Hamilton operator itself rather than in the statistical surface, which can be generated from it. The stationary ground state of the Hamiltonian reproduces the KK profiles of static tonal attraction [10]. We hypothesise that the Hamiltonian encodes further relevant information about tonality. To that end our next steps of investigation are dedicated to a systematic investigation of time developments of various initial states, assuming that the solutions of the associated time-dependent Schrödinger equation could describe dynamic tonal attraction as well [2,9,19]. Specifically, we think that aspects of melodic, contrapuntal and harmonic dynamics could be described within our unified framework. In addition we shall exploit procedures for quantum composition and tonal ambiguity by means of incompatible schematic Hamiltonians.

References

1. Amiot, E.: Music Through Fourier Space, Discrete Fourier Transform in Music Theory. Springer, Cham (2016). https://doi.org/10.1007/978-3-319-45581-5
2. Arthur, C.: When the leading tone doesn't lead: musical qualia in context. Ph.D. thesis, Ohio State University (2016)
3. Blutner, R., beim Graben, P.: Gauge models of musical forces. J. Math. Music **15**(1), 17–36 (2021)
4. Deutsch, D., Feroe, J.: The internal representation of pitch sequences in tonal music. Psychol. Rev. **88**(6), 503–522 (1981)
5. beim Graben, P.: Gauge symmetries of musical and visual forces. J. Math. Arts **17**(3–4), 347–382 (2023)
6. beim Graben, P., Blutner, R.: Quantum approaches to music cognition. J. Math. Psychol. **91**, 38–50 (2019)
7. beim Graben, P., Mannone, M.: Musical pitch quantization as an eigenvalue problem. J. Math. Music **14**(3), 329–346 (2020)
8. beim Graben, P., Noll, T.: Unified musical quantum models for tonal attraction (in prep)
9. Huron, D.: Sweet Anticipation: Music and the Psychology of Expectation. MIT Press, Cambridge (2006)
10. Krumhansl, C.L., Kessler, E.J.: Tracing the dynamic changes in perceived tonal organization in a spatial representation of musical keys. Psychol. Rev. **89**(4), 334–368 (1982)
11. Lerdahl, F.: Tonal Pitch Space. Oxford University Press, New York (2001)
12. McWeeny, R., Coulson, C.A.: Quantum mechanics of the anharmonic oscillator **44**(3), 413–422 (1948)

13. von Neumann, J. https://de.wikipedia.org/wiki/John_von_Neumann
14. Noll, T., beim Graben, P.: Quantum-musical explorations on $\mathbb{Z}_n$. In: Montiel, M., Agustín-Aquino, O.A., Gómez, F., Kastine, J., Lluis-Puebla, E., Milam, B. (eds.) MCM 2022. LNCS, vol. 13267, pp. 369–375. Springer, Cham (2022). https://doi.org/10.1007/978-3-031-07015-0_32
15. Regener, E.: Pitch Notation and Equal Temperament: A Formal Study, Occasional Papers, vol. 6. University of California Press, Berkeley (1973)
16. Russell, S., Norvig, P.: Artificial Intelligence: A Modern Approach, 3rd edn. Pearson (2010)
17. Schrödinger, E.: Der stetige Übergang von der Mikro- zur Makromechanik. Naturwissenschaften **14**(28), 664–666 (1926)
18. Temperley, D.: Music and Probability. MIT Press, Cambridge (2007)
19. Woolhouse, M.: Modelling tonal attraction between adjacent musical elements. J. New Music Res. **38**(4), 357–379 (2009)
20. Yust, J.: Harmonic qualities in Debussy's 'Les sons et les parfums tournent dans l'air du soir'. J. Math. Music **11**(2–3), 155–173 (2017)

Theory and Algorithms for Melodic-Harmonic Analysis and Generation

Melody and Variation Generation Through KAM Theory

Octavio A. Agustín-Aquino(✉) and Alicia Santiago-Santos

Instituto de Física y Matemáticas, Universidad Tecnológica de la Mixteca,
Huajuapan de León, Oaxaca, México
{octavioalberto,alicia}@mixteco.utm.mx
http://www.utm.mx

Abstract. We use the Newton iteration of KAM theory for diffeomorphisms of the circle as a source for melody generation and its variations.

Keywords: Music · melody generation · discrete dynamical systems · KAM theory

1 Introduction

Dynamical systems can be used to generate melodies or its variations, as was shown by the pioneering work of Pressing [9] for certain discrete and discretized systems and Dabby [4] for the case of altering the tonal characteristics of a piece by chaotic trajectories; see [8] for a survey on this and related topics.

In a nutshell, the idea is to associate a sequence of pitches p_k to a sequence of points x_k in a space X such that

$$x_{k+1} = f(x_k)$$

for a given $f : X \to X$ defined by a dynamical system. It is undesirable for the behaviour of the sequence $\{x_k\}_{k\in\mathbb{N}\cup\{0\}}$ to be too "predictable", since it would be uninteresting. On the contrary, if it is too "chaotic" then it would yield melodies indistinguishable from "gibberish", so we would like to have some control between these two extremes.

Here we propose to use KAM theory[1] for the diffeomorphisms of the circle [1,11] as a reasonable alternative not attested in existing literature to accomplish this control, by perturbing an original melody constructed with a simple dynamical system and gradually returning to it. Since the behaviour of the system and its perturbation depends on two easily tunable parameters and the choice of one function, we deem it as a good candidate for melody and variation generation.

In Sect. 2 we establish some notation and terminology that we will use throughout this paper. In Sect. 3 Arnold's theorem is stated after certain preliminaries. This allows us to describe the variation algorithm in Sect. 4 and provide some notes regarding its implementation in Octave. A few examples are given in Sect. 5.

Finally, in Sect. 6, some conclusions and a glimpse of future steps of this work are reached.

[1] KAM stands for Kolmogorov, Arnold and Moser, who developed it.

© The Author(s), under exclusive license to Springer Nature Switzerland AG 2024
T. Noll et al. (Eds.): MCM 2024, LNCS 14639, pp. 175–182, 2024.
https://doi.org/10.1007/978-3-031-60638-0_14

2 Some Basic Notions

In this section we introduce some general notation and the concept of dynamical system, as well as certain terminology we will use frequently. For further details concerning circle mappings, dynamical systems or an introduction to KAM theory, see [5,7] and [11], respectively.

Definition 1. *Given a topological space X and a semigroup $(G, *)$, a* dynamical system *is a continuous function $\psi : G \times X \to X$ such that*

1. *for every $x \in X$ we have $\psi(0, x) = x$, where 0 is the neutral element of G and*
2. *for every $t, s \in G$ and $x \in X$ we have $\psi(t, \psi(s, x)) = \psi(t * x, x)$.*

In particular, we will restrict our attention to *discrete* dynamical systems with $G = (\mathbb{N} \cup \{0\}, +)$ and ψ defined via a continuous function $f : X \to X$ by

$$\psi(k, x) = f^k(x),$$

where f^k is the k-fold composition of f with itself [7, Definition 4.5, p. 33]. Given $x \in X$, we define its *orbit* under f as the set

$$\mathcal{O}(x, f) = \{f^k(x) : k \in \mathbb{N} \cup \{0\}\}.$$

Now we specify f further, taking it as a diffeomorphism of the circle $f : S^1 \to S^1$. There exists a homeomorphism $\phi : \mathbb{R} \to \mathbb{R}$ such that

$$f(\exp(2\pi i t)) = \exp(2\pi i \phi(t))$$

for every $t \in \mathbb{R}$, which is called a *lift* of f.

A fundamental example of a diffeomorphism of the circle is a *rotation*. A lift of a circle rotation r_α by an angle α will be denoted by R_α. In particular, we can take it as

$$R_\alpha(x) = x + \alpha$$

since, as a matter of fact, for any diffeomorphism of the circle f with lift ϕ the limit

$$\rho(f) = \lim_{n \to \infty} \frac{\phi^n(x) - x}{n}$$

exists and it is called the *rotation number* of f. As we will see soon, any diffeomorphism of the circle f is essentially equivalent to $r_{\rho(f)}$.

Moreover, the behaviour of the orbits under a rotation r_α is completely understood: if $\alpha \in \mathbb{Q}$ then $\mathcal{O}(x, r_\alpha)$ is a finite set; otherwise, it is dense in the circle. It is a good moment to mention that an irrational number can be approximated arbitrarily well by some rational number $\frac{m}{n}$, but in general such an approximation cannot get closer than a distance proportional to $\frac{1}{n^2}$. For our purposes we need some quantification on this fact.

Definition 2. *A number $\rho \in \mathbb{R}$ is of type (K, ν) if there exist positive numbers K and ν such that*

$$\left|\rho - \frac{m}{n}\right| > \frac{K}{|n|^{\nu}}$$

for al pairs of integers (m, n).

The number ν from Definition 2 is called the *irrationality measure* of ρ [3, p. 28]. Roth proved that for any irrational algebraic number its irrationality measure is 2 [3, p. 28, Theorem 2.1]; thus, for example, $\varphi = \frac{\sqrt{5}+1}{2}$ is of type $(K, 2)$. In [10] there is a nice proof that e has irrationality measure 2 as well. The irrationality measure of π is not known; as of this writing the best current upper bound on it is approximately 7.103205334 [12].

3 Arnold's Theorem

In this section we deal first with the preliminaries to understand Arnold's theorem statement. We also explain what the *Newton iteration* is in this context.

Let us begin considering diffeomorphisms of the circle with lifts of the form

$$\phi(x) = x + \alpha + \eta(x),$$

where α is an irrational number, $\eta : \mathbb{C} \to \mathbb{C}$ is such that it is real when it is restricted to the real numbers, it is periodic with period 1 and $\frac{d}{dx}\eta(x) > -1$ (so orientation is preserved). Furthermore, we will suppose

$$\sup_{|\mathrm{Im}(z)|<\sigma} |\eta(z)| < \infty.$$

We are looking for a change of variables $x = H(\xi)$ such that

$$H^{-1} \circ \phi \circ H = R_{\alpha},$$

or, equivalently,

$$\phi \circ H = H \circ R_{\alpha}. \tag{1}$$

If we demand from H just to be an homeomorphism, then Denjoy's theorem [5, Theorem 3.4, p. 48] suffices to guarantee its existence. But, if we also require H to be analytical, then Arnold's theorem not only tells us that it exists (under certain relatively mild conditions) but we can also profit from its proof to successively build approximations to H.

Let us elaborate: Suppose H is of the form

$$H(x) = x + h(x). \tag{2}$$

Substituting this in (1) leads to

$$\begin{aligned}\phi(H(x)) = \phi(x + h(x)) &= x + h(x) + \alpha + \eta(x)\\ &= H(R_{\alpha}(x)) = x + \alpha + h(x + \alpha),\end{aligned}$$

thus

$$h(x + \alpha) - h(x) = \eta(x).$$

This implies that the Fourier coefficients of $\hat{\eta}(n)$ satisfy

$$\hat{h}(n) = \frac{\hat{\eta}(n)}{e^{2\pi i n \rho} - 1}$$

for $n \in \mathbb{Z} \backslash \{0\}$. We can now write

$$h(x) = \sum_{n \in \mathbb{Z}, n \neq 0} \frac{\hat{\eta}(n)}{e^{2\pi i n \rho} - 1} e^{-2\pi i n x}. \tag{3}$$

Unless $\hat{\eta}(0) = 0$ this does not completely solve the problem of finding H, but the approximate h obtained this way suffices, for we can now define

$$\phi_1(x) = H^{-1} \circ \phi \circ H$$

so we can calculate an $h_1(x)$ through the Fourier coefficients of $\eta_1(x) = \phi_1(x) - \alpha - x$, and continue inductively with $H_1(x) = x + h_1(x)$ and

$$\phi_n(x) = H_{n-1}^{-1} \circ \phi_{n-1} \circ H_{n-1} \tag{4}$$

for $n \geq 2$. This is the so-called *Newton iteration* for the calculation of the change of coordinates. Now we can state Arnold's theorem.

Theorem 1 (Arnold, 1965, [1, **Theorem 2**]**).** *If*

1. *the number ρ is of type (K, ν),*
2. *the diffeomorphism*

$$\phi(x) = x + \rho + \eta(x)$$

has rotation number ρ, and

3. *for some $\epsilon(K, \nu, \sigma)$ we have $\sup_{|\mathrm{Im}(z)|<\sigma} |\eta(z)| \leq \epsilon(K, \nu, \sigma)$,*

then there is an N such that the change of variables $H = H_1 \circ \cdots \circ H_N$ is analytical and conjugates ϕ to R_ρ.

4 Algorithm and Implementation Details

Our approach to generate a sequence of pitches is to take $x_0 = p_0 = 0$ and then, given a circle diffeomorphism with lift $\phi : \mathbb{R} \to \mathbb{R}$, to calculate

$$p_k = \lfloor 12 \bmod (\phi^k(x_0), 1) \rfloor,$$

thus we obtain a melody where each pitch is codified as semitones from the tonic. Here $\bmod(x, 1)$ denotes the least positive representative of the equivalence class of the relation

$$x \sim y \iff x - y \in \mathbb{Z}.$$

Thus, for example, $\mathrm{mod}(1.5, 1) = 0.5$ and $\mathrm{mod}(-2.3, 1) = 0.7$, and hence

$$\lfloor 12 \bmod (1.5, 1) \rfloor = \lfloor 12 \cdot 0.5 \rfloor = \lfloor 6 \rfloor = 6$$

and

$$\lfloor 12 \bmod (-2.3, 1) \rfloor = \lfloor 12 \cdot 0.7 \rfloor = \lfloor 8.4 \rfloor = 8.$$

It is to be noted that, since in the long run f behaves like a rotation when it comes to the orbit, this tends to stabilize in a pattern that shifts slowly, because of the coarse grain of the 12-tone scale. Of course it is possible to consider other scales that subdivide the octave with a different or finer grain, but we will not pursue such a direction in this paper.

Our diffeomorphism of departure is one with a lift among Arnold's family [2, p. 248]

$$\begin{aligned} \phi : \mathbb{R} &\to \mathbb{R}, \\ x &\mapsto x + \alpha + \frac{\epsilon}{2\pi}\sin(2\pi x), \end{aligned}$$

where ϵ is chosen freely (but small, say, $\epsilon < 1$, so orientation is preserved). The detailed procedure is contained in Algorithm 1. It was implemented[2] in Octave 5.2 as a function of α, ϵ, n, N and M, and of course it is possible to modify the number of samples of step 8. The estimation of Fourier coefficients of step 9 is done with a right-point Riemann sum and the solution step 14 is done with the native Octave function `fsolve`, with the initial estimation chosen as $x_{k-1} + \alpha$. The output is the pure numerical result, before the codification as semitones from the tonic is done.

5 Some Examples

The code to obtain the results of the following three examples can be found in the authors' GitHub directory.

Example 1. If we choose $\alpha = e$ and $\epsilon = 0.5$ in Arnold's diffeomorphism, the first ten pitches generated through ϕ_1^k, $1 \leq k \leq 9$, are

$$0, 8, 4, 1, 11, 7, 3, 0, 9, 5$$

which correspond to the melody seen in Fig. 1 within the octave.

We can calculate exactly the Fourier coefficients of η_1, and thus H_1, which is

$$H_1(x) = x - \frac{\epsilon}{2\pi}\frac{\cos(2\pi x - e\pi)}{2\sin(e\pi)}.$$

By calculating with $\phi_2 = H_1^{-1} \circ \phi_1 \circ H_1$, we obtain the pitches

$$0, 8, 5, 2, 10, 7, 4, 1, 9, 6,$$

namely the melody seen in Fig. 2.

[2] See the authors' GitHub directory https://github.com/octavioalberto/KAM_melody_and_variations/.

Algorithm 1. Melody variation through KAM theory

Require: $\alpha, \epsilon \in \mathbb{R}_+$ with $0 < \epsilon < 1$, $n, M, N \in \mathbb{N}$, where α is irrational, n is the length of the melody, M is the number of variations and N is the number of Fourier coefficients used to approximate the non-linear term of the change of coordinates h.

Ensure: M sets $\{p_k\}$ of cardinality n that correspond to pitches of M variations of a melody.

$\phi \leftarrow x + \alpha + \frac{\epsilon}{2\pi}\sin(2\pi x)$. ▷ A lift of a diffeomorphism within Arnold's family.
$x_0, p_0 \leftarrow 0$
for $1 \le k \le n-1$ **do**
  $x_k \leftarrow \phi(x_{k-1})$.
  $p_k \leftarrow \lfloor 12 \bmod (x_k, 1) \rfloor$. ▷ Original melody.
end for
for $1 \le s \le M$ **do**
  Sample the function $\eta(x) = \phi(x) - \alpha - x$ in $[0, 1]$ at equally spaced points.
  Estimate the Fourier coefficients $\hat{\eta}(j)$ for $-N \le j \le N$ via the sample and a Riemann sum.
  $h \leftarrow \sum_{-N \le j \le N, j \ne 0} \frac{\hat{\eta}(j)}{e^{2\pi i j \alpha} - 1} e^{-2\pi i j x}$. ▷ Truncation of (3).
  $H(x) \leftarrow x + h(x)$. ▷ Change of coordinates (2).
  for $1 \le k \le n-1$ **do**
    $u \leftarrow \phi \circ H(x_{k-1})$.
    Solve $H(x_k) = u$ for x_k.
    $p_k \leftarrow \lfloor 12 \bmod (x_k, 1) \rfloor$. ▷ Perturbed melody by change of coordinates.
  end for
  return $\{p_0, \ldots, p_{n-1}\}$.
  $\phi \leftarrow H^{-1} \circ \phi \circ H$. ▷ Diffeomorphism update (4).
end for

Fig. 1. First melody.

Fig. 2. Second melody, obtained as a variation of the first.

It is also possible to calculate H_2 exactly [1, p. 214], which is

$$H_2(x) = x - \frac{\epsilon}{2\pi}\frac{\cos(2\pi x - e\pi)}{2\sin(e\pi)} + \frac{\epsilon^2}{4\pi^2}\frac{\sin(4\pi x - e\pi)}{4\sin(e\pi)\sin(2e\pi)},$$

but the resulting change with respect to the second variation is nil.

Example 2. If we use the same diffeomorphism of the previous example but we increase the perturbation to $\epsilon = 0.8$, then we obtain the melodies seen in Fig. 3 generated by the exact change of coordinates of two iterations, in this case separated by barlines.

Fig. 3. Three melodies (separated by barlines) obtained taking $\epsilon = 0.8$, $\alpha = e$, $n = 10$ and $M = 3$ in Algorithm 1.

Example 3. If we run the implementation of Algorithm 1 with $\alpha = \pi$, $\epsilon = 0.8$, $n = 10$ and $M = 7$ with 64 equally spaced point samples and keeping 10 Fourier coefficients, then we obtain the following output after the semitone codification, displayed as a matrix,

$$P = \begin{pmatrix} 0 & 1 & 4 & 7 & 8 & 8 & 8 & 8 & 9 & 9 \\ 0 & 0 & 1 & 2 & 4 & 5 & 6 & 6 & 6 & 7 \\ 0 & 2 & 3 & 5 & 7 & 6 & 8 & 8 & 5 & 6 \\ 0 & 1 & 3 & 5 & 6 & 8 & 6 & 8 & 9 & 11 \\ 0 & 1 & 3 & 5 & 6 & 8 & 10 & 0 & 2 & 3 \\ 0 & 1 & 3 & 5 & 6 & 8 & 10 & 11 & 1 & 3 \\ 0 & 1 & 3 & 5 & 6 & 8 & 10 & 11 & 1 & 3 \end{pmatrix}.$$

In this case the convergence towards the rotation is patent.

6 Conclusions and Future Work

As we can readily see the variation obtained by applying a change of coordinates to a circle diffeomorphism in the Arnold family is noticeable yet subtle, similar in spirit to Glass' "repetitive structures" [6, Chapter "Paris"]. The numerical implementation introduces additional relatively small "instabilities", but this could be seen not as a flaw but as a feature!

With the restriction to this type of dynamical system we can only control one aspect of melodic variation (pitch, in this case), but KAM theory also works for integrable Hamiltonian systems, where in principle we can also obtain control over note length or (musical) dynamics, for example. The research in this direction is ongoing.

References

1. Arnold, V.I.: Small denominators. I. Mapping of the circumference onto itself. In: Givental, A., et al. (eds.) Collected Works: Representations of Functions, Celestial Mechanics and KAM Theory, 1957–1965, vol. 1, pp. 152–223. Springer, Heidelberg (2009). https://doi.org/10.1007/978-3-642-01742-1_10

2. Arrowsmith, D.K., Place, C.M.: An Introduction to Dynamical Systems. Cambridge University Press, Cambridge (1990)
3. Bugeaud, Y.: Approximation by Algebraic Numbers. Cambridge Tracts in Mathematics, Cambridge University Press, Cambridge (2004)
4. Dabby, D.S.: Musical variations from a chaotic mapping. Chaos **6**(2), 95–107 (1996)
5. de Faria, E., Farino, P.: Dynamics of Circle Mappings. IMPA, Brazil (2022)
6. Glass, P.: Words Without Music. Liveright Publishing Company, New York (2015)
7. Holmgren, R.A.: A First Course in Discrete Dynamical Systems, 2nd edn. Springer, New York (1996). https://doi.org/10.1007/978-1-4419-8732-7
8. Kaliakatsos-Papakostas, M.A., Epitropakis, M.G., Floros, A., Vrahatis, M.N.: Chaos and music: From time series analysis to evolutionary composition. Int. J. Bifurcat. Chaos **23**(11), 1350181 (2013)
9. Pressing, J.: Nonlinear maps as generators of musical design. Comput. Music. J. **12**(2), 35–46 (1988)
10. Sondow, J.: A geometric proof that e is irrational and a new measure of its irrationality. Am. Math. Mon. **113**(7), 637–641 (2006)
11. Wayne, E.C.: An introduction to KAM theory. In: Givental, A.B., et al. (eds.) Dynamical Systems and Probabilistic Methods in Partial Differential Equations, pp. 3–29. American Mathematical Society, Berkeley (1996)
12. Zeilberger, D., Zudilin, W.: The irrationality measure of π is at most 7.103205334137.... Moscow J. Comb. Number Theory **9**(4), 407–419 (2020)

Modal Pitch Space: A Computational Model of Melodic Pitch Attraction in Folk Music

Gilberto Bernardes(✉) and Nádia Carvalho

University of Porto–Faculty of Engineering and INESC TEC, Porto, Portugal
{gba,nscarvalho}@fe.up.pt

Abstract. We introduce a computational model that quantifies melodic pitch attraction in diatonic modal folk music, extending Lerdahl's Tonal Pitch Space. The model incorporates four melodic pitch indicators: vertical embedding distance, horizontal step distance, semitone interval distance, and relative stability. Its scalability is exclusively achieved through prior mode and tonic information, eliminating the need in existing models for additional chordal context. Noteworthy contributions encompass the incorporation of empirically-driven folk music knowledge and the calculation of indicator weights. Empirical evaluation, spanning Dutch, Irish, and Spanish folk traditions across Ionian, Dorian, Mixolydian, and Aeolian modes, uncovers a robust linear relationship between melodic pitch transitions and the pitch attraction model infused with empirically-derived knowledge. Indicator weights demonstrate cross-tradition generalizability, highlighting the significance of vertical embedding distance and relative stability. In contrast, semitone and horizontal step distances assume residual and null functions, respectively.

Keywords: Modal · Pitch attraction · Folk music · Empirical

1 Introduction

Pitch attraction is a concept that describes the perceptual tendency of one musical pitch to move toward another [13]. It is a pivotal concept in tonal music, namely in establishing pitch hierarchies of stability and tension within a given context or tonality [7]. Various models of pitch attraction have been proposed in the literature, aiming to explain and quantify the cognitive and musical factors that influence the perception of pitch relations. Two prominent models are Lerdahl's Tonal Pitch Space (TPS) [7,8] and Woolhouse's Interval Cycle Model [13]. Both models use mathematical or geometrical representations to account for

This research has been funded by the European Union's Horizon Europe research and innovation programme under the Marie Sklodowska-Curie grant agreement no. 101086338 and by the Portuguese National Funding Agency for Science, Research and Technology under grant agreement no. 2021.05132.BD.

© The Author(s), under exclusive license to Springer Nature Switzerland AG 2024
T. Noll et al. (Eds.): MCM 2024, LNCS 14639, pp. 183–194, 2024.
https://doi.org/10.1007/978-3-031-60638-0_15

pitches, chords, and region's attraction. However, they differ in their underlying assumptions and formalisms: Lerdahl's model is based on the generative theory of tonal music, which derives pitch relations from a set of recursive rules and constraints, while Woolhouse's model is based on the concept of interval cycles, which are circular arrangements of pitches that emphasize certain intervals over others.

Although these models have found successful applications in the analysis of tonal music, particularly within the Western classical music canon, they face significant limitations, mainly when applied in the context of folk music. We identified three challenges: 1) defining a foundational embedding space for diatonic modes, 2) conducting empirical validation, and 3) ensuring computational feasibility.

First, many folk music traditions adopt scales or modes other than major and minor, with a strong emphasis on the seven diatonic modes: Ionian, Dorian, Phrygian, Lydian, Mixolydian, Aeolian, and Locrian. To our knowledge, the only Modal Pitch Space (MPS) proposed so far is by Tsougras [11], itself an extension of Lerdahl's TPS. This extension allows for analyzing modal pitch relations regarding distance, direction, and stability as well as comparing different modes.

Second, one limitation of Tsougras's [11] model and the other models mentioned above is that they are mainly based on theoretical and analytical considerations rather than empirical and experimental evidence. While some studies have tested and validated some aspects of Lerdahl's TPS model [5,10], there is a lack of empirical studies that assess the cognitive or perceptual validity of the other models, especially in the context of modal folk music. A critical aspect is understanding preferences and expectations from different folk musical traditions.

Thirdly, we need frameworks for scaling modal pitch attraction, enabling systematic data-driven and cross-cultural approaches within folk music. This need is particularly emphasized in the context of folk music similarity, as highlighted in the motivation for this work [2]. These frameworks should operate without the necessity of user input context or dependence on automatic context inference, which may be prone to errors. Existing models of pitch attraction, such as Lerdahl's TPS [8] and Tsougras' MPS [11], have some limitations in this regard, as they require chordal context to be known, which may not be readily available nor clearly defined in modal folk music, where the harmonic implications are often ambiguous or variable.

Given the above threefold limitations, we present a computational model to capture melodic pitch attraction in modal music. Our contribution extends and refines the foundational work of Lerdahl [8] and Tsougras [11], emphasizing the modelling of folk modal music with empirical validation and ensuring computational feasibility at scale. The first notable change we introduce and assess is the incorporation of data-driven explicit knowledge to determine the pitch stability in the modal basic embedding space compared to theoretical definitions by Tsougras [11] without the need for local chordal context. Furthermore, based on melodic pitch metrics from Lerdahl's TPS, we define and assess the importance of four quantifiable indicators related to pitch attraction: vertical embedding distance v, horizontal step distance h, semitone interval distance i, and relative stability s. The importance of indicators guides the model weights' definition in computing modal pitch attraction in the context of folk music.

We evaluate the attraction prediction of our model of modal melodic pitch from three Dutch, Irish, and Spanish folk music traditions. Additionally, we analyze the contribution of each indicator to the overall modal pitch attraction and discuss the implications of our findings for understanding modal music cognition and perception. Finally, we measure the degree of generalization of our computational model across folk music traditions.

The remainder of the paper is structured as follows. In Sect. 2, we provide an introduction to Lerdahl's TPS [8] and Tsougras's derived MPS [11]. This section elaborates on existing metrics and mathematical concepts, particularly those pertaining to melodic distance and pitch attraction. In Sect. 3, we present our computational model of melodic pitch attraction, elucidating the melodic pitch indicators and their algorithmic implementations. The empirical evaluation of our model is presented in Sect. 4, and the key findings are discussed in Sect. 5. Finally, Sect. 6 serves as the paper's conclusion, summarizing our contributions.

2 From Lerdahl's Tonal Pitch Space to Tsougras's Modal Pitch Space

Lerdahl's TPS offers a model of tonal hierarchy [1,6], representing an atemporal knowledge schema of pitch relations. This model algebraically defines intuitive relations between pitches, chords, and regions in tonal music as quantifiable distances, which have been shown to roughly align with empirical data from cognitive psychology [8].

Lerdahl's theory of TPS is based on the notion of embeddings, which are hierarchical subsets of pitch classes ordered by their stability. The embeddings are represented by the basic space, which consists of five levels, each containing a different number of pitch classes (or pitch in the case of melodic analysis, as shown in Table 1 for the context of the tonic in Ionian mode). The first level (a) contains only the tonic, the most stable pitch class. The second level (b) adds the dominant, the next most stable pitch class. The third level (c) completes the triad of the tonic chord. The fourth level (d) includes all the pitch classes of the diatonic scale of the current key. The fifth level (e) contains all the chromatic pitch classes. For every chord change, the levels (a-c) must be adapted accordingly; for a key change, level (d) must be adjusted.

From the basic space, Lerdahl derives three measures of pitch distance: vertical distance, horizontal step distance, and tonal pitch attraction. Vertical distance reflects the degree of stability of a pitch class by counting the number of levels from the top of the basic space until the pitch class appears. Horizontal step distance is the number of horizontal steps from one pitch to the next at any level. Note that horizontal distance is measured in terms of actual pitches rather than of pitch classes. Table 1 shows vertical and horizontal step distances of all

Table 1. a) The basic space introduced by Lerdahl [8] within his Tonal Pitch Space delineates the hierarchical salience of each pitch class. Vertical distance is determined by counting the levels from (a) to the occurrence of the pitch class. b) A representation of pitch space demonstrating the calculation of horizontal distance. Arrows indicate three annotated examples with a single horizontal step.

a)

Level a:	0											
Level b:	0							7				
Level c:	0				4			7				
Level d:	0		2		4	5		7		9		11
Level e:	0	1	2	3	4	5	6	7	8	9	10	11
Vertical distance:	0	4	3	4	2	3	4	1	4	3	4	3

b)

	C												(C)
	C							G	⟶	⟶	⟶	⟶	(C)
	C	⟵	⟵	⟵	E			G					(C)
	C	⟵	D		E	F		G		A		B	(C)
	C	C#	D	D#	E	F	F#	G	G#	A	A#	B	(C)
Horizontal distance:	0	1	1	3	1	3	6	1	4	2	2	1	

pitch classes in the context of the tonic in C Ionian. Tonal pitch attraction t reflects the tendency of one pitch class p_1 to resolve to another p_2 by considering both their vertical and semitone distances, such that:

$$t(p_1, p_2) = e_{p_2}/e_{p_1} * 1/i \qquad \text{if } p_1 \neq p_2 \quad , \tag{1}$$

where e_{p_1} and e_{p_2} represent the anchoring strength of pitches p_1 and p_2 in the current configuration of the basic space, and $i = \min(|p_1 - p_2|, 12 - |p_1 - p_2|)$ is the minimal number of semitone intervals between the pitch sequence. The condition $p_1 \neq p_2$ ensures that the denominator is not zero, preventing an undefined result from diving by zero. The higher the value of the t, the higher the attraction between the pitch sequence $p_1 \rightarrow p_2$.

Tsougras' [11] MPS adapts TPS for diatonic modes. The basic space is revised to accommodate different modal scales, adjusting pitch class salience values. Contrary to Lerdahl, who adopts four levels in melodic pitch distance calculations, Tsougras motivates using a basic space with five levels in modal music, reinforcing the saliency of the mode's fifth degree (dominant) in level (b). This revision impacts all distance computations, emphasizing the stability of pitch classes in the context of modal scales and will be referred to hereinafter as the Lerdahl-Tsougras model.

3 Modal Pitch Space: A Computational Melodic Pitch Attraction Model

Following Lerdahl's TPS, our model introduces global design considerations and revises pitch distances to enhance computational feasibility for empirical folk music analysis. Two global design considerations were adopted. The first is assuming a basic space and metric computations on a pitch class space for a more generalized and compact representation of musical structures.

The second is the omission of the chordal context changes in the basic space. This involves assuming an agnostic regional basic space per mode, circumventing the requirement for explicit adaptation to the chordal context over the musical time, which demands higher computational complexity and is prone to error. Hence, the basic space e, capturing the stability of a modal context, is static and corresponds to the mode's most stable condition at the chordal context of the tonic, as shown in Tables 1 and 2. Furthermore, we expanded the discrete levels of the embeddings to a continuous mathematical space to accommodate empirically-derived embeddings from folk music datasets. Its computation is intrinsically linked to our evaluation, detailed in Sect. 5. These two basic spaces will be referred to as *theoretical* and *empirical*.

Table 2. Theoretical basic space for Ionian, Dorian, Mixolydian, and Aeolian diatonic modes – we adopt five levels following Tsougras' [11] modal pitch space.

Mode	*Theorectical* basic space e
Ionian	(5 1 2 1 3 2 1 4 1 2 1 2)
Dorian	(5 1 2 3 1 2 1 4 1 2 2 1)
Mixolydian	(5 1 2 1 3 2 1 4 1 2 2 1)
Aeolian	(5 1 2 3 1 2 1 4 2 1 2 1)

From this representation, we advance a modal pitch attraction metric a, between two given sequential pitch $p_1 \rightarrow p_2$ as the weighted sum of four indicators: 1) vertical embedding distance v, 2) horizontal step distance h, 3) interval (semitone) distance i, and 4) relative stability s, such that:

$$a = \alpha v + \beta h + \gamma i + \delta s \quad , \tag{2}$$

where α, β, γ, and δ are weights defining the relative importance of each indicator. A relevant contribution of our work is the mathematical and algorithmic definition of these indicators, which we detail next.

Vertical embedding distance v adopts the basic space to measure the degree of embedding or stability of the pitch p_2 from a melodic pitch class sequence $p_1 \rightarrow p_2$ on a given modal context. To generalize the metric to any continuous basic space, we define it as the distance to the maximum embedding value max(e), such that $v = \max(e) - e_{p_2}$.

Algorithm 1 computes the horizontal step distance h as the shortest number of steps between the melodic pitch sequence p_{init} and p_{final} at the minimal level of both pitch classes in the basic space. Horizontal steps equal the number of pitch classes between the initial and final pitch whose basic embedding space values are above a threshold, defined as the minimum embedding value of the initial and final pitch indexes, $\min\left(e_{p_{init}}, e_{p_{final}}\right)$. We determine the number of steps with the function `count_steps` twice, with $p_1 \rightarrow p_2$ and $p_2 \rightarrow p_1$, as p_{init} and p_{final}, assessing the steps on an increasing pitch class sequence wrapped modulo 12. The shortest of these two values is the final number of steps h.

Algorithm 1. Horizontal Step Distance h

```
function COUNT_STEPS(basic space e, p_init, p_final)
    count ← 0
    i ← p_init
    threshold ← min(e_{p_init}, e_{p_final})
    while i ≠ p_final do
        if embeddings[i] ≥ threshold then
            count ← count + 1
        end if
        i ← (i + 1) mod length of embeddings array
    end while
    return count
end function
```

The semitone interval distance i represents the shortest distance in semitones between two pitch classes, p_1 and p_2. This calculation incorporates both the direct path connecting the pitch classes and the wrap-around path encountered when traversing across octaves, such that:

$$i = \min\left(\left|p_1 - p_2\right|, 12 - \left|p_1 - p_2\right|\right) \quad . \tag{3}$$

Relative stability s quantifies how the anchoring of p_1 relates to p_2 in the basic space, given their embedding values e_{p_1} and e_{p_2}, respectively. It is computed as the ratio in anchoring strength between the two pitches $s = e_{p_2}/e_{p_1}$.

After examining the range of each indicator, both *theoretical* and *empirical* embedding values are normalized to the $[1, 6]$ range before the computation of the modal attraction value a in 2 as a weighted sum of the indicators. The range was adopted to align all indicators dependent on the basic space with the semitone distance, which reaches a maximum interval distance of six semitones. This adjustment allows for a more balanced interplay between all indicators and retains some transparent degree of interpretation. Figure 1 visually represents the distances between all pitch class pairs for each indicator.[1]

4 Evaluation

Our evaluation has three main objectives: 1) the empirical scrutiny of the established Lerdahl-Tsougras 'baseline' melodic attraction model; 2) the evaluation of both the proposed *empirical* and *theoretical* basic space models, and determine the optimal indicator's weights for modeling pitch attraction within modal folk music; 3) an investigation into the model's generality across various diatonic modes and the three distinct Western folk music traditions.

[1] Please note that the vertical embedding distances are in the [0,5] range, contrary to the remaining indicators in the [0,6] range. We maintain this range differences due to the left-skewed distribution observed in the vertical embedding distance histogram, in contrast to the right-skewed or normal distribution of the remaining indicators, thus flattening the impact of all indicators in the final weighted sum.

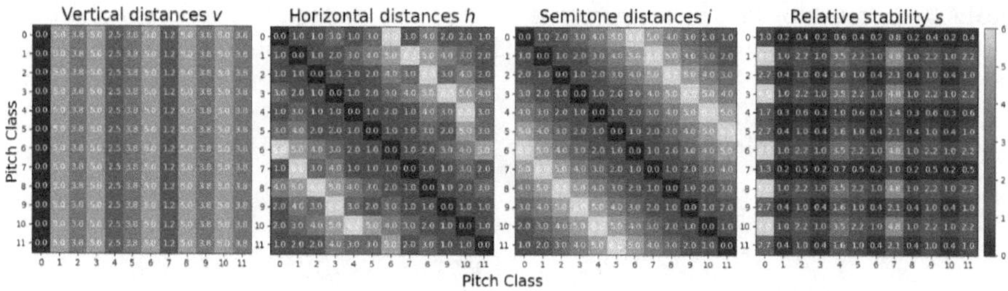

Fig. 1. Visualizing distance indicators between all pitch class distances. From left to right, these matrices illustrate the distances for each indicator: vertical embedding, horizontal step, semitone, and relative stability.

Our empirical ground truth consists of pitch class sequential transition matrices from three representative datasets of folk melodies: The Meertens collection [12], The Session [9], and I-Folk [3] from Dutch, Irish, and Spanish traditions, respectively. We chose these datasets because they have reliable mode annotations. Moreover, while limited to the West, they should provide some insights into how diatonic modes have been adopted in different folk traditions and how generalizable the model is within this geography. We only analyzed four diatonic modes, known to be more representative of folk music, as the remaining modes have few to no examples in the datasets. Table 3 shows the total number of folk melodies per dataset and mode. We note the low number of Dorian and Mixolydian folk melodies in the I-Folk data, which can affect the representativeness of the results.

We parse each folk song in the dataset using the Python library `music21`. Every note in the dataset is mapped according to duration and metric position within a measure, so longer and metrically accented notes, assumingly more stable, have a greater impact on the results. In greater detail, we first convert the note pitch to its corresponding pitch class and then accumulate the product of its duration and metrical position accent in the resulting matrix or vector. Durations are relative to a quarter note as unity (e.g., an eighth and a sixteenth note have .5 and .25 durations, respectively). The metrical position accent is ranked in the $[0, 1]$ range according to the note onset articulation in one of five metrical positions: the measure, half-measure, beat, division, and subdivision pulse. These metrical positions are computed using the `Music21Object.beatStrength` object from `music21` Python library [4] and result in the following values: 1, .5, .25, .125, and .0625.

The first evaluation objective assesses empirically the effectiveness of the baseline Lerdahl-Tsougras model. To this end, we computed the Pearson correlation coefficient r of all pairs of pitch class attraction values from Lerdahl's model in 1 and pitch class transition matrices from the empirical folk music data. We hypothesize that the attraction model should capture pitch transitions in folk music. Therefore, we expect a correlation coefficient close to $r = 1$ $(-1 \leq r \leq 1)$,

Table 3. Number of folk melodies per dataset – Meertens, The Session, and I-Folk, featuring Dutch, Irish, and Spanish traditions – and mode – Ionian, Dorian, Mixolydian, and Aeolian.

Mode	Dataset		
	Meertens	The Session	I-Folk
Ionian	8121	15861	375
Dorian	767	2971	12
Mixolydian	127	1620	54
Aeolian	771	3184	105

indicating a direct linear association between Lerdahl's attraction values and the accumulated pitch transitions in the data.[2]

The second evaluation objective aims to assess the two proposed *empirical* and *theoretical* models, which extend the number of indicators capturing melodic pitch attraction within diatonic modal music. Furthermore, they infer the best weights that align with empirical folk music data. To compute the relative importance of each of the four indicators – vertical embedding distance v, horizontal step distance h, relative stability s, and semitone interval distance i – we measure how well the modal pitch attraction captures sequential pitch transition in the three folk music traditions under study. We use the Pearson correlation coefficient to infer the linear relationship between attraction and pitch transition occurrences. Due to their inverse relationship, the larger the negative value, i.e., closer to -1, the better the model.

We adopted a random 60/40% split per dataset for training and testing. With the training fold, we created our basic embedding space for the *empirical* model and reference pitch transition matrices per mode and dataset for *best weights* computation. We used a grid search to compute the best weights per model, mode, and dataset compared to the reference pitch transition matrices. A total of 14641 weight combinations per dataset and model were assessed, resulting from combinations of 4 values from a pool of 11 numbers in the [0–1] range with .1 equal steps divisions. Adopting the best weights, we used the test fold to validate the alignment of the models with the best weights per dataset and mode.

Finally, we evaluated the generality of the results, namely on what concerns 1) the *empirical* basic embedding spaces per mode across datasets or traditions and 2) the weights across modes per dataset. We inspect their degree of linear relationship using the Pearson correlation coefficient. The results should ultimately inform how generalizable the weights are for the task and how the embeddings differ per mode. The code adopted in the computation of the model and all

[2] Our hypothesis does not link the proposed pitch attraction model to statistical learning or any cognitive theory. Instead, it is grounded in musical practice and seeks to encapsulate the existing folk traditions *praxis*. Future research can delve into the perceptual and cognitive implications of the proposed model.

evaluation procedures can be found online at https://github.com/NadiaCarvalho/MCM-Modal-Pitch-Space.

5 Results

The average Pearson correlation coefficient r for Lerdahl-Tsougras modal against transitions from three datasets – Meertens, The Session, and I-Folk – comprising folk melodies from Dutch, Irish, and Spanish traditions, is $r = .007 \pm .057$. The correlation coefficients were computed per dataset for the four modes under analysis (Ionian, Dorian, Mixolydian, and Aeolian). None of the computed correlations exhibits statistically significant results, as evidenced by all $p > .05$. These findings suggest strong evidence of no discernible linear correlation between Lerdahl's tonal attraction model and the observed pitch class sequential transitions on the selected datasets.

Table 5 presents the Pearson correlation coefficient r between the evaluated *empirical* and *theoretical* models and empirical melodic transition data for each mode and folk music dataset under analysis. Results are shown for the best-computed weights in the grid search method adopted. Table 4 shows the best-computed weights and basic embedding space per mode and dataset.

The most effective model integrates *empirically*-driven basic embedding spaces and weighted indicators, reaching an average Pearson coefficient of $r = -.78 \pm .006$. The lowest correlation is observed in the Dorian mode of the I-Folk dataset ($-.63$), which is potentially attributed to the smaller number of songs in the dataset (12). Adopting *theoretical* basic spaces still provides good results (average of $r = -.71 \pm .078$) and can be adopted in cases where no empirically-driven embeddings exist. All values have strong statistical significance with $p < .001$. All weighted models showcase a substantial improvement over the unweighted strategy and the baseline Lerdahl's [8] tonal attraction model, whose average Pearson correlation coefficient $r = .007 \pm .057$, denotes no linear relationship with empirical folk music data.

For the *empirically-guided* models, the optimal weights ($\alpha, \beta, \gamma, \text{and}\, \delta$) are detailed for each mode and folk tradition in Table 4. Weights are presented in the following sequence: vertical embedding distance v, horizontal step distance, semitone interval distance i, and relative stability s. Notably, the horizontal step distance emerges with a null weight across all modes and traditions, indicating its negligible impact on the model. Moreover, a noteworthy pattern is the higher and similar weights of the vertical embedding distance v and the relative stability s, all dependent on the basic embedding space e. Interestingly, semitone interval distance i does not impact the attraction values as expected due to a common assumption in Western music for favoring parsimonious melodic intervals.[3]

[3] In earlier, unreported tests comparing the linear correlation of Lerdahl's tonal attraction metric t with empirical folk music pitch transitions, we discovered the highest correlation when excluding the inverse intervallic weighting in the denominator from Eq. 1. This adjustment resulted in the separation of indicators in our method and influenced the current evaluation procedure.

Table 4. Weights and basic embeddings space for the highest Pearson correlation coefficient r per mode and dataset under study. Weights (α, β, γ, and δ) order are the following: vertical embedding distance v, horizontal step distance h, semitone distance i, and relative stability s. Weights are normalized to unit sum, i.e., $L1$ norm, to compare the remaining modes and datasets.

		Weights	Basic embeddings space
Meertens	Ionian	(.45 0 .15 .40)	(5.73 1.01 4.96 1.00 5.42 3.52 1.22 6.00 1.03 3.32 1.04 2.86)
	Dorian	(.41 0 .18 .41)	(6.00 1.00 5.20 5.47 1.09 4.21 1.10 5.86 1.90 2.09 3.07 2.23)
	Mixolydian	(.43 0 .17 .39)	(6.00 1.00 4.93 1.07 4.61 5.02 1.09 5.56 1.00 4.11 2.54 1.64)
	Aeolian	(.42 0 .16 .42)	(5.62 1.00 4.51 5.27 1.08 4.51 1.00 6.00 2.66 1.30 2.78 1.83)
The Session	Ionian	(.48 0 .12 .40)	(6.00 1.00 4.34 1.00 5.03 2.88 1.05 5.11 1.00 3.54 1.04 2.75)
	Dorian	(.48 0 .12 .40)	(6.00 1.00 4.35 2.87 1.05 3.96 1.01 4.66 1.03 2.05 4.75 1.06)
	Mixolydian	(.50 0 .06 .44)	(6.00 1.00 3.75 1.06 3.43 3.85 1.00 5.00 1.01 2.74 3.34 1.16)
	Aeolian	(.48 0 .12 .40)	(6.00 1.00 4.08 4.72 1.02 4.21 1.02 5.22 2.04 1.09 3.92 1.17)
I-Folk	Ionian	(.46 0 .14 .40)	(5.16 1.02 3.99 1.06 4.78 3.57 1.00 6.00 1.03 2.75 1.11 2.17)
	Dorian	(.48 0 .16 .36)	(6.00 1.00 4.78 5.09 1.00 4.96 1.18 4.90 2.22 3.38 3.20 1.00)
	Mixolydian	(.49 0 .10 .41)	(6.00 1.10 4.13 1.36 2.54 2.74 1.00 2.05 1.02 2.36 1.87 1.17)
	Aeolian	(.48 0 .14 .38)	(6.00 1.73 4.46 3.65 1.30 3.62 1.00 4.05 1.98 1.46 2.64 1.72)

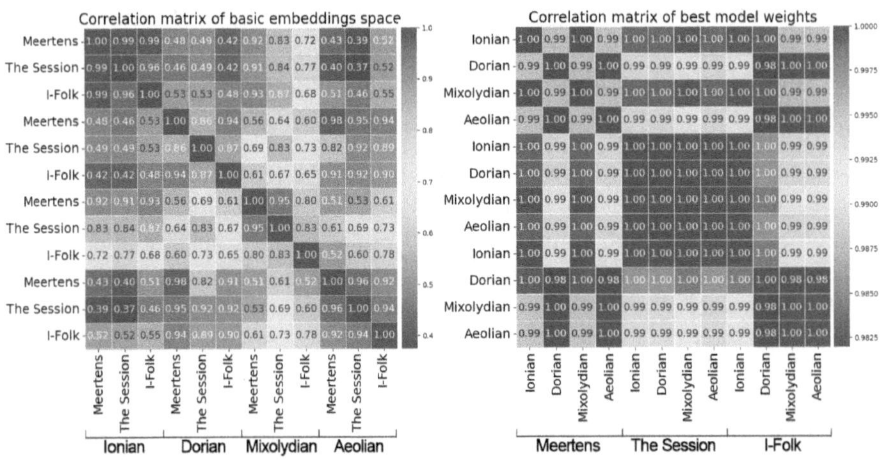

Fig. 2. Correlation matrices depicting the relationships in the basic (embedding) space derived from the empirical model (on the left) and the optimal weights for each mode and dataset (on the right) across the four modes examined in the three datasets. In the basic space matrix, cells are organized by mode, while in the weights matrix, cells are arranged by dataset.

Table 5. Pearson correlation coefficients r between the *empirical* and *theoretical* attraction models and empirical transition data from folk diatonic modal music. Statistical tests are conducted with three folk music datasets – Meertens, The Session, and I-folk – and four diatonic modes – Ionian (Ion), Dorian (Dor), Mixolydian (Mix), and Aeolian (Aeo). The highest correlation coefficient per mode and dataset are shown in bold.

		Meertens				The Session				I-Folk			
		Ion	Dor	Mix	Aeo	Ion	Dor	Mix	Aeo	Ion	Dor	Mix	Aeo
Empirical	Unweighted	−.61	−.57	−.58	−.58	−.58	−.54	−.55	−.59	−.63	−.60	−.53	−.65
	Weighted	**−.81**	**−.73**	**−.74**	**−.75**	**−.82**	**−.80**	**−.81**	**−.82**	**−.82**	**−.63**	**−.79**	**−.79**
Theoretical	Unweighted	−.57	−.50	−.50	−.54	−.51	−.45	−.51	−.51	−.57	−.50	−.50	−.64
	Weighted	−.77	−.65	−.70	−.69	−.79	−.69	−.79	−.76	−.77	−.56	−.58	−.72

Figure 2 shows the correlation of the best-computed weights (on the left) and basic embedding spaces (on the right) for the *empirical* model (listed in Table 4). The basic embedding space is ordered by mode and the weights by dataset or musical tradition. Results show a high correlation between basic embedding space per mode irrespective of dataset or musical tradition. A simplified model is then derived by computing an average basic embedding space per mode. Additionally, we note a reasonably anticipated linear solid correlation between Ionian and Mixolydian modes and between Dorian and Aeolian modes as they align as major and minor $\hat{3}$ degree scales, respectively.

The Pearson correlation between all weights results in a $r \geq .99$ across almost all modes and datasets. A few exceptions exist in the I-Folk dataset, whose fewer examples may drive some inconsistencies. However, the results suggest the possibility of adopting a unique set of weights for all modes and folk traditions under study. Hence, the resulting average $L1$ norm weights equal $(.45, 0, .13, .42)$.

6 Conclusions

This paper addresses three key challenges in analyzing pitch attraction in modal folk music: the absence of a comprehensive modal embedding, the lack of empirical validation, and the computational feasibility of models. The authors introduce a novel computational model that extends and refines existing work, focusing on folk modal music with empirical validation and ensuring scalability. Incorporating data-driven explicit knowledge to determine pitch stability in modal embeddings without needing local chordal context represents a significant contribution. The model introduces four quantifiable indicators related to pitch attraction: vertical embedding distance, horizontal step distance, relative stability, and semitone distance.

The empirical evaluation, conducted on Dutch, Irish, and Spanish folk music traditions, demonstrates the model's superiority over existing ones. The results reveal a substantial improvement over Lerdahl-Tsougras model, which exhibits no statistically significant correlation with empirical folk music data. The proposed model achieves a Pearson correlation coefficient of approximately −.8, indicating a robust prediction of the examined data from melodic transitions in folk music. Moreover, a streamlined melodic pitch attraction model is presented by computing average basic embedding spaces per mode and applying unified weights across all traditions, driven by their high correlation. This approach enhances the model's generalizability for the modal folk music traditions under study. Finally, this work markedly advances cognition of modal music pitch attraction. Emphasis is placed on the significance of the regional basic embedding space, from which two pivotal indicators are derived: vertical embedding distance and relative stability. It is notable that semitone interval distances and horizontal step distance have demonstrated residual to negligible impact, respectively, in the model.

References

1. Bharucha, J.J.: Event hierarchies, tonal hierarchies, and assimilation: a reply to Deutsch and Dowling. J. Exp. Psychol. Gen. **113**(3), 421–425 (1984). https://doi.org/10.1037/0096-3445.113.3.421
2. Carvalho, N., Diogo, D., Bernardes, G.: Computational similarity of portuguese folk melodies using hierarchical reduction. In: Proceedings of the 10th International Conference on Digital Libraries for Musicology, pp. 22–29 (2023)
3. Carvalho, N., Gonzalez-Gutierrez, S., Merchan Sanchez-Jara, J., Bernardes, G., Navarro-Caceres, M.: Encoding, analysing and modeling i-folk: a new database of iberian folk music. In: 8th International Conference on Digital Libraries for Musicology, pp. 75–83 (2021)
4. Cuthbert, M.S., Ariza, C.: music21: a toolkit for computer-aided musicology and symbolic music data (2010)
5. Keller, P.E., Stevens, C.J.: Constraints on the use of global instructions in melody perception. Music Percept. Interdisc. J. **18**(4), 465–496 (2001)
6. Krumhansl, C.L., Cuddy, L.L.: A theory of tonal hierarchies in music. In: Riess Jones, M., Fay, R.R., Popper, A.N. (eds.) Music Perception. Springer Handbook of Auditory Research, pp. 51–87. Springer, New York (2010). https://doi.org/10.1007/978-1-4419-6114-3_3
7. Lerdahl, F.: Tonal pitch space. Music Percept. Interdisc. J. **5**(3), 315–349 (1988). https://doi.org/10.2307/40285402
8. Lerdahl, F.: Tonal Pitch Space. Oxford University Press, Oxford (2001)
9. Session, T.: Traditional Irish music on the session (2021). https://thesession.org/
10. Temperley, D.: Music and Probability. MIT press, Cambridge (2007)
11. Tsougras, C.: Modal pitch space-a theoretical and analytical study. Music Sci. **7**(1), 57–86 (2003)
12. Van Kranenburg, P., de Bruin, M., Grijp, L.P., Wiering, F.: The meertens tune collections. Meertens Online Reports **2014**(1) (2014)
13. Woolhouse, M.: Modelling tonal attraction between adjacent musical elements. J. New Music Res. **38**(4), 357–379 (2009)

Persistent Homology and Harmonic Analysis

Riccardo C. Gilblas[1,2](✉)

[1] University of Padua, Padua, Italy
riccardo.gilblas@math.unipd.it
[2] University of Strasbourg, Strasbourg, France

Abstract. In this paper, we apply persistent homology, one of the main tools in Topological Data Analysis, to the study of harmonic complexity of musical pieces. A directed weighted graph of chords provides a flexible structure that allows to adapt the distances between chords to a training corpus and to retain the asymmetry of human perception of chord progressions. Each musical piece of the corpus is modelled as a sequence of chords, which is used to compute the persistent homology of the simplicial complexes obtained by the Dowker filtration. The values associated to persistence barcodes are then studied to link the geometric properties of the simplicial complexes to the harmonic characteristics of the musical pieces.

1 Introduction

Topological Data Analysis. Persistent homology is one of the main tools in Topological Data Analysis, a domain that is quickly acquiring importance in data analysis thanks to the deep geometrical information it provides. Its popularity among data analysts has been growing since the early 2000's, and it has been employed to face classification problem and feature extraction in several scientific domains. Persistent homology is at the core of TDA and it allows to represent geometrical and topological features of conveniently represented datasets. More specifically, given a filtration of simplicial complexes, which can approximates the underlying geometrical structure of a set of points, the lifespans of homological classes throughout the filtration are represented via barcodes or persistence diagrams. This allows to gather topological information from the dataset and features from the barcodes that can be further used in Machine Learning algorithms for clustering and classification purposes.

The main reference for persistent homology and its current implementation is [25], which presents the mathematical setting and an efficient algorithm for computing persistent homology, while a wide introduction to its use in topological data analysis can be found in [10] and [18]. An interesting review of Machine Learning techniques applied to persistent barcodes is [17]. Persistent homology has been successfully used in classification problems, from fingerprint classification [13] to general image recognition [1].

© The Author(s), under exclusive license to Springer Nature Switzerland AG 2024
T. Noll et al. (Eds.): MCM 2024, LNCS 14639, pp. 195–209, 2024.
https://doi.org/10.1007/978-3-031-60638-0_16

Persistent Homology and Music. More recently, persistent homology has been used also in math-music research, with the aim of linking the topological features of barcodes to the musical properties of datasets. In particular, in [4] persistent homology is used in conjunction with the Tonnetz for automatic style classification. In [6], Vietoris-Rips filtration is used to compute persistent homology of a musical score and to visualise thematic features of musical pieces.

In this paper we focus on harmony or, more precisely, on harmonic density. Similarly to [4], the aim is to recognise the harmonic properties of sequences of chords associated to musical pieces. To do so, we make use of a rich musical database [8], which contains the harmonic analysis of several classical music corpora. With respect to [4], we abandon the Tonnetz in favour of a more flexible object: a weighted directed graph of chords, whose weights are extracted from the database itself. This construction allows us to freely adjust chord relations and also to consider not only distances but more general, non-symmetric, similarity functions between chords. The possibility of having non-symmetric chords relations plays, in our opinion, a central role in harmonic progressions, as it enables to distinguish for example the chord progression (Cmajor $\rightarrow$ Gmajor) from (Gmajor $\rightarrow$ Cmajor). This allows to model the human perception of harmony while retaining this asymmetry, whose importance has been highlighted in several studies (see [3,16]).

There are several ways to associate simplicial complexes to digraphs (see [5] for a complete analysis). Here we make use of the Dowker filtration, whose benefits have been shown in [2,7].

2 Persistent Homology and Chord Notation

Persistent Homology

We avoid to properly introduce persistent homology here: it would result in a compressed presentation that would be boring for experts and not sufficiently clear for novices to algebraic topology. The interested reader can find a formal introduction to persistent homology in [25], which is also the main reference for the computational aspects. Here we present it visually using Vietoris-Rips filtration, which is maybe the most intuitive one.

The following example shows the filtration of simplicial complexes obtained from the set

$$S = \{(1,2),(1,3),(2,2),(2,4),(3,1),(4,3)\} \subset \mathbb{R}^2$$

using the euclidean distance on $\mathbb{R}^2$ and Vietoris-Rips filtration.

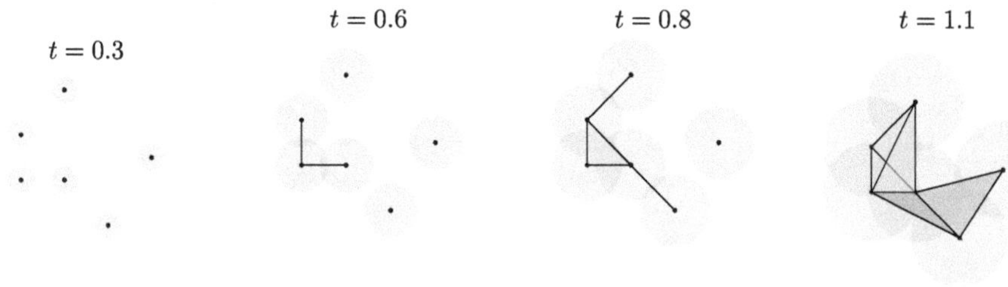

For radius $t = 0.3$, the balls centred in the points of S have empty pairwise intersections, hence the simplicial complex obtained is just the set S itself, whose points are called 0-simplices, having dimension 0. For radius $t = 0.6$, there are non-empty intersections and we draw the two lines (called 1-simplices) that connect the points where the intersecting balls are centred. For radius $t = 0.8$, there are new intersections and the corresponding lines are added. More, the triplet of points $(1,2), (1,3), (2,2)$ are the centres of balls which all have pairwise non-empty intersection, hence we draw a triangle (2-simplex) between the corresponding points. For radius $t = 1.1$, simplices are similarly added. Notice that a tetrahedron (3-simplex) connecting the points $(1,2), (1,3), (2,2), (2,4)$ appears.

Each value of t gives a simplicial complex, which is the collection of the simplices of all the dimensions, together with a derivation function, that associates to each n-simplex an algebraic combination of the $(n-1)$-simplices lying on its border. The collection of simplicial complexes for every t gives a filtration, since the simplicial complex associated to t can be included in the simplicial complex associated to t' if $t < t'$.

To give a naive idea of the information provided by persistent homology on a set, we use another example. Take a finite set of points $\{p_i\}_i$ on the unit circle $\mathbb{S}^1 \subset \mathbb{R}^2$ and imagine to consider them just as points in R^2, forgetting the underlying variety $\mathbb{S}^1$. One would be interested in reconstructing this variety in order to understand the topological properties of the points. This can be done using persistent homology.

Indeed consider the filtration of simplicial complexes as shown above, where $t_1 < t_2 < t_3 < t_4 \in \mathbb{R}$ are the values of the radius of the balls centred in the points. We omit to represent the simplices of dimension greater than 1 to keep the drawing intelligible.

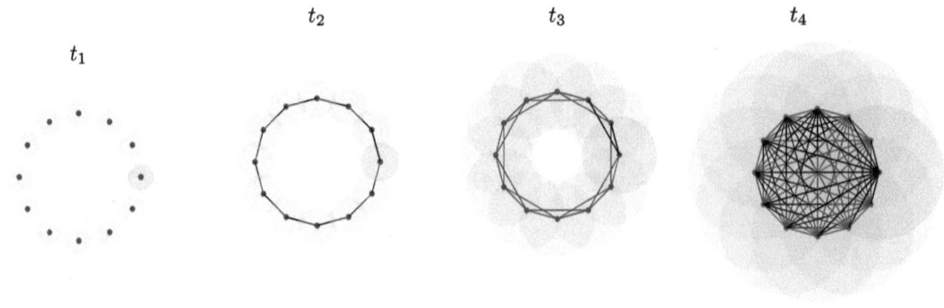

In this example, one of the main topological features of the original variety $\mathbb{S}^1$ is the (1-dimensional) hole, which mathematically is formalised by the homology group in degree 1 having rank 1. As one can notice, this hole appears as well in the filtration of simplicial complexes, which somehow approximate the unit circle. Roughly, the hole appears for $t = t_2$ and it disappears for $t = t_4$. These values are called respectively the time of *birth* and *death* of the hole in the filtration.

Persistent diagrams and barcodes represent the birth and death time of the holes of the various dimensions appearing in the filtration, encoding the evolution of the topological features of the simplicial complexes. More specifically, persistent homology highlights the holes that last longer in the filtration: they are usually the most important ones as they reveal more accurately the underlying topological structure of the set of points. We finish with the definition of entropy for persistent diagrams, which will be used later:

Definition 1. *Consider a persistent diagram* $X = \{(x_i, y_i) \mid x_i < y_i, x_i, y_i \in \mathbb{R}\}$*, where* x_i *(resp.* y_i*) represents the birth (resp. death) time of the generators. If we denote* $S = \sum(y_i - x_i)$*, the entropy of* X *is defined as:*

$$\varepsilon_X = \sum_i -\frac{y_i - x_i}{S} \log\left(\frac{y_i - x_i}{S}\right).$$

Dowker Filtration

We briefly present here the Dowker filtration, following [7]. This construction associates to a directed weighted graph G a filtration of simplicial complexes, which can be further analysed using persistent homology. Firstly, let $\omega_G : G \times G \to \mathbb{R}$ denote the weights function associated to the digraph G. We define the function:

$$\begin{aligned} \bar{\omega}_G : G \times G &\longrightarrow \mathbb{R} \\ (v, v') &\longmapsto \max\{\omega_G(v, v'), \omega_G(v, v), \omega_G(v', v')\}. \end{aligned}$$

Given a parameter $t \geq 0$, we consider $R_{t,G} := \{(v, v') \in G \times G \mid \bar{\omega}_G(v, v') \leq t\}$. Clearly if $t_1 \leq t_2$, one has $R_{t_1,G} \subset R_{t_2,G}$. We define the *Dowker t-sink simplicial complex* $\mathfrak{D}^{\mathrm{si}}_{t,G}$ to be:

$$\mathfrak{D}^{\mathrm{si}}_{t,G} = \{\sigma = (v_0, \ldots, v_k) \mid \exists\, v' \in G \text{ s.t. } (v_i, v') \in R_{t,G} \text{ for all } v_i\}.$$

If v' satisfies the condition above for the simplex σ, we say that it is a *t-sink* for σ. The filtration $(R_{t,G})_{t\geq 0}$ gives rise to the filtration $(\mathfrak{D}^{\mathrm{si}}_{t,G})_{t\geq 0}$, which is a filtration of simplicial complexes whose persistent homology can be computed.

The dual construction is the Dowker source filtration, defined as follows:

$$\mathfrak{D}^{\mathrm{so}}_{t,G} = \{\sigma = (v_0, \ldots, v_k) \mid \exists\, v' \in G \text{ s.t. } (v', v_i) \in R_{t,G} \text{ for all } v_i\}.$$

Here v' is called a *t-source* of σ. In what follows, we will only say *Dowker filtration* without specifying if we are referring the sink or the source construction, as their respective homologies have been proven to be equivalent:

Theorem 1. *[7, Theorem 17] Let us denote $Dgm_k^{si}(G)$ (resp. $Dgm_k^{so}(G)$) the persistence diagram associated to the k-th homology groups of the sink (resp. source) Dowker filtration of the weighted directed graph G. Then one has:*

$$Dgm_k^{si}(G) \simeq Dgm_k^{so}(G).$$

Remark 1. Observe that the previous theorem does not imply that the information carried by Dowker filtration reduces to the symmetric case. In [7] it is shown that Dowker filtration remains sensible to asymmetry of data.

Chord Notation

We denote the set of pitch classes with a very common notation in math-music research:

$$\mathcal{P} := \{C, C\sharp, D, D\sharp, E, F, F\sharp, G, G\sharp, A, A\sharp, B\} \simeq \mathbb{Z}_{12}.$$

To maintain a unique correspondence between labels and pitches, we decided to use only sharp notation, so $E\flat$ will be always written as $D\sharp$.

Definition 2. *An non-degenerate* abstract chord *is a subset of $\mathcal{P}$ of cardinality ≥ 3. If $\mathbf{C} = \{P_1, P_2, P_3\}$ is an abstract chord and $n \in \mathbb{Z}_{12}$, we say that $\{P_1 + n, P_2 + n, P_3 + n\}$ is the translation of the chord $\mathbf{C}$ by n and we denote it by $\mathbf{C} + n$.*

We will consider the following group of chords:

- *Major chords*: $\mathfrak{M} = \{\{C, E, G\} + n \mid n \in \mathbb{Z}_{12}\}$. If P is the fundamental pitch class (i.e. it corresponds to the translation of C) of a major chord, we shortly denote the chord by PM.
- *Minor chords*: $\mathfrak{m} = \{\{C, D\sharp, G\} + n \mid n \in \mathbb{Z}_{12}\}$, shortly Pm.
- *Major seventh chords*: $\mathfrak{M}_7 := \{\{C, E, G, B\} + n \mid n \in \mathbb{Z}_{12}\}$, shortly $P\Delta$.
- *Dominant chords*: $\mathfrak{do} := \{\{C, E, G, A\sharp\} + n \mid n \in \mathbb{Z}_{12}\}$, shortly $P7$.

- *Minor seventh chords*: $\mathfrak{m}_7 := \{\{C, D\sharp, G, A\sharp\} + n \mid n \in \mathbb{Z}_{12}\}$, shortly $Pm7$.
- *Major-seventh minor chords*: $\mathfrak{m}_{\bar{7}} := \{\{C, D\sharp, G, B\} + n \mid n \in \mathbb{Z}_{12}\}$, shortly $PmM7$.
- *Diminished chords*: $\mathfrak{di} := \{\{C, D\sharp, F\sharp, A\} + n \mid n \in \mathbb{Z}_{12}\}$, shortly $P\circ$. Notice that there are only 3 types of diminished chords, as $P\circ = (P+3)\circ = (P+6)\circ = (P+9)\circ$.
- *Half-diminished chords*: $\mathfrak{di}_{\natural} := \{\{C, D\sharp, F\sharp, A\sharp\} + n \mid n \in \mathbb{Z}_{12}\}$, shortly $P\%$.
- *Augmented chords*: $\{\{C, E, G\sharp\} + n \mid n \in \mathbb{Z}_{12}\}$, shortly P^+. Notice that there are only 4 types of diminished chords, as $P^+ = (P+4)^+ = (P+8)^+$.
- *4-suspended chords*: $\{\{C, F, G\} + n \mid n \in \mathbb{Z}_{12}\}$, shortly $Psus4$.
- *2-suspended chords*: $\{\{C, D, G\sharp\} + n \mid n \in \mathbb{Z}_{12}\}$, shortly $Psus2$.
- *augmented sixths*: there are several types of augmented sixths. We introduce the so-called *German*, *Italian* and *French*.

$$\begin{aligned} Ger :=& \{\{C, E, G, A\sharp\} + n \mid n \in \mathbb{Z}_{12}\}. \\ It :=& \{\{C, E, A\sharp\} + n \mid n \in \mathbb{Z}_{12}\}. \\ Fr :=& \{\{C, E, F\sharp, A\sharp\} + n \mid n \in \mathbb{Z}_{12}\}. \end{aligned}$$

Remark 2. Notice that the sets *Ger* and $\mathfrak{do}$ coincide. While this may seem bizarre from the mathematical point of view, it is very useful in practice to have two possible labels for the same type of chords. Indeed while coinciding pitch-wise, German augmented sixths and dominants have very different roles in functional tonal harmony [20] and it is convenient to distinguish them. From the formal point of view, this does not originate any problem, as in the end we are treating chords just as labels associated to the vertices of a digraph.

Remark 3 (The role of time and harmonic density). If the aim is to model the human perception of harmony, it seems important to take into account time when looking at chord sequences. Indeed it is natural to assume that human perception of harmony is strongly influenced by the number and the variety of chords in a given time span: the same chord progression (for example, $CM \to D\sharp m$) can be perceived more or less unusual (or strange, or complex) depending on the time span it occupies and the presence of chords between them that form a sort of *tonal bridge*. In what follows, we chose not to consider the duration of each chord and also to ignore repetitions of the same chord, while focusing on distinguishing whether two chords are next to each other or they are separated by others in the sequence.

Indeed *repetitions* can be considered not to have a huge impact from the point of view of harmonic perception; the choices we made (see Eq. (1)) reflect this, so that persistent homology ignores subsequent repetitions of the same chord. Regarding *duration*, we decided to simplify it to a standard unit duration for every chord, which in the end corresponds to the index of the chord in the sequence. This may seem simplistic, but it allows to retain the fundamental information given by two chords being close to each other in time or separated by other chords that may act as tonal bridges, hence reducing the perceived harmonic complexity of the progression. Also, the computational complexity of the algorithms is conveniently reduced.

So, we will consider the chords of a piece as a sequence $(\mathbf{C}_i)_{0\leq i\leq k}$ and define *time-chords* to be the couples $(\mathbf{C}_i, i)$. Then we define a distance function that takes the time i into consideration.

3 Digraph Associated to a Corpus

Definition 3. *A musical corpus $\mathcal{M}$ is a set of chord sequences of the type $S = (\mathbf{C}_i)_{1\leq i\leq \tau_S}$ where $\tau_S \in \mathbb{N}$ is the number of chords in S and for every i, $\mathbf{C}_i$ is an abstract chord. In this case we write $\mathbf{C}_i \in \mathcal{M}$ and the total number of chords in $\mathcal{M}$ is denoted as $\tau_{\mathcal{M}} := \sum_{S\in\mathcal{M}} \tau_S$.*

For every pair of chords $\mathbf{C}_i, \mathbf{C}_j \in G$, we write $(\mathbf{C}_i, \mathbf{C}_j) \in \mathcal{M}$ if there exists $S \in \mathcal{M}$ that contains the chord progression $(\mathbf{C}_i, \mathbf{C}_j)$. If $(\mathbf{C}_i, \mathbf{C}_j) \in \mathcal{M}$, we define $\nu(\mathbf{C}_i, \mathbf{C}_j)$ to be the total number of instances of the chord progression in $\mathcal{M}$, divided by $\tau_{\mathcal{M}}$.

Definition 4. *Given a musical corpus $\mathcal{M}$, we define the* associate directed weighted graph *$G_{\mathcal{M}}$ as follows:*

- *as vertices, we consider the set of all the chords contained in $\mathcal{M}$.*
- *As directed edges, we consider the set of all the couples $(\mathbf{C}_i, \mathbf{C}_j) \in M$.*
- *As weights, we take $\omega(\mathbf{C}_i, \mathbf{C}_j) = \frac{1}{\nu(\mathbf{C}_i, \mathbf{C}_j)}$.*

We denote by $\bar{G}_{\mathcal{M}}$ the extended graph *associated to $G_{\mathcal{M}}$, where we consider vertices as in $G_{\mathcal{M}}$ and edges:*

$$(\mathbf{C}_i, \mathbf{C}_j) \textit{ if } (\mathbf{C}_i + n, \mathbf{C}_j + n) \textit{ is edge of } G_{\mathcal{M}}, \quad \exists n \in \mathbb{Z}_{12}.$$

Remark 4. Notice that the graph $G_{\mathcal{M}}$ retains the precise information about the chords used in the corpus $\mathcal{M}$, hence including the choice of tonalities. The extended graph $\bar{G}_{\mathcal{M}}$ flattens down the data about the tonalities and which chords are used, and preserves only the informations about the progressions of chords as mutual degree (see [20] for details). This turns out to be very useful in practice, as it allows to analyse a certain musical piece using a corpus $\mathcal{M}$ even if the piece and the corpus don't possess chords in common.

Classical Music Database

The database we used has been created by the Digital and Cognitive Musicology Lab at École Polytechnique Fédérale de Lausanne (see [8]). This database contains several music corpora, with a complete analysis of the measures, chords and harmonies.

The Composers. We focused specifically on the following:

- Annotated Beethoven Corpus, containing all Beethoven's string quartets.
- Beethoven's Piano Sonatas.
- Chopin's Mazurkas.
- Corelli's Trio Sonatas.
- Debussy's *Suite Bergamasque.*
- Liszt's *Pelegrinage.*
- Mozart's Piano Sonatas.
- Schumann's *Kinderszenen* op. 15.
- Tchaikovsky's *Seasons* op. 37a.

In particular, we used the analysis from the *harmonies* section of the database. It is worth stressing out that this contains the harmonic annotations of the musical pieces instead of the sequence of chords. This is a non-trivial difference from the music theory point of view, but we preferred to ignore these aspects and consider the analysis conducted from Lausanne Lab as sequences of chords. We are sure that even the most careful reader agrees that harmonic annotations are at least an acceptable approximation of chords progressions. More in detail, the notation used in the database, based on global/local tonalities and scale degrees, was converted in the simpler notation presented above, with a number $n \in \mathbb{Z}_{12}$ expressing the fundamental and a short string (such as "*M*", "*m*", "Δ") to denote the type of chord. We denote by $\mathcal{M}$ the musical corpus of the chord sequences obtained by the union of the musical compositions listed above. We used $\mathcal{M}$ both to construct the graph of chords $G_{\mathcal{M}}$ and as a testing set of chord sequences.

Frequencies Extraction and Composers' Balance. When extracting the frequencies related to chord progression, we took into account the remarkable difference of dimension of the various composers' corpora. For example, the corpus of Beethoven's piano sonatas (resp. string quartets) has a total of 18928 (resp. 24303) chords, while the corpus of Debussy's *Suite Bergamasque* only contributes with 920 chords. We decided to balance the contribution of each composer by multiplying the frequencies deriving from each composer by a suitable correction factor. We preferred to apply this correction instead of just restricting the total number of chords of each composer to a fixed (forcedly small) upper-bound in order to retain the maximum amount of information from the corpora.

The possibility of adjusting the balance of the various contributions to $\mathcal{M}$ highlights the flexibility of the construction: it is very simple to modify how much a certain composer *influences* the corpus by changing the correspondent correction factor.

4 Analysis

The construction above provides the graph $\bar{G}_{\mathcal{M}}$ of chords and relations extracted from the database $\mathcal{M}$. We denote by d the minimal-path weight function that associate to every couple of chords $(\mathbf{C}_i, \mathbf{C}_j)$ the weight of the minimal directed path in $\bar{G}_{\mathcal{M}}$ from $\mathbf{C}_i$ to $\mathbf{C}_j$. Remember that here d is not a distance, as it lacks symmetry. In Table 1 we report some of the significant values of d.

Table 1. Some significant weights of $\bar{G}_{\mathcal{M}}$.

Chord progression	Weight
$CM \to GM$	25.72
$CM \to G7$	21.47
$CM \to FM$	17.57
$C7 \to FM$	13.11
$CM \to Am$	118.39

Remark 5. As one can notice, the values are quite small for very common chord progressions, but quickly increase for rarer ones. For example, the weight of the progression $C\Delta \to BmM7$ is 10737.29, as major-seventh minor chords are quite scarce in the corpus. It is interesting to observe also some unexpected results: the progression $CM \to Am$, which one could expect to be very common, has already weight 118.39, bigger than the tritone $CM \to G\sharp M$, which stops at 105.47.

In order to build a testing set, we divided each musical piece in the corpus in several parts having the same number of chords, in order to have comparable results. To obtain this, we used thresholds τ equal to 5, 10 and 20 chords and obtained chord sequences of the type $S = (\mathbf{C}_i)_{1 \leq i \leq \tau}$.

We included the index of the chord as a time element, so to have a local description of the harmonic content rather than the overall harmonic complexity, as discussed in Remark 3. To do so, we define the function $\bar{d}$ on time-chords $(\mathbf{C}_i, i)$ as follows:

$$d((\mathbf{C}_i, i), (\mathbf{C}_j, j)) = \begin{cases} 0 & \text{if } i = j \\ \log(|j-i|^\alpha) & \text{if } \mathbf{C}_i = \mathbf{C}_j \\ \max\{1, \log(d(\mathbf{C}_i, \mathbf{C}_j)|j-i|^\alpha)\} & \text{otherwise.} \end{cases} \tag{1}$$

Here the parameter $\alpha \in \mathbb{R}$ allows to tweak the importance of time in the computation: by increasing its value one gets a more local harmonic description, while $\alpha = 0$ gives the overall harmonic complexity. In what follows, we take $\alpha = 1$.

Then, for every music piece S, we built a filtration of simplicial complexes using the Dowker filtration and the function d and we computed the persistent homology. To do so, we used the package SIMPLICIAL, available in JULIA. For every chord sequence S we obtained barcodes of degree 0 and 1.

We associated to each barcode a set of values (usually called *features* in data analysis) that allows to conduct further analysis on the barcodes. In particular, restricting to finite length bars of the barcode, we considered: maximal length, second maximal length, average length and variance length. One can consider barcode entropy as well as a feature.

In Table 2 we depict some of these values of barcodes obtained from musical pieces having chord threshold $\tau = 20$. More precisely, one can read the features **0-ML**, **0-A**, **0-V**, **1-ML**, **1-A** and **1-V**, corresponding to maximal length, average length and variance length for degrees 0 and 1. For maximal length and average, we wrote both the mean and the maximum (with respect to the corpus), while for variance we wrote only the mean.

Table 2. Values associated to the barcodes from each corpus.

Corpus	0-ML	0-A	0-V	1-ML	1-A	1-V
Beeth. Quart.	39.6, 125	9.6, 24	152	8, 51	4.8, 31.5	12.5
Beeth. Son.	41.5, 112	9.4, 23.9	159	7.8, 69	4.6, 26.3	10.9
Chopin	34.3, 106	8.5, 25.4	125	7, 41	4.3, 21	8.9
Corelli	43.8, 138	10.6, 27.9	187	9.7, 49	5.1, 26	20.6
Debussy	46.8, 106	11.0, 22.1	200	11.3, 34	5.9, 18	23.4
Liszt	38.2, 99	10.0, 22.7	156	7.7, 40	4.3, 21.5	12.7
Mozart	42.7, 138	9.9, 25.1	169	7.5, 48	4.5, 28	10.6
Schumann	40.6, 100	10.3, 15.6	142	8.6, 17	4.7, 8.5	8.7
Tchaikovsky	43.6, 106	10.2, 20.3	176	10.3, 40	6, 29	19.6

Machine Learning Approach

Support Vector Machine. Support Vector Machine is one of the most common techniques for classification in machine learning. We briefly present here the basic definitions. Consider a training set of parameters $\{\mathbf{x}_i\}_i \subset \mathbb{R}^n$ with a corresponding set of values $\{y_i\}_i \in \mathbb{R}^n$. Here typically $y_i \in \{-1, 1\}$ and represents the property function that one wants to predict. For example, in our case $\mathbf{x}_i$ is the vector of features associated to a barcode of a musical piece and y_i has value 1 if the musical piece has a fixed property, for example being composed by Beethoven, and -1 otherwise. We say that $\mathbf{x}_i$ is a *positive* (resp. *negative*) training point if the corresponding $y_i = 1$ (resp. $y_i = -1$. The aim of the algorithm is to find an hyperplane $\mathbb{H}$ of $\mathbb{R}^n$ (or an algebraic hyper-surface, more in general) that divides the set of positive training points from the one of negative training points. Then one can use $\mathbb{H}$ to estimate the value $y_j \in \{-1, 1\}$ of a testing point $\mathbf{x}'_j \in \mathbb{R}^n$. More formally, the goal is to find $w \in \mathbb{R}^n$ and $b \in \mathbb{R}$ solving the following problem:

$$\min_{w,b,\zeta} \frac{1}{2} w^T w + C \sum_{i=1}^{s} \zeta_i$$

$$y_i(w^T \phi(\mathbf{x_i}) + b) \geq 1 - \zeta_i$$

$$\zeta_i \geq 0 \text{ for } i = 1, \ldots, s$$

where s is the cardinality of the training set, ϕ is a linear transformation, ζ_i is the error we admit on the training point i and C is a penalty term that forces to stay next to the margin boundary. Indeed with a perfect prediction, i.e. if the two subsets of positive and negative training points are actually perfectly separable by an hyperplane, one has $y_i(w^T\phi(\mathbf{x_i}) + b) \geq 1$. In applications, this situation is very rare, and one usually accepts to miss-classify the point $\mathbf{x}_i$ by introducing the parameter ζ_i, which expresses the distance of the point from the boundary of the hyperplane $\mathbb{H}$.

We used SVM to attack the problem of automatic composer classification. We divided the barcodes obtained in the previous section in two subsets: Beethoven's chord sequences and non-Beethoven ones. Using the previous notation, we took as parameters $\mathbf{x}_i$ the vectors

$$(\mathbf{0\text{-}E}, \mathbf{0\text{-}ML}, \mathbf{0\text{-}A}, \mathbf{0\text{-}V}, \mathbf{1\text{-}E}, \mathbf{1\text{-}ML}, \mathbf{1\text{-}A}, \mathbf{1\text{-}V}) \in \mathbb{R}^8$$

where **0-E** (resp. **1-E**) denotes the entropy of the barcode of degree 0 (resp. degree 1) as in Definition 1. As anticipated, we set $y_i = 1$ if $\mathbf{x}_i$ is a positive point (i.e. corresponds to a Beethoven's chord sequence) and $y_i = 0$ otherwise.

To perform the computations, we used the package SVM provided by SciKit [19] in Python. In particular, we used the function StandardScaler to properly scale the coefficients and the built-in SVC algorithm. However, we did not obtain particularly interesting results: we never managed to obtain a score higher than 0.56, where 0.5 denotes the score of a random classification function. This remained the case even applying some tweaks to the input: we modified the chord threshold of the chord sequences used to extract the barcodes ($\tau = 5, 10, 20$), and the best result (a score of 0.558) was obtained with $\tau = 20$. We also tried to use an enlarged database composed of the barcodes associated to chord sequences having all three possible thresholds, but this did not significantly improve the accuracy. There are still many tweaks that can be explored, as adjusting the parameter α of time in the computation of time-chord distances, or manually modify the graph of chords to suit it to an expected outcome.

Multi-linear Regression. We obtained slightly better results using another machine learning tool: linear regression. The mathematical setting is similar to SVM: given training set $(\mathbf{x}_i, y_i) \in \mathbb{R}^{n+1}$, one wants to determine $\beta_0 \in \mathbb{R}$ and $\beta \in \mathbb{R}^n$ such that

$$y_i = \beta_0 + \beta^T\mathbf{x}_i + \epsilon_i \qquad \forall\, i$$

where ϵ_i denotes an error variable. This allows to use (β_0, β) to predict the value y_j of a test point $\mathbf{x}_j$.

Again, we used SciKit [19] in Python with its Linear Regression models. Similarly to SVM, we define the features vector as:

$$\mathbf{x}_i = (\mathbf{0\text{-}E}, \mathbf{0\text{-}ML}, \mathbf{0\text{-}A}, \mathbf{0\text{-}V}, \mathbf{1\text{-}E}, \mathbf{1\text{-}ML}, \mathbf{1\text{-}A}, \mathbf{1\text{-}V}) \in \mathbb{R}^8$$

and the corresponding value $y_i = 1$ if the chord sequence is composed by Beethoven, $y_i = 0$ otherwise. We considered several possibilities for the chord sequences constituting the database: we tried with chord threshold $\tau = 5, 10, 20$ and even all the three choices combined. When considering a positive database given by all Beethoven's compositions and a negative database composed by all non-Beethoven compositions, we did not get particularly interesting results: the best score was obtained with $\tau = 20$ and it was equal to 0.010, where 0 is the random prediction and 1 is the perfect prediction. Better results were obtained by restricting the dataset. For example, we reduced the positive database to Beethoven's Sonatas only and the negative database to Debussy's, Schumann's and Tchaikovsky's compositions. We made a random selection of Beethoven's barcodes in order to have an even cardinality of positive and negative part. With these modifications, the score was consistently around 0.2 with a peak of 0.27. Of course these results are still too low to get accurate predictions, but they suggest that it is possible to fine tweak the model to increase the accuracy of the regression.

Discussion

Musical Choices

Much debate can originate from the musical choices associated to harmonic analysis. Here we preferred to adhere to the annotations in [8] and simply convert the notation to the one presented above. Also, we decided to ignore the added note to chords (like ninths, elevenths, etc.) and consider every chord without a given mode as major. A first improvement can be using chord analysis rather than harmonic annotation, which would better suit the purpose of describing harmonic complexity. Unfortunately we were not able to find a sufficiently big and coherent database of chord analysis, yet it would be possible to build one using automatic harmonic analysis algorithms (see for example [24]). On the other hand, it is possible to introduce added notes to fundamental chords just by considering new classes of chords in the digraph, which is straight forward both from the formal and the computational point of view. This highlights again the flexibility of the construction.

Features and Composers' Style

As expected, to higher values (**0-ML**, **0-A**, **1-ML**, **1-A**) correspond more harmonically complex corpora. In the case of Debussy (which has the highest values) and Tchaikovsky, the high complexity probably comes from the wide range of chords used in the corpus, and in particular for Debussy also from the frequent use of non-standard chord progressions from the tonal point of view. Notice that Debussy has high mean values, both in degree 0 and 1, while keeping relatively low maximal values, somehow showing a high average harmonic complexity throughout the corpus without many harmonic density peaks.

Corelli's corpus also has interestingly high values, this time probably caused by the considerable density of harmony and the quick changes of tonality (so called *modulations*), typical of Baroque period and polyphonic music, rather than tonally unusual chord progressions. This interpretation is supported by Corelli's high scores in maximal values, which would suggest the presence of very dense harmonic sections in the corpus.

On the other side of the spectrum, Chopin's corpus are characterised by very low scores, which well describe the limited overall harmonic complexity and the low density of harmonic changes of Mazurkas.

Non surprisingly, Beethoven's corpora have very similar scores and both occupy the middle of the spectrum with respect to almost every value. Also Schumann can be placed more or less in the middle. Liszt's corpus seems more complex to analyse: it has quite low scores in all the indicators except **0-A**. In any case, it is worth noticing that *Pelegrinage* is a very heterogeneous collection of pieces, with remarkably different artistic choices and musical structures.

Machine Learning Improvements

The attempt of using machine learning techniques on persistent diagrams features seems quite natural and it has been successfully done for other classification problems (see [1,13,17]). Here we obtained only partial results and we did not manage to get a sufficiently accurate classification. In any case, there is ample room for improvements, as it is possible to fine-tune the numerous parameters that come into play in the computations. A first one could be to tweak manually the weights in the graph $\bar{G}_{\mathcal{M}}$, in order to fix unexpected or unwanted values, as observed in Remark 5. Also modifying the balance between chord distance and time in Eq. (1) highly affects persistent homology and consequently the values of the features used in machine learning algorithms. This can be done by adjusting the value of α, a parameter that allows to modify the impact of time in the computation of $\bar{d}$ and hence affects how local is considered the harmonic complexity. Again in Eq. (1), the choice of applying the logarithm on chord distance seems natural, given the steep increase in the weights of $\bar{G}_{\mathcal{M}}$ corresponding to less common chord progressions, but clearly this can be adjusted by using different functions.

5 Conclusions

In this work we focused on harmony and in particular on harmonic complexity, with the aim of classifying musical pieces and possibly composers' styles using persistent homology. To do so, we introduced the idea of a weighted digraph associated to a musical corpus, whose vertices are the chords and the edges with their weights represent the probability of the corresponding chord progressions in the corpus. Then we computed the persistent homology of the Dowker filtration associated to each musical piece. Finally, we extracted from the barcodes a set of features that have been studied with machine learning algorithms.

While the use of machine learning has much room for improvement and the classification results are only partial, we believe that the construction of the digraph associated to a corpus offers a useful and flexible tools for harmonic analysis, on its own as well as in tandem with persistent homology. With respect to the Tonnetz, it allows to model asymmetry and it makes easy to add new types of chords and adjust the weights of the chord progressions.

References

1. Asaad, A.T.: Persistent Homology Tools for Image Analysis, PhD Thesis, University of Buckingham (2020)
2. Atkas, M.E., Akbas, E., El Atmaoui, A.: Persistent homology of networks: methods and applications. Appl. Netw. Sci. **4**, 61 (2019)
3. Bharucha, J., Krumhansl, C.L.: The representation of harmonic structure in music: Hierarchies of stability as a function of context. Cognition **13**(1), 63–102 (1983)
4. Bergomi, M.G., Baratè, A.: Homological persistence in time series: an application to music classification. J. Math. Music **14**, 204–221 (2020)
5. Bergomi, M.G., Ferri, M.: Exploring graph and digraph persistence. Algorithms **16**(10), 465 (2023)
6. Callet, V.: Persistent Homology on Musical Bars. Math. Comput. Music, MCM (2022)
7. Chodhury, S., Mémoli, F.: Persistent homology of asymmetric networks: an approach based on Dowker filtrations, arXiv:1608.05432
8. Digital and Cognitive Musicology Lab. https://github.com/DCMLab, École Polytechnique Fédérale de Lausanne (EPFL)
9. Bas De Haas, W., et al.: Automatic functional harmonic analysis. Comput. Music. J. **37**(4), 37–53 (2013)
10. Edelsbrunner, H., Harer, J.: Persistent homology – a survey. Discr. Comput. Geom. **453** (2009)
11. Edelsbrunner, H., Harer, J.: Computational topology: an introduction. Appl. Math. Am. Math. Soc. (2010)
12. Ghrist, R.: Computing persistent homology. Bull. Ame. Math. Soc **45**, 61–75 (2009)
13. Giansiracusa, N., Giansiracusa, R., Moon, C.: Persistent homology machine learning for fingerprint classification, pp. 1219–1226 (2019). https://doi.org/10.1109/ICMLA.2019.00201
14. Harte, C.: Towards Automatic Extraction of Harmony Information from Music Signals, PhD thesis, Queen Mary, University of London (2010)
15. Centre for Digital Music, Annotations: The Beatles. isophonics.net/content/reference-annotations-beatles
16. Milne, A.J., Holland, S.: Empirically testing Tonnetz, voice-leading, and spectral models of perceived triadic distance. J. Math. Music (2016)
17. Pun, C.S., Lee, S.X., Xia, K.: Persistent-homology-based machine learning: a survey and comparative study. Art. Intell. Rev. **55**, 5169–5213 (2022)
18. Oter, N., Porter, M.A., Tillmann, U., et al.: A roadmap for the computation of persistent homology. EPJ Data Sci. **6**, 17 (2005)
19. Pedregosa, F., et al.: Scikit-learn: machine learning in python. JMLR **12**, 2825–2830 (2011)
20. Schönberg, A.: Structural Functions of Harmony. Ernest Benn Limited, London (1969)
21. Tymoczko, D.: The generalized tonnetz J. Music Theory, **56** (2012)

22. Vietoris, L.: Über den höheren Zusammenhang kompakter Räume und eine Klasse von zusammenhangstreuen Abbildungen. Math. Ann. **97**, 454–472 (1927)
23. Yust, J.: Generalized Tonnetze and Zeitnetze, and the topology of music concepts. J. Math. Music **14**, 170–203 (2020)
24. Yust, J., Lee, J., Pinsky, E.: A clustering-based approach to automatic harmonic analysis: an exploratory study of harmony and form in mozart's piano sonatas, transactions of the international society for music. Inf. Retrieval **5**, 113–128 (2022)
25. Zomorodian, A., Carlsson, G.: Computing persistent homology. Discrete Comput. Geom. **33**, 249–274 (2005)

Melodic Contour Generation with Spline Models of Cycles

Matt Klassen[1(✉)] and Paul Lanthier[2]

[1] DigiPen Institute of Technology, Redmond, WA, USA
mklassen@digipen.edu
[2] Unilasalle Rouen, Rouen, France

Abstract. In this paper we present techniques to generate melodies which follow the contours coming from segments of audio signal graphs. We refer to these segments as cycles, which we model with cubic splines, focusing on the case where the audio signal has recognizable fundamental frequency. These spline models of cycles are sampled in various ways to give a sequence of values which are used to define logarithmic pitch deviations from some chosen central pitch. Melodic fragments are constructed using cycles extacted from recorded wave forms, and these same cycles are also used to construct the tones of the melodic fragments as audio segments with given pitch and duration. In this way, the musical elements of pitch, duration, and timbre are captured in the context of spline modeling of cycles.

Keywords: Spline · Audio · Signal · Melody · Cycle

1 Introduction

Central to most forms of music composition is the notion of a sequence of pitches, or melody, chosen from some set of possible pitches, or pitch space. Pitch is often equated with fundamental frequency, which we adopt in this work, and which we refer to as f_0. We do this with the understanding that in the case of musical instrument samples f_0 is not well defined, but can be approximated by various methods, such as the Discrete Fourier Transform. This can be done since the waveforms of such instrument samples are approximately periodic, and so an approximate period or cycle length can be identified. The non-periodicity should be understood not as an accumulation of error in physical measurement, but rather as a fundamental non-linearity in the generation of sound with physical objects such as strings, plates, and resonating cavities.

Pitch spaces are often discrete, as is the case with the modern (equal-tempered) piano whose 88 keys take on one of the fundamental frequencies given by:

$$f_0(k) = 440 * 2^{\frac{k}{12}}, \quad k = -48, \ldots 39.$$

Supported by DigiPen Institute of Technology.

© The Author(s), under exclusive license to Springer Nature Switzerland AG 2024
T. Noll et al. (Eds.): MCM 2024, LNCS 14639, pp. 210–223, 2024.
https://doi.org/10.1007/978-3-031-60638-0_17

If we think of duration of musical notes along an x axis, (with an appropriate minimal subdivision) and the above values of k along a y axis, we can plot melodies on the piano as dots on a grid. If we replace this regular grid with continuous time and pitch, then we can interpret curves as giving continuous time and pitch graphs. We do this with graphs of cubic splines which come from representations of cycles for waveforms. By cycle we mean loosely an interval along the time axis of approximate length given by $1/f_0$ where f_0 is an approximate value of fundamental frequency, together with the sample values of the waveform on that interval. It is convenient to define cycles with endpoints given by zero crossings, but this is not required. (See [1] for discussion.) Since our methods are dependent on computing f_0, we devote Sect. 8 to computation of f_0, in particular to those techniques which use the notion of cycles.

In addition to pitch sequences, melodies also imply some choice of note durations. As with pitch space, we can use simple regular subdivision of some fixed time duration or we can compute time durations in other ways. In this work we use the spline models of cycles to compute a sequence of time durations of notes, either with y values of the spline, in a similar way that we arrive at a sequence of pitches or f_0 values, or with stationary points of the spline, where the first derivative is zero. In the latter case, we use the y values for pitch and the intervals along the x axis between stationary points for the time durations.

Generation of melodies in continuous pitch space suggests that intervals between pitches may be microtonal, which raises the question of whether or not the pitches generated are following any particular scale or discrete pitch system. In this work we replace the traditional notion of scale system with the generation of pitches with spline graphs. Since the spline graphs are coming from cycles of waveforms, we are connecting the generation of melody to the generation of sound. Melody typically occurs at the time scale of seconds, whereas one cycle of length $1/f_0$ occurs at the time scale of milliseconds (for example 1/440 is less than 3 ms). See Sect. 7 for a discussion of the connection between timbre and the generation of sound samples with cycle interpolation as an embedding into high dimensional space.

In this work we present examples of melodies which have a certain *self-reflective property*, which we can state as follows: The individual notes in the melody are waveforms which can be described as each having some fundamental frequency (or pitch) $f_0(k)$ for some sequence of indices $k = 0, \dots, m$, such that the graph or contour of these pitches $f_0(k)$ matches the graph or contour of the cycles inside these waveforms. To be more precise, suppose also that each notes lasts for some time duration t_k and the first note starts at time $t = 0$. Then we can plot the points (x_k, y_k) where $x_{k+1} - x_k = t_k$ and

$$y_k = \log_2 \left(\frac{f_0(k)}{f_0(0)} \right).$$

If we sketch a curve through these points we can say that we have a graph of the "*melodic contour*" represented by this sequence of pitches. Now suppose we examine the waveform of one of the notes in the melody. As we zoom in on the waveform we see a repeating pattern of cycles which occur at the approximate rate of $f_0(k)$ cycles per second. If we focus on one cycle, we see that the shape of that cycle appears to be quite similar to the melodic contour described above, up to scaling of the axes. Since there is a strong connection between local cycle representations of a waveform and the resulting timbre of the sound of that waveform, we have exhibited a way to link melody and timbre in music in a mathematical or computational way, which we do not find in existing musical examples. We do find that graphical representations of existing melodies tend to have less variation than random pitch sequences. Also, we can say that it is possible to perceive more variation in a melodic contour in a similar way that we can hear more variation in a repeating cycle of a waveform which results in a richer spectrum and timbre.

The innovation of crafting melodies from the fundamental structure of a single waveform is a contemporary extension of a time-honored tradition in Western music. Historically, composers have often taken a motif, a subject, or a theme – a musical "DNA" – and explored its transformative potential through variation, development, and elaboration. This thematic development is akin to a fractal process, where a simple figure yields infinite complexity and beauty through self-similar expansion across different scales and contexts. Our methodology infuses this concept with a modern perspective by treating waveforms not just as sonic building blocks, but as embryonic ideas containing within their oscillations the seeds of elaborate musical expressions.

2 Related Work

Timbre has been defined by the Acoustical Society of America in (ASA) Acoustical Terminology definition 12.09 which describes it as "that attribute of auditory sensation which enables a listener to judge that two nonidentical sounds, similarly presented and having the same loudness and pitch, are dissimilar", adding, "Timbre depends primarily upon the frequency spectrum, although it also depends upon the sound pressure and the temporal characteristics of the sound". (See [5] and [6] for more discussion.) Timbre is thus considered to be independent of pitch, duration, and loudness. Each of the latter three musical elements has a simple proxy, even though it is in each case not the full story. For example, in the case of pitch, the most commonly associated quantity is fundamental frequency f_0 measured in Hz. Though pitch is a psychoacoustic property, with dependencies on existence of overtones and harmonicity in the spectrum, we use the proxy f_0 freely. There is, however, no such proxy for timbre, at least not in the form of a single number.

Timbre is typically related to spectrum, and in the work of Sethares (see [7]) timbre is also related to the notion of scale system. He points out the relationship between dissonance, beats, and harmonics, and how the harmonic spectrum can be used to make choices for intervals in a scale system. Although these are very interesting relationships, there is not a direct connection between timbre and generation of melody.

A different direction, beginning with Xenakis' work on UPIC and extended to UPISketch (see [8]) focuses on graphical generation of pitch with computer software. Our work is inspired by this, and can be thought of as replacing the system of user-defined graphs, or sketches, with a system of graphs generated by audio signals. In the same vein, Collins' work on SplineSynth [9] allows the user to change the timbre of a sound being generated by one spline cycle through the use of control points, but this is not extended to melody generation.

In a third direction, splines are often used to generate envelopes for synthesized sound, in works such as [10]. The notion of envelope plays an important role in timbre, since the user can often distinguish instruments partly due the natural attack and decay coming from the physical sound production process. Envelope can be seen as distinct from spectrum, since the shape of individual cycles occurs at a smaller time scale. Although envelope occurs at a similar time scale to melody (in seconds) it is distinct from melody since the latter implies a sequence of pitch changes, whereas envelope is typically applied to one pitch.

3 Summary of Spline Modeling of Audio Signals

For more details and background to this section, please refer to [1,3] and [4].

As in [1], we normalize one cycle to be given as a function on the interval $[0, 1]$. We work with C^2 cubic splines f on $[0, 1]$, which are defined by a sequence of cubic polynomials $p_i(t)$ on a sequence of k subintervals $[u_i, u_{i+1}]$ with *breakpoints* $0 = u_0 < u_1 < u_2 < \cdots < u_{k-1} < u_k = 1$. The dimension of the vector space V of such spline functions $f(t)$ is $n = k + 3$. Each f is then a linear combination of n B-spline basis functions:

$$f(t) = c_0\mathcal{B}_0(t) + \cdots + c_{n-1}\mathcal{B}_{n-1}(t).$$

Choosing n data points on $[0, 1]$ to be sufficiently spread out, we can then solve for such an f as an interpolating spline, according to the Schoenberg-Whitney Theorem (see [4]). In the case of cycles for an audio signal $x(t)$, we choose points (t, s) with inputs $t \in [0, 1]$ and outputs $x(t) \in [-1, 1]$ obtained from the audio data function. The B-spline basis is determined uniquely by choice of a *knot sequence* containing the subinterval breakpoints $u_1, \ldots, u_{k-1}$ and another $d + 1 = 4$ values on either end. See [3] for details on bases and knot sequences. Here we use the common knot sequence

$$\mathbf{t} = \{t_0, \ldots, t_N\} = \{0, 0, 0, 0, \frac{1}{k}, \frac{2}{k}, \ldots, \frac{k-1}{k}, 1, 1, 1, 1\}$$

which has the property that endpoint values and first derivatives are encoded by the first and last pairs of B-spline coefficients. It is natural to place interpolation points at the subinterval endpoints, which gives $k + 1$ conditions. Two more are needed for a unique interpolant, since the dimension is $n = k+3$. We place these conditions at the midpoints of the first and last subintervals. We summarize the choices of data (see [1]) for one cycle spline model here:

- degree d = 3
- number of subintervals k
- subinterval sequence: $[u_i, u_{i+1}] = [\frac{i}{k}, \frac{i+1}{k}]$, $i = 0, \ldots, k-1$
- knot sequence: $\mathbf{t} = \{t_0, \ldots, t_N\} = \{0, 0, 0, 0, \frac{1}{k}, \frac{2}{k}, \ldots, \frac{k-1}{k}, 1, 1, 1, 1\}$
- dimension of vector space V of B-spline functions: $n = k + 3 = N - 3$
- B-spline basis functions: $\mathcal{B}_0(t), \mathcal{B}_1(t), \ldots, \mathcal{B}_{n-1}(t)$
- interpolating spline: $f(t) = c_1\mathcal{B}_1(t) + \cdots + c_{n-2}\mathcal{B}_{n-2}(t)$ $(c_0 = c_{n-1} = 0)$
- input values: $s_0 = 0, s_1 = \frac{1}{2k}$, $s_i = u_{i-1}$, $i = 2, \ldots, n-2$, $s_{n-2} = 1 - \frac{1}{2k}$, $s_{n-1} = 1$
- target (signal) values: $x(s_0), x(s_1), \ldots, x(s_{n-1})$

Rather than working with a spline model of each cycle, it is more efficient to save particular cycles which give a good representation of the signal locally. This means that cycles may be placed irregularly for a good match, often more densely at the beginning of an instrument sample. These are called *key cycles.* To reconstruct an approximation of an audio waveform, simple linear interpolation between B-spline coefficients of cycles can be used.

The use of cubic splines for cycle modeling is motivated from several perspectives. First, there is the variational property which is also called the *smoothest interpolation property*, which states that among all C^2 functions g on some interval $[a, b]$ satisfying n interpolation conditions (on say a sequence of regular $n-2$ subinterval endpoints, plus a first derivative at a and b) the C^2 cubic spline achieves the minimum possible value of the energy integral: (see [11], p. 320)

$$\int_a^b g''(x)^2 \, dx.$$

Second, the cubic spline has the advantage of being easy to compute with, using the DeBoor algorithm which is nested linear interpolation using 4 B-spline coefficients determined by the placement of input t and the overlapping supports of the nearest B-spline basis functions containing t. Third, cubic polynomials have a full spectrum of Fourier coefficients c_m which decay as $1/\text{m}^3$, so with minimal data can provide a rich basis for audio signal approximation.

4 Melodic Generation with One Sinusoidal Cycle

To fix ideas, we first describe the method of pitch sequence generation from a waveform which is sinusoidal, say at frequency 440 Hz. One cycle then has the shape of one period of the sine function. One could say that this waveform is devoid of timbre, with spectrum having only one value at frequency 440 Hz. For the melodic generation we normalize the cycle to have length 1. In order to generate a melody we use the contour of the cycle graph to represent a sequence of pitches at sampled points. Sampling the cycle can be done at regularly spaced subintervals, for example with x values 0, 1/4, 1/2, 3/4, 1. Next we consider scaling to obtain both time durations and pitch range. For example, suppose that we use time duration of each note in the melody to be 1/4 s. Then the time durations will correspond directly to the distance between successive x values of the chosen sampled points. Next, suppose we scale the y axis to obtain a two octave spread around 440 Hz. In other words $y = 0$ corresponds to $f_0 = 440$ Hz, $y = 1$ is 880 Hz, and $y = -1$ is 220 Hz. This is obtained by using the exponential mapping of the interval $[-1, 1]$ given by

$$f_0 = 440 * 2^y.$$

Since note durations are interpreted as intervals on the x axis, we can use the interpretation of *'sample and hold'* from signal processing which gives a piecewise constant graph anchored at the left endpoint of each subinterval. This sampling process results in the pitch sequence given in the following staff, where the quarter note is $MM = 240$ bpm, or 4 notes per second (Fig. 1):

Fig. 1. sinusoidal pitch sequence

If we change the sampling on the x axis to have 8 subintervals, we get some new pitches. In particular, the y value at $x = 1/8$ is $\sin(\pi/4) = \sqrt{2}/2$, which yields

$$f_0 = 440 * 2^{\sqrt{2}/2}.$$

This corresponds to an interval with approximate cent value:

$$\frac{1200}{\ln 2} \ln\left(2^{\sqrt{2}/2}\right) = \frac{1200}{\sqrt{2}} = 848.53$$

which is known as a "*neutral sixth*" since it is about half way between a minor sixth and a major sixth. We can also give an approximate version of this with music notation, but we indicate the neutral sixth simply with a hat symbol to indicate that the given note is raised by a quarter tone, or about 50 cents. Since we are using regular subintervals, and in general we allow for time duration scaling, the time duration is not significant, so here we choose to stick with the same tempo and use eighth notes with duration 1/8 second (Fig. 2).

This process can of course be continued, with further subdivision yielding other intervals. For example, if we use equal division of $[-1, 1]$ into six parts, we have the y value $\sin(\pi/3) = \sqrt{3}/2$,

$$f_0 = 440 * 2^{\sqrt{3}/2}.$$

Fig. 2. sinusoidal pitch sequence with neutral sixth

This in turn corresponds to an interval with approximate cent value:

$$\frac{1200}{\ln 2} \ln\left(2^{\sqrt{3}/2}\right) = 1200\frac{\sqrt{3}}{2} = 1039.23$$

giving a type of neutral seventh. In general, the cent value deviation from the central pitch at $y = 0$ is simply given as $1200y$.

From these examples using only one cycle of a sinusoid, with regular subdivision along the x axis, we see that there are already some microtonal intervals appearing in the derived sequence of pitches. To have an interesting interplay between pitch, duration, and timbre, however, we will need to use a cycle which is not simply sinusoidal.

5 Melodic Generation with One Spline Cycle

For a slightly more involved example, we use a cycle selected from a recorded waveform. In order to avoid something too similar to a sinusoid, we recorded a human voice for about 0.45 s pronouncing "zhzh". The recording uses sample rate 44100 and bit depth 16. We compute a weak f_0 with STFT followed by calculation of cycles based on zero crossings. The f_0 approximation is done on segments of 2048 samples with FFT size 1024 and step size 128. There are 9 segments, with approximate f_0 around 110 Hz. We chose the fourth cycle from the third segment which has length 406 samples. This cycle is graphed here in Fig. 3 with an interpolating cubic spline through $n = 25$ points from the audio data. Most of these numerical choices are standard in audio signal processing and have no particular significance for our methods. The dimension n is however an important parameter and does play a role in the musical elements we generate. The smaller the value of n, the fewer points in the audio graph of one cycle are interpolated by the cubic spline, hence the lower the level of accuracy in the cycle approximation. We think of n as a free parameter to be used in the generation process. Small values of n will produce spline curves which have smaller total variation in the first derivative, and as n approaches the number of samples in the cycle, the curve approaches the audio sample graph. We think of (and compute with) the audio sample graph as piecewise linear, so it is possible to choose more interpolation points for the spline than the number of sample points, but we do not do this in general. On the other hand, we may choose to focus on points in the cycle which are not at audio sample points, such as stationary points, where the first derivative is zero.

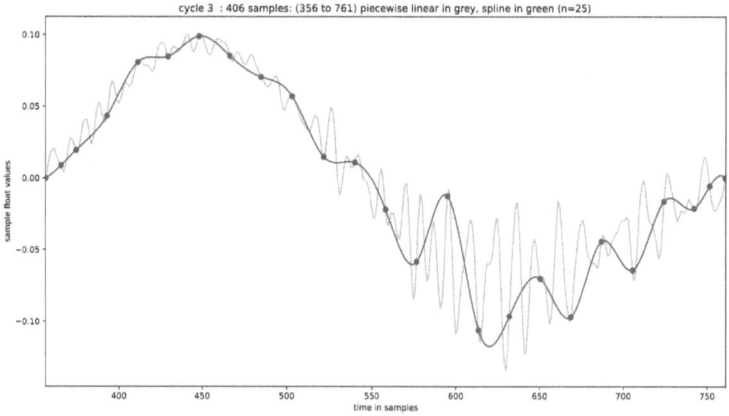

Fig. 3. cubic spline approximation of one cycle

Such a graph is shown in Fig. 4 where we compute approximate first derivative and plot the stationary points in red. For practical reasons we set a lower bound on note duration and delete stationary points which are within 0.02 of the previous one, which happens three times in this example. The resulting set of points (left endpoint plus stationary points) is then:

$$(0,0), (0.156, 0.084), (0.230, 0.099), (0.428, 0.011), (0.539, -0.060), (0.587, -0.012), (0.652, -0.117),$$

$$(0.721, -0.069), (0.768, -0.098), (0.822, -0.043), (0.861, -0.064), (0.918, -0.014), (0.950, -0.022)$$

In Fig. 5 we plot these stationary points and the piecewise constant graph obtained by the sample and hold method. Each line segment then represents the relative note duration (in the x direction) and the relative pitch of that note (in the y direction).

Next, we generate audio waveforms for each of the pitches and time durations in the graph. This requires a choice of central, or starting, pitch which we take to be $f_0 = 220$ Hz. The waveforms are generated in this example using only this one cycle together with a simple envelope. In addition to the melody represented in this graph, we also generate transforms, such as inversion, retrograde, and retrograde-inversion. All of this is currently being done with python scripts. In Fig. 6 we display audio waveforms corresponding to the original melody and the three transforms.

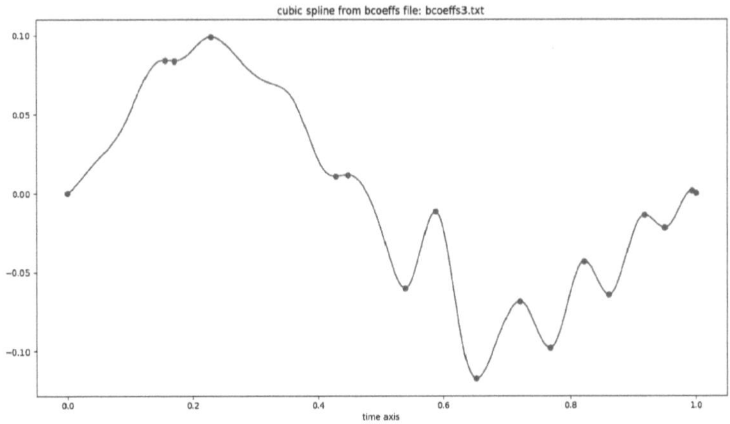

Fig. 4. cubic spline for one cycle with stationary points

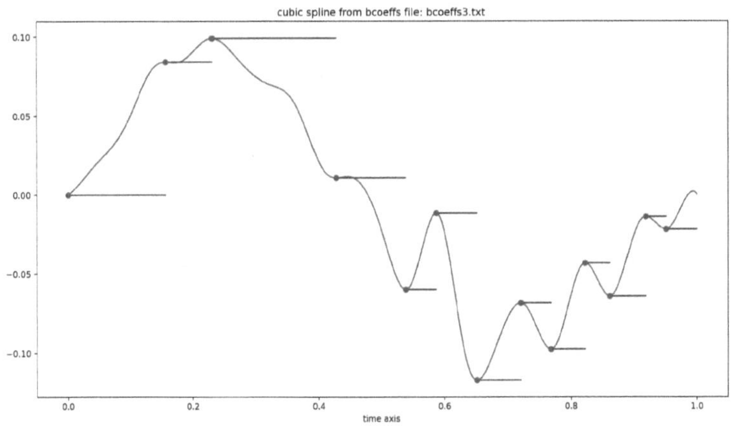

Fig. 5. sample and hold for one cycle with stationary points

Once we have the melody and its three transforms, we have enough basic musical material to arrive at the stage of musical composition. In addition to these fragments, we can also modify these in standard ways such as transposition, time stretching, and pitch stretching. Before going further, we pause to speak about implementation details and how these can fit into a music composition pipeline.

The generative process we have described is implemented with python scripts, which starts with an audio sample and computes an approximate f_0, cycles corresponding to the predicted period length, and B-spline coefficients for each cycle. The next phase is then to choose a cycle (or several cycles) for the generation of melody. Graphs of all the cycles and their B-spline approximations are generated to assist in the choice. The melodic fragment is then generated with

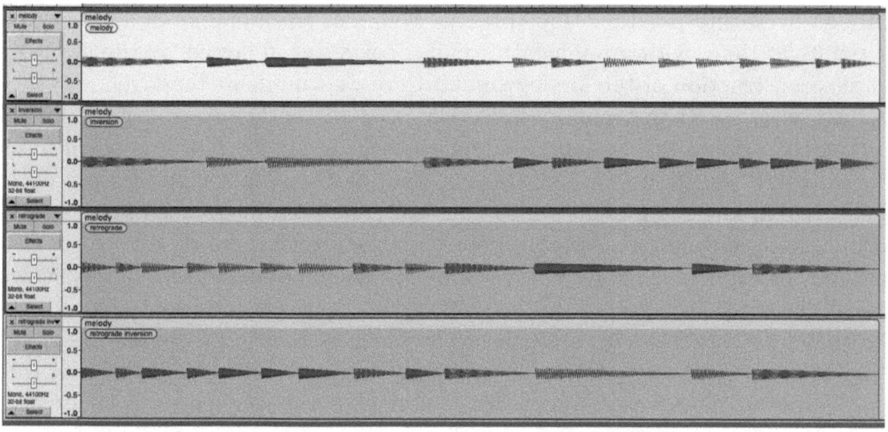

Fig. 6. audio waveforms of melody with three transforms

a chosen cycle by using its B-spline model. The output melody consists of notes which have the correct pitches to match the contour, but also which have the cycle shape matching the same contour. Thus, the melody has the self-reflective property referred to earlier. In our python scripts we also have options to generate each of the three transforms as well as starting pitch, and scaling factors. The scaling factors can be applied to pitch or time duration. The scaling of pitch is applied to y which has the default interpretation that $y = 1$ is an octave above the starting pitch at $y = 0$. The scaling of time duration is based on the first note in the melody which can simply be assigned a time duration in seconds, which scales the entire melody appropriately. In the example in Fig. 4 the maximum y value is around 0.1, so for a melody that extends up to around one octave we use a scale factor of 10. This is also the value used in the transforms which appear in Fig. 6. Combinations of these fragments can be assembled in a DAW (digital audio workstation) to form the skeleton of a musical composition.

Up to this point we have used a single cycle to generate various melodic fragments with transforms. If all of these are based on one cycle, then they all effectively have the same timbre, with varying pitch and time duration of notes. In the next section we address the issue of varying timbre.

6 Melodic Generation with Several Spline Cycles

In the spline models of an instrument sample, one key cycle can be used as a local representation of timbre. To model one-second we found that it is good to have about 16 cycles. Since the first part of the sound might represent an attack phase, it is natural to concentrate more of the key cycles in this part, and fewer towards the end or tail of the sample where cycles are more stable. The change in spectrum over the course of one second is modeled by the change in the shape of the key cycles. The audio sample can be approximately reconstructed through cycle interpolation. This process is similar to sampling an audio signal at points in time, with an analog to digital converter, followed by the approximate reconstruction of the analog signal. Here we sample at the level of cycles, and we select cycles in a non-uniform way. This method captures timbre for an instrument sample as a sequence of numbers, namely the B-spline coefficients of each cycle. We have found that using 33 such B-spline coefficients per cycle is effective in forming models which can easily distinguish between instruments such as flute and guitar. We note here that the sequence of 16 key cycles also contains an implicit envelope, since scaling the B-spline coefficients simply scales the amplitude of the cycles. This is a much more accurate representation of timbre than that obtained by a single cycle which is then repeated as a strictly periodic function, as we did in the previous section. Next we indicate how this representation can be used to construct more elaborate sequences of melodies.

Just as the sequence of key cycles generates a *coherent* audio waveform through cycle interpolation, we can also generate a *coherent* sequence of melodic fragments with the same techniques we have already described applied to each cycle. In this way, the self-reflective property can be seen at the structural level of music composition. This raises the question of what plays the role of interpolation in the case of melodic fragments. It is possible to carry the analogy through in a strict way, by gradually evolving the melodic fragments which then occur at a much larger time scale. For example, an instrument sample at $f_0 = 220$ with melodic fragment of length say 4 s would expand into a sequence of gradually evolving melodic fragments lasting a total of $4 * 220 = 880$ s, or about 14.6 minutes. Instead of this somewhat mechanical process, a composer can take a sequence of melodic fragments derived from key cycles, and apply transforms to those, then arrange these fragments with other supporting musical material. In either case, one could say that in the resulting composition timbral elements are related to melodic elements in a nontrivial way.

Another direction is to use sound samples of distinct timbres within the same composition, say for example two guitar plucks, one with timbre *sul tasto* and one with timbre *sul ponticello*. Or two different instruments, such as flute and cello. Various melodic fragments can be generated and then realized with notes using one of those timbres. If the melody played is directly related to the timbre of the waveforms in the notes, we call this a melodic fragment of *cycle type I*. If the melody is related to the timbre of waveforms occuring in other musical material in the composition then we call this a melodic fragment of *cycle type II*. A special case is a composition in which the featured melodic fragments occur as counter-melodies in pairs, each being derived from the other's waveforms. A third case is a melodic fragment derived from some other source waveform not being used in the composition, of *cycle type III*.

Finally, it is possible to also use the cycle graphs to influence other musical elements, such as structure or even form. For example, the entrances of voices can occur at times derived from the y values in the cycle graph, or the lengths of melodic fragments can be scaled according to the lengths of line segments which were also used to determine note duration values.

Examples of compositions using these ideas as well as many melodic fragments and the python scripts used to generate them are available on the authors' website.

7 Timbral Model with Splines

In [1] and [2] it was demonstrated that instrument samples of length approximately one second could be modeled by a sequence of 16 "*key cycles*, each with $n = 33$ B-spline coefficients. Further, the sounds generated using the model for instruments such as guitar and flute were clearly recognizable to a human listener, and sounds could be blended by simple linear interpolation of B-spline coefficients in the 16 key cycles. This work requires further testing on a large number of instrument models and a rigorous study to measure human recognition of sounds. Here we make a few definitions and ask some questions.

First, define the abstract manifold T to be the set of one second audio samples of human recognizable pitched sounds. To fix ideas, one can assume that these are labelled files (pitch plus description) and that we use a standard audio sample rate of $N = 44100$. This gives a realization of T as a set of points in $\mathbb{R}^N$.

One obvious question is: How big is T in terms of dimension? Further, we can ask what is the convex hull of points in T? To study these questions one would need data, so we can also ask what is the size of currently available data? Given a large amount of data representing points in T, we propose to study T by the embedding into the smaller space of dimension $16 * 33 = 528$ given by using spline models of the audio samples with cycle interpolation. We also note that in this case the convex hull of this embedded space is precisely the set of reconstructed audio samples using linear cycle interpolation (of B-spline coefficients).

8 Computation of Fundamental Frequency f_0

In this paper we have demonstrated an approach to the generation of melody which is dependent on the computation of f_0 and the construction of cycles which are representative of the local timbre of waveforms. Historically there are many different approaches to computation of f_0, and it is interesting to note that some of them use the notions of cycles and zero crossings. It is important to keep in mind that the waveforms we use are typically recorded instrument samples, and are only approximately periodic. So there is no "ground truth" f_0. The methods of Fourier Analysis are only approximately relevant. The use of discrete frequency domain transforms are subject to the uncertaintly principle and are plagued by lack of good frequency resolution. For example, we use the STFT (Short Time Fourier Transform) to compute a weak f_0 with the argmax of magnitude frequency bins, with fft size 1024 and step size 128 samples. For sample rate 44100 this gives a frequency resolution of $44100/1024 = 43$ Hz which is quite indadequate for computational purposes. But as an initial guess, in our computation of cycles, this works reasonably well.

Our process can be described as: 1) Begin with a "weak f_0" computation, and then 2) Proceed to construct cycles. Once we arrive at the sequence of cycles, we are less interested in declaring a new or enhanced value of f_0, but an available candidate can be defined as the reciprocal of the average of cycle lengths. Here we provide a few more details of this process.

In [1] the audio signal was partitioned into cycles based on zero crossings, where a cycle is an interval $[a, b]$ on the time axis such that the piecewise linear function f given by the audio samples satisfies $f(a) = f(b) = 0$. If a (or b) is between consecutive time samples t_1 and t_2 then this is equivalent to assuming that $f(t_1)$ and $f(t_2)$ have opposite sign. The cycles were formed iteratively by simply starting with some zero, say z_0, and then finding the next zero crossing z_1 closest to $z_0 + P$ where P is the "weak period length" predicted by the weak f_0 as $P = 1/f_0$ s. This process gives in many cases a coherent partition into cycles with various lengths close to P.

In this work, we modify and extend these computations in several ways. First, we make a simplifying assumption that the zero crossings are positive, by which we mean that $f(t_1) < 0$ and $f(t_2) > 0$ when a zero crossing a satisfies $t_1 < a < t_2$ for consecutive samples t_1 and t_2. Second, we compute cycles at each available positive zero crossing, allowing for overlaps. This is done in order to maximize the number of "local cycle representatives" which can be used for either local cycle models (such as key cycles) or to compute an average cycle length for some stretch of the audio sample. In summary, this method can be used to compute an enhanced f_0, given a weak f_0 obtained from some initial guess. If the initial guess comes from Fourier methods like STFT, as we describe above, then we can call this enhanced f_0 computation a hybrid time-frequency method.

9 Future Work

It would be interesting to reverse the process described in this paper by taking examples of melodies from existing music and converting these to cycles for the construction of waveforms. Since melodies can be represented as piecewise constant graphs based on f_0 values of pitches in the melody, then these graphs could be interpolated in various ways, including the use of stationary points or regular subdivision into k subintervals with the dimension $n = k+3$ of the C^2 cubic splines matching the number of notes in the selected melody. Once a spline is attached to a melody it is possible to generate waveforms with the associated timbre and also to resample the waveform to create related melodies with different pitch characteristics. Music compositions using these techniques as well as python code can be found at the following link: http://azrael.digipen.edu/research.

References

1. Klassen, M.: Spline modeling of audio signals and cycle interpolation. In: Montiel, M., Agustín-Aquino, O.A., Gómez, F., Kastine, J., Lluis-Puebla, E., Milam, B. (eds.) Mathematics and Computation in Music. LNCS, vol. 13267, pp. 155–167. Springer, Cham (2022). https://doi.org/10.1007/978-3-031-07015-0_13

2. Klassen, M., Lanthier, P. : Design of timbre with cellular automata and B-spline interpolation. In: Michon, R., Pottier, L., Orlarey, Y. (eds.) Proceedings of the 19th Sound and Music Computing Conference (SMC 2022). SMC Network, pp. 485–492 (2022). ISBN: 978-2-9584126-0-9 EAN: 9782958412609
3. Klassen, M.: Lecture notes on Bezier curves and polynomial splines. http://azrael.digipen.edu/notes/
4. de Boor, C.: A Practical Guide to Splines, revised Springer, New York (1980)
5. https://en.wikipedia.org/wiki/Timbre
6. Acoustical Society of America Standards Secretariat: Acoustical Terminology ANSI S1.1-1994 (ASA 111–1994). ANSI / Acoustical Society of America, American National Standard (1994)
7. Sethares, W.A.: Tuning Timbre Spectrum Scale, 2nd edn. Springer, London (2004)
8. Bourotte, R., Kanach, S.: UPISketch: the UPIC idea and its current applications for initiating new audiences to music. Organised Sound **24**(3), 252–260 (2019)
9. Collins, N.: SplineSynth: an interface to low-level digital audio. In: Proceedings of the Diderot Forum on Mathematics and Music, pp. 49–61. Vienna (1999). ISBN 3-85403-133-5
10. Ardaillon, L., Degottex, G., Roebel, A.: A multi-layer F0 model for singing voice synthesis using a B-spline representation with intuitive controls. Interspeech 2015, Sep 2015, Dresden, Germany. hal-01251898v2
11. Davidson, K.R., Donsig, A.P.: Real Analysis with Real Applications, Prentice Hall (2002)

Geometric Approaches to Musical Algorithms and Microtonality

I-Shaped Tiles in the *Tonnetz*

M. A. Coury-Hall(✉)

New York City, NY, USA
mh2468@aol.com

Abstract. An I-shaped tile proposed in this paper is used to construct an enharmonically conformed *Tonnetz* space that preserves more music-theoretic information related to tonality, arguably the original purpose of a *Tonnetz*, than the standard parallelogram tiles currently used in the literature. The resulting tiling is monohedral, monomorphic and enantiomorphic, giving music-theoretic meaning in particular to the non-horizontal tile boundaries that cross fifths-related regions.

Keywords: *Tonnetz* · Tiling · Diatonic · Tonality · Myhill's property · Spelling Consistency

1 Introduction

Candace Brower [1] uses the mathematical concept of tiling in the *Tonnetz* to explore the geometries of triadic pitch-class space through image schema. As discussed previously by this author [10,11], the full algebraic properties of Brower's tiling is undermined by music theory's reliance on parallelogram-shaped regions within the *Tonnetz* which Brower organizes into three collections of tiles: expanded key space, hexatonic space, and octatonic space. In her essentially geometric approach, tile boundaries lose musical meaning, a feature endemic to tile collections that form parallelograms because these tilings are not intrinsically monomorphic in the plane without some additional assumptions such as edge-to-edge tiling. This paper considers a new construction of the *Tonnetz* in a spelled pitch-class space based on an I-shaped tile whose geometry gives music-theoretic and mathematical meaning to these regions through the tile's shape and boundaries.

2 An I-Shaped Tile: A Music-Theoretic Geometry

Figure 1 proposes a new tile shape for tiling the *Tonnetz*.[1] The new tile, an I-shaped tile, has the following geometry in the plane:

[1] Tilings in the *Tonnetz* are considered in detail by Brower [1]. Hook [14] also considers the topic in the context of exploring musical spaces. For histories of the *Tonnetz*, see Cohn [5,6]. The specific version used in this paper, a *Tonnetz* with tilted rather than rectangular axes, follows Tymoczko [17] and originates with Otakar Hostinský (1879).

M. A. Coury-Hall—Independent Scholar.

© The Author(s), under exclusive license to Springer Nature Switzerland AG 2024
T. Noll et al. (Eds.): MCM 2024, LNCS 14639, pp. 227–241, 2024.
https://doi.org/10.1007/978-3-031-60638-0_18

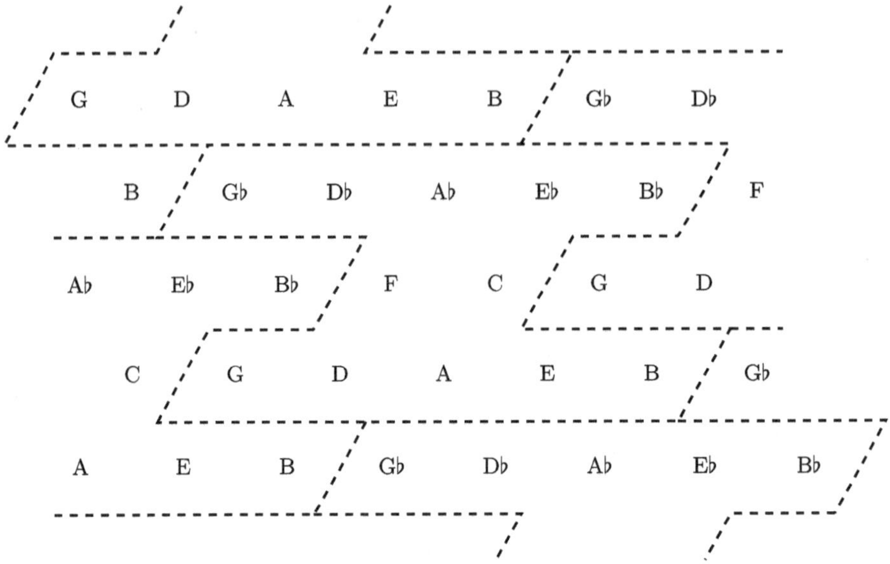

Fig. 1. Sheared *Tonnetz* space: an enharmonically conformed *Tonnetz*

1. **Dodecagon.** A simple concave polygon with twelve sides. Six sides are horizontal; six are 30° off vertical, denoted in this paper as "tilted." Horizontal sides alternate with tilted sides around the perimeter.
2. **Interior Angles and the Four Equivalence Classes of Quotients.** Among the interior angles, there are four reflex angles ($180° < \theta < 360°$), four obtuse angles ($90° < \theta < 180°$), and four acute angles ($0° < \theta < 90°$). Two reflex angles are 240° and two are 300°; all obtuse angles are 120°; and all acute angles are 60°.[2] All interior angles are divisible by 60°, and the quotients belong to one of four congruence classes (mod 6).[3]
3. **First-Level Triangulation: Equilateral Triangle Partitions.** There are 24 equilateral triangles that are congruent (equal by translation or reflection) that partition the *I*-shaped tile as follows: ten in the upper crossbar; ten in the lower crossbar; and four in the middle section. All equilateral triangles

[2] Starting in the upper right corner moving clockwise, the interior angles are: 60°, 120°, 240°, 300°, 60°, 120°, 60°, 120°, 240°, 300°, 60°, and 120° respectively. The sum of all the interior angles is 1800°; the same sum as the interior angles of a regular dodecagon.

[3] When the interior angles are partitioned further based on their rotational symmetry (Item 7.), they belong to one of eight congruence classes (mod 12). The SUM classes of the consonant triads also belong to the same congruence classes (mod 12).

have a horizontal base with half of the equilateral triangles oriented up from the horizontal and half oriented down.[4]

4. **Calissons Partitions.** The first-level regular triangular tessellation described above can be organized into twelve pairs of equilateral triangles such that each pair forms a rhombus containing a 60° angle, called a calisson after the French candy. See Fig. 2. These calissons are congruent and composed of oppositely oriented equilateral triangles. The calisson center is labeled with the *Tonnetz* graph node circled.[5]
5. **Second-Level Triangulation and the *Tonnetz*** *Graph.* Each equilateral triangle can be partitioned further into four smaller equilateral triangles, each of whose bases is half the base of the larger triangle, forming another regular triangular tessellation. The *Tonnetz* graph is a subgraph of this additional partitioning, coinciding with alternating horizontal and alternating SW-NE boundary lines. See dashed lines in Fig. 3 for a subgraph of this *Tonnetz.* Shading in this instance distinguishes calissons from neighboring ones.
6. **Jigsaw Puzzle Tabs and Blanks.** There are four tabs and two blanks, following the nomenclature of jigsaw puzzle pieces.[6] Unlike a standard puzzle, the interlocking of the *I*-shaped tiles do not form a vertical-horizontal grid but represent a sheared lattice.
7. **Rotational Symmetry of Order 2.** There exists a center point around which the *I*-shaped tile has rotational symmetry. The angle of rotation is 180°, so the rotation is an involution with the center of rotation acting as a fixed point. Six calissons define an axis of symmetry.[7]

The motivation for developing this *I*-shaped tile and its *Tonnetz*-related decompositions is to construct a geometric object that restricts possible translations and reflections of the figure in order to better represent musical properties

---

[4] These triangles are offset by translation from the equilateral triangles that form chords. For comparison, see pp. 28 and 93 in Hook [14] for standard examples of triangular tessellation in the *Tonnetz* that represent chords. The geometry of this first-level partitioning will be used to discuss: 2/3-8/9 axis of symmetry, functional music-theoretic boundaries, and shadings within tonal regions.

[5] In the example, two angles of the G-node calisson from Fig. 2 form interior angles of the *I*-shaped tile given in Fig. 1: $\measuredangle XYZ = 60°$ (a defining angle of the calisson) and $\measuredangle YZW = 120°$.

[6] Mathematicians describe a polygonal "ear" as a triple of successive vertices x_{i-1}, x_i, x_{i+1} of a simple polygon P such that $[x_{i-1}, x_{i+1}]$ is a diagonal that lies in the interior of the polygon. If the diagonal $[x_{i-1}, x_{i+1}]$ is in the exterior, then it is called the polygonal "mouth." Successive ears form tabs; and successive mouths form blanks.

[7] With this 2/3-8/9 axis of symmetry, the interior angles can be partitioned further into two equivalence classes. By convention, interior angles of simple polygons are measured clockwise, so they are always positive with the arc pointing clockwise as shown in Fig. 2. There are music-theoretic reasons based on the *I*-shaped tile's rotational symmetry, to measure the interior angles in reference to the horizontal rays uniquely associated to each interior angle. Such measurements result in some negative-valued interior angles but this paper follows the standard convention.

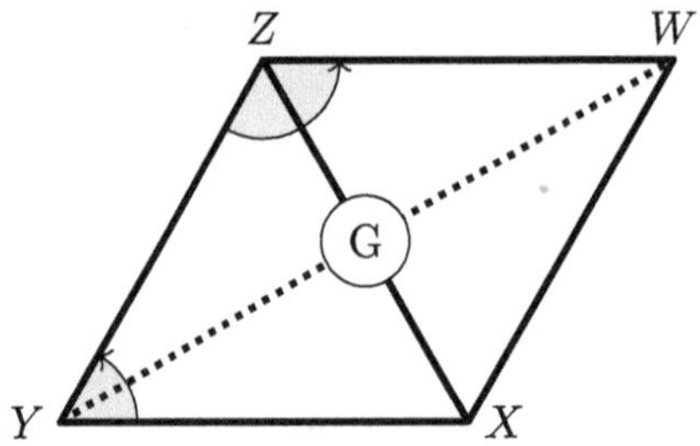

Fig. 2. Example of calisson sub-tile from the I-shaped tile in Fig. 1 composed of two oppositely oriented equilateral triangles. Calisson center labeled with spelled pitch-class *Tonnetz* node G.

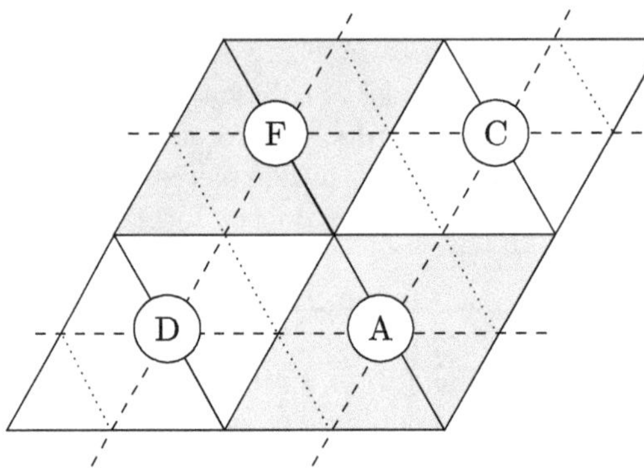

Fig. 3. *Tonnetz* subgraph (dashed lines) positioned with respect to calissons. *Tonnetz* graph nodes are circled. First-level (solid lines) and second-level (solid, dashed, and dotted lines) partitioning shown. *Tonnetz* graph edges are the dashed lines that coincide with second-level partitioning.

in the *Tonnetz* and to preserve these properties as invariants under rigid transformations in the plane. Thus, twelve vertices and twelve edges of the tile can be used to organize music-theoretic properties such as pitch classes, major triads, minor triads, etc. In the next section, a discussion of tiling basics can help understand how the I-shaped tiles "fit together" to result in a tiling of the plane that preserves the *Tonnetz*. After this discussion, a spelled pitch-class space will be discussed followed by some basic musical properties that apply to the musical example.

3 An I-Shaped Tile: Tiling Basics

The I-shaped tile in Fig. 1, contained within the dotted lines, results in a monohedral and monomorphic tiling. Grünbaum and Shephard [12], a standard reference on tiling, provide examples of this kind of tiling, and their results are utilized throughout this paper. In order to make the results more rigorous, the following definitions are noted from Grünbaum and Shephard: Let a **tile** T be a closed topological disk in the plane, and further let a tiling, denoted $\mathcal{T}$, be a countable collection of tiles T_i with disjoint interiors whose union is the entire plane. Hence, $\mathcal{T} = \{T_1, T_2, \dots\}$ and $\bigcup_{i \in \mathbb{N}} T_i$ a plane.

If each $T_i \in \mathcal{T}$ is of the form g_i for some planar isometry g_i, which defines these tiles as being congruent to each other, then the tiling $\mathcal{T}$ is called **monohedral**. Let $\mathcal{S}(\mathcal{T})$ be the set of shapes in a tiling $\mathcal{T}$ such that no two shapes are congruent to each other in $\mathcal{S}(\mathcal{T})$ and every $T_i \in \mathcal{T}$ is congruent to one of the shapes in $\mathcal{S}(\mathcal{T})$. The elements of $\mathcal{S}(\mathcal{T})$ are called **prototiles**. The set of prototiles for a tiling $\mathcal{T}$ is not unique but the cardinality of this set is well defined. Hence, a monohedral tiling has one and only one prototile in the set of shapes. A tile that admits a monohedral tiling is said to be **monomorphic** provided that any two tilings are directly or reflective congruent and all such tilings are considered to be equal.[8]

4 Spelled Pitch-Class Space and the Marked *I*-Shape Tile

As used in this paper, a spelled pitch-class is a named note (e.g., a letter from A to G) with its accidentals belonging to an unconformed space generated by fifths, called the line-of-fifths, following Hook [14], in which every note that is separated by twelve positions on the line is enharmonically equivalent. See Fig. 4. "This space organizes all standard note names (without register designations) into a single fifth-generated line, 'sharper' notes appearing in one direction, 'flatter' in the other."[9] When enharmonically equivalent pitches are identified, the space is often "rolled up" to form an enharmonically conformed fifths space such as the circle of fifths.

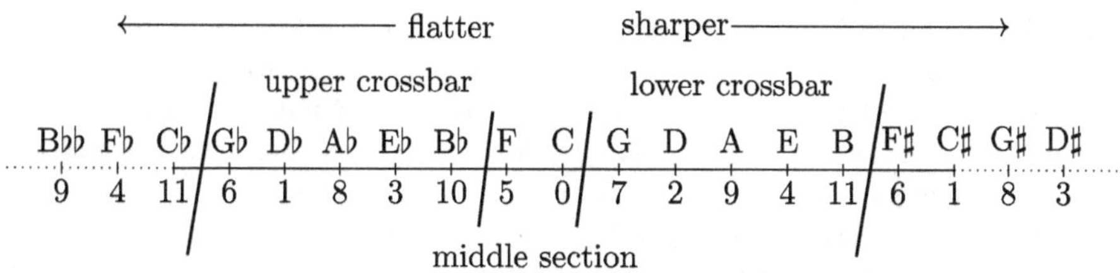

Fig. 4. Line-of-fifths: unconformed spelled pitch-class space (upper list) with standard integer pitch-classes modulo 12 (lower list) and tilted *I*-shaped tile boundaries across fifths-related regions.

In an *I*-shaped tile, the twelve spelled pitch-classes again represent twelve consecutive positions on the line-of-fifths. The tiling in Fig. 1 contains one set of

[8] Equal tilings are defined by Grünbaum and Shephard [12] in Sect. 1.1 in which there exists a similarity transformation of the plane that maps one tiling onto the other.

[9] For further discussion, see Sect. 1.3 and Fig. 1.3.3 in Hook [14]. This space is spelling consistent, so generic intervals such as a fifth reflect ordering of letters by a fifth. B moves to F♯, not G♭.

these representatives.[10] See Fig. 4 for the "unpacking" of these representatives onto the line-of-fifths: the first five of the consecutive spelled pitch-classes correspond to the pitches from the upper crossbar of the I-shaped tile; the next two, from the middle section; and the last five, from the lower crossbar. Let Fig. 1 given by the I-shaped tile represent a small portion of the tiling $\mathcal{T}$. With the representatives labeled on each tile, the tiling $\mathcal{T}$ is a special case of a marked tiling.[11] Each tile in Fig. 1 connects to the next set of twelve spelled pitch-classes geometrically so as to preserve the network with this process repeated throughout the tiling. See Sect. 2 for precise locations of nodes within calissons. Thus, the tiling $\mathcal{T}$ monomorphically constructs the *Tonnetz* network in its entirety. The edges between *Tonnetz* nodes are omitted in order to avoid clutter in the diagram. The line-of-fifths is preserved horizontally from left to right on each row, and the lines of major and minor thirds are also preserved in their respective directions. With the *Tonnetz* labels marking the tiles, there are four distinct tilings that preserve the *Tonnetz* and infinitely many that do not. These four tilings are **enantiomorphic**, i.e., mirror images, so each can be paired with its mirror image. All tilings, however, are congruent by rotation or reflection, so the tiling $\mathcal{T}$ is considered to be monomorphic.[12]

5 A Brief Comparison: Parallelogram Tiles

For comparison, Hook divides his *Tonnetz* space, labeled in his work [14] as Fig. 1.4.2, into partitions of three-by-four rectangular modules that he calls tiles. See Fig. 5. Hook is following the standard mathematical approach for defining the shape of a fundamental domain which, by convention, requires the set to be "reasonably nice" topologically; made mathematically rigorous as either smooth or polyhedral. Hook's tile is an example of a fundamental polygon, one of the "nice" choices. A fundamental polygon can be defined for every compact Riemann surface C of genus greater than zero, so it is a natural choice from a mathematical viewpoint.[13] For example, for the case of genus one (a surface conformally equivalent to a torus C/Λ for lattice Λ in C), the fundamental polygon for this lattice is either a period parallelogram or a centrally symmetric hexagon once the polygon is assumed to be convex.[14] By assuming convexity, Hook has to choose a tile that takes one of two shapes in the *Tonnetz*, forming his lattice Λ in C.

[10] These twelve representatives are repeated through each tile to produce a conformed spelled pitch-class space which also corresponds to a conformed chromatic space.

[11] A tiling in which there is a marking or motif on each tile. See p. 28 of Grünbaum and Shephard [12].

[12] Monomorphic tilings are characterized by tile shapes, not tile labelings.

[13] But not necessarily from a music-theoretic viewpoint. This paper argues that there is another choice of polygon: one that is concave.

[14] A result first proved by Fedorov in 1885. See Marjorie Sennechal and R. V. Galiulin [15] for a thorough analysis and summary in French and English of Federov's 1885 work "Nachala Ucheniya o Figurakh" (An Introduction to the Theory of Figures).

B	D	F	A♭	B	D
G	B♭	C♯	E	G	B♭
E♭	F♯	A	C	E♭	F♯
B	D	F	A♭	B	D
G	B♭	C♯	E	G	B♭

Fig. 5. Enharmonically conformed parallelogram tile over an unsheared *Tonnetz* space, modeled after Fig. 1.4.2 in Hook [14]. Note the lines-of-fifths are SW-NE, not horizontal.

The problem with this choice of a convex fundamental polygon is immediately acknowledged by Hook: the boundaries do not have any musical meaning. The terminology "false boundary," used by Hook, is slightly misleading with respect to tiling and tessellation, especially as Hook's discussion moves between different types of spaces: an infinite plane, a tiled plane, and a quotient space (Figs. 1.4.1, 1.4.2, and 1.4.3 respectively in Hook [14]). The tile boundaries, which define the tiling of the plane being explored in this paper, are not expected to be "true" boundaries in the quotient space which has the well-known topology of the torus. In the quotient topology, the boundaries of the fundamental domain can be understood informally as "false" boundaries of some tile where they become the "seams from gluing together" points that belong to the same equivalence classes, a mapping used to construct the new topological space. The space of the quotient construction in this case is mathematically distinct from the space used for tiling: the Euclidean plane; and as a result, each space defines these boundaries differently (or not at all) based on the context.

6 Some Music-Theoretic Features of the *I*-Shaped Tile

This section covers some of the music-theoretic properties represented by the *I*-shaped tiles befitting an introduction. This discussion and its heuristics should not be considered exhaustive but represents a few music-theoretic properties that will be used in the analysis of the musical example (Sect. 7).

The Lower 0-Accidentals and Upper 5-Flat Parts of the Diatonic Collection. In the example given in Fig. 1, the lower crossbar and the middle section of the *I*-shaped tile contains the spelled pitch-classes of the C major scale. These two lower tile rows represent the subset of the diatonic collection denoted as DIA_0 in Straus' notation [16]: hence, C Ionian (major), D Dorian, E Phrygian, F Lydian, G Mixolydian, A Aeolian (natural minor) and B Locrian. These are all the diatonic scales (Forte 7–35) that contain no sharps and no flats in the key signature, which Straus' notation organizes through his indexing.[15]

Because the tiling with the I-shaped tile is monomorphic, the geometry of the boundaries of any one of the tiles defines the properties of the tiling. Figures 6 and 7 are introduced. Both figures use shading as a visual aid with two adjacent tiles outlined in order to study the interlocking of individual tiles like puzzle pieces. For Fig. 6, it is helpful to think of the keys of the piano. The white keys of the piano are unshaded in the figure, which represent the spelled pitch-classes of C major. The black keys are shaded. The white keys belong to the lower crossbar and the middle of the tile, so these regions will be called the lower major scale or, more formally, the part of the diatonic collection associated to the lower region of the I-shaped tile. In this example, the lower major scale is C major, and the lower part of the diatonic collection is DIA_0. The black keys belong to the upper crossbar and form a complementary pentatonic scale (Forte 5–35).[16]

In Fig. 7, the shading of the upper and lower crossbars is exchanged.[17] If a calisson for the associated spelled pitch-class in either region (upper or lower) in Fig. 6 is shaded, then it will be unshaded in Fig. 7 and vice versa. The middle section's F and C calissons, however, remain unshaded throughout the exchange. These regions will be called the upper major scale or, more formally, the part of the diatonic collection associated to the upper region of the I-shaped tile. In this example, the newly unshaded regions now contain the spelled pitch-classes of the D♭ major scale, and the part of the diatonic collection associated to the upper region of the I-shaped tile is DIA_{-5}.[18]

Striped Tiling and Myhill's Property. In Figs. 6 and 7, unshaded and shaded regions of the tiling can be extended indefinitely in either direction, visually forming jagged-edged stripes in the plane. These stripes alternate between unshaded and shaded, representing one part of the diatonic collection (unshaded) and its complementary part of the pentatonic collection (shaded) as they repeat throughout the tiling.[19] These stripes can be understood as visual representations

15 Recall that the diatonic collection can be partitioned into twelve equivalence classes with respect to key signatures: $DIA/\sim_k$ with $k \in \mathbb{Z}$ representing the number of accidentals (mod 12). See the Table in Example 5–14 on p. 246 in Straus [16]. This paper modifies Straus' index notation in order to be more mathematical with negative indices ($k < 0$) indicating the number of flats in the key signature and positive indices ($k > 0$) indicating the number of sharps. Straus uses the symbols ♭ and ♯ in his indexing along with positive integers. See Hook [14] for a different mathematical notation underpinned by signature transformations.

16 In the general case, this lower part will be designated DIA_k with $k \in \mathbb{Z}$, possibly (mod 12), indicating the number of accidentals in the key signature.

17 In the musical example (Sect. 7), this upper/lower exchange represents the world of pure imagination (D♭ major) in contrast to an unspecified scale from DIA_0 that contains of the Fmaj^9 chord, a scale of unadorned reality with no accidentals.

18 In the general case, this upper part will be designated DIA_{k-5} with $k \in \mathbb{Z}$, possibly (mod 12), as before.

19 These tile regions represent two collections that are maximally even, a property first described by Clough and Douthett [3]. The scales in these collections have Myhill's property and are said to be a well-formed scale as defined by Carey and Clampitt [2].

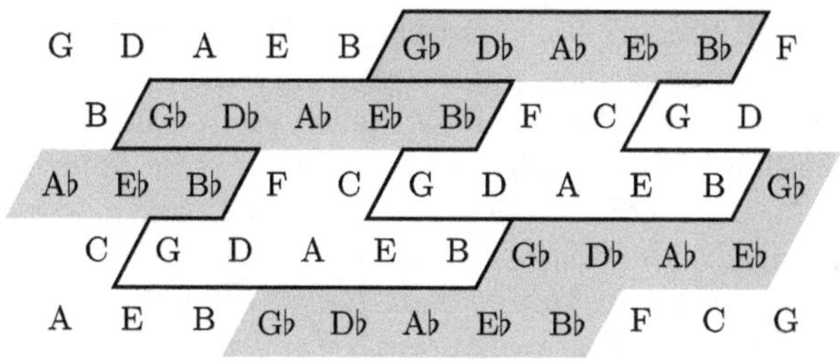

Fig. 6. Connected DIA_0 regions in tiling $\mathcal{T}$ (Shading #1): The tonic-dominant organization of the major scale. Tiling with an *I*-shaped tile is shown in this example using shading to represent the white (unshaded) and black (shaded) keys on the piano. Connected unshaded regions represent skew alleys.

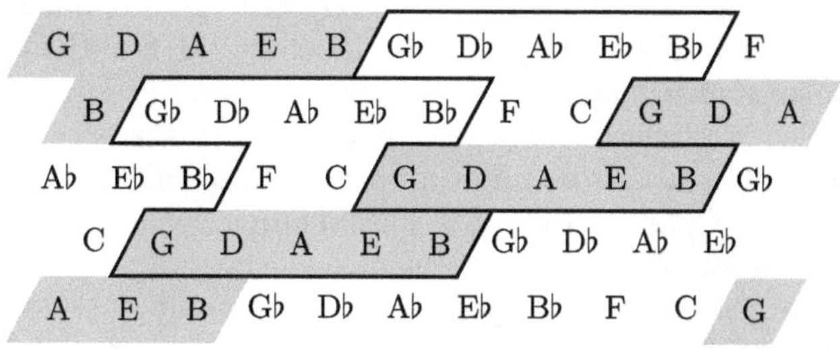

Fig. 7. Connected DIA_{-5} regions in tiling $\mathcal{T}$ (Shading #2): The submediant-mediant organization of the major scale. Tiling with an *I*-shaped tile is shown in this example that exchanges the shading of the upper and lower crossbars of Fig. 6.

in the *Tonnetz* of collected, regionalized sets of (spelled) pitch-classes having Myhill's property; thus, these shadings can be called Myhill's stripes.[20] The *Tonnetz* organizes the spelled pitch-classes into Myhill's stripes, alternating between pentatonic and diatonic stripes, because of how fifths (and indirectly seconds and thirds) are distributed in the *Tonnetz*. Compare the formally equivalent conditions of symmetry and closure for well-formed scales discussed by Carey and Clampitt [2] to the tile distributions. The tile stripes are visual representations of these conditions.

The angle of these stripes is determined by the fifth-relations of the pitch classes of the diatonic collection appearing within the tile. Six of the pitch classes have a perfect fifth located above them within the collection. One pitch class does not. This disparity, represented by the staggered fitting of the *I*-shaped tiles, breaks the shading within the horizontal line-of-fifths which leads to the

[20] Recall that Myhill's property describes collections in which every non-zero generic interval has two specific intervals. The property was first described by Clough and Myerson [4] and named after mathematician John Myhill.

jaggedness of the striping.[21] Note that the middle section's F and C have half steps that precede these spelled pitch-classes in scale order for DIA_0 while F and C have half steps that follow these spelled pitch-classes in scale order for DIA_{-5}. The location of these half-step intervals in scale order is invariant for the middle section in all constructions of lower DIA_k and upper DIA_{k-5} parts of a tile exhausting all possibilities for the middle section.

Skew Alleys: Syntonic Image and Key Signature Pitch Center. In Fig. 6, a straight line can be drawn through *Tonnetz* nodes with spelled pitch-class D, never touching any shaded calissons.[22] This line is the center of a skew alley, representing a music-theoretic structure within C major and within DIA_0 itself. In general, the center of the skew alley is a balance point of the line-of-fifths representation of the k-th part of the diatonic collection: there are three pitch classes below and three above when generated by a perfect fifth.[23] Informally, this center can be called the center of a scale written as a line-of-fifths. It is independent of mode. For example, D is the center of F-C-G-D-A-E-B. This center can be constructed for both the lower DIA_k and the upper DIA_{k-5}. In Fig. 6, the spelled pitch-class D is the center for the unshaded DIA_0; in Fig. 7, the spelled pitch-class E♭ is the center for the unshaded DIA_{-5}.

For comparison, Cohn [6] discusses the "syntonic image" as part of a contrasting formalism for the *Tonnetz* of regions which is also represented by this center.[24] Cohn's discussion is of interest for the present work as the analytic problems that arise from the various *Tonnetze* of regions are directly related to the tile shapes of those regions. As parallelograms, none of the shapes in Cohn's discussion support a monomorphic tiling without some extrinsic information unrepresented by the tile shape.[25] The I-shaped tile, in contrast, has

---

21 The work of Marek Žabka [18] first explored well-formedness in two dimensions in the *Tonnetz*. The reader should note how parallelogram-shaped tiles are used by Žabka as a simplifying structure in his theory's use of just intonation with the theorist selecting "representatives in accordance with the tiling implied by the commas." (p. 2).

22 This is an infinite subset of all *Tonnetz* nodes with spelled pitch-class D.

23 The center in Fig. 1 is a special case of this balanced scale representation discussed by Hook [14] on p. 20; where he identifies the spelled pitch-class D as 0 because D is "centrally located among the seven unsigned notes (white keys) on the line of fifths, so taking this note as the origin for spc [spelled pitch-class] numbers results in a system that exactly balances sharps and flats." In the general case of the I-shaped tiles, accidentals are not balanced by this center, only the number of fifth-generated pitch classes are balanced.

24 As Cohn acknowledges, this viewpoint represents a competing strain of thought, not his own. This viewpoint emerged from eighteenth-century thoroughbass treatises via Gottfried Weber (1821–1824) and Francois-Joseph Fétis (1844) where "tonal regions are primarily represented by scales rather than chords." Cohn [6] uses this contrasting description on p. 329 to motivate his own theoretic work and discussion of Lerdahl's tonal pitch space.

25 A *Tonnetz* of regions can be reconciled with Cohn's *Tonnetz* of triads but requires a different set of tile shapes, specifically serrated alley tiles in one of three orientations based on homotopic equivalence classes from Tymoczko [17] discussed in Coury-Hall [10,11].

three levels of horizontal orientation (two levels related to diatonic structure) resulting in the tabs and blanks that eliminate the ambiguity in the tiling.

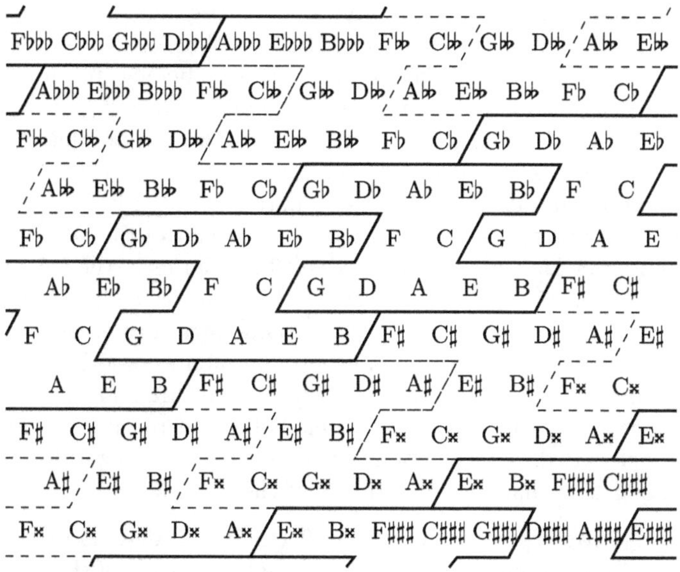

Fig. 8. Spelling-consistent *Tonnetz*: an enharmonically conformed space systematically constructed from each set of twelve consecutive representatives in the line-of-fifths; enharmonic conformity restricted to one skew alley of *I*-shaped tiles per each set of representatives.

An Alternate Tiling: The Spelling-Consistent *Tonnetz*. The skew alleys suggest another approach to constructing the enharmonically conformed *Tonnetz*. Rather than requiring the entire plane to be enharmonically conformed, this construction only requires the tiles forming a skew alley to be enharmonically conformed. This tiling, a portion of which is shown in Fig. 8, results in the plane being tiled with all possible spelled pitch-classes (organized discretely in groups of twelve consecutive pitches in the line-of-fifths) but instead of each unconformed tile having a unique set of representatives in the plane, each skew alley and its associated tiles has a unique set of representatives used to enharmonically conform the skew alley only. These skew alleys then add representatives systematically alley by alley. If the line-of-fifths in Fig. 4 is extended in each direction, every set of twelve consecutive spelled pitches is uniquely a set of representatives for a skew alley. The advantage of this tiling is that all *Tonnetz* triads are spelling consistent; thus, the consonant chords are represented in the sheared *Tonnetz* as equilateral triangles constructed from the neighboring spelled pitch-class nodes: two in the same line-of-fifths and one node either above or below that line indicating chord quality. Also under this construction, there is no spelling inconsistency at any boundaries which reveals more of the underlying structure of the *Tonnetz* graph itself.[26]

Tonic-Dominant and Submediant-Mediant Organizations of the *I*-Shaped Tile. The shaded tilings in Figs. 6 and 7 organize their respective diatonic parts of the *Tonnetz* into two rows of fifths within a single tile: either lower form or upper form. The tonic-dominant organization belongs to the lower region of the *I*-shaped tile, and the submediant-mediant organization belongs to the upper region of the tile. The names of these organizations are derived from the (tilted) tile boundaries of the major scales: in Fig. 6, the C and G tilted boundary separates the tonic and dominant of the C major scale; in Fig. 7, the B♭ and F tilted boundary separates the submediant and mediant of the D♭ major scale.[27] The tonic-dominant and submediant-mediant boundaries in the context of their key signatures orient the spelled pitch-classes in terms of the half steps that precede (or follow), again defining a property without regard to mode related to the symmetry of the well-formed scales and the essential automorphism that defines them as well-formed.[28] As a result, there is one line-of-fifths form in each shaded tiling for the respective part of the diatonic collection running across two tiles with the named tilted vertical boundary unequally dividing this line-of-fifths region and having no shading on each side of the boundary.

Rotational Symmetry and the 2/3-8/9 Inversional Axis of Symmetry. The *Tonnetz* graph is preserved by the rotational symmetry of the *I*-shaped tile: nodes map to nodes and edges map to edges; however, the *Tonnetz* labels for the nodes change. See Fig. 9. Thus, there is an isomorphism of graphs with relabeling that accompanies the tile's rotational symmetry. The rotation represents an inversion of pitch classes about this axis of symmetry (dotted line), so the rotation has musical meaning. See Fig. 10. Pitch classes are inverted through the 2/3-8/9 axis (Sum 5) in the notation of Straus [16]. A tilted line drawn in the *I*-shaped tile through the center of rotation and parallel to the tilted lines of the *I*-shape forms the axis of symmetry.[29] The pitches next to the axis

[26] These more advanced mathematical ideas are explored by this author (Coury-Hall [10,11]) in terms of another *I*-shaped tiling structure in the *Tonnetz* whose lower structure is the result of a transformational groupoid embedded in the schritt-wechsel group, called the Riemannian groupoid, acting on the consonant triads.

[27] Any part of the diatonic collection can be organized into either form through tile transposition/translation, or similarly through transposing/translating the twelve consecutive positions in the line-of-fifths of Fig. 4. The matching diatonic counterpart of the collection, however, will be different for each organization. For example, if DIA_0 is in the lower region, then DIA_{-5} is in the upper region. If DIA_0 is in the upper region, then DIA_5 is in the lower region.

[28] See p. 196 of Carey and Clampitt [2] for the formal definition of well-formed scales in a context separate from tiling.

[29] Six calissons border this axis (2/3: D and E♭; 8/9: A♭ and A) and poles (F and C). The axis-bordering edges of the calissons form a dotted line connecting the boundary of the convex-to-convex fittings (tab-to-tab) of the tiles immediately above and below and can be extended indefinitely.

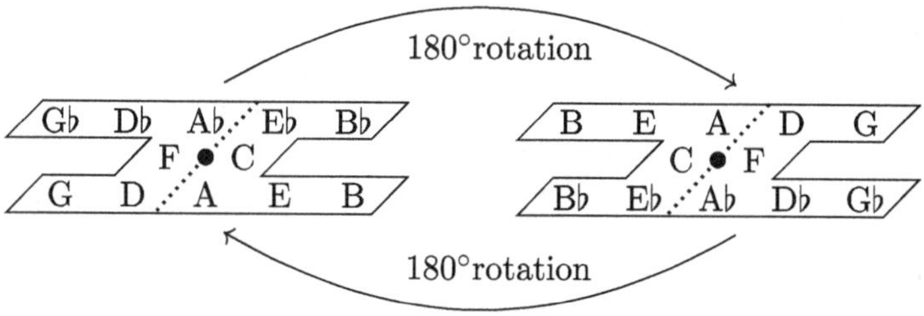

Fig. 9. Rotational symmetry around center of the *I*-shaped tile (black dot); floating nodes used for visual clarity to avoid nodes with upside down labels.

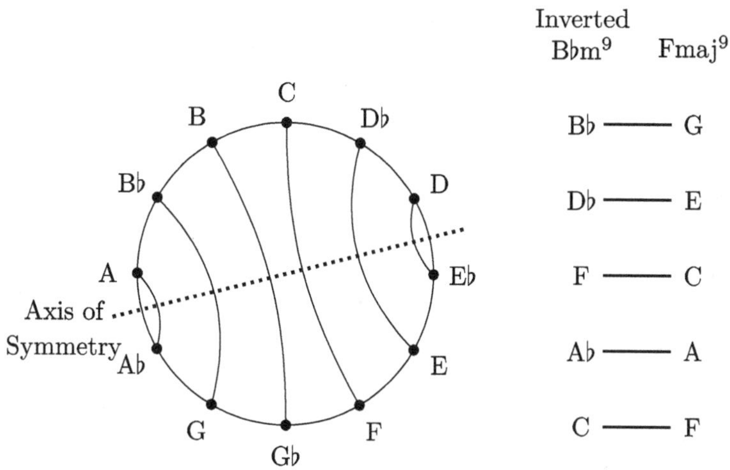

Fig. 10. The 2/3-8/9 inversional axis of symmetry with example B♭m^9 inverted to Fmaj^9

line in the upper/lower crossbars are the pitch classes that name the axis in the notation of Straus: 2/3 and 8/9. Node locations of all inverted pitches are preserved rotationally. This inversional symmetry is used in the next section's analysis with the *I*-shaped tile providing the ontology and epistemology for understanding the sonic worlds established in the text-setting.

7 Music Analysis: "Pure Imagination" by Leslie Bricusse and Anthony Newley (1971)

In the 1971 film "Willy Wonka and the Chocolate Factory," the song "Pure Imagination" by Anthony Newley and Leslie Briscusse sets the text of "defy explanation" with the dominant A♭^{9} chord followed by an Fmaj^9, reminiscent of the jazz style.[30] Although the Fmaj^9 chord does not belong to the song's key of D♭ major, the Fmaj^9 acts as a tonic substitution in the text-setting, mimicking an earlier tonic resolution in the phrasing. The I-shaped tile from Fig. 1 perfectly contextualizes this choice. The Fmaj^9 chord has five of the pitch classes of DIA_0 with only the pitches B and D missing. By containing the pitch classes of the A minor triad and the C major triad, Fmaj^9 harmonizes an octave jump of C in the melody with one of the scales of the 0-accidental collection (DIA_0), providing the contrast to the 5-flat collection (DIA_{-5}) of the *D*♭ major scale and the hinted *B*♭ minor sonority used in the introduction to the song. The phrase is set up for a deceptive cadence with a B$^\flat$m^9 at "defying explanation" but the surprise arrives with the innovative choice of the Fmaj^9 chord, the inversion of that possible deceptive resolution along the 2/3-8/9 axis of symmetry.[31] See Fig. 10. The B♭m^9, stacked top down, is inverted by the 2/3-8/9 axis of symmetry to the Fmaj^9, stacked bottom up.

The I-shape tile explains how the 2/3-8/9 axis of symmetry is used in the text-setting of the piece. The upper crossbar of the I-shaped tile belongs to the world of "pure imagination" filling in the gaps of a more grounded reality in the lower crossbar with the middle section's shared F and C mediating between the two worlds: one of pure imagination (D♭ major scale or DIA_{-5}) and one of practicality, an unexpectedly innovative world within a mass-producing factory (C major scale/A minor scale or DIA_0). The I-shaped tile captures Wonka's entrepreneurial spirit in its creativity and execution. By using the I-shaped tile in analysis, the magician-chocolatier's harmonic trick, instead of defying explanation in the music, is revealed with the music theorist's tools.

8 Conclusion

The I-shaped tile provides a powerful tool for understanding and re-organizing the *Tonnetz* through examples such as skew alleys and Myhill's stripes or the spelling-consistent *Tonnetz*. The author has also presented several works on topics that can be developed further using this technique. See Coury-Hall [7–11]. Tile boundaries distinguish tonal function in the tonic-dominant and submediant-mediant organizations of the major scales; thus, endowing the tile boundaries with music-theoretic meaning across the lines-of-fifths that fits these tiles together like puzzle pieces. These boundaries can be generalized to reflect symmetries that result from Myhill's property and apply to certain partitions of the diatonic collection based

---

30 The upper extensions of world of "pure imagination" have been reduced at this point from an earlier A♭^{13} to form a less ethereal A♭^{9} in order to match the Fmaj^9 that follows. This reduction in the upper extensions places the harmony at the level of his guests' (lesser) imaginations, referenced as "your imagination" in the song, which is an octave lower than the setting for Wonka's own "pure imagination.".

31 Recall that Willy Wonka is testing the character of the young golden ticket winners unbeknownst to them (or the audience) until the end of the story, so a deceptive cadence is certainly appropriate in the text-setting. The choice of Fmaj^9, however, is more subtle and engaging in its deception.

on key signatures. The next step is to develop the mathematical theory of tiling with *I*-shaped tiles beyond the heuristics presented in this paper to open up the *Tonnetz* to more algebraic and geometric properties, extending the mathematical framework initiated by Brower [1] through tilings of the *Tonnetz.*

Acknowledgements. This author wishes to thank the anonymous reviewers for many insightful comments.

References

1. Brower, C.: Paradoxes of pitch space. Music. Anal. **27**(1), 51–106 (2008)
2. Carey, N., Clampitt, D.: Aspects of well-formed scales. Music Theory Spectrum **11**, 187–206 (1989)
3. Clough, J., Douthett, J.: Maximally even sets. J. Music Theory **35**, 93–173 (1991)
4. Clough, J., Myerson, G.: Musical scales and the generalized circle of Fifths. Am. Math. Mon. **93**(9), 695–701 (1986)
5. Cohn, R.: Neo-riemannian operations, parsimonious trichords, and their "Tonnetz" representations. J. Music Theory **41**(1), 1–66 (1997)
6. Cohn, R.: Tonal pitch space and the (Neo-)Riemannian Tonnetz. In: Gollin, E., Rehding, A. (eds.) The Oxford Handbook of Neo-Riemannian Music Theories, pp. 322–48. Oxford University Press, New York (2011)
7. Coury-Hall, M. A.: The Myhill property: the strong vs. the weak. In: Annual Meeting of Society for Music Theory. Conference Poster. Peer-reviewed. Columbus, Ohio (2019)
8. Coury-Hall, M. A.: Reconsidering negative harmony: melodic dualism in Bárdos' scalar schemata. In: Annual meeting of Society for Music Theory. Conference Poster. Peer-reviewed. Virtual (2020)
9. Coury-Hall, M. A.: Rogue symmetry: the groupoid of Riemannian UTTs. In: Annual meeting of Society for Music Theory. Conference Presentation. Peer-reviewed. Virtual (2021)
10. Coury-Hall, M. A.: The "Colors" of Parsimony in Cohn's reinterpreted *Tonnetz*. In: Annual Meeting of Music Theory Midwest. Conference Presentation. Peer-reviewed. University of Manitoba, Winnipeg, Canada (2023)
11. Coury-Hall, M. A.: The "Colors" of parsimony in Cohn's reinterpreted *Tonnetz*. In: Annual meeting of Society for Music Theory. Conference Poster. Peer-reviewed. Denver, Colorado (2023)
12. Grünbaum, B., Shephard, G.C.: Tilings and Patterns, 2nd edn. Dover Publications, New York (2016)
13. Hook, J.: Spelled Heptachords. In: Agon, C., Andreatta, M., Assayag, G., Amiot, E., Bresson, J., Mandereau, J. (eds.) MCM 2011. LNCS (LNAI), vol. 6726, pp. 84–97. Springer, Heidelberg (2011). https://doi.org/10.1007/978-3-642-21590-2_7
14. Hook, J.: Exploring Musical Spaces: A Synthesis of Mathematical Approaches (Oxford Studies in Music Theory). Oxford University Press, Oxford (2022)
15. Senechal, M., Galiulin, R.V.: An introduction to the theory of figures: the geometry of E.S. Fedorov. Struct. Topol. **10**, 5–22 (1984)
16. Straus, J.N.: Introduction to Post-Tonal Theory, 4th edn. W. W. Norton & Company, New York (2016)
17. Tymoczko, D.: Why topology? J. Math. Music **14**(2), 114–169 (2020)
18. Žabka, M.: Well-formedness in two dimensions: a generalization of Carey and Clampitt's theorem. J. Math. Music **4**(1), 1–30 (2010)

Advanced Polyphonic Music Pattern Matching Algorithms with Timing Invariances

Antti Laaksonen(✉) and Kjell Lemström

Department of Computer Science, University of Helsinki, Helsinki, Finland
ahslaaks@cs.helsinki.fi

Abstract. This paper considers polyphonic pattern matching in symbolically encoded music under transposition invariance. We show how, in two specific retrieval problems, the worst-case time complexities can be improved by an order of magnitude by applying a relatively straightforward algorithm design technique. Moreover, the technique applies to most of the previously best-known algorithms for the other cases of the problem category, giving comparable worst-case running times. In general, given two point sets, a musical work T and a pattern P, the task is to find possibly transposed occurrences of P in T. The occurrences may be full or partial, with or without time-warping or time-scaling. The technique uses a merge-like approach to incrementally build pattern occurrences using sorted lists of notes. The algorithms search for incomplete matches that are subsequently extended as far as possible.

Keywords: Music pattern matching · geometric algorithms · timing invariances

1 Introduction

Content-based music-retrieval (CBMR) in symbolically encoded music is a classical music information retrieval related pattern matching problem that was first addressed with string matching techniques (see, e.g. [4,5,9,17]). However, linear strings do not capture the essence of polyphonic CBMR problems where the polyphony may be dense, without any instrument information, or the pattern occurrences are scattered across the instruments. In such cases, attempting to form all possible linear strings of the musical work would quickly result in a combinatorial explosion.

Later, several authors proposed different geometric modellings of music that allow the natural handling of polyphony and *translation invariance*, that is, transposition and time-shifting invariances [12] with different algorithmic techniques [13,18–20]. In this paper, we use the piano-roll-like representation, or point-set representation, suggested by Meredith et al. [16], where a note is represented as a point in a two-dimensional space such that its *pitch* (the perceived height of the note) is given by the vertical position and the *onset* time (note-on) information by the horizontal position of the point (see Fig. 1).

© The Author(s), under exclusive license to Springer Nature Switzerland AG 2024
T. Noll et al. (Eds.): MCM 2024, LNCS 14639, pp. 242–254, 2024.
https://doi.org/10.1007/978-3-031-60638-0_19

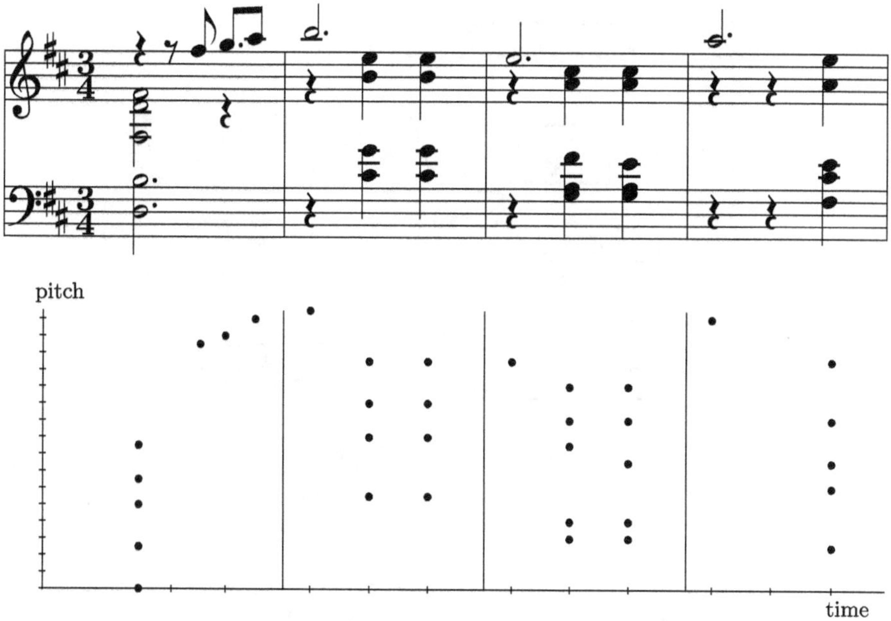

Fig. 1. An excerpt of Sibelius' Kuusi is displayed with common music notation above and a synchronized geometric point-set representation below. Each note is represented as a point, with the x-coordinate as the onset time and the y-coordinate as the pitch.

When dealing with performed music, the rhythmical information is not accurate: the overall tempo between performances varies (the case of a uniform time-scaling), and, more gravely, individual notes are next to never played precisely in time (the case of local time warping). For instance, such features would be needed in score following or autonomous accompaniment applications. The original geometric algorithms suffered from the lack of coping with any changes or errors in the rhythm[1].

In this paper, we present new geometric algorithms designed to address inaccuracies in rhythm information. We show how an algorithm design technique can be applied to any algorithm within the problem category we are considering, which involves identifying occurrences of a given pattern P of length m within a musical work T of length n, represented as geometric point sets. More precisely, the problems of this category are as follows:

1. **Full matching**: Find plain/time-scaled/time-warped translations of P such that all the points in P match against the points in T.
2. **Partial matching**: Find plain/time-scaled/time-warped translations of P where a subset of the points in P match against the points in T.

[1] They tolerate rhythmical noise by simply ignoring any event out of the rhythm.

Fig. 2. (a) A theme from Bach's fugue (BWV 871) in *Das Wohltemperierte Klavier.* (b) A plain transposed, (c) a time-scaled, and (d) a time-warped occurrence of it.

Typically, the pattern is a melody whose occurrences we want to find in a musical work. For example, Fig. 2 shows a theme from Bach's fugue and its plain, time-scaled and time-warped occurrences.

The related problem category of finding recurring patterns within a given music database has been under active study also in very recent years (see, e.g. [3,7,8,14,15]). Although this is not the case for our problem category, there are still several solutions to these problems in the literature. For instance, there are algorithms for the pattern matching problem for the plain translated case [18,20], time-scaled and translated case (where the temporal distance of the notes in the occurrence are multiplied by a positive uniform time-scale factor of those of the pattern) [1,2,6,10,18], and time-warped and translated case (where the time-scale factor has to be positive but not uniform) [6,11]. We have summarised these problems with their best-known solving algorithms concerning the worst-case time complexities in Table 1.

Although the invariance needed for the time-warped problems is more robust than that of time-scaled problems [12], the time-scaled problems appear harder than time-warped problems. This is because, in time-warped problems, the subproblems are independent, making techniques like dynamic programming applicable to efficient solutions. In contrast, time-scaled problems are influenced by a global scaling factor, preventing the utilization of any such approach.

The algorithms based on the geometric representation, both in the sequel and generally in the literature, assume that the underlying musical work T is given in a lexicographic order. To that end, let $t \in T$ be a note (a point in our two-dimensional representation), and let us denote by $t.x$ and $t.y$ its onset time and pitch, respectively. We are interested in two lexicographic orderings, or permutations, of T. The permutation T^o is based on the onset times. In T^o any two points $(t_i.x, t_i.y)$ and $(t_j.x, t_j.y)$ are ordered by onset times, denoted as $(t_i.x, t_i.y) \prec^o (t_j.x, t_j.y)$, that is, $t_i.x < t_j.x$, or $t_i.x = t_j.x$ and $t_i.y < t_j.y$. The

Table 1. A taxonomy of musical pattern matching problems under consideration. The "Previous work" and "This paper" columns give the worst-case time complexities for the previous best-known and our algorithms, respectively. In the highlighted area, our new algorithms outperform the previously best-known solutions.

Problem	Previous work	This paper
Full		
Plain	$O(nm)$ [20]	$O(nm)$
Time-scaled	$O(n^2m)$ [6]	$O(n^2m)$
Time-warped	$O(nm)$ [6]	$O(nm)$
Partial		
Plain	$O(nm \log m)$ [20]	$O(nm^2)$
Time-scaled	$O(n^2m^2 \log n)$ [10]	$O(n^2m^2)$
Time-warped	$O(n^2m^2 \log n)$ [11]	$O(nm^2)$

permutation T^p based on pitched values is defined accordingly. T^o and T^p are generated only once in a preprocessing phase, taking $O(n \log n)$ time.

The technique presented in this paper can be used to solve several types of problems. The technique is instrumental in partial time-scaled and time-warped problems where we can create more efficient algorithms than the best previously known ones. We introduce the technique in Sect. 2 in the context of time-warped problems that are easier to solve. After that, in Sect. 3, we show how the technique can also be extended to time-scaled problems, and in Sect. 4, we give a link to our implementations. Finally, Sect. 5 concludes the paper.

2 Time-Warped Problems

This section deals with time-warped pattern-matching problems. First, we discuss a variant of the algorithm presented in [6] that solves the full time-warped pattern matching problem in $O(nm)$ time (after preprocessing), which is a starting point for our technique. After that, we extend the idea of the algorithm to solve the partial time-warped pattern matching problem in $O(nm^2)$ time.

2.1 Full Pattern Matching

In the full time-warped pattern matching problem, the task is to determine, for each note $t \in T$, whether there is a time-warped occurrence of P in T whose last note is t. More precisely, the task is to determine whether there are m notes

$$t_1, t_2, \ldots, t_m$$

where $t_m = t$ and, for each $1 \leq i \leq m-1$, $t_i.x < t_{i+1}.x$ and $t_{i+1}.y - t_i.y = p_{i+1}.y - p_i.y$, i.e., the onset times increase, and each pitch interval corresponds to the pattern.

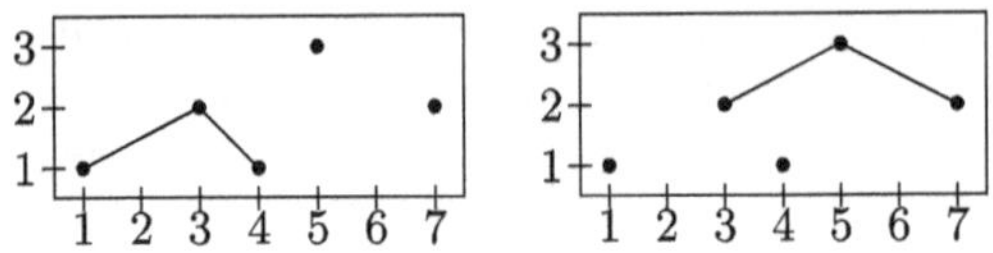

Fig. 3. Two time-warped pattern occurrences for $P = \{(1,1),(2,2),(3,1)\}$.

To exemplify the problem, let us now consider the following simple, monophonic case (any polyphonic case would be treated analogously):

$$T^o = \{(1,1),(3,2),(4,1),(5,3),(7,2)\}; \quad P^o = \{(1,1),(2,2),(3,1)\}.$$

In this case, the algorithm should determine two full occurrences of the pattern that end at points $(4,1)$ and $(7,2)$. Figure 3 shows the occurrences: the first occurrence consists of points $(1,1),(3,2),(4,1)$, and the second occurrence consists of points $(3,2),(5,3),(7,2)$.

Note that the number of possible occurrences may be vast in a time-warped search. For example, if every note in T and P has the same pitch, then every subset of m notes in T is a partial time-warped occurrence of P. However, when focusing on ending notes of the occurrences, there are at most m occurrences.

The core of the time-warped algorithm is expressed as pseudocode in Algorithm 1. In the preprocessing phase, the algorithm creates a list $U = T^p$ consisting of the notes in T ordered in pitch first manner. After that, the algorithm executes m rounds. On round i, it creates a list L_i consisting of notes where a partial pattern occurrence $p_1, p_2, \ldots, p_i$ can end, and L_m corresponds to the algorithm's output. The first round initiates $L_1 = U$ because any note in T is a partial occurrence with only one note. Then, on round i, for $1 < i \leq m$, a merge-like approach is used to create L_i based on L_{i-1} and U. The algorithm goes through each note $u \in U$ from left to right and maintains a pointer z to L_{i-1}, initially pointing to the first element of L_{i-1}. The algorithm must check if a note in L_{i-1} can be the previous note in a pattern occurrence whose ith note is u. To that end, it uses a pointer z. As long as $u.y - L_{i-1}[z].y > p_i.y - p_{i-1}.y$, it increases z. After that, if $u.y - L_{i-1}[z].y = p_i.y - p_{i-1}.y$ and $L_{i-1}[z].x < u.x$; thus, u is advanced at this point, and the algorithm adds it to L_i.

As each round takes $O(n)$ time because the pointer z moves $O(n)$ steps, the total time complexity of the algorithm is $O(nm)$.

In our simple example scenario, the algorithm first creates the list

$$U = \{(1,1),(4,1),(3,2),(7,2),(5,3)\}.$$

In the example, the pattern consists of three notes $(1,1),(2,2),(3,1)$, so the algorithm performs three rounds. In round 1, the algorithm creates the list

$$L_1 = U = \{(1,1),(4,1),(3,2),(7,2),(5,3)\}$$

which consists of all notes where an occurrence of the first pattern note can end. Then, in round 2, the algorithm uses the lists L_1 and U to create the list

$$L_2 = \{(3,2),(7,2),(5,3)\}.$$

Algorithm 1. Time-warped algorithm

$L_1 \leftarrow U \leftarrow T^P$
Create empty lists $L_2, L_3, \ldots, L_m$
for $i \leftarrow 2$ to m **do**
  If L_{i-1} is empty, terminate the loop
  $z \leftarrow 1$
  for $u \in U$ **do**
    while $(z+1 \leq |L_{i-1}|) \wedge ((u.y - L_{i-1}[z].y) > (p_i.y - p_{i-1}.y))$ **do**
      $z \leftarrow z+1$
    end while
    if $(u.y - L_{i-1}[z].y = p_i.y - p_{i-1}.y) \wedge (L_{i-1}[z].x < u.x)$ **then**
      Add u to L_i
    end if
  end for
end for
return L_m

For example, note $(3,2)$ corresponds to the partial pattern occurrence $(1,1),(3,2)$. Finally, in round 3, the algorithm uses the lists L_2 and U to create the list

$$L_3 = \{(4,1),(7,2)\}.$$

This corresponds to the following pattern occurrences:

$$(1,1),(3,2),(4,1) \text{ and } (3,2),(5,3),(7,2).$$

2.2 Partial Pattern Matching

A partial time-warped pattern matching algorithm finds the maximum partial occurrence size ending at each note $t \in T$. Let $d(t)$ denote the maximum occurrence size for note t. If $d(t)=k$, there must be k notes

$$t_1, t_2, \ldots, t_k$$

and a monotonically increasing injection

$$f : \{1,2,\ldots,k\} \rightarrow \{1,2,\ldots,m\}$$

such that $t_k = t$ and, for each $1 \leq i \leq k-1$, $t_i.x < t_{i+1}.x$, $p_{f(i)}.x < p_{f(i+1)}.x$ and $t_{i+1}.y - t_i.y = p_{f(i+1)}.y - p_{f(i)}.y$. Here, f is an increasing function determining which pattern notes are included in the partial pattern occurrence. A full pattern occurrence ends at note t exactly when $d(t) = m$.

For example, when

$$T^o = \{(1,1),(3,2),(4,1),(5,3),(7,2)\}; \quad P^o = \{(1,1),(2,2),(3,1)\}.$$

the algorithm should determine the following values: $d((1,1)) = 1, d((3,2)) = 2, d((4,1)) = 3, d((5,3)) = 2, d((7,2)) = 3$. For example, $d((3,2)) = 2$ because

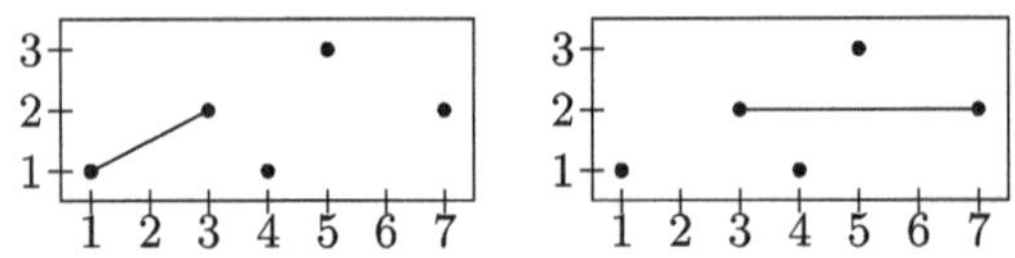

Fig. 4. Two partial time-warped pattern occurrences. The first occurrence is a maximum occurrence that ends at note $(3, 2)$. The second occurrence is not maximum because it is still extensible by adding note $(5, 3)$.

there is a partial pattern occurrence $(1, 1), (3, 2)$ which is maximum. In this case $k = 2$, $f(1) = 1$ and $f(2) = 2$. Figure 4 shows two partial pattern occurrences. The first occurrence is $d((3, 2)) = 2$. The second occurrence is not maximum because it only contains 2 notes while $d((7, 2)) = 3$.

We can modify the algorithm described above to support partial pattern matching. In partial pattern matching, there can be gaps in the pattern, which means that when the algorithm processes a note u on round i, the previous occurrence note is not necessarily in L_{i-1}. Instead, it can be in any previous lists $L_1, L_2, \ldots, L_{i-1}$.

As in the full pattern matching algorithm, the algorithm creates a sorted list $U = T^p$ and then performs m rounds that create the lists $L_1, L_2, \ldots, L_m$. To count the maximum number of notes in partial patterns, the algorithm stores, for each note u, additional values of the form $d_i(u)$: the maximum number of notes in a partial pattern occurrence whose last note is u and u corresponds to the ith pattern note. On round i, the algorithm calculates the value $d_i(u)$ for each note u. The algorithm maintains $i - 1$ pointers $z_1, z_2, \ldots, z_{i-1}$ that point to lists $L_1, L_2, \ldots, L_{i-1}$, respectively. Each pointer initially points to the first element of the corresponding list. Then, to calculate the value $d_i(u)$, the algorithm looks for the previous pattern notes in lists $L_1, L_2, \ldots, L_{i-1}$ using the same trick used in the full pattern matching algorithm. The algorithm chooses the previous note whose partial pattern occurrence size is maximum and extends that occurrence by adding note u. $O(m)$ pointers are moved $O(n)$ times on each round. Since the number of rounds is m, the total time complexity of the rounds is $O(nm^2)$.

Consider calculating the value $d_3((7, 2))$ in our example scenario. We can choose the previous note either from L_1 or L_2. First, we can choose note $(3, 2)$ from L_1. We know that $d_1((3, 2)) = 1$, so extending that occurrence would give a new occurrence of size 2. However, we can also choose note $(5, 3)$ from L_2. This is a better choice because $d_2((5, 3)) = 2$, and we get an occurrence of size 3 by extending it. Thus, $d_3((7, 2)) = 3$ gives the maximal partial match for $t = (7, 2)$ that happens to be also a full occurrence.

2.3 Time-Warped vs Plain Problems

Notably, minor adjustments to the time-warped algorithms mentioned earlier would suffice to address the plain search problem. In the previously described algorithms, the pointers are shifted to identify two notes with intervals matching

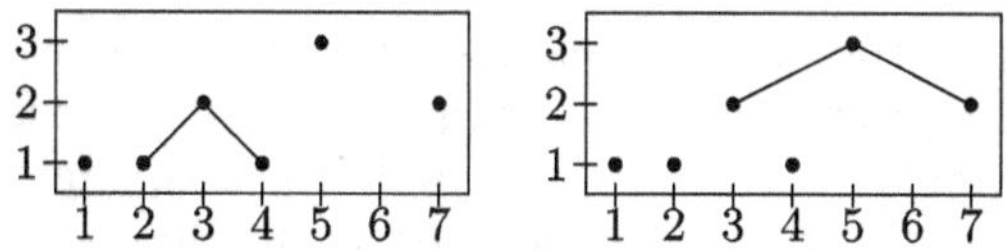

Fig. 5. Two time-scaled pattern occurrences for $P = \{(1,1),(2,2),(3,1)\}$. The α values are 1 and 2 in the first and second occurrence, respectively.

the pattern, and the only additional condition is the preservation of pattern onset times in an occurrence. However, we could extend this requirement to match the difference between onset times, transforming the algorithm's focus from time-warped occurrences to plain occurrences.

This yields $O(nm)$ and $O(nm^2)$ time algorithms for full and partial plain search problems. Yet, the $O(nm)$ algorithm is more complex than the existing algorithm of the same complexity. Also, the $O(nm^2)$ algorithm is less efficient than the known $O(nm \log m)$ algorithm using a heap structure (see Table 1).

3 Time-Scaled Problems

Let us next focus on the time-scaled pattern-matching problems. First, we use the ideas discussed in the previous section to show how to solve the full-time-scaled pattern matching problem in $O(n^2m)$ time. The resulting algorithm is an alternative to the earlier algorithm [6] of the same complexity. Then, we present an $O(n^2m^2)$ time algorithm for solving the partial time-scaled matching problem.

3.1 Full Pattern Matching

In full-time-scaled pattern matching, the task is to determine, for each note $t \in T$, whether there is a time-scaled occurrence of P in T whose last note is t. More precisely, the task is to determine whether there are m notes

$$t_1, t_2, \ldots, t_m$$

such that $t_m = t$ and, for each $1 \le i \le m-1$, $(p_{i+1}.x - p_i.x)\alpha = t_{i+1}.x - t_i.x$, where α is a positive real number, and $t_{i+1}.y - t_i.y = p_{i+1}.y - p_i.y$. The difference between time-scaled and time-warped search is the scaling factor α. Note that α may vary between any pattern occurrences.

Let us consider an example where

$$T^o = \{(1,1),(2,1),(3,2),(4,1),(5,3),(7,2)\}; \quad P^o = \{(1,1),(2,2),(3,1)\}.$$

In this case, the algorithm should determine that two occurrences end at points $(4,1)$ and $(7,2)$. Figure 5 shows the occurrences: the first occurrence consists of points $(2,1),(3,2),(4,1)$, the second of points $(3,2),(5,3),(7,2)$. The α values are 1 and 2 in the first and second occurrence, respectively.

Like in the time-warped algorithms, the algorithm first creates a list $U = T^p$ that consists of the pairs in T. The algorithm sorts the list based on the pitches of the notes. If two points have the same pitch, the algorithm sorts them based on their onset times. This first step takes $O(n \log n)$ time.

After that, the algorithm iterates $m-1$ rounds. To find time-scaled occurrences, it maintains, for each round $i = 2, \ldots, m$ and note u, a list $L_i(u)$ containing all scaling factors corresponding to partial time-scaled pattern occurrences that consist of the first i pattern notes and end at note u. Initially, all lists are empty. When the algorithm processes note u on round i, it considers all notes $u' \in U$ for which $u'.x < u.x$ and $u.y - u'.y = p_i.y - p_{i-1}.y$. The scaling factor for a pattern occurrence that contains u and u' would be

$$\alpha = \frac{u.x - u'.x}{p_i.x - p_{i-1}.x}.$$

On round 2, the algorithm adds α to $L_2(u)$ because the two first notes in a pattern occurrence determine the scaling factor. On round i where $i > 2$ the algorithm adds α to $L_i(u)$ if α also appears in $L_{i-1}(u')$. This extends a partial pattern with scaling factor α. Finally, each list $L_m(u)$ contains scaling factors corresponding to full-time-scaled pattern occurrences ending at note u. An occurrence ends at note u exactly when $L_m(u)$ is not empty.

The number of rounds is $O(m)$, and the algorithm processes $O(n)$ notes on each round. In addition, there can be $O(n)$ ways to choose the previous note. To efficiently check if $L_m(u)$ contains α, the lists are generated so that the scaling factors are sorted by processing the u' values from right to left. Implementing the check using binary search would add an extra logarithmic factor to the time complexity. Instead, we use a pointer that moves from left to right for each list and the same trick as in the time-warped search. Using such pointers, the time complexity of the rounds is $O(n^2 m)$ because there are $O(nm)$ pointers, and each pointer moves $O(n)$ times.

In our example scenario, $L_2((3,2))$ contains two scaling factors 1 and 2 which correspond to partial pattern occurrences $(2,1),(3,2)$ and $(1,1),(3,2)$, respectively. The pattern occurrence with scaling factor 1 can be extended to a full pattern occurrence $(2,1),(3,2),(4,1)$, and for this reason $L_3((4,1))$ also contains 1. In addition, $L_2((5,3))$ and $L_3((7,2))$ contain the scaling factor 2 which corresponds to the pattern occurrence $(3,2),(5,3),(7,2)$.

3.2 Partial Pattern Matching

Finally, let us consider the partial time-scaled pattern matching problem, in which the task is to determine, for each note $t \in T$, the maximum size of a partial time-scaled pattern occurrence whose last note is t. Let $d(t)$ denote the maximum occurrence size for note t. If $d(t) = k$, there must be k notes

$$t_1, t_2, \ldots, t_k$$

and an injection

$$f : \{1, 2, \ldots, k\} \to \{1, 2, \ldots, m\}$$

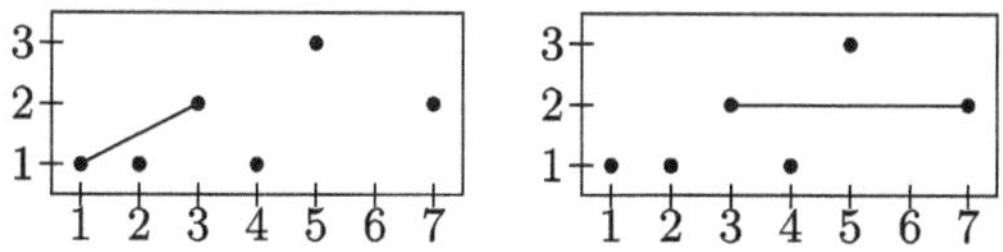

Fig. 6. Two time-scaled pattern occurrences for $P = \{(1,1),(2,2),(3,1)\}$. The scaling factor in both occurrences is 2.

such that $t_k = t$ and, for each $1 \leq i \leq k-1$, $t_i.x < t_{i+1}.x$, $(p_{f(i+1)}.x - p_{f(i)}.x)\alpha = t_{i+1}.x - t_i.x$, where α is a positive real number, and $t_{i+1}.y - t_i.y = p_{f(i+1)}.y - p_{f(i)}.y$. As in the corresponding time-warped problem, f determines which pattern notes are included in the partial pattern occurrence. A full pattern occurrence ends at note t if and only if $d(t) = m$.

For example, when

$$T^o = \{(1,1),(2,1),(3,2),(4,1),(5,3),(7,2)\}; \quad P^o = \{(1,1),(2,2),(3,1)\},$$

the algorithm should determine the following values: $d((1,1)) = 1, d((2,1)) = 1, d((3,2)) = 2, d((4,1)) = 3, d((5,3)) = 2, d((7,2)) = 3$. For example, $d((3,2)) = 2$ because there are partial pattern occurrences $(1,1),(3,2)$ and $(2,1),(3,2)$. In those cases $k = 2$, $f(1) = 1$ and $f(2) = 2$. Figure 6 shows two partial pattern occurrences. The first occurrence is $d((3,2)) = 2$. The second occurrence is not maximum because it only contains 2 notes while $d((7,2)) = 3$.

Only two modifications are needed to the full-time-scaled pattern-matching algorithm so that it also finds partial pattern occurrences. First, instead of plain scaling factors, each list $L_i(u)$ contains pairs of the form (α, c) where α is the scaling factor, and c is the size of a partial pattern occurrence that ends at note u corresponding to the scaling factor α. Second, on round i, after choosing the note u', the algorithm considers all lists $L_2(u'), \ldots, L_{i-1}(u')$ and looks for partial patterns that may contain gaps. In addition, it also considers the case where u' is the first note in the pattern occurrence. Due to this modification, there will be an additional $O(m)$ factor in the time complexity because, for each round i and note u, the algorithm goes through $O(nm)$ ways to select the previous note and round. Thus, the total time complexity of the rounds is $O(n^2m^2)$.

Consider calculating the value $d((7,2))$ in our example scenario. Two ways exist to create a partial pattern occurrence with note $(7,2)$. First, we can create a partial pattern occurrence consisting of two notes: $(3,2)$ and $(7,2)$. In this case, there is a gap in the occurrence. Second, list $L_2((5,3))$ contains $(2,2)$, which means that there is a partial pattern occurrence with scaling factor 2 that consists of 2 notes. We can extend this occurrence by adding $(7,2)$. The latter creates an occurrence of 3 notes, giving the maximum in this case.

4 Implementation

We have implemented the six problems discussed in this paper using the proposed algorithmic technique. The implementations are available on GitHub (a link will be given in the final version). We have tested that the implementations work correctly and can be used to process real-world data sets.

5 Conclusions

This paper presented a CBMR technique for pattern matching in symbolically encoded, polyphonic music. The matching algorithms seek either *full* or *partial* matches under a given set of allowed invariances of a user-provided pattern in an underlying musical work.

A clear advantage of the proposed technique is its flexibility: Although the algorithms are not the most efficient in all the problem cases, they can be applied to all the CBMR problems under consideration (see Table 1). In the full and partial *plain* cases, which allow only translation (transposition and pitch-shift) invariance, we expect that the two case-specialised algorithms by Ukkonen et al. [20] outperform our algorithms: In the partial plain case, their heap-based algorithm has a better worst-case time-complexity than that of ours ($O(nm \log m)$ vs $O(nm^2)$). We have the same worst-case time complexity $O(nm)$ in the full plain case, but their algorithm runs in $O(n)$ expected time. However, their algorithms do not apply to the other cases considered here.

When applying the translation invariance with the full *time-scaled* (using a uniform, positive time-scaling factor) and full *time-warped* (applying any sequence of positive time-scaling factors) translation-invariant cases, our algorithms are on par with the previously best-known solutions, $O(n^2m)$ and $O(nm)$, respectively. In the partial time-scaled and time-warped cases (with translation invariance), our algorithms have at least an order of magnitude better worst-case time complexities than the previously best-known solutions: in the partial time-scaled case, we improve the worst-case time-complexity from $O(n^2m^2 \log n)$ to $O(n^2m^2)$, in the partial time-warped case the margin of improvement is even more considerable: from $O(n^2m^2 \log n)$ to $O(nm^2)$.

Considering the robustness of our time-warped partial matching algorithm in the geometric framework, we anticipate its significant potential in applications such as score following and autonomous accompaniment. Previous solutions have predominantly operated in the audio domain, but we posit that superior solutions may be attainable within the symbolic domain due to the algorithm's resilience to noise. The validation of this hypothesis remains a subject for future research.

References

1. Arzt, A., Böck, S., Widmer, G.: Fast identification of piece and score position via symbolic fingerprinting. In: Proceedings International Society for Music Information Retrieval Conference, pp. 433–438 (2012)
2. Collins, T., Arzt, A., Flossmann, S., Widmer, G.: SIARCT-CFP: improving precision and the discovery of inexact musical patterns in point-set representations. In: Proceedings of International Conference on Music Information Retrieval, pp. 549–554 (2013)
3. Collins, T., Arzt, A., Frostel, H., Widmer, G.: Using geometric symbolic fingerprinting to discover distinctive patterns in polyphonic music corpora. In: Meredith, D. (ed.) Computational Music Analysis, pp. 445–474. Springer, Cham (2016). https://doi.org/10.1007/978-3-319-25931-4_17
4. Crawford, T.: String matching techniques for musical similarity and melodic recognition. Computing in Musicology **11**, 73–100 (1998)
5. Ghias, A., Logan, J., Chamberlin, D., Smith, B.: Query by humming - musical information retrieval in an audio database. In: ACM Multimedia, pp. 231–236 (1995)
6. Laaksonen, A.: Efficient and simple algorithms for time-scaled and time-warped music search. In: Proceedings of International Symposium on Computer Music Multidisciplinary Research, pp. 621–630 (2013)
7. Laaksonen, A., Lemström, K.: Discovering distorted repeating patterns in polyphonic music through longest increasing subsequences. J. Math. Music **15**(2), 99–111 (2021)
8. Laaksonen, A., Lemström, K., Björklund, O.: Transposition and time-scaling invariant algorithm for detecting repeated patterns in polyphonic music. In: Proceedings of Mathematics and Computation in Music, pp. 168–179 (2022)
9. Lemström, K.: String Matching Techniques for Music Retrieval. Ph.D. thesis, University of Helsinki (2000)
10. Lemström, K.: Towards more robust geometric content-based music retrieval. In: Proceedings of International Society for Music Information Retrieval Conference, pp. 577–582 (2010)
11. Lemström, K., Laitinen, M.: Transposition and time-warp invariant geometric music retrieval algorithms. In: International Conference on Multimedia and Expo, pp. 1–6 (2011)
12. Lemström, K., Wiggins, G.: Formalizing invariances for content-based music retrieval. In: Proceedings of the 10th International Society for Music Information Retrieval Conference, pp. 591–596 (2009)
13. Lubiw, A., Tanur, L.: Pattern matching in polyphonic music as a weighted geometric translation problem. In: Proceedings of International Conference on Music Information Retrieval, pp. 289-296 (2004)
14. Meredith, D.: RECURSIA-RRT: recursive translatable point-set pattern discovery with removal of redundant translators. In: Proceedings International Workshop on Machine Learning and Music, pp. 485–493 (2019)
15. Meredith, D.: Understanding and compressing music with maximal transformable patterns. In: Rauterberg, M. (ed.) HCII 2023. LNCS, vol. 14035, pp. 309–325. Springer, Cham (2023). https://doi.org/10.1007/978-3-031-34732-0_24
16. Meredith, D., Lemström, K., Wiggins, G.: Algorithms for discovering repeated patterns in multidimensional representations of polyphonic music. J. New Music Res. **31**(4), 321–345 (2002)

17. Mongeau, M., Sankoff, D.: Comparison of musical sequences. Comput. Humanit. **24**, 161–175 (1990)
18. Romming, C.A., Selfridge-field, E.: Algorithms for polyphonic music retrieval: the Hausdorff metric and geometric hashing. In: Proceedings of International Conference on Music Information Retrieval, pp. 457–462 (2007)
19. Typke, R., Giannopoulos, P., Veltkamp, R.C., Wiering, F., Van Oostrum, R.: Using transportation distances for measuring melodic similarity. In: Proceedings of International Conference on Music Information Retrieval (2003)
20. Ukkonen, E., Lemström, K., Mäkinen, V.: Geometric algorithms for transposition invariant content-based music retrieval. In: Proceedings of International Conference on Music Information Retrieval, pp. 193–199 (2003)

Tonnetze and Tori for the 19-, 31-, and 53-Tone Equal Temperaments

Luis Nuño(✉)

Communications Department, Polytechnic University of Valencia, Camino de Vera S/N, 46022 Valencia, Spain
lnuno@dcom.upv.es, harmonicwheel@gmail.com

Abstract. The Tonnetz is a graphical representation of consonant intervals and chords, together with their corresponding parsimonious transformations, which are the basis of neo-Riemannian theory. It has proved to be useful for analyzing some chromatic passages of nineteenth century, as well as for composition. Furthermore, it has led to several generalizations and variants, which have been applied to different musical styles. In most cases, it implements the ordinary 12-tone equal temperament or 12-TET, which has a finite number of notes and allows to transform it into a torus. However, regarding consonances, they are only exactly matched in just intonation, which has infinite notes. Therefore, other finite tuning systems were developed to better approximate it, among them 19-, 31-, and 53-TET, which in turn gave rise to related new instruments. In this paper, the Tonnetze and tori for these tuning systems are developed, which are intended to be helpful when using them.

Keywords: Tonnetz · Torus · 12-TET · 19-TET · 31-TET · 53-TET · Neo-Riemannian theory

1 Introduction

Historically, the concept of consonance, in music, is normally associated to the impression of stability, repose, and pleasure produced on a listener by certain combinations of sounds [1]. According to music theory [2, 3], there are 6 types of consonances, which are 4 intervals, namely octave, perfect fifth, major and minor thirds; and 2 triads, namely major and minor (which, in fact, consist of consonant intervals).

All these consonances can be represented graphically on a triangular lattice, where the notes are at the vertices, the consonant triads on the triangles, and the edges link the notes by consonant intervals. This is the (Oettingen/Riemann) Tonnetz, which also has a dual version known as "chicken-wire torus" [4, 5]. In both cases, the two triads connected by the edges are related by P, L, and R operations, which are special types of *parsimonious transformations* [5] (see Sect. 3). These characteristics are the basis of neo-Riemannian theory, which has proved to be useful for analyzing some chromatic passages of nineteenth century, as well as for composition [6, 7]. The Tonnetz has also led to several generalizations and variants, which have been applied to different musical

© The Author(s), under exclusive license to Springer Nature Switzerland AG 2024
T. Noll et al. (Eds.): MCM 2024, LNCS 14639, pp. 255–267, 2024.
https://doi.org/10.1007/978-3-031-60638-0_20

styles [8–11]. Normally, the Tonnetz implements the 12-tone equal temperament or 12-TET, so although being infinite in two dimensions (a plane), it is periodic in three directions (the three types of intervals). Therefore, it can easily be transformed into a torus, which is finite and more compact, but three-dimensional [4, 5].

Regarding consonances, they can only be "exact" if some of the first harmonics of the notes match each other, mainly the 3rd or 5th, apart from the octaves (powers of 2), which are inherent to most tuning systems (including those considered here). But both the 3rd and 5th are only matched in *just intonation.*[1] In *Pythagorean tuning*, only the 3rd (perfect fifth);[2] in *1/4-comma meantone temperament*, the 5th (major third);[3] and in *1/3-comma meantone temperament*, the ratio of the 3rd to the 5th (minor third)[4] [12, 13]. These systems, however, have the drawback of being "divergent", in the sense that the number of notes within an octave tends to infinity. Thus, some finite tuning systems are used to approximate them, such as 53-, 31-, 19- and, of course, 12-TET, where the octave is divided into a finite number of equally-sized parts (in a logarithmic frequency scale), here called *steps*. The Zarlino's cembalo (1558) is an example of 19-TET keyboard, while the Vicentino's archicembalo (1555), the Collona's Sambuca Lincea (1618) and the Fokker organ (1950) are examples of 31-TET keyboards [12, 13].

In [14], mathematical criteria for selecting relevant subsets from the infinite Tonnetz in just intonation are given, which include 12-, 19-, 31-, and 53-tone systems. Other authors followed different criteria, not necessarily related to harmonics, for obtaining specific N-TET systems, together with their associated Tonnetze. For example, [15] focuses on the chromatic (melodic), diatonic (tonal), and triadic (harmonic) structures, leading to $N = k(k+1), k \in \mathbb{N}$, and analyzes the cases N = 12, 20, 30, and 42. And [6] considers both the "abstract" (generic) and "parsimonious" Tonnetze, the latter yielding $N = 3(x+1), x \in \mathbb{N}$, and analyzes the cases N = 12, 18 (third-tone steps), and 24 (quarter-tone steps).

2 The 12-, 19-, 31-, and 53-Tone Equal Temperaments

Although the 12-, 19-, 31-, and 53-TET are well known, it is worth reviewing now how they can be obtained from the above-mentioned tuning systems that match different harmonics. In fact, this can be done by simple reasoning as follows:

1. First, considering only the 2nd and 3rd harmonics (octave and perfect fifth), as in Pythagorean tuning, there must be a whole number of notes from C to G, frequency ratio 3/2, and from G to the next C, 4/3. The quotient of these intervals, in cents,[5] is $701.955/498.045 = 1.40942 \cong 1.4 = 7/5$. This means that a perfect fifth can consist of 7 steps (in this case, semitones) and a perfect fourth of 5. Thus, an octave will consist of $7 + 5 = 12$ steps and, consequently, 12 notes. This is the 12-TET.

[1] Here, we refer to the 5-limit just intonation, which involves the harmonics 2, 3, and 5.

[2] This system is a sequence of perfect fifths, frequency ratio 3/2. Then, starting from C, the difference between note E and the 5th harmonic, frequency ratio 5/4, is the *syntonic comma*.

[3] This is achieved by shortening the perfect (Pythagorean) fifths by 1/4 of syntonic comma.

[4] This is achieved by shortening the perfect (Pythagorean) fifths by 1/3 of syntonic comma.

[5] Given the frequency ratio of two notes, b/a, the interval in cents is given by $1200\log_2(b/a)$.

2. By also including the 5th harmonic (thus giving rise to the major and minor thirds), as in just intonation, if we start, for example, from C, there will be two different E's: one corresponding to the 5th harmonic, frequency ratio 5/4, and the other obtained by perfect fifths, $(3/2)^4/4$, which will be equal to 2 tones, each with frequency ratio $(3/2)^2/2$. Thus, the intervals from these two E's to F (4/3 from C) will give rise to two types of semitones.[6] Their quotient, in cents, is $111.731/90.225 = 1.2384 \cong 1.25 = 5/4$, giving 5 steps for the greater (chromatic) semitone and 4 steps for the lesser (diatonic). By considering a tone as the sum of these two types of semitones, it will consist of 9 steps. And, because an octave consists of 5 tones plus 2 diatonic semitones, this makes $5 \cdot 9 + 2 \cdot 4 = 53$ steps and notes. This is the 53-TET.
3. Considering now the 1/4-comma meantone temperament, there must be a whole number of notes from C to E and from E to G. In this system, the quotient of these intervals, in cents, is $386.314/310.265 = 1.24511 \cong 1.25 = 5/4$ or, better, 10/8, because it is required an even number of notes for a major third to be equal to 2 tones. Thus, 10 steps for a major third (5 steps per tone) and 8 steps for a minor third. Because from G to the next C there is one tone plus a minor third, an octave will consist of $(10 + 8) + 5 + 8 = 31$ steps and notes. This is the 31-TET. Here, a diatonic semitone consists of 3 steps and a chromatic one of 2.
4. And, considering the 1/3-comma meantone temperament, there must be a whole number of notes from C to E♭ and from E♭ to G. In this system, the quotient of the latter interval to the former (the greater over the lesser), in cents, is $379.145/315.641 = 1.20119 \cong 1.2 = 6/5$. So, 5 steps for a minor third and 6 steps for a major third, which will be equal to 2 tones, each with 3 steps. Because from G to the next C there is one tone plus a minor third, an octave will consist of $(5 + 6) + 3 + 5 = 19$ steps and notes. This is the 19-TET. Here, a diatonic semitone consists of 2 steps and a chromatic one of 1.

Table 1 shows the pitches of all notes in these tuning systems, in cents, with respect to C, where some enharmonic notes are also indicated. The note names are always assigned following perfect fifth or fourth intervals, which, in 19-, 31-, and 53-TET, correspond, respectively, to 11 or 8, 18 or 13, and 31 or 22 steps. This causes that, in 53-TET, a (pure) major third (386.314 cents) is not the interval from C to E, but from C to F♭; and a (pure) minor third (315.641 cents) is not the interval from C to E♭, but from C to D♯. Additionally, to avoid multiple accidentals and simplify the notation, the symbols "+" and "–" are used, which stand, respectively, for raising or lowering the note by a step in this system (22.642 cents), which is called *Holdrian comma* (from W. Holder). It is important to highlight that all note names are consistent within their own tuning systems, but their relations among other systems are not generally obvious. For instance, A𝄪 in 31-TET is related to B♭ in 19-TET (enharmonic to A𝄪 in this system), while its pitch is nearer to B. And this fact is more striking when the 53-TET is involved. Thus, for instance, F♯! in 31-TET is related to G♭– and G♭ in 53-TET, but not to F♯![7] Fig. 1 shows basic keyboards for 19-, 31-, and 53-TET, where, in the last two cases, the notes on the grey and black keys raise upward.

[6] The two E's are a syntonic comma apart, frequency ratio 81/80, that is, 21.506 cents.

[7] Note that, for example, the D major triad in 53-TET is formed by notes D, G♭, A.

Table 1. Notes of 12-, 19-, 31-, and 53-TET, and their pitches, in cents, with respect to C.

12-TET	19-TET	31-TET	53-TET
C 1200	**C 1200**	**C 1200**	**C 1200**
	B♯ → C♭ 1136.8	B♯ 1161.3	D♭♭ 1177.4
			A×+ → D♭♭- 1154.7
		C♭ 1122.6	A× 1132.1
B 1100	**B 1073.7**	**B 1083.9**	**B 1109.4**
A♯ → B♭ 1000	B♭ 1010.5	C♭♭ → A× 1045.2	C♭ 1086.8
			C♭- 1064.2
		B♭ 1006.5	A♯+ 1041.5
			A♯ 1018.9
	A♯ 947.4	A♯ 967.7	B♭ 996.2
			C♭♭ → B♭- 973.6
		B♭♭ 929.0	G×+ 950.9
			G× 928.3
A 900	**A 884.2**	**A 890.3**	**A 905.7**
G♯ = A♭ 800	A♭ 821.1	G× 851.6	B♭♭ 883.0
			B♭♭- 860.4
		A♭ 812.9	G♯+ 837.7
			G♯ 815.1
	G♯ 757.9	G♯ 774.2	A♭ 792.5
			A♭- 769.8
		A♭♭ 735.5	F×+ 747.2
			F× 724.5
G 700	**G 694.7**	**G 696.8**	**G 701.9**
G♭ → F♯ 600	G♭ 631.6	F× 658.1	A♭♭ 679.2
			A♭♭- 656.6
		G♭ 619.4	E× → F♯+ 634.0
			F♯ 611.3
	F♯ 568.4	F♯ 580.6	G♭ 588.7
			G♭- 566.0
		E× → G♭♭ 541.9	E♯+ 543.4
			E♯ 520.8
F 500	**F 505.3**	**F 503.2**	**F 498.1**
	F♭ → E♯ 442.1	E♯ 464.5	G♭♭ 475.5
			G♭♭- → D×+ 452.8
		F♭ 425.8	D× 430.2
E 400	**E 378.9**	**E 387.1**	**E 407.5**
D♯ → E♭ 300	E♭ 315.8	F♭♭ → D× 348.4	F♭ 384.9
			F♭- 362.3
		E♭ 309.7	D♯+ 339.6
			D♯ 317.0
	D♯ 252.6	D♯ 271.0	E♭ 294.3
			F♭♭ → E♭- 271.7
		E♭♭ 232.3	C×+ 249.1
			C× 226.4
D 200	**D 189.5**	**D 193.5**	**D 203.8**
D♭ → C♯ 100	D♭ 126.3	C× 154.8	E♭♭ 181.1
			E♭♭- 158.5
		D♭ 116.1	B× → C♯+ 135.8
			C♯ 113.2
	C♯ 63.2	C♯ 77.4	D♭ 90.6
			D♭- 67.9
		B× → D♭♭ 38.7	B♯+ 45.3
			B♯ 22.6
C 0	**C 0**	**C 0**	**C 0**

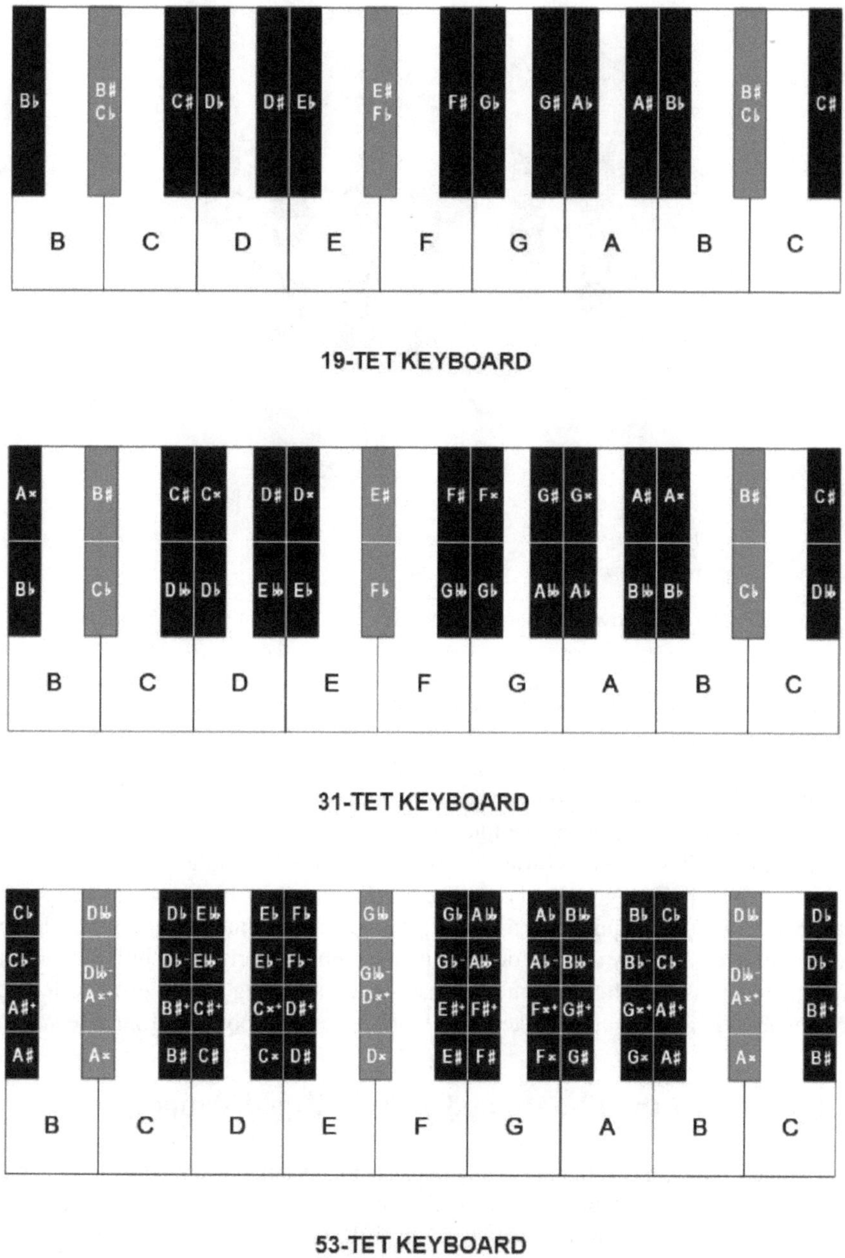

Fig. 1. Basic keyboards for 19-, 31-, and 53-TET.

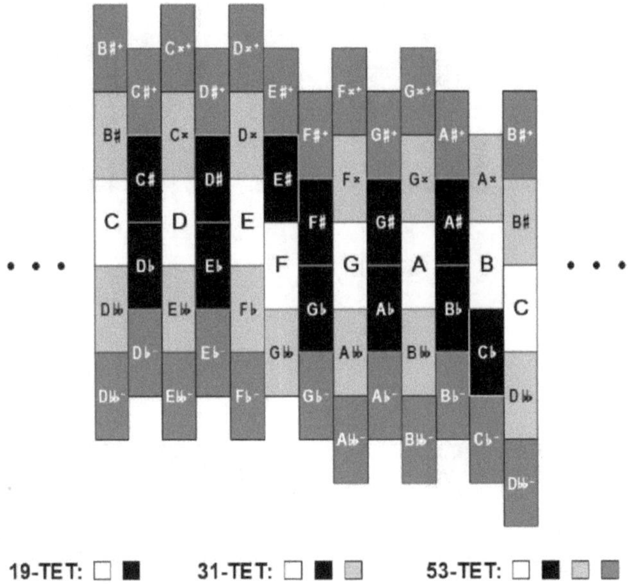

Fig. 2. Alternative keyboards for 19-, 31-, and 53-TET ("Fokker style").

Alternatively, Fig. 2 shows other keyboards (in "Fokker style"), where these three systems are superimposed to show how the different notes are progressively added: 19-TET comprises the white and black keys, 31-TET adds the light grey keys, and 53-TET also adds the dark grey keys, where the notes are assigned the symbols "+" and "–", which makes the keyboard simpler and more uniform. Note that, in the vertical direction, the pitches raise downward in 19- and 31-TET (D♭ higher than C♯), but raise upward in 53-TET (C♯ higher than D♭). The block of notes in this figure can be repeated horizontally (with the required vertical shift) to increase the number of octaves, but can also be repeated vertically (up and down, with the required horizontal shifts) to facilitate playing some passages (but without increasing the pitch range). The great differences between 53-TET and the other systems will affect the corresponding Tonnetze and tori.

3 Tonnetze for the 19-, 31-, and 53-Tone Equal Temperaments

As explained above, the Tonnetz usually consists in a triangular lattice where the notes are at the vertices, the consonant triads (major and minor) on the triangles, and the edges link the notes by consonant intervals (perfect fifth, major and minor thirds). Figure 3 shows such a lattice for the 31-TET. In this case, the horizontal lines correspond to the perfect fifths and the two oblique lines to the major and minor thirds. These intervals also relate the two triads linked by them by *P*, *L*, and *R* operations[8] [4]. These are special

[8] *P*, *L*, and *R* stand for Parallel, Leading-tone exchange, and Relative, which are basic operations between major and minor triads. For example, they map, respectively, C major to C minor, C major to E minor, and C major to A minor; and vice versa.

types of *parsimonious transformations*[9] [5], where 2 notes are sustained and the third one is moved by 1 or 2 semitones. A region defined by a thick solid line contains the notes and chords only once, except the note A𝄪, which is repeated at the four corners.

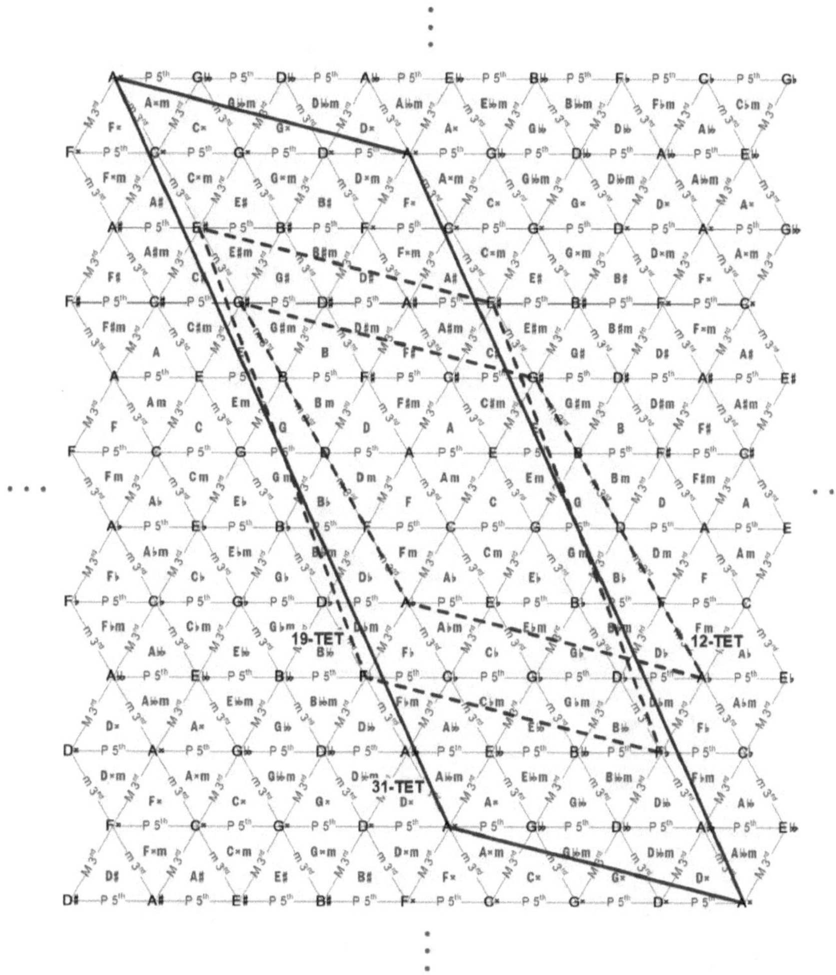

Fig. 3. The 31-TET Tonnetz, together with the 12- and 19-TET Tonnetze.

A dashed line defines the Tonnetz for the 19-TET, which is a subset of 31-TET (the note F♭ at the two lower corners is enharmonic, in 19-TET, to E♯). And another dashed line defines the Tonnetz for the 12-TET, which in turn is a subset of 19-TET (the note A♭ at the two lower corners is enharmonic, in 12-TET, to G♯). Moreover, in this case, the

[9] Transformation between two chords, with the same cardinality, performed by sustaining the common notes and moving the rest of them by one or two semitones, but normally no more than two semitones in all.

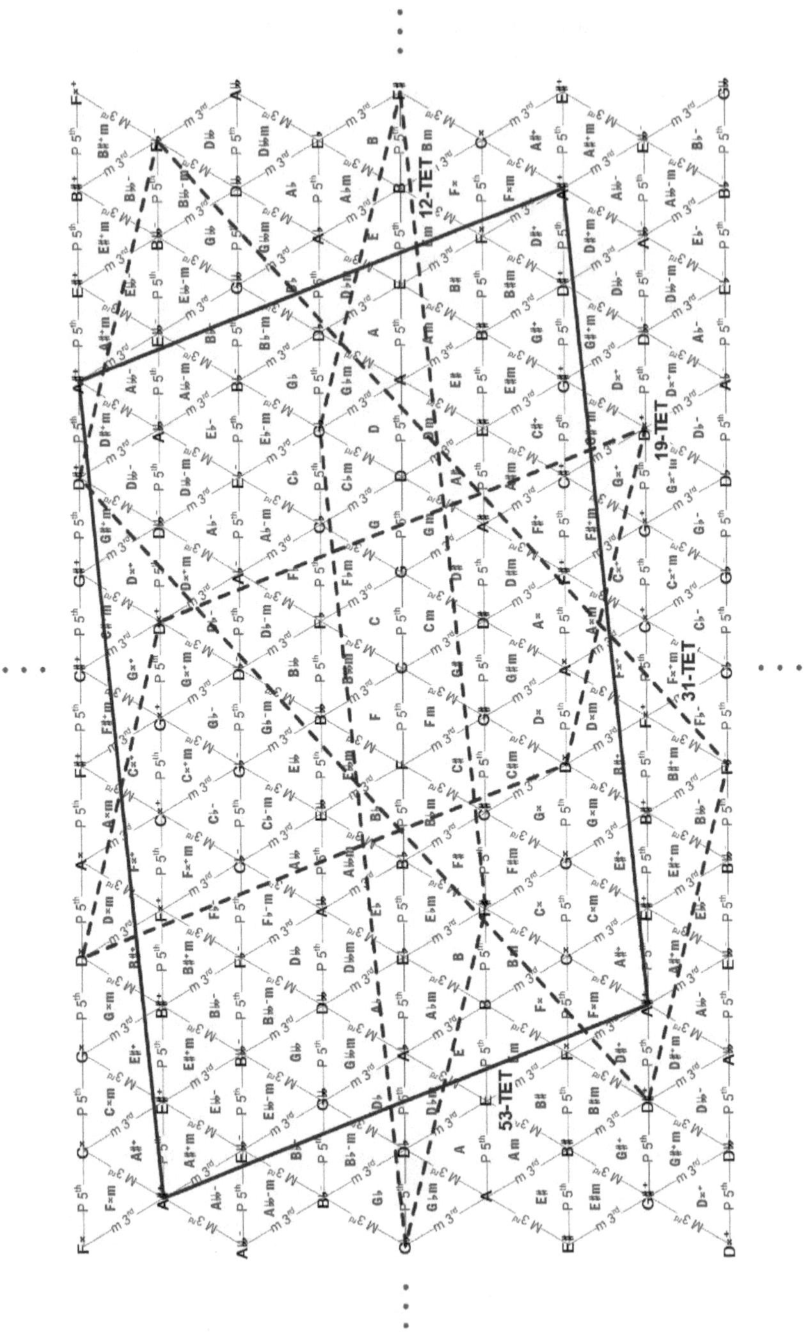

Fig. 4. The 53-TET Tonnetz and the approximated 12-, 19-, and 31-TET Tonnetze.

four notes at the left side are repeated at the right one. A correlation is evident between these three Tonnetze and the corresponding groups of notes in Fig. 2.

Additionally, in these Tonnetze, if we start, for example, from C and follow the series of perfect fifths (G, D, A, E,...), the third note (A) forms a minor third with it, and the fourth note (E), a major third, so these two notes are linked to C because they form consonant intervals with it. This is the reason why the three Tonnetze have similar widths (three different notes along each horizontal line). However, the Tonnetz for the 53-TET will be different, as notes A and E in this tuning system do not form consonant intervals with C (a minor third below C is B♭♭, and a major third above, F♭). In this case, the first notes linked to C in the series of perfect fifths are the eighth (G♯, a major third below C) and ninth (D♯, a minor third above). The result is shown in Fig. 4, where, to allow for comparisons, also the 12-, 19-, and 31-TET are included. But, because in these cases most of the note names will not coincide with the real ones, they are represented as close as possible to their actual pitches, considering the values in Table 1.

4 Tori for the 19-, 31-, and 53-Tone Equal Temperaments

The different spaces resulting from N-TET Tonnetze with arbitrary trichords are studied in [16]; and, in our cases, they are single tori. In fact, as can be seen, each of the 12-, 19-, and 31-TET Tonnetze in Fig. 3, and the 53-TET Tonnetz in Fig. 4, has the same note at the four corners (considering the enharmonic notes in 12- and 19-TET). Therefore, each one can easily be transformed into a torus, which is a finite and more compact geometry, although three-dimensional. As an example, Fig. 5 shows the 12-TET Tonnetz, changed into a rectangle, and indicates how to bend it (steps 1 and 2) to obtain the torus. The result is shown in Fig. 6, which, for clarity reasons, only includes the notes, the lines of minor thirds and half of the line of perfect fifths.

Angles α and β in these figures start from note C, and their values can be expressed as functions of r, a variable that follows the line of perfect fifths, so that $r = 0$ corresponds to note C, $r = 1$ to note G, and so on. Thus, when r completes a cycle (the 12 notes), β completes 4 turns while α completes -1 turn (that is, 1 turn in the opposite direction), so these angles, in degrees, for the 12-TET, are given by

$$\alpha_{12} = -r \cdot 360°/12 = -r \cdot 30°, \beta_{12} = r \cdot 4 \cdot 360°/12 = r \cdot 120° \quad (1)$$

Similarly, Figs. 7, 8 and 9 show the tori for 19-, 31-, and 53-TET, where the corresponding angles α and β are given by

$$\alpha_{19} = -r \cdot 360°/19 \cong -r \cdot 18.95°, \beta_{19} = r \cdot 6 \cdot 360°/19 \cong r \cdot 113.68° \quad (2)$$

$$\alpha_{31} = -r \cdot 360°/31 \cong -r \cdot 11.61°, \beta_{31} = r \cdot 10 \cdot 360°/31 \cong r \cdot 116.13° \quad (3)$$

$$\alpha_{53} = r \cdot 360°/53 \cong r \cdot 6.79°, \beta_{53} = r \cdot 6 \cdot 360°/53 \cong r \cdot 40.75° \quad (4)$$

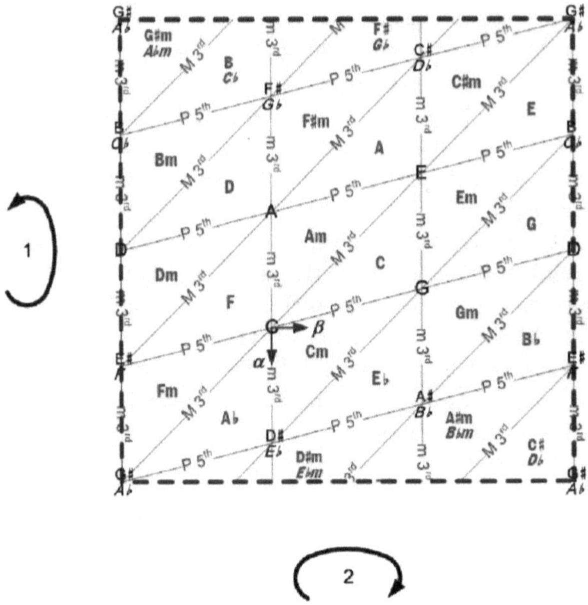

Fig. 5. Transformation of the 12-TET Tonnetz into a torus.

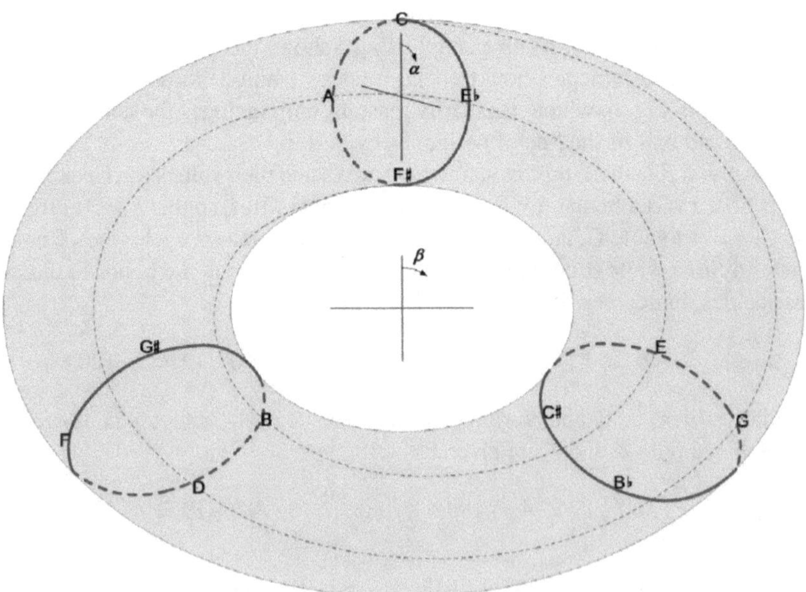

Fig. 6. 12-TET torus. Minor-third lines (thick) and half of the perfect-fifth line (thin).

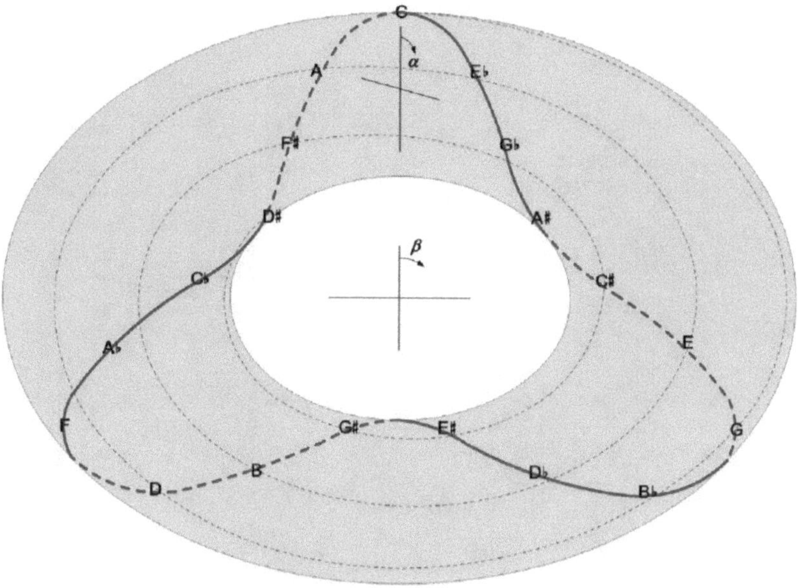

Fig. 7. 19-TET torus. Minor-third line (thick) and half of the perfect-fifth line (thin).

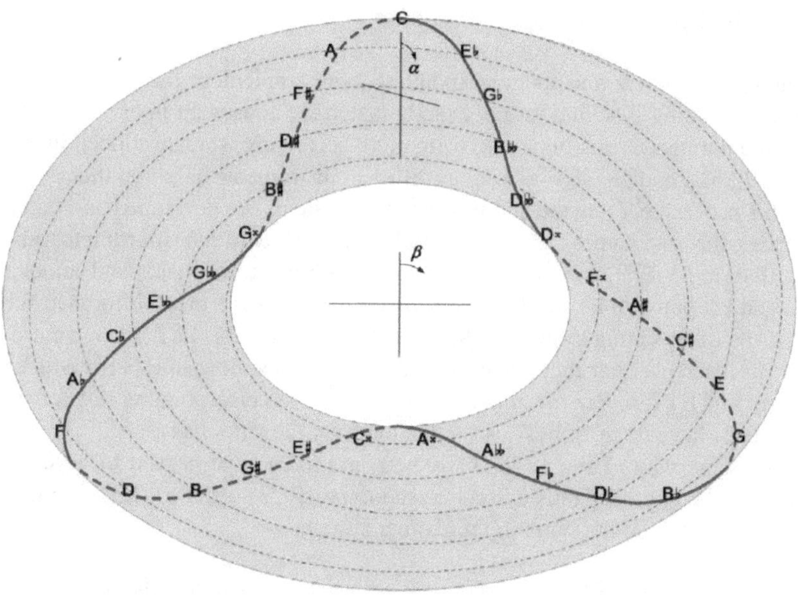

Fig. 8. 31-TET torus. Minor-third line (thick) and half of the perfect-fifth line (thin).

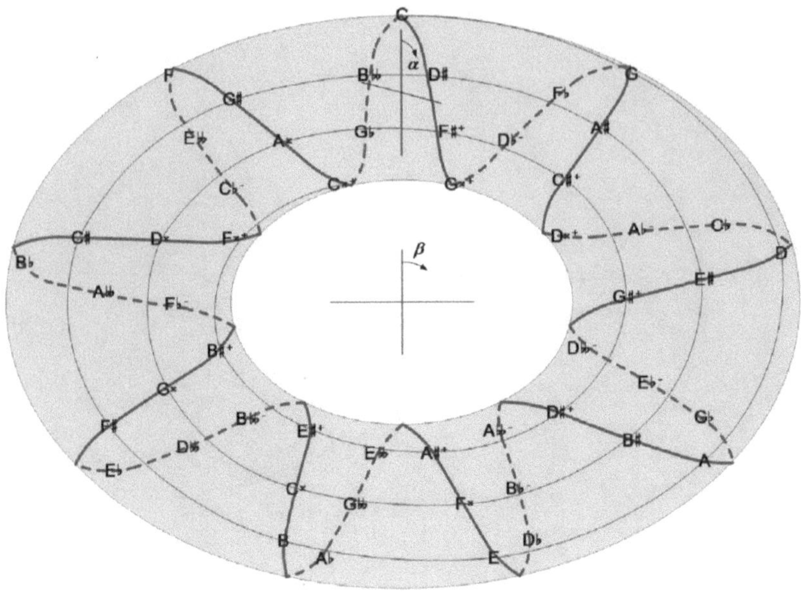

Fig. 9. 53-TET torus. Minor-third line (thick) and half of the perfect-fifth line (thin).

5 Conclusions

Normally, the Tonnetz and its corresponding torus implement the ordinary 12-TET. Because this tuning system does not exactly match the consonant intervals and chords, other finite tuning systems matching them more accurately were developed in the past, such as 19-, 31-, and 53-TET, as well as different instruments based on them.

In this paper, the Tonnetze and tori for these systems are developed and compared. In all cases, the note names are assigned following perfect fifth or fourth intervals. This causes that, in 53-TET, for example, the note C do not form major and minor thirds with E and E♭, but with F♭ and D♯, respectively. This makes a great difference between this and the other tuning systems. Thus, the 12-, 19-, and 31-TET Tonnetze are quite similar, in the sense that they have three different notes along the perfect-fifth lines, while the 53-TET Tonnetz has nine. Consequently, the corresponding tori for the first three systems have three "lobes", while the torus for the latter has nine.

In the same way as the 12-TET Tonnetz and torus have proved to be useful for analyzing and composing music in this system, the 19-, 31-, and 53-TET Tonnetze and tori are intended to do the same in these other systems.

Disclosure of Interests. The author has no competing interests to declare that are relevant to the content of this article.

References

1. Tenney, J.: A History of 'Consonance' and 'Dissonance.' Excelsior Music Publishing, New York (1988)

2. Schönberg, A.: Theory of Harmony, 3rd edn. University of California Press, Berkeley (1983)
3. Piston, W.: Harmony, 5th edn. W. W. Norton and Co., New York (1988)
4. Crans, A.S., Fiore, T.M., Satyendra, R.: Musical actions of dihedral groups. Amer. Math. Monthly **116**(6), 479–495 (2009)
5. Douthett, J., Steinbach, P.: Parsimonious graphs: a study in parsimony, contextual transformations, and modes of limited transposition. J. Music Theory **42**(2), 241–263 (1998)
6. Cohn, R.: Neo-Riemannian operations, parsimonious trichords, and their 'Tonnetz' representations. J. Music Theory **41**(1), 1–66 (1997)
7. Cohn, R.: Audacious Euphony: Chromatic Harmony and the Triad's Second Nature. Oxford University Press, New York (2012)
8. Gollin, E.: Some aspects of three-dimensional 'Tonnetze.' J. Music Theory **42**(2), 195–206 (1998)
9. Tymoczko, D.: The generalized Tonnetz. J. Music Theory **56**(1), 1–52 (2012)
10. Briginshaw, S.B.P.: A Neo-Riemannian approach to jazz analysis. Nota Bene Can. Undergraduate J. Musicol. **5**(1), 57–87 (2012)
11. Nuño, L.: Representation of harmonies on the harmonic wheel. In: 6th International Conference on Technologies for Music Notation and Representation (TENOR), pp. 24–32, Hamburg University for Music and Theater, Hamburg (2020). https://doi.org/10.5281/zenodo.4105918
12. Barbour, J.: Tuning and Temperament: A Historical Survey. Dover Publications, Inc., Mineola (2004 [1951])
13. Goldáraz, J.: Afinación y temperamento en la música occidental (Tuning and temperament in Western music). Alianza Editorial, S. A., Madrid (1998)
14. Žabka, M.: The Minkowski geometry of numbers applied to the theory of tone systems. In: Yust, J., Wild, J., Burgoyne, J.A. (eds.) MCM 2013. LNCS (LNAI), vol. 7937, pp. 226–240. Springer, Heidelberg (2013). https://doi.org/10.1007/978-3-642-39357-0_18
15. Balzano, G.: The group-theoretic description of 12-fold and microtonal pitch systems. Comput. Music. J. **4**(4), 66–84 (1980)
16. Catanzaro, M.J.: Generalized Tonnetze. J. Math. Music **5**(2), 117–139 (2011)

A Model of Scores as Abstract Syntactic Trees

Gonzalo Romero-García[1(✉)], Carlos Agón[1], and Isabelle Bloch[2]

[1] Sorbonne Université, CNRS, IRCAM, STMS, Paris, France
{romero,agonc}@ircam.fr
[2] Sorbonne Université, CNRS, LIP6, Paris, France
isabelle.bloch@sorbonne-universite.fr

Abstract. This paper deals with the structure of a musical piece. The score is modeled as an Abstract Syntactic Tree (AST) to account for the hierarchy of its elements. Formal definitions of harmony, texture and instrumentation are proposed and constitute the main components of the model. Concatenation and parallelization operators are then proposed to combine these components and organize them in a tree structure. This approach is illustrated on some examples.

Keywords: Harmony · Texture · Instrumentation · Abstract Syntactic Tree · Hierarchical score modeling

1 Introduction

Through centuries, Western classical music has been engraved in scores with staff notation. This way of laying down music is very well suited for sharing music and allows mutual comprehension between the composer and the interpreter. However, staff notation does not account, in general, for the hierarchical nature of music, nor for the structure that underlies a musical piece.

In this article, we propose to model a score as an *Abstract Syntactic Tree* (AST). This way, we can express the structure of a musical piece in a hierarchical way, which has proven to be very useful for analyzing and composing music [5,11–15].

To organize a score into a tree has already been proposed, for instance through the MusicXML [8] format that uses XML notation and thus a tree structure. However, this format is a pure translation of the score information, and the hierarchy is mainly structured in bars. In our model, we aim to go beyond the bars decomposition into a more high level one.

Modeling a score as an AST (as in [16]) requires making several key decisions, and two fundamental questions must be answered: What are the most basic materials? And, how can they be combined? In computational terms, these questions translate to: What are the leaves of the tree, and what are the nodes? These questions have profound philosophical implications, as their answers are essentially equivalent to asking: What is music made of, and how is it constructed?

© The Author(s), under exclusive license to Springer Nature Switzerland AG 2024
T. Noll et al. (Eds.): MCM 2024, LNCS 14639, pp. 268–279, 2024.
https://doi.org/10.1007/978-3-031-60638-0_21

It is a common approach to say that music is made by combining *frequency* with *time*. To this, we add a third component: *timbre*. Using these three dimensions, we will construct our computational model. More precisely, we will identify three musical notions—*harmony*, *texture*, and *instrumentation*—and provide them with formal mathematical definitions. These elements will be shown to be the essential materials of our model.

To combine these materials, we introduce two fundamental operators: *concatenation* and *parallelization*. These operators are analogous to those proposed in [9,10]. They complete our grammar and allow for the construction of ASTs.

Once we have constructed an AST, we can compile it into various musical representations such as a MIDI file, MusicXML, piano roll, audio file, and others. In this article, we focus on the compilation into a MIDI file. Note that audio and piano rolls can be generated from the MIDI file.

The strength of our model is as follows: constructing our AST allows for a direct interpretation of the structure and hierarchy of a musical piece and the relationships between its components. Successfully expressing common repertoire pieces as ASTs enables us to assert that we have conducted an analysis of the piece. This has deep implications, since it transforms the analysis problem into the construction of an AST. In this sense, our work is close to the one proposed in [18].

Additionally, our approach offers an alternative to traditional score notation for music composition. While the staff notation is the standard method for engraving Western classical music, this process typically requires a learning curve that can span years. By contrast, composing through an AST may be more intuitive for newcomers, especially when presented interactively, as in a Computer Assisted Composition (CAC) tool, like OpenMusic [1,4]. Currently, we offer a Python library named `harmtex` (that can be found in the code that accompanies this article) and an XML Schema for expressing the AST as an XML document.

The article is structured as follows. We first expose the main three materials: harmony, texture and instrumentation. These are introduced in mathematical terms to provide a robust framework that extends beyond mere computational implementation. Then, we define the three operators: tensor contraction, parallelization and concatenation, which complete the grammar necessary for our AST model. Finally, we present a computational implementation in the form of a Python library and XML Schema.

All the code and examples from this article are accessible in the public repository that accompanies this article: https://github.com/Manza12/MCM-2024.

2 Harmony, Texture and Instrumentation

The concepts of *harmony*, *texture*, and *instrumentation* are subject to various interpretations and hold different meanings across different domains. In this article, we provide mathematical definitions for each, tailored specifically to our model yet capturing a significant part of their traditional meaning.

2.1 Harmony

Harmony is a term that is very used in musicology, commonly referring to the distribution and relation between pitches.

In this article, we consider that a *harmony* is simply a list of chords. In our framework, a *chord* is a finite set of pitches. More formally, if we identify the musical pitches (..., C3, C♯3, D3, ..., C4, ...) with the integer numbers[1] denoted by $\mathbb{Z}$, a **chord** is a finite subset of $\mathbb{Z}$, that we denote by $C \in \mathcal{P}_{\mathrm{F}}(\mathbb{Z})$, where $\mathcal{P}_{\mathrm{F}}(\mathbb{Z})$ denotes the finite subsets of $\mathbb{Z}$. For instance, $\{\mathrm{C3}, \mathrm{E3}, \mathrm{G3}\} = \{60, 64, 67\}$ is a C major chord. Notice that, with this definition, the empty set is a chord that corresponds to the absence of pitch and thus to a rest, a single note is a chord of cardinal 1, an interval is a chord of cardinal 2, a scale has varying cardinality, usually from 5 to 8, etc. We denote by $\mathcal{C}$ the set of all chords.

Since a **harmony** is a list of chords, we can formally say that $H \in \mathcal{C}^N$ where N is a finite set of indices, i.e. $|N| < \infty$. In general, we choose $N = \{0, 1, ..., n-1\}$, for some $n \in \mathbb{N}$, and think of H as an ordered list.

This definition of harmony is simple yet powerful; for instance, when we talk about the *harmony* of a piece, we may use the list of chords of the piece[2]. In a similar way, the *harmony* of a jazz standard is the ordered list of chords that forms this standard.

In this work, we use harmonies in a particular way, which is in combination with *textures*. Let us then define textures.

2.2 Texture

The notion of *texture* has a large amount of interpretations, depending on the domain. In this article, we refer to what is commonly called *symbolic* or *compositional texture* [7], which is rather recent compared to the study of harmony [6].

In this framework, we aim to give a precise definition of symbolic texture; it is the analogue to harmony in time: instead of pitches, we use what we call *hits*, which are pairs of onsets and durations; instead of chords, we use *rhythms*, that are sets of hits; and a *texture* is a list of rhythms.

While saying that a texture is simply a list of rhythms might appear as oversimplification, we will see in the next section that combining a single texture with different harmonies captures the essence of a symbolic texture paired with a chord progression.

In order to be mathematically precise, we need to choose a model for time and define hits, rhythms and textures. Since our main target is to model scores, we consider that time inside a score is modeled through the use of rational numbers, denoted by $\mathbb{Q}$, where we associate 1 to 𝅝, $\frac{1}{2}$ to 𝅗𝅥, $\frac{1}{4}$ to ♩, and so on[3].

---

1 The standard way to do that is to associate each pitch with its corresponding MIDI number, i.e., C3 → 60, C♯3 → 61, D3 → 62, C4 → 72, etc.

2 The question that naturally arises is how these chords are chosen, but this will be the subject of the next sections.

3 We may model ties by the sum of two rational numbers and tuplets by using different numbers in the denominator.

Then, the duration of a hit can be modeled by a positive rational number. It is a bit more delicate how to model the onset; we say that there is a 0 that acts as origin and then the number associated to the onset is the time (expressed in rationals) elapsed between this origin and the onset. In the case of notes of a piece, the origin would be the start of the piece[4]. With this formalization, a **hit** $h = (o, d)$ is an element of $\mathbb{Q} \times \mathbb{Q}^{+}$.

Let us move on to the definition of rhythms. We define a **rhythm** R as a finite set of hits, thus an element of $\mathcal{P}_{\mathrm{F}}(\mathbb{Q} \times \mathbb{Q}^{+})$. For instance, the rhythm , where the | symbol accounts for the origin, is associated with the set $\{(0, \frac{1}{4}), (\frac{1}{2}, \frac{1}{8}), (\frac{5}{8}, \frac{1}{8}), (\frac{3}{4}, \frac{1}{4})\}$. We denoted by $\mathcal{R}$ the set of all rhythms.

Similarly to the case of harmony, a **texture** is a list of rhythms, that is, an element of $\mathcal{R}^N$, where N is a finite set.

2.3 Instrumentation

Instrumentation has been an important domain of music composition and musicology already developed by Berlioz [3], and is today an essential skill for music composition. As in the case of harmony and texture, we expose it here in a rigorous mathematical way.

Let us consider a set of instruments I. We call a **section**[5] S a finite subset of instruments, i.e. an element of $\mathcal{P}_{\mathrm{F}}(I)$. We call $\mathcal{S}$ the set of all sections[6]. Now, we can define an **instrumentation** O[7] as a list of sections, i.e. $O = (S_n)_{n \in N}$, where N is a finite set and S_n is a section for each n in N.

We now have a formal setting for harmony, texture and instrumentation, that is summarized in Table 1.

Table 1. Main terminology employed through this article for frequency, time and timbre. We consider that N is a finite set, i.e. $|N| < \infty$.

FREQUENCY ELEMENTS

Term	Notation	Belonging to
pitch	p	$\mathbb{Z}$
chord	C	$\mathcal{C} = \mathcal{P}_{\mathrm{F}}(\mathbb{Z})$
harmony	H	$\mathcal{C}^N$

TIME ELEMENTS

Term	Notation	Belonging to
hit	h	$\mathbb{Q} \times \mathbb{Q}^{+}$
rhythm	R	$\mathcal{R} = \mathcal{P}_{\mathrm{F}}(\mathbb{Q} \times \mathbb{Q}^{+})$
texture	T	$\mathcal{R}^N$

TIMBRE ELEMENTS

Term	Notation	Belonging to
instrument	i	I
section	S	$\mathcal{S} = \mathcal{P}_{\mathrm{F}}(I)$
instrumentation	O	$\mathcal{S}^N$

4 We may even allow for negative numbers for the case of anacrusis.
5 While the term section usually involves instruments of the same family, in this case a section means the instruments that play the same pattern in a fragment.
6 The dependence on I is omitted to simplify notations.
7 We use the O for instrumentation since I is already used and it refers to the term *orchestration*, usually taken as synonym.

3 Composition Operators

We have now our three main mathematical objects (harmony, texture and instrumentation) that we are using as leaves of our AST. Let us now expose the three main operators that we are using to combine them: tensor contraction, parallelization and concatenation.

3.1 Tensor Contraction

Music in a score is made of *notes*. Each note has a lot of parameters attached to it, but we are keeping four of them: the *pitch*, the *onset*, the *duration* and the *instrument*. These are, as presented before, elements of $\mathbb{Z}$, $\mathbb{Q}$, $\mathbb{Q}^+$ and I, which makes a note an element of $\mathbb{Z} \times \mathbb{Q} \times \mathbb{Q}^+ \times I$, that we will call the space of notes and denote it by $\mathcal{N}$.

We call the finite power set of $\mathcal{N}$, $\mathcal{P}_{\mathrm{F}}(\mathcal{N})$, the **musical space** and we denote it by $\mathcal{M}$. With these definitions, a musical piece might be defined as an element of $\mathcal{M}$.

The aim of tensor contraction is to merge a harmony, a texture and an instrumentation into a set of notes belonging to $\mathcal{M}$, and thus forming a musical piece (or fragment). Let us present the mathematical formalism.

Let N be a finite set of indices. Let $H = (C_n)_{n\in N} \in \mathcal{C}^N$ be a harmony, $T = (R_n)_{n\in N} \in \mathcal{R}^N$ be a texture and $O = (S_n)_{n\in N} \in \mathcal{S}^N$ be an instrumentation. We define the **tensor contraction**[8] generated by H, T and O, denoted by $H \otimes T \otimes O$ as the union of the Cartesian products between C_n, R_n and S_n, i.e.,

$$H \otimes T \otimes O = \bigcup_{n\in N} C_n \times R_n \times S_n \tag{1}$$

$$= \bigcup_{n\in N} \{(p, o, d, i) \in \mathbb{Z} \times \mathbb{Q} \times \mathbb{Q}^+ \times I : p \in C_n,\ (o, d) \in R_n,\ i \in S_n\}. \tag{2}$$

Notice that we require H, T and O to have the same set of indices N, such that we can pair the chords with the rhythms and with the sections.

Let us show how we can use this operator to build music fragments. To this end, we use as example the first 7 measures of Mozart's Symphony No.40 in G minor, K.550, depicted in Fig. 1.

Let us focus first in the accompaniment, that is, the pattern

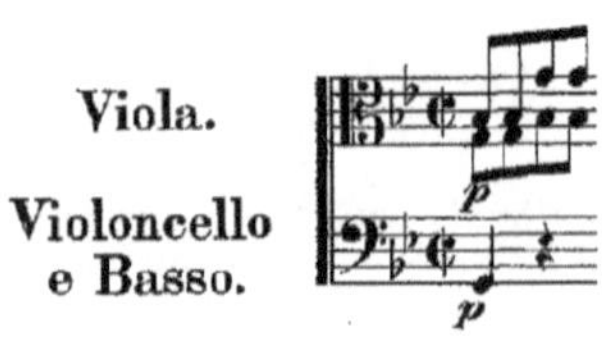

8 The term *tensor contraction* comes from the similarity of this operation with tensor contraction, where the contraction operator is the union instead of the sum.

Fig. 1. First 7 measures of Mozart's Symphony No.40 in G minor, K.550.

that can be expressed as the set of notes

$$
\begin{aligned}
A_0^{\mathrm{a}} = \{&(\mathrm{G2}, 0, \tfrac{1}{4}, \textbf{Vlc.}), (\mathrm{G2}, 0, \tfrac{1}{4}, \textbf{Cb.}),\\
&(\mathrm{G3}, 0, \tfrac{1}{8}, \textbf{Vla.}), (\mathrm{B}\flat 3, 0, \tfrac{1}{8}, \textbf{Vla.}), (\mathrm{G3}, \tfrac{1}{8}, \tfrac{1}{8}, \textbf{Vla.}), (\mathrm{B}\flat 3, \tfrac{1}{8}, \tfrac{1}{8}, \textbf{Vla.}),\\
&(\mathrm{B}\flat 3, \tfrac{2}{8}, \tfrac{1}{8}, \textbf{Vla.}), (\mathrm{G4}, \tfrac{2}{8}, \tfrac{1}{8}, \textbf{Vla.}), (\mathrm{B}\flat 3, \tfrac{3}{8}, \tfrac{1}{8}, \textbf{Vla.}), (\mathrm{G4}, \tfrac{3}{8}, \tfrac{1}{8}, \textbf{Vla.})\}.
\end{aligned}
$$

We can use the following harmony, texture and instrumentation:

$$
\begin{aligned}
H_0^{\mathrm{a}} &= (\{\mathrm{G2}\}, \{\mathrm{G3}, \mathrm{B}\flat 3\}, \{\mathrm{B}\flat 3, \mathrm{G4}\}) = (\{43\}, \{55, 58\}, \{58, 67\})\\
T^{\mathrm{a}} &= (|\text{♩}, |\text{♪♪}, |\text{𝄽♪♪}) = (\{(0, \tfrac{1}{4})\}, \{(0, \tfrac{1}{8}), (\tfrac{1}{8}, \tfrac{1}{8})\}, \{(\tfrac{2}{8}, \tfrac{1}{8}), (\tfrac{3}{8}, \tfrac{1}{8})\})\\
O^{\mathrm{a}} &= (\{\textbf{Vlc.}, \textbf{Cb.}\}, \{\textbf{Vla.}\}, \{\textbf{Vla.}\})
\end{aligned}
$$

and we have

$$
A_0^{\mathrm{a}} = H_0^{\mathrm{a}} \otimes T^{\mathrm{a}} \otimes O^{\mathrm{a}}. \tag{3}
$$

We can actually express the rest of the accompaniments by only changing the harmony, that is by choosing

$$
\begin{aligned}
H_1^{\mathrm{a}} &= (\emptyset, \{\mathrm{G3}, \mathrm{B}\flat 3\}, \{\mathrm{B}\flat 3, \mathrm{G4}\}) & H_2^{\mathrm{a}} &= (\{\mathrm{G3}\}, \{\mathrm{G3}, \mathrm{B}\flat 3\}, \{\mathrm{B}\flat 3, \mathrm{G4}\})\\
H_4^{\mathrm{a}} &= (\{\mathrm{G2}\}, \{\mathrm{G3}, \mathrm{B}\flat 3\}, \{\mathrm{D4}, \mathrm{G4}\}) & H_5^{\mathrm{a}} &= (\emptyset, \{\mathrm{G3}, \mathrm{B}\flat 3\}, \{\mathrm{D4}, \mathrm{G4}\})\\
H_6^{\mathrm{a}} &= (\{\mathrm{G3}\}, \{\mathrm{G3}, \mathrm{B}\flat 3\}, \{\mathrm{B}\flat 3, \mathrm{D4}\}) & H_8^{\mathrm{a}} &= (\{\mathrm{G2}\}, \{\mathrm{A3}, \mathrm{E}\flat 4\}, \{\mathrm{E}\flat 4, \mathrm{A4}\})\\
H_9^{\mathrm{a}} &= (\emptyset, \{\mathrm{A3}, \mathrm{E}\flat 4\}, \{\mathrm{E}\flat 4, \mathrm{A4}\}) & H_{10}^{\mathrm{a}} &= (\{\mathrm{G3}\}, \{\mathrm{A3}, \mathrm{E}\flat 4\}, \{\mathrm{E}\flat 4, \mathrm{A4}\})\\
H_{12}^{\mathrm{a}} &= (\{\mathrm{F}\sharp 2\}, \{\mathrm{A3}, \mathrm{D4}\}, \{\mathrm{D4}, \mathrm{C5}\}) & H_{13}^{\mathrm{a}} &= (\emptyset, \{\mathrm{A3}, \mathrm{D4}\}, \{\mathrm{D4}, \mathrm{C5}\})
\end{aligned}
$$

and $H_7^{\mathrm{a}} = H_3^{\mathrm{a}} = H_1^{\mathrm{a}}$, $H_{11}^{\mathrm{a}} = H_9^{\mathrm{a}}$.

This means that each half bar[9] A_i can be expressed as

$$
A_i^{\mathrm{a}} = H_i^{\mathrm{a}} \otimes T^{\mathrm{a}} \otimes O^{\mathrm{a}}. \tag{4}
$$

A similar approach could be used to model the melody; let us consider the main motive:

[9] A_i corresponds to the bar $\lfloor \frac{i}{2} \rfloor + 1$ and to the first half if i is even and to the second half if i is odd.

We can model it through the texture

$$T^{\mathrm{m}} = (\text{♪𝄾}|, \text{♪}|, |\text{♩}, |\text{𝄽♩}). \tag{5}$$

Notice that, in this case, we will have negative time values, since the motive contains an anacrusis.

The harmonies will depend on the instrument we consider; in order to avoid a too large AST in Fig. 3, we consider that both violins are a single instrument[10], **Vln.**, and then we have the instrumentation $O^{\mathrm{m}} = (S_{\mathbf{Vln.}}, S_{\mathbf{Vln.}}, S_{\mathbf{Vln.}}, S_{\mathbf{Vln.}})$ with $S_{\mathbf{Vln.}} = \{\mathbf{Vln.}\}$. The corresponding harmonices would be

$$\begin{aligned}
H_0^{\mathrm{m}} &= (\{\mathrm{E}\flat 4, \mathrm{E}\flat 5\}, \{\mathrm{D4}, \mathrm{D5}\}, \{\mathrm{D4}, \mathrm{D5}\}, \emptyset) \\
H_2^{\mathrm{m}} &= (\{\mathrm{E}\flat 4, \mathrm{E}\flat 5\}, \{\mathrm{D4}, \mathrm{D5}\}, \{\mathrm{D4}, \mathrm{D5}\}, \{\mathrm{B}\flat 4, \mathrm{B}\flat 5\}) \\
H_3^{\mathrm{m}} &= (\{\mathrm{B}\flat 4, \mathrm{B}\flat 5\}, \{\mathrm{A4}, \mathrm{A5}\}, \{\mathrm{G4}, \mathrm{G5}\}, \emptyset) \\
H_4^{\mathrm{m}} &= (\{\mathrm{G4}, \mathrm{G5}\}, \{\mathrm{F4}, \mathrm{F5}\}, \{\mathrm{E}\flat 4, \mathrm{E}\flat 5\}, \emptyset) \\
H_5^{\mathrm{m}} &= (\{\mathrm{E}\flat 4, \mathrm{E}\flat 5\}, \{\mathrm{D4}, \mathrm{D5}\}, \{\mathrm{C4}, \mathrm{C5}\}, \{\mathrm{C4}, \mathrm{C5}\}) \\
H_6^{\mathrm{m}} &= (\{\mathrm{D4}, \mathrm{D5}\}, \{\mathrm{C4}, \mathrm{C5}\}, \{\mathrm{C4}, \mathrm{C5}\}, \emptyset) \\
H_8^{\mathrm{m}} &= (\{\mathrm{D4}, \mathrm{D5}\}, \{\mathrm{C4}, \mathrm{C5}\}, \{\mathrm{C4}, \mathrm{C5}\}, \{\mathrm{A4}, \mathrm{A5}\})
\end{aligned}$$

and $H_1^{\mathrm{m}} = H_0^{\mathrm{m}}$, $H_7^{\mathrm{m}} = H_6^{\mathrm{m}}$.

We have now a way of combining harmonies, textures and instrumentation through the tensor contraction. Let us now expose how we can combine these tensor contractions to build bigger ones. To that end, we propose two operators: *parallelization* and *concatenation*.

3.2 Parallelization

Parallelization is a way of combining tensor contractions such that they start at the same time. To do that, we align the origins of the textures, represented by the 0. This means that the operation is equivalent to the union. In mathematical terms, let X_1 and X_2 be two tensor contractions. Then, the **parallelization** of them, denoted by $X_1 \parallel X_2$, is the set

$$X_1 \parallel X_2 = X_1 \cup X_2. \tag{6}$$

In fact, the parallelization of two tensor contractions is a tensor contraction: indeed, if we have $X_1 = H_1 \otimes T_1 \otimes O_1$ and $X_2 = H_2 \otimes T_2 \otimes O_2$ with set of indices N_1 and N_2 respectively, we can choose $H = (H_1, H_2) := (C_n)_{n \in N_1 \sqcup N_2}$,

[10] This is only a simplification, the model can handle several instances of an instrument like **Vln. I** and **Vln. II**.

$T = (T_1, T_2) := (R_n)_{n \in N_1 \sqcup N_2}$ and $O = (S_1, S_2) := (S_n)_{n \in N_1 \sqcup N_2}$, where $\sqcup$ is the disjoint union[11] and then we have

$$X_1 \parallel X_2 = H \otimes T \otimes O \tag{7}$$
$$= (H_1, H_2) \otimes (T_1, T_2) \otimes (O_1, O_2) \tag{8}$$
$$= \big(H_1 \otimes T_1 \otimes O_1\big) \cup \big(H_2 \otimes T_2 \otimes O_2\big). \tag{9}$$

Notice that this operator inherits all its properties from union, and in particular it is associative and commutative. The graphic representation of the parallelization is shown in Fig. 2b.

3.3 Concatenation

The case of the concatenation requires slightly more work; first, we need to define the endpoint of a texture; we define the **endpoint** of a texture as the maximum of the ends of the hits, i.e., if T is a texture with indices N, its endpoint $e(T)$ is defined by

$$e(T) = \max_{n \in N} \{\max\{o + d \in \mathbb{Q} : (o, d) \in R_n\}\}. \tag{10}$$

Now, to concatenate two tensor contractions, we shift all the onsets of the second one by the endpoint of the texture of the first one and then we take the parallelization. Mathematically, let X_1 and X_2 be two tensor contractions with $X_j = H_j \otimes T_j \otimes H_j$. The **concatenation** of them, denoted by $X_1 \to X_2$, is the set

$$X_1 \to X_2 = X_1 \parallel \big(H_2 \otimes (T_2 + e(T_1)) \otimes O_2\big) \tag{11}$$

where $T_2 + e(T_1) = (R_n + e(T_n))_{n \in N_2}$ with $R_n + e(T_1) = \{(o + e(T_1), d) \in \mathbb{Q} \times \mathbb{Q}^+ : (o, d) \in R_n\}$.

Similarly as in the case of parallelization, the concatenation of two tensor contractions is also a tensor contraction.

The concatenation is also associative, but it is not commutative. The graphic representation of the concatenation is shown in Fig. 2c.

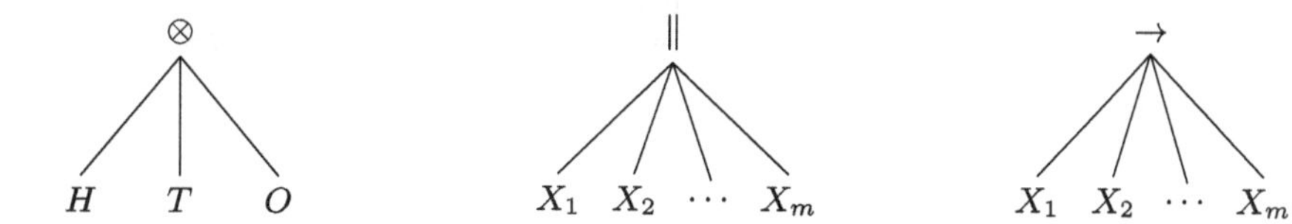

(a) Tensor contraction (b) Parallelization (c) Concatenation

Fig. 2. Graphic representation of the operators of the AST.

[11] We need to use the disjoint union for the case where there are common members between N_1 and N_2.

Let us show how to model the previous example as an AST. We use the harmonies, textures and instrumentations previously introduced. Since the size of the AST grows very fast, we only expose the first three measures.

We chose to split the first bar apart since it is an introduction. We concatenate that with the parallelization of the accompaniment and the melody; the accompaniment is made up from four tensor contractions but the melody has only three repetitions of the motive.

If we wanted to include the subsequent bars, we can map the same distribution of the parallelization simply changing the harmonies.

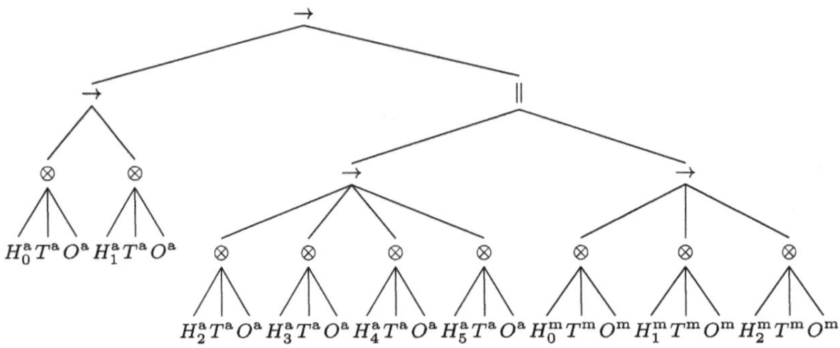

Fig. 3. A proposition of AST of the first three measures of Mozart's Symphony No.40 in G minor, K.550.

4 Computational Implementation

In order to compose through this model, we have made an implementation as a Python library that allows us to create harmonies, textures and instrumentations, and use the three operators $\otimes$, $\|$ and $\rightarrow$ to build ASTs.

In addition to this library, we have specified a XML Schema that allows us to write XML files that can be compiled with the Python library into MIDI files. The compilation is done by transforming the AST into a Python object representation, and then by converting this representation into a MIDI file using `pretty_midi` [17].

The interface and data structures used are shown in the following listing; the tensor contraction is an object and the parallelization and concatenation operators are denoted by `|` and `-`, respectively. They are implemented as overridden operator and thus presented with the `__or__` and `__sub__` methods, respectively.

```
Hit: {
    onset: frac
    duration: frac
}
Rhythm: {
```

```
    hits: Set[Hit]
}
Texture: {
    rhythms: List[Rhythm]
}
Pitch: {
    number: int
}
Chord: {
    pitches: Set[Pitch]
}
Harmony: {
    chords: List[Chord]
}
Instrument: {
    name: str
}
Section: {
    instruments: Set[Instrument]
}
Instrumentation: {
    sections: List[Section]
}
TensorContraction: {
    texture: Texture
    harmony: Harmony
    __or__(self, other: TensorContraction) -> TensorContraction
    __sub__(self, other: TensorContraction) -> TensorContraction
}
```

Regarding the XML Schema, we used as root a tag `score` that includes some metadata, and that has an `ast` tag inside. We also allowed in the metadata section the declaration of harmonies, textures and instrumentations, such that they can be reused in the AST through an id.

5 Conclusions and Future Work

Through this article, we have seen how we can model a score as an Abstract Syntactic Tree (AST). This representation makes the structure of the music explicit and organizes it in a hierarchical way. Structuring musical information in such a way sheds light on the different relations between elements and might be useful both for analysis and composition.

One of the main contributions of this article is the formal definition of key musical notions, namely harmony, texture and instrumentation. While these definitions shall not to be taken as the full description of these notions, they capture a significant part of their essence, especially when combined together.

For combining these three elements and building bigger musical structures, we presented three operators; the tensor contraction allows for the mix of harmony, texture and instrumentation, and transforms them into a set of notes, and parallelization and concatenation are the necessary operators to arrange tensor contractions in time.

These six concepts are the elements that constitute the basic grammar of our language and enable the construction of an AST. While the AST can be compiled into a MIDI file, it might serve several other purposes; for instance, we can change some harmonies or textures from the AST for performing harmony or texture transfer. Even more practically, we can change some instrumentations to make arrangements.

We think that one of the most promising way of using this model is for pedagogical purposes; an exercise for a student might be to provide an AST for a music fragment. This can be followed by a transformation of certain elements showing up their impact on the global result.

Several avenues of research could be explored in order to enrich the expressiveness of the AST. First of all, there are important parameters contained inside a score that we have dropped, mainly dynamics and articulation. It is a question whether we could include them in the texture or rather as a separate component of the contraction. We could also extend the number of operators available in the language: it would be interesting to have a transposition operator that can be applied to harmonies, or even include some other operators like the octave duplication. Another important improvement would be to implement a compiler that transforms the AST into a MusicXML file. Note that we do not use the word compiler as a metaphor but we talk about a real compiler as defined in [2]. This will require to make several choices, for instance the way of engraving the different rhythms of a texture into a single voice. Regarding the interactive part, we aim to implement this model in the Computer Assisted Composition software OpenMusic, as we think that it is well suited to its functional approach and most of the elements are already implemented (like the harmony, that can be mapped to the `ChordSeq` object).

References

1. Agón, C.: OpenMusic: un langage visuel pour la composition musicale assistée par ordinateur. Ph.D., Paris 6 (1998)
2. Aho, A.V., Sethi, R., Ullman, J.D.: Compilers: Principles, Techniques, and Tools. Addison-Wesley, Boston (1986)
3. Berlioz, H.: Grand traité d'instrumentation et d'orchestration modernes. Schonenberger, Paris (1844)
4. Bresson, J., Agon, C., Assayag, G.: OpenMusic: visual programming environment for music composition, analysis and research. In: Proceedings of the 19th ACM International Conference on Multimedia, New York, NY, USA, pp. 743–746 (2011)
5. Carnovalini, F., Rodà, A., Harley, N., Homer, S.T., Wiggins, G.A.: A new corpus for computational music research and a novel method for musical structure analysis. In: Proceedings of the 16th International Audio Mostly Conference, New York, NY, USA, pp. 264–267 (2021)

6. Couturier, L., Bigo, L., Levé, F.: Annotating symbolic texture in piano music: a formal syntax. In: Proceedings of Sound and Music Computing 2022, pp. 577–584 (2022)
7. Couturier, L., Bigo, L., Levé, F.: A dataset of symbolic texture annotations in Mozart Piano Sonatas. In: Proceedings the 23rd International Society for Music Information Retrieval Conference, pp. 509–516 (2022)
8. Good, M.: MusicXML for notation and analysis. The virtual score: representation, retrieval, restoration **12**(113–124), 160 (2001)
9. Hudak, P.: An algebraic theory of polymorphic temporal media. In: Jayaraman, B. (ed.) PADL 2004. LNCS, vol. 3057, pp. 1–15. Springer, Heidelberg (2004). https://doi.org/10.1007/978-3-540-24836-1_1
10. Hudak, P., Janin, D.: Tiled polymorphic temporal media. In: 2nd ACM SIGPLAN International Workshop on Functional Art, Music, Modeling & Design (FARM), Gothenburg, Sweden, pp. 49–60 (2014)
11. Koelsch, S., Rohrmeier, M., Torrecuso, R., Jentschke, S.: Processing of hierarchical syntactic structure in music. Proc. Natl. Acad. Sci. **110**(38), 15443–15448 (2013)
12. Lerdahl, F.: A Generative Theory of Tonal Music. MIT Press, Cambridge (1983)
13. Lerdahl, F., Jackendoff, R.: An overview of hierarchical structure in music. Music Perception Interdisc. J. **1**(2), 229–252 (1983)
14. Marsden, A., Hirata, K., Tojo, S.: Towards computable procedures for deriving tree structures in music: context dependency in GTTM and Schenkerian theory. In: Proceedings of the Sound and Music Computing Conference 2013, pp. 360–367 (2013)
15. Orio, N., Roda, A.: A measure of melodic similarity based on a graph representation of the music structure. In: Proceedings of the 10th International Society for Music Information Retrieval Conference, Kobe, Japan, pp. 543–548 (2009)
16. Orlarey, Y., Fober, D., Letz, S., Bilton, M.: Lambda calculus and music calculi. In: International Conference on Mathematics and Computing (1994)
17. Raffel, C., Ellis, D.P.: Intuitive analysis, creation and manipulation of MIDI data with pretty_midi. In: 15th International Society for Music Information Retrieval Conference Late Breaking and Demo Papers, pp. 84–93 (2014)
18. Tymoczko, D.: Tonality: An Owner's Manual. Oxford Studies in Music Theory. Oxford University Press, Oxford (2023)

Piston Words

Shawn Michael Westmoreland(✉)

College of San Mateo, San Mateo, CA 94402, USA
westmorelands@smccd.edu

Abstract. We mathematize the notion of a "Piston word" (i.e. a diatonic chord sequence that uniquely determines key) and seek a formal language method of constructing all triadic Piston words. Since these triadic words are (simplified) superstrings of atomic triadic words, it suffices to just construct the language $\mathcal{L}$ of the atomic triadic words. To this end, we construct a minimal DFA (deterministic finite automaton) for this language $\mathcal{L}$. Interestingly, the states of this DFA, which are the Myhill-Nerode classes of $\mathcal{L}$, correspond to classical harmonic building blocks: silence, the tonic, the subdominant, the dominant, the diminished triad, the perfect cadence, the plagal cadence, the deceptive cadence, and a fourth type of cadence that we call the "deceptive plagal cadence".

Keywords: diatonic harmony · formal languages · automata theory

1 Introduction and Overview

Walter Piston (Piston [1] page 55) writes "... groups of two or three chords with distinct meaning as to key, might be regarded as musical 'words' ... " An example of a musical word, in any major key, is IV-V. As Piston himself states, IV-V "... cannot be interpreted in more than one tonality, without chromatic alteration..." (Piston [1] page 54). By contrast, V-I is equivalent to I-IV in the key of the dominant (Piston [1] page 54). If a suitable third chord is added to V-I however, then some instances of Piston's musical words can be formed. For example, V-I-IV and V-I-ii are musical words. (Roman numeral conventions in the present paper are the same as in e.g. Kostka, Payne, and Almén [2].) Piston advises his students "...to write as many of these formulae [musical words] as possible in a notebook where they will be available for consultation, beginning with the commonest, like IV-V-I, I-IV-V, etc. They should then be studied for their individual qualities, their strength of tonality and rhythmic feeling, and notice should be taken of their frequency of use in music and of the varied forms in which they may appear." (Piston [1] page 55). Piston clearly regards these musical words as fundamentally important to real diatonic music: "The study of harmony deals with the variants of these fundamental formulae, or 'words.' It will be found that they exist in the most complex texture, and from the very beginning much attention should be given to harmonic analysis of standard works, to bring out the endless capacity for varied treatment of these harmonic structural elements" (Piston [1] page 56).

© The Author(s), under exclusive license to Springer Nature Switzerland AG 2024
T. Noll et al. (Eds.): MCM 2024, LNCS 14639, pp. 280–288, 2024.
https://doi.org/10.1007/978-3-031-60638-0_22

The main objective of the present paper is to generalize Piston's concept of musical words in a mathematically precise way, and to devise a method of constructing all Piston words over the triads (triadic words). To achieve this objective we must first derive a few basic results. The paper begins with general considerations that apply to all diatonic sequences, not just triadic sequences. In Sect. 4 we specialize to triadic sequences and we construct a formal language $\mathcal{L}$ of atomic triadic words such that each triadic word is a (simplified) superstring of an element of $\mathcal{L}$. Figure 1 depicts a minimal DFA for $\mathcal{L}$.

While this paper was under review, the author was informed that its topic overlaps with earlier work by Mazzola on cadences (see Mazzola [3] Chap. 26, or Mazzola [4] for an earlier source). Our contribution supplements Mazzola's work with an independently developed (and apparently novel) formal language approach. Points of convergence with Mazzola will be highlighted as they arise.

2 Preliminaries

In the present paper, we assume octave equivalence (i.e. notes are identified with their pitch classes). As for the twelve-note chromatic scale, we assume equal temperament. A *diatonic scale* can be thought of as any seven-note scale that can be expressed in one of the fifteen standard key signatures without using accidentals. We shall always refer to each key signature K by its major mode name and, for simplicity, we shall assume that any diatonic scale in K (i.e. any seven-note scale expressible in K without accidentals) is ordered in accordance with the major mode. We do this regardless of whether the major mode is actually being emphasized musically. It is just a convenient naming and labeling convention that suits the purposes of the present paper.

Some key signatures, such as B major and $C\flat$ major, are enharmonically equivalent. In the present paper, *key* can be identified with key signature up to enharmonic equivalence.

By a *diatonic chord* we simply mean a nonempty set of notes from a diatonic scale. Thus, in the present paper, an arbitrary diatonic chord can have at minimum one note and at maximum seven notes. This definition may be a bit nonstandard, but it suits our present purposes. By octave equivalence, changing the chord voicing or applying an inversion does not change the chord as far as the present paper is concerned. Moreover, since chords are sets, we can perform set-theoretic operations on chords, such as unions (e.g. see Definitions 3 and 4).

By a *diatonic sequence* we mean a finite sequence s of diatonic chords all of which are from the same diatonic scale K (see also Kostka et al. [2] page 646). In this case, s can be expressed without accidentals in any key signature corresponding to K and hence we say that s is *in the key of* K. We represent a diatonic sequence s of length $n \geq 1$ by a string $s = \alpha_1\alpha_2 \cdots \alpha_n$, where each α_k is a diatonic chord from some fixed scale. A sequence of length zero is represented by the empty string ε. When chords are written in Roman numeral notation, concatenation by juxtaposition is impractical so we opt for dashes (e.g. V^7-I-IV).

By a *simplified diatonic sequence* we mean a diatonic sequence that does not have consecutive repeats of the same chord. We are apt to use simplified sequences because, as far as the present paper is concerned, a progression progresses only when the chords change. For example, I-V-V-I is not simplified, and we are apt to regard it more simply as I-V-I.

We generalize Piston's concept of musical words as follows (compare/contrast this with Mazzola's notion of a *cadence*, Mazzola [3] pages 453–454):

Definition 1. *A* **Piston word** *(or alternatively herein, a* **word***) is a simplified diatonic sequence s that is in one and only one key K. In this case, we say that s is a* **word in the key of** K *or, put another way, s* **establishes the key of** K.

For us, words can be arbitrarily long, but are always finite. In the present paper, "scales," "chords," "sequences," etc. specifically refer to diatonic scales, diatonic chords, diatonic sequences, etc. The term "word" herein is short for *Piston word* and the term "string" is our generic term for a finite sequence.

The property of being a Piston word is *permutation invariant up to simplification.* For example, I-IV-I-V is a Piston word, but the permutation I-I-IV-V is technically not a Piston word because it is not simplified. Of course, when I-I-IV-V is simplified it becomes the Piston word I-IV-V.

Plainly then, chord order does not affect whether a sequence establishes key and so one might wonder whether Piston words would be better defined as sets rather than as sequences of chords. The reason why we define Piston words in terms of sequences is because we are interested in actual ordered sequences possessing the Piston word property. That said, we are indeed also interested in the property itself. The distinction between that of objects possessing a property and that of the property itself plays out in the difference herein between Sect. 4 and Sect. 3. When studying the Piston word property itself we are indeed apt to eschew sequential information and simply analyze sets of chords (cf. Chap. 26 of Mazzola [3]). This is effectively our approach in Sect. 3. On the other hand, when it comes to systematically composing specific types of Piston words (e.g. words that only use triads, as in Sect. 4), it then makes sense to work with finite sequences (i.e. strings) of chords.

Chord order can affect emotional expression, voice-leading, harmonic function, and more. In particular, chord order can influence the relative timing of key establishment. For example, both IV-V-I and IV-I-V are words using the same chords, but the former establishes key at the second chord change whereas the latter does so at the third chord change.

Of particular interest herein is the notion of an atomic word (compare/contrast this with Mazzola's notion of *minimal cadential families*, Mazzola [3] page 455):

Definition 2. *An* **atomic word** *is a word ℓ such that no proper substring of ℓ is also a word.*

For example, IV-I-V is an atomic word but IV-V-I is not because it has the word IV-V as a proper substring. Thus, the property of being an atomic word is not permutation invariant. However, it should be mentioned that atomic words are *reversal invariant*: if $\alpha_1\alpha_2\alpha_3\cdots\alpha_n$ is an atomic word then so is $\alpha_n\cdots\alpha_3\alpha_2\alpha_1$. The reversal invariance of atomic words is reflected in the results of Theorem 2 and in the symmetrical structure of our DFA (Theorem 3/Fig. 1).

3 The Tritone Theorem

In this section, our focus is on the Piston word property itself. Hence, the sequential ordering of chords is irrelevant, and so it is useful to flatten out any given sequence to just a set of notes. This leads to the following definition:

Definition 3. *Let $s = \alpha_1\alpha_2 \cdots \alpha_n$ be a diatonic sequence. The* **note set** *of s, denoted $\mathcal{N}(s)$, is the set of all notes used by the chords of s.*

$$\mathcal{N}(s) = \bigcup_{k=1}^{n} \alpha_k$$

If $s = \varepsilon$ then $\mathcal{N}(s) = \emptyset$ (the empty set).

Each key has exactly one tritone, but each tritone appears in exactly two tonally distinct keys: K and K'. These keys, K and K', are a tritone apart (see e.g. Butler [5]). By presenting a tritone together with a third note not included in the tritone however, a unique key is established (Butler [5]). This fact is important to the present paper, and so we incorporate it as a theorem:

Theorem 1. *(The Tritone Theorem) A diatonic sequence s is a Piston word if and only if $\mathcal{N}(s)$ contains a tritone and at least one other note.*

Proof. The "if" direction is already implied by the paragraph preceding the theorem, so it suffices to prove the "only if" direction. To this end, suppose s is a word in the key of K. By transposition, we can assume without loss of generality that this key is C major. We claim that the note set $\mathcal{N}(s)$ contains the tritone $\{B, F\}$. For indirect proof, assume otherwise. This gives two cases: (1) the note set $\mathcal{N}(s)$ does not have B, or (2) the note set $\mathcal{N}(s)$ does not have F. In case (1), the word s could be interpreted in the key of F major. In case (2), the word s could be interpreted in the key of G major. In either case, we contradict s being a word in the key of C major, so $\{B, F\} \subseteq \mathcal{N}(s)$. Moreover, $\mathcal{N}(s)$ must also contain a note distinct from B and F. Otherwise, s could be interpreted in the key of $F\sharp$ or $G\flat$. □

The tritone theorem implies that, in general, in order to build Piston words one must utilize chords in such a way that the note set contains a tritone together with a third note. Hence, we define a concept of *utilizable chords* as follows:

Definition 4. *Let $\mathcal{U}$ denote a set of diatonic chords all of which are from the same diatonic scale. We call $\mathcal{U}$ a set of* **utilizable chords** *if $\bigcup \mathcal{U}$ satisfies the tritone theorem (i.e. if $\bigcup \mathcal{U}$ contains a tritone and a third note).*

If $\mathcal{U}$ is a set of utilizable chords, then it is possible to construct a word using only chords from $\mathcal{U}$. The set of all words constructed thusly is denoted $\mathcal{W}_{\mathcal{U}}$.

Definition 5. *Chord α is said to be* **relative** *to chord β when, for all keys K, chord α is in the key of K if and only if chord β is in the key of K.*

On any set $\mathcal{U}$ of chords (whether $\mathcal{U}$ is utilizable or not), *relative* is clearly an equivalence relation. Taking the specific case where $\mathcal{U}$ is the (utilizable) set of diatonic triads, we get exactly four relative classes: (1) the *tonic relative class* $\{\mathrm{I}, \mathrm{vi}\}$, (2) The *subdominant relative class* $\{\mathrm{IV}, \mathrm{ii}\}$, (3) the *dominant relative class* $\{\mathrm{V}, \mathrm{iii}\}$, and (4) the *tritone relative class* $\{\mathrm{vii}^\circ\}$. We call triads I and vi *tonic relatives*, IV and ii *subdominant relatives*, and V and iii *dominant relatives.* Our tonic, subdominant, and dominant relatives correspond to Hugo Riemann's *parallel clangs* (Riemann [6] pages 71, 193–194).

If one were to work with a different set of chords, then of course the relative classes could be different. For example, if one were to take $\mathcal{U}$ as the set of diatonic triads together with chords V^7 and $\mathrm{I}^{\mathrm{M}7}$, then the tritone relative class would become $\{\mathrm{vii}^\circ, \mathrm{V}^7\}$ and there would be a fifth relative class $\{\mathrm{I}^{\mathrm{M}7}\}$.

4 The Language of Atomic Triadic Words

In the previous section, the type of diatonic chords used (and their order of appearance in a sequence) was unimportant, but henceforth we specialize to diatonic sequences built from triads only. In traditional Roman numeral notation, the triads in a key K are I, ii, iii, IV, V, vi, vii°, where I denotes the tonic triad of K. (We number chords in accordance with the major mode out of convenience. If needed, one can always translate our labels to suit other modes.) The set of all seven triads in the key of K is, of course, a set of utilizable chords in the sense of Definition 4.

The set of all chord sequences, in Roman numeral notation, can be thought of as a formal language whose alphabet is a certain set of Roman-numeral-based symbols. Since Roman numerals make for unwieldy strings, we will henceforth use Arabic numerals, $\Sigma = \{1, 2, 3, 4, 5, 6, 7\}$. We will also use Arabic numerals to specify, not only chords, but also key. For example, triad 4 is equivalent to triad 1 in key 4, and it is equivalent to triad 5 in key ♭7. Hence, triad 4 is expressible in keys 4 and ♭7 (herein, "expressible" always means without accidentals).

The *star* of a set $\mathcal{A}$, denoted $\mathcal{A}^*$, is the set of all finite strings (finite sequences) that can be formed using the elements of $\mathcal{A}$ (see e.g. Sipser [7] page 41). If $\Sigma = \{1, 2, 3, 4, 5, 6, 7\}$, then the set Σ^* includes strings such as ε, 13, 7, 1776, etc. However, we need the option to simplify away consecutively repeated symbols so that e.g. 1776 becomes 176. Let $\mathcal{S}$ denote the set of all such *simplified sequences*, or *simplified strings.* Of course, each element of $\mathcal{S}$ is expressible in key 1, but an element of $\mathcal{S}$ is a triadic word (i.e. an element of $\mathcal{W}_\Sigma$) if and only if it is not expressible in any other key.

Definition 6. *The* **language of atomic triadic words** *is* $\mathcal{L} = \{\ell \in \mathcal{W}_\Sigma :$ *no proper substring of* ℓ *is in* $\mathcal{W}_\Sigma\}$.

As per our remarks in Sect. 2, $\mathcal{L}$ is reversal invariant, but not permutation invariant. For example, $3612 \in \mathcal{L}$, $2163 \in \mathcal{L}$, but $3261 \notin \mathcal{L}$. The elements of $\mathcal{L}$ are the *atomic triadic words*, but to simplify our prose we will henceforth call them "atomic words" or "atoms" even though $\mathcal{L}$ is a proper subset of all atomic words.

If we had all the elements of $\mathcal{L}$ then we could easily construct all the elements of $\mathcal{W}_\Sigma$ by forming (simplified, but otherwise unrestricted) superstrings of the elements of $\mathcal{L}$. Indeed, we can enumerate the elements of $\mathcal{L}$ as follows:

Theorem 2. *(The Enumeration Theorem)*

(a) There are no atomic words of length zero.
(b) There is a unique atomic word of length one. It is the word 7.
(c) There are 8 atomic words of length two: 23, 25, 43, 45, and their reverses: 32, 52, 34, 54.
(d) For each $n \geq 3$, there are 16 atomic words of length n. These are of the form $\alpha\gamma\beta$ where $\alpha, \beta \in \Sigma$ and $\alpha\beta$ is an atomic word of length two and γ is a simplified string of length $n-2$ that only involves tonic relatives. (I.e. $\gamma \in \{1,6\}^* \cap \mathcal{S} \setminus \{\varepsilon\} = \{1, 6, 16, 61, 161, 616, \cdots\}$.)

In part (d), by reversal invariance, any eight of the 16 atoms are just the reverses of the other eight. Parts (b) and (c) rediscover Mazzola's minimal cadential families for the major mode triads (Mazzola [3] page 455). Since Mazzola's approach ignores chord order, part (d) does not arise therein.

Proof. Part (a) holds because $\mathcal{N}(\varepsilon)$ $(= \emptyset)$ does not contain a tritone. Hence, by Theorem 1, the empty string is not a word, let alone an atomic word.

Part (b) holds because 7 is the only member of Σ whose note set satisfies the tritone theorem and moreover the only proper substring of 7 is ε.

To get (c), suppose that $\alpha\beta$ is a string of length two, where $\alpha, \beta \in \Sigma$. In order for $\alpha\beta$ to be an atomic word, the set $\mathcal{N}(\alpha\beta) = \alpha \cup \beta$ must satisfy the tritone theorem and yet neither α nor β can be 7. Hence, the set $\{\alpha, \beta\}$ must consists of a subdominant relative and a dominant relative. So part (c) holds.

To get (d), first observe that if $\alpha\beta$ is an atomic word (with $\alpha, \beta \in \Sigma$), and if $\gamma \in \{1,6\}^* \cap \mathcal{S} \setminus \{\varepsilon\}$ is a string of tonic relatives only, then $\alpha\gamma\beta$ will of course be a word in $\mathcal{W}_\Sigma$. Moreover, each nonempty proper substring of $\alpha\gamma\beta$ is expressible in either key 4 or key 5. Thus, $\alpha\gamma\beta$ is an atomic word, and so any string of length $n \geq 3$ having the form described in (d) will be an atomic word. Moreover, since $\gamma \in \{1,6\}^* \cap \mathcal{S} \setminus \{\varepsilon\} = \{1, 6, 16, 61, 161, 616, \cdots\}$, it follows that for each $n \geq 3$ there are exactly two choices for which string (of length $n-2$) the substring γ can be. Thus, since there are 8 atomic words of length two, it follows that there are $8 \cdot 2 = 16$ atomic words of length $n \geq 3$ having the form as described in (d).

To complete the proof of (d), we show that any atomic word of length $n \geq 3$ necessarily has the form described in (d). Let $\alpha\gamma\beta \in \mathcal{L}$, where $\alpha, \beta \in \Sigma$ and $\gamma \in \Sigma^* \setminus \{\varepsilon\}$. Since $\alpha\gamma\beta \in \mathcal{L}$, we get that $\alpha \neq 7$, $\beta \neq 7$, and γ cannot have a 7. We must prove two claims: (1) $\alpha\beta \in \mathcal{L}$ and (2) $\gamma \in \{1,6\}^* \cap \mathcal{S} \setminus \{\varepsilon\}$. To prove (1), assume $\alpha\beta \notin \mathcal{L}$. Then we get conditionals $\alpha \in \{2,4\} \to \beta \notin \{3,5\}$ and $\alpha \in \{3,5\} \to \beta \notin \{2,4\}$. So, if $\alpha \in \{2,4\}$, then γ has a 3 or 5 to ensure $\alpha\gamma\beta$ isn't in key 4, but we also then get that $\alpha\gamma$ is a word, contradicting $\alpha\gamma\beta \in \mathcal{L}$. A similar contradiction ensues if $\alpha \in \{3,5\}$. If $\alpha \in \{1,6\}$, then $\gamma\beta$ has a 2 or 4 to ensure $\alpha\gamma\beta$ isn't in key 5. Moreover, $\gamma\beta$ has a 3 or 5 to ensure $\alpha\gamma\beta$ isn't in key 4. Thus, $\gamma\beta$ involves a subdominant relative and a dominant relative, making it a word, which contradicts $\alpha\gamma\beta \in \mathcal{L}$. To prove (2), assume $\gamma \notin \{1,6\}^* \cap \mathcal{S} \setminus \{\varepsilon\}$. Since $\gamma \neq \varepsilon$ and $\alpha\gamma\beta \in \mathcal{L}$, we get $\gamma \in \mathcal{S} \setminus \{\varepsilon\}$, and so $\gamma \notin \{1,6\}^*$. Hence, γ (which does not involve a 7) involves a subdominant relative or a dominant relative. So, since $\alpha\beta$ is an atomic word of length 2, we get that $\alpha\gamma$ or $\gamma\beta$ is a word, contradicting $\alpha\gamma\beta \in \mathcal{L}$. □

A standard way to construct a DFA that accepts $\mathcal{L}$ is to use the Myhill-Nerode Theorem (e.g. Lewis and Papadimitriou [8] page 96, Shallit [9] page 77). The *Myhill-Nerode equivalence relation*, corresponding to $\mathcal{L}$, is a relation $\sim$ on strings in Σ^* such that $x \sim y$ if and only if $\forall z \in \Sigma^*, xz \in \mathcal{L} \leftrightarrow yz \in \mathcal{L}$ (see e.g. Shallit [9] page 78). For each $x \in \Sigma^*$, let $\overline{x} = \{y \in \Sigma^* : x \sim y\}$ denote the *Myhill-Nerode class* of x. According to the Myhill-Nerode Theorem, a language is regular if and only if it only has finitely many Myhill-Nerode classes, and in this case a DFA for the language can be constructed such that each state is identified with a Myhill-Nerode class and for each $\alpha \in \Sigma$ the transitions are given by $\alpha : \overline{x} \mapsto \overline{x\alpha}$. The initial state of the DFA is identified with $\overline{\varepsilon}$. The accepting states are identified with classes $\overline{x}$ where x is in the language (see e.g. Shallit [9] pages 78–81). This DFA is minimal in the sense that it uses the minimum number of states necessary to accept the language, and moreover it is the unique minimal DFA for the language up to relabeling the states (e.g. Shallit [9] pages 81–82, Hopcroft and Ullman [10] pages 65–68). For the language $\mathcal{L}$, we get:

Theorem 3. *The language $\mathcal{L}$ has exactly nine Myhill-Nerode classes, namely: $\overline{\varepsilon}$, $\overline{1}$, $\overline{4}$, $\overline{5}$, $\overline{7}$, $\overline{41}$, $\overline{46}$, $\overline{51}$, and $\overline{56}$. Moreover, $\overline{\varepsilon} = \{\varepsilon\}$, $\overline{1} = \overline{6}$, $\overline{4} = \overline{2}$, $\overline{5} = \overline{3}$, $\overline{41} = \overline{21}$, $\overline{46} = \overline{26}$, $\overline{51} = \overline{31}$, $\overline{56} = \overline{36}$, $\overline{7} = \mathcal{L}$, and the minimal DFA is given by Fig. 1.*

Proof. Let $y \in \overline{\varepsilon}$. Then $\forall z \in \Sigma^*, yz \in \mathcal{L} \leftrightarrow z \in \mathcal{L}$. Taking $z = 7$, we infer $y7 \in \mathcal{L}$, which implies $y = \varepsilon$. Hence, $\overline{\varepsilon} = \{\varepsilon\}$ (the inclusion $\{x\} \subseteq \overline{x}$ is true for all x since $\sim$ is reflexive). Since $\varepsilon\alpha = \alpha$, we get that $\alpha : \overline{\varepsilon} \mapsto \overline{\alpha}$ for each $\alpha \in \Sigma$. To get that $\overline{1} = \overline{6}$, $\overline{2} = \overline{4}$, and $\overline{3} = \overline{5}$, we can use Theorem 2 (which lists out $\mathcal{L}$) to observe that if $\alpha, \beta \in \Sigma$ are relative then $\forall z \in \Sigma^*, \alpha z \in \mathcal{L} \leftrightarrow \beta z \in \mathcal{L}$. Thus, if α and β are relative then $\overline{\alpha} = \overline{\beta}$ (and hence $\sim$ generalizes Definition 5 to strings).

To get that $\overline{41} = \overline{21}$, $\overline{46} = \overline{26}$, $\overline{51} = \overline{31}$, and $\overline{56} = \overline{36}$, observe that if $\alpha, \beta \in \Sigma$ are relative and γ is a tonic relative, then $\forall z \in \Sigma^*, \alpha\gamma z \in \mathcal{L} \leftrightarrow \beta\gamma z \in \mathcal{L}$. Thus, when α and β are subdominant or dominant relatives, we get $\overline{41} = \overline{21}$, $\overline{46} = \overline{26}$, $\overline{51} = \overline{31}$, and $\overline{56} = \overline{36}$. Moreover, the Myhill-Nerode Theorem tells us that our DFA has transitions $1 : \overline{4} \mapsto \overline{41}$, $6 : \overline{4} \mapsto \overline{46}$, $1 : \overline{5} \mapsto \overline{51}$, and $6 : \overline{5} \mapsto \overline{56}$.

To show $\overline{7} = \mathcal{L}$, observe that for each $y \in \mathcal{L}$, we get that $\forall z \in \Sigma^*$, $yz \in \mathcal{L} \leftrightarrow z = \varepsilon \leftrightarrow 7z \in \mathcal{L}$. Hence, for each $y \in \mathcal{L}$, we have $\overline{y} = \overline{7}$. Put another way, $\overline{7} = \mathcal{L}$.

We thus have nine Myhill-Nerode classes, or DFA states. There is only one accepting state and it is $\overline{7}$ because $\overline{7}$ is the only class containing members of $\mathcal{L}$.

To verify the remaining DFA transitions, let $\alpha \in \Sigma$. Using Theorem 2, no member of $\mathcal{L}$ begins with 1. Hence, $\forall z \in \Sigma^*$, $1z \in \mathcal{L} \leftrightarrow 1\alpha z \in \mathcal{L}$ and so $\overline{1\alpha} = \overline{1}$ (i.e. $\alpha : \overline{1} \mapsto \overline{1}$ for all α). Also, $\forall z \in \Sigma^*$, $7\alpha z \in \mathcal{L} \leftrightarrow 1z \in \mathcal{L}$ and so $\overline{7\alpha} = \overline{1}$ (i.e. $\alpha : \overline{7} \mapsto \overline{1}$ for all α). Similarly, if $\alpha, \beta \in \Sigma$ are relative, or if $\beta = 7$, then no member of $\mathcal{L}$ begins with $\alpha\beta$. Hence, $\forall z \in \Sigma^*$, $\alpha\beta z \in \mathcal{L} \leftrightarrow 1z \in \mathcal{L}$ and so $\overline{\alpha\beta} = \overline{1}$ (i.e. $\beta : \overline{4} \mapsto \overline{1}$ if $\beta \in \{2,4,7\}$ and $\beta : \overline{5} \mapsto \overline{1}$ if $\beta \in \{3,5,7\}$).

If α is a subdominant relative and β is a dominant relative then $\alpha\beta \in \mathcal{L}$ and hence $\overline{\alpha\beta} = \overline{7}$ and so $\beta : \overline{\alpha} \mapsto \overline{7}$. Furthermore, if $\gamma \in \Sigma$ is a tonic relative, then $\alpha\gamma\beta \in \mathcal{L}$ and hence $\overline{\alpha\gamma\beta} = \overline{7}$ and so $\beta : \overline{\alpha\gamma} \mapsto \overline{7}$. By reversal invariance, the analogous results hold if α is a dominant relative and β is a subdominant relative. Hence, $\beta : \overline{4} \mapsto \overline{7}$, $\beta : \overline{41} \mapsto \overline{7}$, $\beta : \overline{46} \mapsto \overline{7}$, for $\beta \in \{3, 5\}$ and similarly $\beta : \overline{5} \mapsto \overline{7}$, $\beta : \overline{51} \mapsto \overline{7}$, $\beta : \overline{56} \mapsto \overline{7}$, for $\beta \in \{2, 4\}$.

If α is a subdominant relative, and γ is a tonic relative, and either $\beta = \gamma$ or $\beta = 7$ or β is a subdominant relative then, using Theorem 2, $\alpha\gamma\beta z \notin \mathcal{L}$ for all $z \in \Sigma^*$ and hence $\forall z \in \Sigma^*, \alpha\gamma\beta z \in \mathcal{L} \leftrightarrow 1z \in \mathcal{L}$, so $\beta : \overline{\alpha\gamma} \mapsto \overline{1}$ (the same holds if α is a dominant relative and either $\beta = \gamma$ or $\beta = 7$ or β is a dominant relative). Hence, if $\beta \in \{1, 2, 4, 7\}$ then $\beta : \overline{41} \mapsto \overline{1}$. If $\beta \in \{2, 4, 6, 7\}$ then $\beta : \overline{46} \mapsto \overline{1}$. If $\beta \in \{1, 3, 5, 7\}$ then $\beta : \overline{51} \mapsto \overline{1}$. If $\beta \in \{3, 5, 6, 7\}$ then $\beta : \overline{56} \mapsto \overline{1}$. If α is either a subdominant relative or a dominant relative, and if γ_1 and γ_2 are distinct tonic relatives, then using part (d) of Theorem 2 we get that $\forall z \in \Sigma^*$, $\alpha\gamma_1\gamma_2 z \in \mathcal{L} \leftrightarrow \alpha\gamma_2 z \in \mathcal{L}$, so $\gamma_2 : \overline{\alpha\gamma_1} \mapsto \overline{\alpha\gamma_2}$. Hence, $1 : \overline{46} \mapsto \overline{41}$, $6 : \overline{41} \mapsto \overline{46}$, $1 : \overline{56} \mapsto \overline{51}$, and $6 : \overline{51} \mapsto \overline{56}$. Thus, the transitions shown in Fig. 1 are correct.

Moreover, given any input string $z \in \Sigma^*$, the DFA will eventually stop in one of our nine states. Therefore, there are no additional states beyond these nine. Moreover, our nine states are distinct because each has a string that it shares with no other state (for $\overline{x}$, such a string is x itself). □

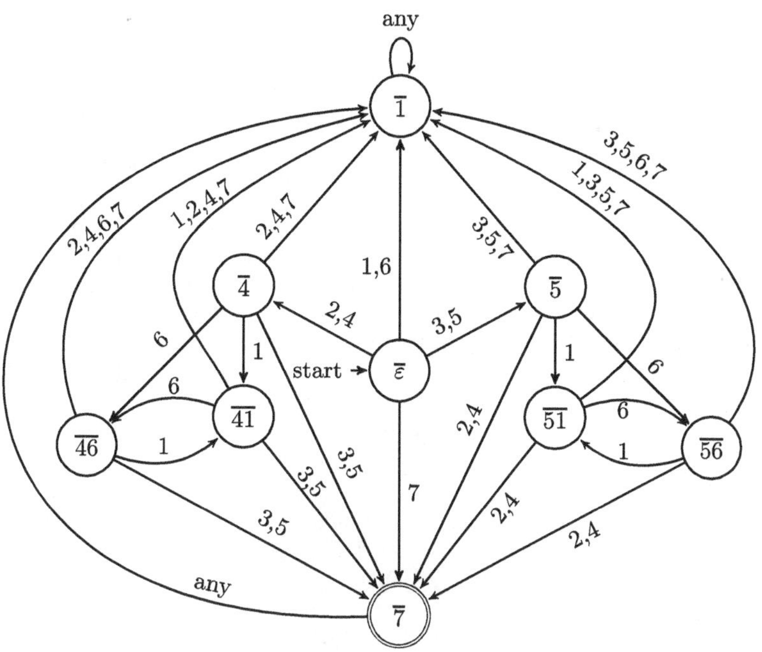

Fig. 1. Minimal DFA for $\mathcal{L}$. State $\overline{7}$ (double-circled) is the only accepting state. The symmetric structure of the DFA is expected since $\mathcal{L}$ is invariant under reversal.

5 Conclusion and Discussion

The elements of $\mathcal{W}_{\Sigma}$ can easily be constructed as the simplified superstrings of the elements of $\mathcal{L}$. (Herein Σ is the set of diatonic triads. One expects that the analogous problem for a different utilizable set can be solved similarly.)

Walter Piston was perhaps not as rigid about words as is Definition 1. For example, the deceptive cadence V-vi is equivalent to I-ii in key V, but Piston says that V-vi is "strongly indicative of a single tonality" (Piston [1] page 54).

Indeed, $\overline{56}$ is one of our Myhill-Nerode classes (i.e. DFA states). These classes generalize Riemann's notion of tonic, subdominant, and dominant parallel clangs from chords to strings, and the states of our DFA go beyond the tonic, subdominant, dominant, and diminished, to include silence (as $\overline{\varepsilon}$), three *classical cadential classes* (the perfect $\overline{51}$, the plagal $\overline{41}$, the deceptive $\overline{56}$), and a fourth *exceptional cadential class* (the "deceptive plagal" $\overline{46}$). Interestingly, recognizable ingredients of common practice are derived herein from purely structural considerations.

Future work might use a formal language approach to open up Mazzola's cadential theory in a syntactic way (i.e. by working with strings, not just sets).

Acknowledgements. The author thanks Thomas Noll and the reviewers for their valuable feedback.

References

1. Piston, W.: Harmony, 4th edn. Revised and expanded by M. DeVoto. Norton, New York (1978)
2. Kostka, S., Payne, D., Almén, B.: Tonal Harmony with an Introduction to Twentieth Century Music, 7th edn. McGraw Hill, New York (2013)
3. Mazzola, G.: The Topos of Music I: Theory, 2nd edn. Springer, Cham (2017)
4. Mazzola, G.: Gruppen und Kategorien in der Musik. Heldermann, Berlin (1985)
5. Butler, D.: The initial identification of tonal centres in music. In: Rogers, D., Sloboda, J.A. (eds.) The Acquisition of Symbolic Skills, pp. 251–261. Springer, Boston (1983). https://doi.org/10.1007/978-1-4613-3724-9_29
6. Riemann, H.: Harmony Simplified or the Theory of the Tonal Functions of Chords, 2nd edn. Augener Limited, London (1895)
7. Sipser, M.: Introduction to the Theory of Computation, Preliminary PWS Publishing Company, Boston (1996)
8. Lewis, H.R., Papadimitriou, C.H.: Elements of the Theory of Computation, 2nd edn. Prentice-Hall, Upper Saddle River (1998)
9. Shallit, J.: A Second Course in Formal Languages and Automata Theory. Cambridge University Press, New York (2009)
10. Hopcroft, J.E., Ullman, J.D.: Introduction to Automata Theory, Languages, and Computation. Addison-Wesley, Reading (1979)

Fourier Analysis for Music

DFT and Persistent Homology for Topological Musical Data Analysis

Victoria Callet(✉)

IRMA, UMR 7501, CNRS, Université de Strasbourg, Strasbourg, France
victoria.callet@math.unistra.fr

Abstract. There are several works that already exist in the context of persistent homology for Topological Musical Data Analysis, and we can cite [2] and [3] among others: in each one of these works, the main problem is to find how we can associate a point cloud with a musical score, that is a set of points with a metric. This paper proposes to combine persistent homology with a symbolic representation of musical structures given by the Discrete Fourier Transform to answer this question: the points are the musical bars and the metric is given by the DFT in dimension two. We start with the mathematical background, and the main goal of this paper is thus to support the use of the DFT in this context, by extracting barcodes from artificially constructed scores based on Tonnetze, and then recovering topological features.

Keywords: Filtered simplicial complex · Persistent homology · Barcodes · Discrete Fourier Transform · Topological Data Analysis · Musical analysis · Tonnetz

1 Context, Definitions and Problematic

1.1 Persistent Homology

A finite filtered simplicial complex $K = (K^i)_{i=0}^N$ is a nested sequence of simplicial complexes, such as

$$\emptyset = K^{-1} \subset K^0 \subseteq K^1 \subseteq \ldots \subseteq K^N = K$$

We call $\{0, 1, \ldots, N\}$ the set of **filtration times**. For instance, Fig. 1 represents a filtered complex with six filtration times.

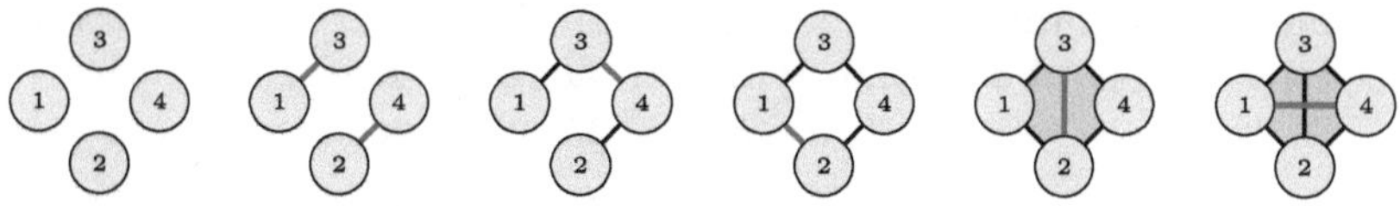

Fig. 1. A filtered simplicial complex K with six times of filtration.

© The Author(s), under exclusive license to Springer Nature Switzerland AG 2024
T. Noll et al. (Eds.): MCM 2024, LNCS 14639, pp. 291–304, 2024.
https://doi.org/10.1007/978-3-031-60638-0_23

For any filtered complex K and any pair of integers $(i, p) \in \mathbb{N}^2$, there exists a natural inclusion map $\eta^{i,p} : K^i \hookrightarrow K^{i+p}$ which induces a homological morphism

$$\eta_*^{i,p} : H_*(K^i) \to H_*(K^{i+p})$$

Therefore, for each degree d, we get the following sequence of dth-homology groups:

$$0 \longrightarrow H_d(K^0) \xrightarrow{\eta_d^{0,1}} H_d(K^1) \xrightarrow{\eta_d^{1,1}} ...H_d(K^{N-1}) \xrightarrow{\eta_d^{N-1,1}} H_d(K^N) = H_d(K)$$

Each one of the maps $\eta_d^{i,p}$ sends a d-homological class at time i to the one containing it at time $i + p$ of the filtration. If the cycle has become a boundary, then it sends the class to 0. Thus, the image of each map $\eta_*^{i,p}$ will describe the homology evolution during the filtration: more precisely, the p-**persistent homology group** associated with K^i in degree d is given by the image of each map $\eta_*^{i,p}$:

$$H_d^{i,p}(K, \mathbb{F}_2) = H_d^{i,p}(K) := \operatorname{im} \eta_d^{i,p} : K^i \to K^{i+p}.$$

From [12], we have the following structure theorem of persistent homology:

Theorem 1. *Let K be a finite filtered simplicial complex. In any degree n, we have the following isomorphism of $\mathbb{F}_2[t]$-module:*

$$\bigoplus_{i=0}^{N} H_n(K^i) \cong \bigoplus_{j=0}^{N} t^{b_j} \cdot \mathbb{F}_2[t] / (t^{c_j})$$

This isomorphism naturally leads to the construction of graphics named barcodes, which mesures the lifetime evolution of homology classes in a filtration and provide a natural visualization of persistent homology.

Definition 1. *Let K be a finite filtered complex and $d \in \mathbb{N}$ be a fixed degree. The* ***barcode*** *$BC_d(K)$ associated with $H_d(K)$ is the graph where the x-axis describes the time of filtration and each generator of $H_d(K)$ corresponds to a bar whose start and end are given by Theorem's equivalence 1:*

- ✱ *a class that was born at time a_i and never dies is a bar that starts at the abscissa point a_i and never stops*
- ✱ *a class that was born at time b_j and died at time $b_j + c_j$ is a bar that starts and ends at abscissa points b_j and $b_j + c_j$, respectively.*

An illustration of this definition is given in Fig. 2, where we have computed the corresponding barcodes family with filtration from Fig. 1.

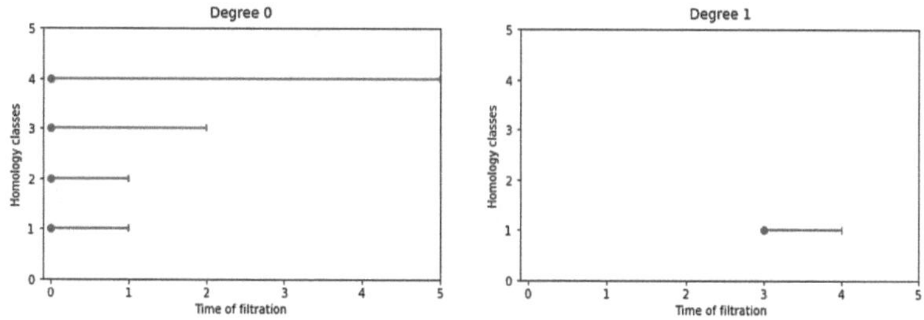

Fig. 2. The associated barcodes with filtration from Fig. 1.

1.2 Topological Data Analysis

Persistent homology is an algebraic tool that is commonly used in Topological Data Analysis context. In practice, we have a starting object that we want to analyze using its hidden topological structure. For this purpose, we transform it into a **point cloud**, that is to say into a finite metric set, and from this point cloud, we build a filtered simplicial complex using the Vietoris-Rips algorithm, which is presented in Definition 2 below. Once this is done, it is possible to compute persistent homology and then extract the associated barcodes family. Therefore, we can use this family as a **topological fingerprint** to characterize the starting object. In our case, our starting object is given by a score, symbolically represented by a MIDI file. The general TDA process is summarized in Fig. 3.

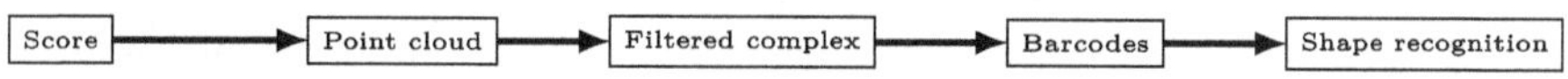

Fig. 3. Topological Data Analysis process.

Definition 2. *Let $X = \{x_1, ..., x_N\}$ be a point cloud, that means a collection of points in a metric space, and let $\epsilon \geq 0$ be a non-negative parameter. The* ***Vietoris-Rips complex*** *$\mathcal{R}_\epsilon(X)$ is the simplicial complex where:*

* *X is the set of vertices*
* *$\sigma = \{x_1, ..., x_n\}$ is a n-simplex if and only if the vertices it contains are pairwise close, i.e. if $d(x_i, x_j) \leq \epsilon$ for all pairs x_i, x_j of σ.*

Figure 4 shows the classical construction of a Vietoris-Rips complex starting from a given point cloud X and a parameter ϵ. The filtration is obtained by changing the parameter ϵ. In other words, this process consists of adding n-simplices to the point cloud as a certain parameter increases. The smaller the parameter is, the more separated the points are, while conversely, the larger the parameter is, the more trivial and topologically equivalent to a single point the resulting complex is. The point is therefore to analyze the complex for the "right" parameter, i.e. when the topological features are correctly represented. This construction method is presented in the article [7] by R. Ghrist.

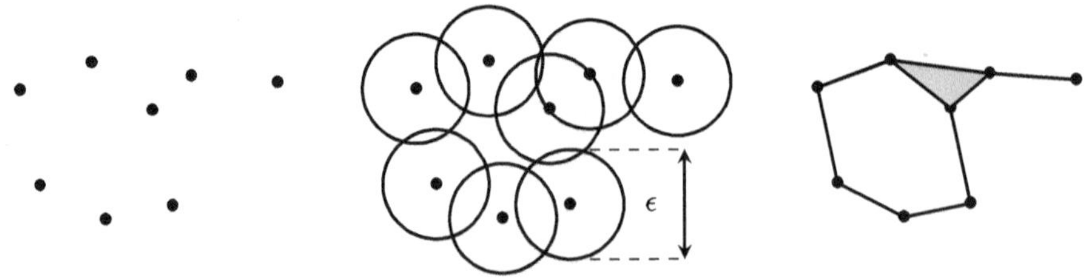

Fig. 4. The Vietoris-Rips method.

1.3 General Problematic

The essential question that remains incomplete in the TDA process is how to extract a point cloud from a musical score. There are many arbitrary choices to be made, and we already tried to answer this question in [4] by considering that a musical score is given by the set of its musical bars. Then, each bar was a subset of $\mathbb{R}^3$, so the metric was given by the Hausdorff distance. In this paper, we propose to keep the representation of a musical score by the set of its musical bars, but we consider the metric given by the two-dimensional Discrete Fourier Transform. In fact, we start by making explicit the transition from a musical bar to a subset of $\mathbb{Z}/t\mathbb{Z} \times \mathbb{Z}/p\mathbb{Z}$ and then define the notion of DFT-distance on the set of musical bars. This construction allows us to obtain a filtered complex from a MIDI file, and then to extract a barcodes family using persistent homology. The second part focuses on the question of musical analysis using this new topological signature. In this paper, our main motivation is to support the use of the DFT in the context of persistent homology and TDA, so we propose an experiment on different famous musical structures from which we expect to capture topological features. More precisely, the point cloud will be extracted from sets of chords based on two-dimensional Tonnetze.

2 Persistent Homology Using Discrete Fourier Transform

Associating a filtered complex with a piece of music requires extracting a point cloud, i.e. a collection of points in a metric space. As mentioned in the previous paragraphs, the points are given by the musical bars of a score, and we propose here to use the two-dimensional Discrete Fourier Transform as the metric.

2.1 A Model Based on the Two-Dimensional DFT

The DFT is a mathematical tool introduced into the context of music analysis by D. Lewin in 1959, and used mainly for modeling simple musical sequences, such as scales or collections of rhythms (see the work of E. Amiot and J. Yust, for instance [1] and [10]).

In this paper, we are primarily interested in modeling a musical sequence using the DFT: if $\mathcal{M} = (m_1, \ldots, m_N)$ is a musical sequence in $\mathbb{Z}/n\mathbb{Z}$ (for instance a sequence of pitch-classes or rhythmic data), then it can be associated with a list of n Fourier coefficients $\big(\mathcal{F}_{\mathcal{M}}(0), \ldots, \mathcal{F}_{\mathcal{M}}(n-1)\big)$ using the corresponding characteristic map. For instance, the pitch-classes set from Fig. 5 is given by $\mathcal{P} = \{9, 4, 5, 2\} \in \mathbb{Z}/12\mathbb{Z}$, and its characteristic map is given by

$$\mathbb{1}_{\mathcal{P}} : \{9, 4, 5, 2\} \mapsto (0, 0, 1, 0, 1, 1, 0, 0, 0, 1, 0, 0).$$

In the same way, using the quaver note as the unit of time, $\mathcal{T} = \{0, 1, 3, 5\}$ becomes the set of **onsets**, and the associated characteristics map is given by

$$\mathbb{1}_{\mathcal{T}} : \{0, 1, 3, 5\} \mapsto (1, 1, 0, 1, 0, 1, 0, 0)$$

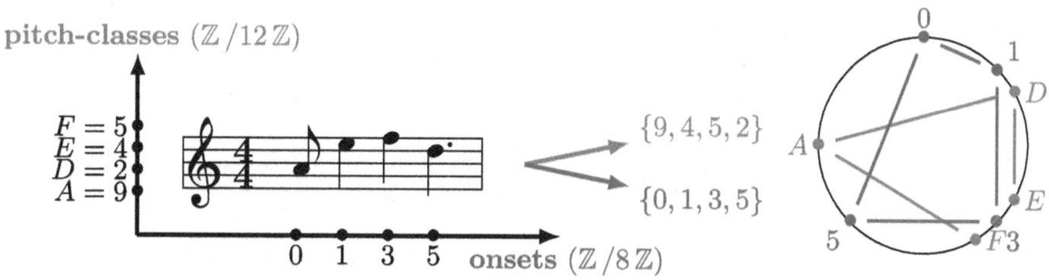

Fig. 5. A musical bar where the x-axis and y-axis represent time and pitches, respectively.

Now we can compute the **Discrete Fourier Transform** (DFT) of $\mathcal{M}$ ($\mathcal{P}$ or $\mathcal{T}$), which is simply given by the DFT of its characteristic map $\mathbb{1}_{\mathcal{M}} : \mathbb{Z}/n\mathbb{Z} \rightarrow \{0, 1\}$ with the usual definition, that is:

$$\begin{aligned} \mathcal{F}_{\mathcal{M}} = \widehat{\mathbb{1}_{\mathcal{M}}} : \mathbb{Z}/n\mathbb{Z} &\longrightarrow \mathbb{C} \\ x &\longmapsto \sum_{k \in \mathcal{M}} \exp\left(\tfrac{-2i\pi kx}{n}\right) \end{aligned}$$

Therefore, each musical structure can be associated with a family of **Fourier coefficients** that are given by the n-tuple of the DFT values on $\mathbb{Z}/n\mathbb{Z}$:

$$\Big(\mathcal{F}_{\mathcal{M}}(0), \mathcal{F}_{\mathcal{M}}(1), \ldots, \mathcal{F}_{\mathcal{M}}(n-1)\Big)$$

The DFT can be applied to any abelian group, so the previous definition can naturally be generalized ot the two-dimensional case. More precisely, instead of considering onsets and pitch-classes separately, we combine both representations and thus see $\mathcal{T}$ and $\mathcal{P}$ as a subset of $\mathbb{Z}/8\mathbb{Z} \times \mathbb{Z}/12\mathbb{Z}$, and we will call it a **musical bar**, denoted by $\mathcal{B}$. For instance, the description of the musical bar from Fig. 5 is given by:

$$\mathcal{B} = (\mathcal{T}, \mathcal{P}) = \{(0, 9), (1, 4), (3, 5), (5, 2)\} \subset \mathbb{Z}/8\mathbb{Z} \times \mathbb{Z}/12\mathbb{Z}$$

In general, if t is a **unit of time** and p is an **ambitus**, both fixed, we say that a **musical bar** $\mathcal{B}$ with n elements is a subset of $\mathbb{Z}/t\mathbb{Z} \times \mathbb{Z}/p\mathbb{Z}$ given by

$$\mathcal{B} = \{(t_1, p_1), \ldots, (t_n, p_n)\}$$

where an element (t_j, p_j) of $\mathcal{B}$ is called a **note** which is characterized by its **onset** t_j modulo t and its **pitch-class** p_j modulo p. Therefore, a **musical score** $\mathcal{S}$ is the non-ordered set of its N distinct musical bars modulo (t, p):

$$\mathcal{S} = \{\mathcal{B}_1, \mathcal{B}_2, \ldots, \mathcal{B}_N\} \text{ with } \mathcal{B}_i \subset \mathbb{Z}/t\mathbb{Z} \times \mathbb{Z}/p\mathbb{Z} \text{ and } \mathcal{B}_i \neq \mathcal{B}_j \text{ if } i \neq j$$

Thus, we get a description of any musical bar of a given score using a matrix of Fourier coefficients.

Definition 3. *Let B be a musical bar in $\mathbb{Z}/t\mathbb{Z} \times \mathbb{Z}/p\mathbb{Z}$.*

1. *The associated* ***characteristic map*** *$\mathbb{1}_{\mathcal{B}} : \mathbb{Z}/t\mathbb{Z} \times \mathbb{Z}/p\mathbb{Z}$ is given by:*

$$\mathbb{1}_{\mathcal{B}} : (x, y) \mapsto \begin{cases} 1 \textit{ if } (x, y) \in \mathcal{B} \\ 0 \textit{ otherwise.} \end{cases}$$

2. *The associated* ***DFT*** *is the DFT of the characteristic map $\mathbb{1}_{\mathcal{B}}$:*

$$\begin{aligned} \mathcal{F}_{\mathcal{B}} = \widehat{\mathbb{1}_{\mathcal{B}}} : \mathbb{Z}/t\mathbb{Z} \times \mathbb{Z}/p\mathbb{Z} &\longrightarrow \mathbb{C} \\ (x, y) &\longmapsto \sum_{(k,l) \in \mathcal{B}} \exp\left(\tfrac{-2i\pi kx}{t}\right) \exp\left(\tfrac{-2i\pi ly}{p}\right) \end{aligned}$$

3. *The associated* ***Fourier coefficients*** *are given by the matrix $\widehat{M}_{\mathcal{B}} \in \mathcal{M}_{t \times p}(\mathbb{C})$:*

$$\widehat{M}_{\mathcal{B}} = \Big(\mathcal{F}_{\mathcal{B}}(x, y)\Big)_{(x,y) \in \mathbb{Z}/t\mathbb{Z} \times \mathbb{Z}/p\mathbb{Z}}$$

2.2 From the DFT to a Point Coud

We denote by $\mathfrak{B}_{t,p}$ the set of all the musical bars in $\mathbb{Z}/t\mathbb{Z} \times \mathbb{Z}/p\mathbb{Z}$. Now, using Definition 3, we can define a metric based for computational reasons on the 1-norm to compare musical bars together. We call it the DFT-distance, which we will use to extract a point cloud from a musical score.

Definition 4. *Let $\mathcal{B}$ and $\mathcal{B}'$ be two elements of $\mathfrak{B}_{t,p}$. The* ***DFT-distance*** *between $\mathcal{B}$ and $\mathcal{B}'$ is given by the 1-norm between their respective Fourier coefficients matrices:*

$$d_{\mathrm{DFT}}(\mathcal{B}, \mathcal{B}') = \| \widehat{M}_{\mathcal{B}} - \widehat{M}_{\mathcal{B}'} \|_1 = \sum_{x=1}^{t} \sum_{y=1}^{p} |\mathcal{F}_{\mathcal{B}}(x, y) - \mathcal{F}_{\mathcal{B}'}(x, y)|$$

We denote by $(\mathfrak{B}_{t,p}, d_{\mathrm{DFT}})$ the metric space of all the musical bars of $\mathbb{Z}/t\mathbb{Z} \times \mathbb{Z}/p\mathbb{Z}$ equipped with the DFT-distance.

We can now turn any musical piece into a point cloud, i.e. a subset of a metric set. More precisely, if $\mathcal{S}_{\mathfrak{P}} = \{\mathcal{B}_1, \ldots, \mathcal{B}_N\}$ is a musical score representing a musical piece $\mathfrak{P}$, then $\mathcal{S}_{\mathfrak{P}}$ is a subset of the metric space $(\mathfrak{B}_{t,p}, \mathrm{d}_{\mathrm{DFT}})$ and thus becomes a point cloud constructed in the following way:

* each **point** is a musical bar $\mathcal{B}_i$
* the **distance** between $\mathcal{B}_i$ and $\mathcal{B}_j$ is given by the DFT-distance.

Remark 1. In the following, we will apply a normalization on the distances computed with the DFT-distance in order to have all the barcodes on the same scale: in fact, our goal is to obtain barcodes that we can compare with each other, and we are more interested in the distance ratios between musical bars than in the distance values themselves. Therefore, by taking a percentage of the maximum distance for a given score, we consider the set of times of filtration to be $\{0, 1, \ldots, 100\}$. Furthermore, we will say that we are looking at the filtration with a **scaling parameter** of $\epsilon\%$, with $\epsilon \in \{0, 1, \ldots, 100\}$.

2.3 An Illustration of the Model

As an illustration of these definitions, let us consider the score $\mathcal{S}$ from Fig. 6 which is an excerpt from the music piece *One Summer's Day*, composed by Joe Hisaishi in 2001 for the movie *Spirited Away.*

Fig. 6. The score $\mathcal{S}$ extracted from *One Summer's Day* written by Joe Hisaishi.

The above score is divided into five distinct musical bars: $\mathcal{S} = \{\mathcal{B}_1, \ldots, \mathcal{B}_5\}$. Furthermore, the shortest note here is a quaver one, so the time unit is $t = 8$, and the ambitus can be taken as $p = 24$, with $C_4 = 0 = C_6$, $C_5 = 12$ and $B_5 = 23$. With these observations, each musical bar $\mathcal{B}_i$ of the score $\mathcal{S}$ is a subset of $\mathbb{Z}/8\mathbb{Z} \times \mathbb{Z}/24\mathbb{Z}$. We extract the corresponding point cloud, as shown in Fig. 7.

$\mathcal{B}_1 = \{(6,9), (7,11)\}$

$\mathcal{B}_2 = \{(0,12), (1,12), (2,14), (3,12), (4,11), (6,4), (7,7)\}$

$\mathcal{B}_3 = \{(0,9), (1,9), (2,7), (3,5), (4,7), (7,7)\}$

$\mathcal{B}_4 = \{(0,7), (1,5), (2,5), (3,3), (4,5), (6,0), (7,5)\}$

$\mathcal{B}_5 = \{(0,2), (0,7), (0,11)\}$

	$\mathcal{B}_1$	$\mathcal{B}_2$	$\mathcal{B}_3$	$\mathcal{B}_4$	$\mathcal{B}_5$
$\mathcal{B}_1$	0	86	81	82	63
$\mathcal{B}_2$		0	91	100	91
$\mathcal{B}_3$			0	95	87
$\mathcal{B}_4$				0	82
$\mathcal{B}_5$					0

Fig. 7. The point cloud associated with the score from Fig. 6. The points are the musical bars in $\mathbb{Z}/8\mathbb{Z} \times \mathbb{Z}/24\mathbb{Z}$ (left) and the corresponding distance ratios are computed from the DFT-distance (right).

From this point cloud we can easily apply the Vietoris-Rips method, and the resulting filtration is presented in Fig. 8. Notice that this illustration only shows the graphs filtration (to avoid overloading the figure): in particular, the triangles are all filled (always the case with the Vietoris construction).

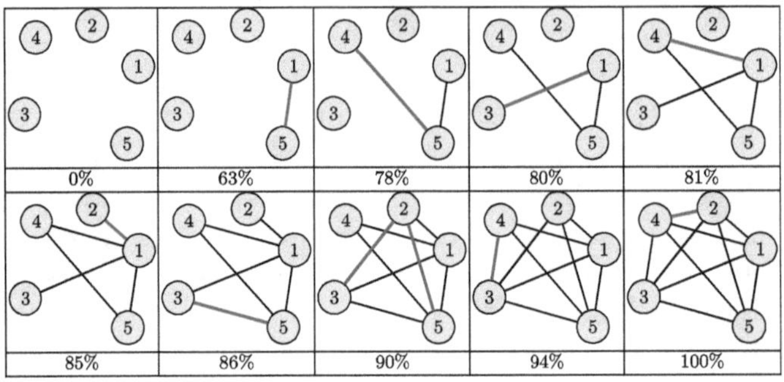

Fig. 8. The filtered simplicial complex associated with *One Summer's Day*.

For each scaling parameter $\epsilon \in \{0, 1, \ldots, 100\}$, the corresponding pairs of musical bars form edges. Therefore, we can look at the evolution of homology and then compute the associated barcodes family, as shown in Fig. 9. Here and throughout this paper, we will focus only on barcodes in degree 0 and degree 1: indeed, the intuition of a musical interpretation goes along with these low dimensions, as for example we think of a cycle in dimension 1 as a musical loop. Of course, higher dimensions could be interpreted in future work. We can also note that there is no element of homology in degree 1, which is not so surprising considering the number of musical bars in the score.

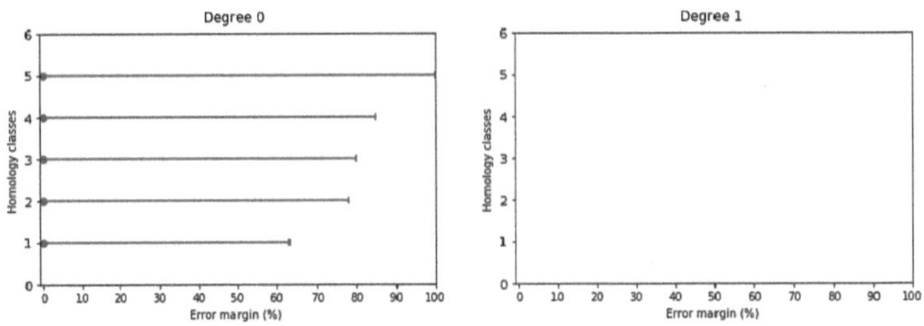

Fig. 9. The barcodes family associated with *One Summer's Day*.

3 Experiment on Artificial Scores

The goal of this section is to support the use of the two-dimensional DFT in the context of persistent homology, and in particular to answer the question of whether this new metric makes musical sense. To do this and in order to support this new model, we will try it on some artificially created musical scores: more precisely, we will construct a musical score consisting of very simple musical bars containing only one triad from a given family of chords, such as the major and minor chords (Euler's Tonnetz).

3.1 A Point Cloud from the Set of Minor and Major Chords

Let us consider the score based on the twenty-four major and minor chords, that means a score with twenty-four musical bars, each one containing a 3-chord, as shown in Fig. 10. The point cloud thus contains twenty-four points, and can be identified with Euler's Tonnetz $T[3,4,5]$ of minor and major triads. A representation of this Tonnetz as a simplicial complex is given in Fig. 11, and we refer to [5] for a formalized definition of the two-dimensional Tonnetze of the form $T[a,b,-(a+b)]$.

Notice that the representation from Fig. 11 provides a topological structure for Euler's Tonnetz, which is given by a torus. We also see the action of the PLR-group of Neo-Riemannian transformations P (Parallel), L (Leading-tone) and R (Relative), which transforms a major (or minor) triad into the nearest minor (or major) triad. Note that this action is simply transitive, and we refer to [8] for more details on this subject. These two observations are exactly the kind of properties that we want to emphasize with our model. Indeed, it will support the use of the two-dimensional DFT in this particular context.

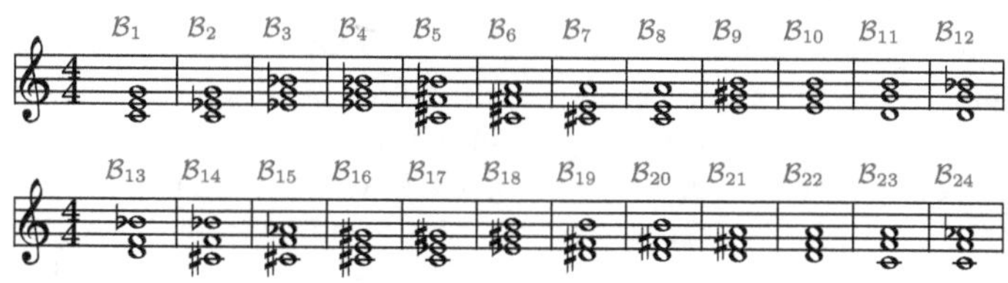

Fig. 10. The score of twenty-four musical bars based on the major and minor chords.

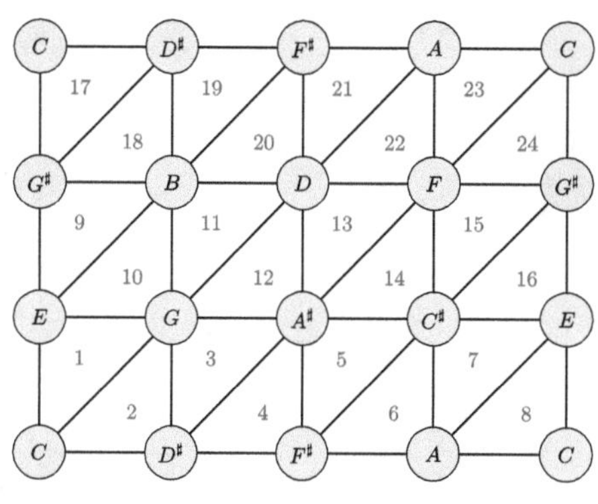

Fig. 11. Euler's Tonnetz as a simplicial complex $T[3,4,5]$.

3.2 The Results: PLR-Group and One-Dimensional Cycles

Now that we have the point cloud, we can compute the associated barcodes family, as presented in Fig. 12.

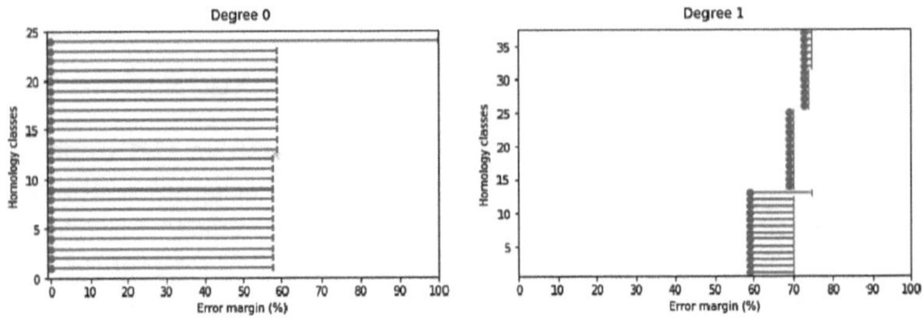

Fig. 12. The associated barcodes family with Euler's Tonnetz $T[3, 4, 5]$.

Let us analyze these two barcodes, starting with degree 0: at 58% of the filtration, each chord is connected to its relative, which means that the DFT understands this proximity. With a scaling parameter of 59%, the complex looks like in Fig. 13 (left), where we have shown only the graphs (with vertices in $\{1, 2, \ldots, 24\}$ from the score 10): in this illustration, we have chosen to show the complexes at 59% and 69% because the filtration stagnates at 59% (the first moment where it is connected) and remains the same until it becomes as in 69%. There are several things to note here: first, in both cases we recover the shape of a torus, with a duality relation since a triangle in $T[3, 4, 5]$ is a vertex in our graph. Secondly, each vertex of the complex at 59% has exactly three neighbors: for instance, C-major triad (musical bar $\mathcal{B}_1$, vertex $\{1\}$) is connected to musical bars $\mathcal{B}_2$, $\mathcal{B}_8$ and $\mathcal{B}_{10}$. Looking back at the simplicial representation of Fig. 11, we see that it corresponds to the minor chords C_m, A_m and E_m respectively, the three neighbors of C given by its parallel P, its relative R and its the leading-tone L, which are exactly the three minor chords that minimize the DFT-distance from a given triad. This observation is illustrated in Fig. 14 and is formalized in Theorem 2.

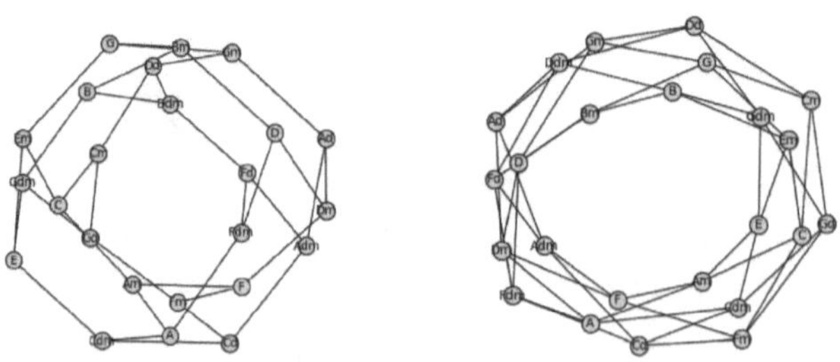

Fig. 13. The associated filtration with the Tonnetz $T[3, 4, 5]$ with a scaling parameter of 59% (left) and 69% (right).

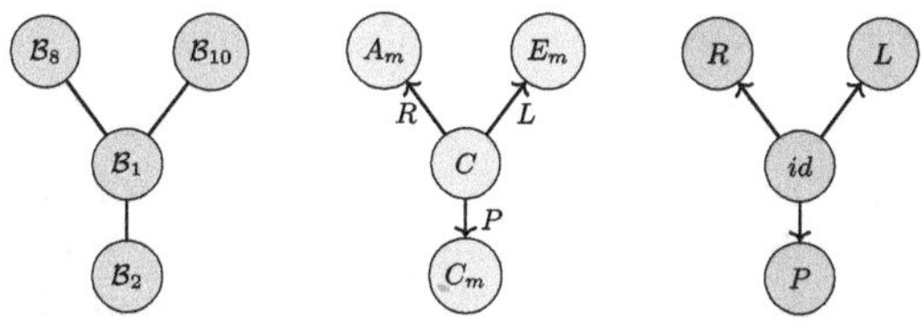

Fig. 14. A zoom on the graph at 59% of the filtration: each chord has exactly three neighbors that are given by the three basic transformations P, L and R.

Theorem 2. *The graph associated with the twenty-four major and minor chords of the Tonnetz* $T[3,4,5]$, *given by the filtration below with a scaling parameter of* 59%, *is exactly the Cayley graph of the* PLR*-group generated by the three transformations* P, L *and* R.

Proof. The action of the PLR-group on the set of the twenty-four major and minor chords is simply transitive, so there is a bijection between the elements of the PLR-group and the chords of the Tonnetz. Therefore, using the illustration from Fig. 14, the left graph from Fig. 13 is exactly the corresponding Cayley graph.

The barcode in degree 0 seems to capture the topological structure of the Euler's Tonnetz, and there is a strong relationship between the filtration and the PLR-group. On the other hand, the cycles in degree 1 also seem to emphasize this property: in fact, we can classify them into three types, all given by the transformation P, L and R. To illustrate these elements of 1-homology, we draw a representative of each type of one-dimensional cycle that we can construct from C-Major triad (but it is important to note that we can get these cycles from any chosen chord). These cycles are obtained respectively with transformations $(PRL)^2$, $(LP)^3$ and $(PR)^4$, as shown in Fig. 15. Also notice that these cycles have been pointed out in [6] and it is musically consistent to recover such results.

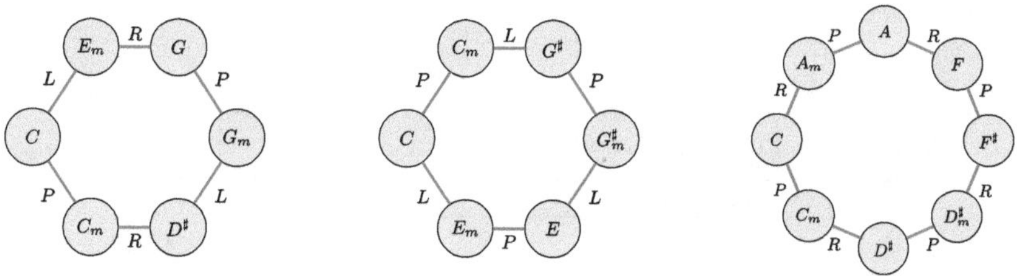

Fig. 15. The one-dimensional cycles obtain from the major and minor chords and the DFT as a metric: they are given respectively by transformations $(PRL)^2$, $(LP)^3$ and $(PR)^4$ on a given chord (here C-Major).

3.3 A Study of the Two-Dimensional Tonnetze

As an extension of this work, we apply this new model to the eleven other sets of triads based on the two-dimensional Tonnetze of the form $T[a, b, -(a+b)]$. This analysis, which is fully described in [5], leads to a summary classification of the different filtrations for each point cloud, which is presented in Table 1.

The barcodes in degree 0 capture the topological structure of each Tonnetz. As for Euler's Tonnetz, we recover the shape of the different $T[a, b, -(a+b)]$ in the filtration, such as the tori, the cylinders, or even the several connected components for $T[2, 2, 8]$, $T[2, 4, 6]$, $T[3, 3, 6]$ and $T[4, 4, 4]$. In addition, barcodes of degree 1 allow us to manually generate one-dimensional cycles. Each Tonnetz provides specific types of generators of H_1, and it is interesting to compare them: for some Tonnetz we have cycles of length 4, while for others it can go up to 12. Table 1 classifies the different types of cycles we can find in these two-dimensional Tonnetze.

Table 1. A classification of the twelve two-dimensional Tonnetz $T[a, b, -(a+b)]$ by their topological structure and their types of cycle in H_1 using the DFT.

Point cloud	Topology	Most relevant stage of the filtration	Length of cycles (degree 1)				
			4	6	8	10	12
$T[1, 2, 9]$, $T[2, 3, 7]$	Torus	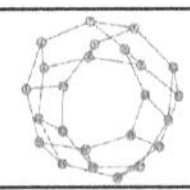		×	×		×
$T[1, 3, 8]$, $T[3, 4, 5]$				×	×		
$T[1, 4, 7]$				×			
$T[1, 1, 10]$, $T[2, 5, 5]$	Cylinder	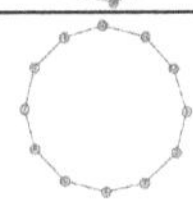					×
$T[1, 5, 6]$	Necklace of six tetrahedra	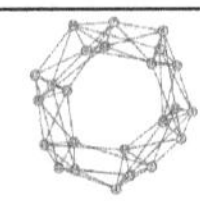				×	
$T[2, 2, 8]$	Two cylinders	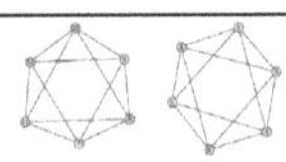		×			
$T[2, 4, 6]$	Two necklaces of three tetrahedra	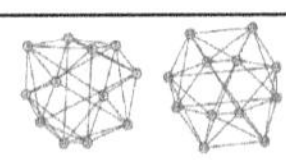	×	×			
$T[3, 3, 6]$	Three tetrahedra	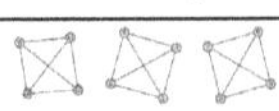	×				
$T[4, 4, 4]$	Four triangles						

4 Conclusion and Perspective for Future Work

We begin this paper by introducing a new method for extracting a filtration and more precisely a point cloud from a given score. In order to support this approach, we wanted to consider the inverse problem, that is, to apply the DFT together with persistent homology to artificially constructed musical scores and show that the DFT is a reasonable choice of metric.

The score we choose is given by triads from a specific space, specifically the Euler's Tonnetz, or more generally any two-dimensional Tonnetz of the classical form $T[a, b, -(a+b)]$. The barcodes in degrees 0 and 1 seem to reveal consistent musical features: in particular, we find back the topological structure associated with the Tonnetze, such as the torus in the case of $T[3, 4, 5]$. This construction also allows us to emphasize the strong relationship with PLR-group of fundamental transformations of Euler's Tonnetz, as illustrated by Theorem 2. Finally, we recover some famous cycles with barcode in degree 1, which is musically worthwhile. This confirms that the metric constructed by means of the DFT, together with persistent homology, seems to be a reasonable tool for understanding known musical structures.

However, there are several other artificial musical scores on which we might want to test our model. As a natural starting point, we could try artificial scores based on scales, as in [5]. We could also try the metric on the set of all the triads in $\mathbb{Z}/12\mathbb{Z}$: for a given triad, the closest would still be its relative, then its leading-tone and its parallel, but this will allow us to classify all the 220 triads with this DFT, and this will also give a meta-classification between the twelve Tonnetze. Another work could be to compare the DFT with other metrics, such as the Hausdorff distance which provide much less interesting results (see [5]). Finally, the DFT-distance could also be applied to n-chords, or simply to higher dimensional Tonnetze, as presented in [9]. To go further, we could also imagine trying out our metric on musical objects that focus more on rhythms, such as the general *Zeitnetz* that is introduced in [11].

Finally, we have proposed here a new model for associating a filtration of simplicial complexes with a musical score, and thus a topological signature given by barcodes. If this method seems to provide strong and encouraging results in applications on artificial scores, we also apply it as a natural work in the context of more general problems, such as automatic classification of musical styles. These other kinds of applications of this model based on the DFT and persistent homology are presented in [5].

Acknowledgments. The author would like to thank the anonymous referees of this paper for their valuable comments and suggestions. We also would like to acknowledge the support by the Interdisciplinary Thematic Institute CREAA, as part of the ITI 2021-2028 program of the Université de Strasbourg, the CNRS, and the Inserm (funded by IdEx Unistra ANR-10-IDEX-0002, and by SFRI-STRAT'US ANR-20-SFRI-0012 under the French Investments for the Future Program).

References

1. Amiot, E.: Music Through Fourier Space. Computational Music Science, Springer, Cham (2016). https://doi.org/10.1007/978-3-319-45581-5
2. Bergomi, M.G., Baraté, A., Di Fabio, B.: Towards a topological fingerprint of music. In: Bac, A., Mari, J.L. (eds.) CTIC 2016. LNCS, vol. 9667, pp. 88–100. Springer, Cham (2016). https://doi.org/10.1007/978-3-319-39441-1_9

3. Bigo, L., Andreatta, M.: Filtration of pitch-class sets complexes. In: Montiel, M., Gomez-Martin, F., Agustín-Aquino, O.A. (eds.) MCM 2019. LNCS, vol. 11502, pp. 213–226. Springer, Cham (2019). https://doi.org/10.1007/978-3-030-21392-3_17
4. Callet, V.: Persistent homology on musical bars. In: Montiel, M., Agustín-Aquino, O.A., Gómez, F., Kastine, J., Lluis-Puebla, E., Milam, B. (eds.) MCM 2022. LNCS, vol. 13267, pp. 349–355. Springer, Cham (2022). https://doi.org/10.1007/978-3-031-07015-0_29
5. Callet, V.: Modélisation topologique des structures et processus musicaux. Theses, Université de Strasbourg (UNISTRA) (2023). https://theses.hal.science/tel-04322624
6. Cohn, R.: Neo-Riemannian operations, parsimonious trichords, and their "Tonnetz" representations. J. Music Theory **41**(1), 1–66 (1997). http://www.jstor.org/stable/843761
7. Ghrist, R.: Barcodes: the persistent topology of data. Bull. Am. Math. Soc. (N.S.) **45**(1), 61–75 (2008)
8. Popoff, A., Andreatta, M., Ehresmann, A.: Relational poly-Klumpenhouwer networks for transformational and voice-leading analysis. J. Math. Music. **12**(1), 35–55 (2018). Mathematical and Computational Approaches to Music Theory, Analysis, Composition and Performance
9. Tymoczko, D.: The generalized Tonnetz. J. Music Theory **56**, 1–52 (2012). https://doi.org/10.2307/41508604
10. Tymoczko, D., Yust, J.: Fourier phase and pitch-class sum. In: Montiel, M., Gomez-Martin, F., Agustín-Aquino, O.A. (eds.) MCM 2019. LNCS, vol. 11502, pp. 46–58. Springer, Heidelberg (2019). https://doi.org/10.1007/978-3-030-21392-3_4
11. Yust, J.: Generalized *tonnetze* and *zeitnetze*, and the topology of music concepts. J. Math. Music. **14**(2), 170–203 (2020). Mathematical and Computational Approaches to Music Theory, Analysis, Composition and Performance
12. Zomorodian, A., Carlsson, G.: Computing persistent homology. Discret. Comput. Geom. **33**(2), 249–274 (2005)

Fourier (Common-Tone) Phase Spaces are in Tune with Variational Autoencoders' Latent Space

Nádia Carvalho(✉) and Gilberto Bernardes

Faculty of Engineering of University of Porto & INESC TEC, Porto, Portugal
{nscarvalho,gba}@up.pt

Abstract. Expanding upon the potential of generative machine learning to create atemporal latent space representations of musical-theoretical and cognitive interest, we delve into their explainability by formulating and testing hypotheses on their alignment with DFT phase spaces from $\{0,1\}^{12}$ pitch classes and $\{0,1\}^{128}$ pitch distributions – capturing common-tone tonal functional harmony and parsimonious voice-leading principles, respectively. We use 371 J.S. Bach chorales as a benchmark to train a Variational Autoencoder on a representative piano roll encoding. The Spearman rank correlation between the latent space and the two before-mentioned DFT phase spaces exhibits a robust rank association of approximately $.65 \pm .05$ for pitch classes and $.61 \pm .05$ for pitch distributions, denoting an effective preservation of harmonic functional clusters per region and parsimonious voice-leading. Furthermore, our analysis prompts essential inquiries about the stylistic characteristics inferred from the rank deviations to the DFT phase space and the balance between the two DFT phase spaces.

Keywords: Variational Autoencoders · Latent Space · Discrete Fourier Transform · Musical Harmony Analysis

1 Introduction

The enigmatic nature of how the human brain encodes and represents musical properties continues to challenge researchers. Despite advancements in cognitive psychology, the intricate mechanisms underlying music neural representation remain elusive. Nonetheless, many fields, including music information retrieval [12,15], composition and analysis [13], and generative music [5,8], have advanced knowledge by incorporating cognitive-led frameworks.

A promising avenue for conceptualizing atemporal representations of music structures lies in leveraging artificial neural networks (ANNs) [2,3]. These computational models provide a biologically inspired approach to comprehending

This research has been funded by the Portuguese National Funding Agency for Science, Research and Technology [2021.05132.BD].

© The Author(s), under exclusive license to Springer Nature Switzerland AG 2024
T. Noll et al. (Eds.): MCM 2024, LNCS 14639, pp. 305–316, 2024.
https://doi.org/10.1007/978-3-031-60638-0_24

cognitive processes. They can serve as valuable tools in exploring the representation of music within the brain, notwithstanding the interpretational challenges posed by their intricate structures. In a groundbreaking insight, Dawson and other co-authors [11,16] have shown remarkably high correlations between ANNs' weights and the phase of the discrete Fourier transform (DFT) space employed to represent musical sets. This unexpected correlation sparks intriguing questions about the relationship between network learning rules and formal musical analysis. While groundbreaking, the experiment was conducted on an ANN trained to classify simple musical structures such as intervals and diatonic triads, thus excluding the temporal manifestation of surface elements in music.

Amid the rapid progress in generative machine learning, state-of-the-art deep learning Variational Autoencoders (VAEs) have shown great potential in generating meaningful and interpretable music structures in lower-dimension latent spaces [14,19]. VAEs have emerged as a powerful tool for capturing cognitive distances within the temporal dimension of music through their distinctive latent spaces. VAEs excel in modeling complex relationships and structures inherent in music by transforming musical data into a lower-dimensional representation latent space, which results from encoding input data into a probabilistic distribution, optimizing for reconstruction accuracy and regularization during training. This dual-objective optimization yields a latent space where similar data points cluster, reflecting their cognitive proximity.

Carvalho and Bernardes [10] have shown that a flat baseline VAE derives latent spaces aligned with cognitive pitch spaces and tonal hierarchies at the regional (key) space. Similarly, Turker et al. [21] showcase that disentangled directions commonly display monotonic and global traits within a well-organized and regularized latent space. These directions encode fundamental musical characteristics like colorfulness, speed, and repetitiveness, influencing the style and features of the input music. Bryan-Kinns et al. [9] integrate latent space regularization (LRS) into MeasureVAE and AdversarialVAE models. Their findings illustrate the feasibility of improving the interpretability of music VAE models by introducing semantically meaningful attributes, such as note density, range, rhythmic complexity, and average interval jump. Importantly, the scope of their research is confined to monophonic melodies, and no exploration is undertaken regarding harmony. They further stress the need for future studies examining the influence of genre diversity, dataset variations, sizes, musical attributes, and training methods on the performance of explainable AI models, aiming to advance the field and augment the adaptability of explainable AI across diverse musical scenarios.

Our work expands previous research by relying on advanced MusicVAEs to unlock the structuring of musical harmony from a commonly-used Western music corpus of J.S. Bach chorales, spatially organized in latent spaces. We hypothesize that VAEs' latent spaces correlate with DFT phase spaces, similarly explored in music theory to the harmonic paths in the *Tonnetz* with a wider range of harmonic objects [23]. DFT phase spaces from $\{0,1\}^{12}$ pitch classes and $\{0,1\}^{128}$ pitch distributions represent harmony according to common tone principles, and privilege either tonal functional harmony or voice-leading spaces [22], respectively. Possible deviations to this space should inform the stylistic traits of a representative corpus or composer, which naturally uses different degrees of note retention and voice-leading to elaborate the musical time. Ultimately, we aim to analyze deviations from a purely common-tone DFT phase space and the degree to which they promote the notion of style within a corpus.

In Sect. 2, we introduce VAE models and elaborate on constructing the latent space within the MusicVAE architecture. Section 3 introduces and discusses the properties of DFT phase spaces. Section 4 delineates our methodology for assessing the correlation between latent and DFT phase spaces. The subsequent Sect. 5 presents and delves into the results obtained. Finally, Sect. 6 presents conclusions from our study and outlines potential directions for future research.

2 MusicVAE

Utilizing VAEs with bidirectional Long Short-Term Memory (LSTM) networks for encoder and decoder components, MusicVAE [19] is an advanced generative model designed to learn intricate patterns in musical data. This innovative approach enables the model to generate diverse and expressive musical sequences, fostering creative exploration for musicians and composers. Notably, MusicVAE stands out for its proficiency in smoothly interpolating between different musical latent spaces, providing a seamless means to explore the vast landscape of musical possibilities.

In this architecture, the encoder learns to compress musical input sequence, X, into a latent space by parameterizing a Gaussian distribution with mean (μ) and standard deviation (σ). This is expressed as $q_\phi(Z|X) = \mathcal{N}(\mu_\phi(X), \sigma_\phi(X))$, where ($\phi$) represents the encoder parameters. The training objective involves maximizing the evidence lower bound (ELBO), balancing the reconstruction term $\mathbb{E}_{q_\phi(Z|X)}[\log p_\theta(X|Z)]$ (decoder's ability to reconstruct input) with the regularization term $\mathrm{KL}[q_\phi(Z|X)||p(Z)]$ (ensuring a structured latent space). The decoder, $p_\theta(X|Z)$, models the likelihood of reconstructing X given a latent point Z, and KL measures the divergence between the encoder's distribution and a chosen prior distribution.

3 Pitch Fourier Phase Space

Within music theory and music information retrieval (MIR), Fourier phase spaces are typically obtained by applying the DFT to a pitch class set from a binary 12-element distribution $\{0, 1\}^{12}$, where activation (i.e., 1) denotes the presence of a pitch class. From the resulting 12 complex values (i.e., DFT coefficients), only the DFT coefficients one to six are relevant, as the others are redundant due to the symmetry. From the complex values from the DFT analysis, one can compute magnitude and phase information [18], the first relating to harmonic qualities and the second to pitch organization within regional sets [23].

Using the periodic-shaped phase values of fewer coefficients as coordinates results in spaces shaped like a torus, with clear links to the *Tonnetz*, namely, the phases of the third and fifth coefficients [1,24]. Drawn from this link, the preservation of its common-tone basis has been observed. However, due to its continuous geometrical space, a larger number of harmonic objects within a wider range of hierarchies (e.g., pitch, chords, and regions) can be projected. In the pitch class DFT phase space, the arrangements of pitches within chords are disregarded, holding no significance. Hence, chord inversions and intervallic displacement across octaves are reduced to an agnostic pitch class space ('chroma').

Adopting L_1 pitch class distributions normalization, we can compare and analyze different hierarchies [4], aligning with many music-theoretical and cognitive spaces of interest. Regional spaces align with the circle of fifths and preserve major and minor relationships. Derived from the common-tone property of the space, functional harmonic clusters can be observed in the phase space per region. These relationships can be quantified by both Euclidean and cosine distances across pitch class sets.

Yust [23] adopts the third and fifth coefficients to draw a two-dimensional phase space to study the Trio and Adagio from Schubert's String Quintet (D. 956, Op. posth. 163), namely by expanding and refining the application of neo-Riemannian theory concepts to the analysis of chromatic harmony, enlightening triadic voice-leading and tonal regions relationships. Bernardes et al. [5] adopt the common-tone phase space to design generative music strategies for tonal music by navigating in the clustered functional harmonic functions within regional spaces. In [6,7], DFT phase spaces are expanded to the audio domain and adopted to find compatible audio loops that share controllable degrees of common-tone content within the same regional space to navigate and retrieve musical loops from large datasets for musical mashup creation.

Having been subject to much less theoretical discussion, phase spaces derived from expanding the 12-element pcset distribution to the 128 pitch range in $\{0,1\}^{128}$ have been explored in [10]. In contrast to the pitch class DFT phase space, this exploration considers pitch arrangements, accounting for chord inversions and octave displacements. The study's results indicate that pitch distribution DFT VAE latent spaces closely correspond to cognitive spaces, forming a common-tone space where overlapping objects within a key create fuzzy clusters. This configuration establishes a well-defined order of structural significance or stability, essentially resembling a tonal hierarchy, notably following the circle of fifths.

Figures 1 and 2 show the distances between major and minor regions (represented by their diatonic scale) in both DFT phase spaces from pitch class and pitch distributions, enhancing the circle of fifth relations. Furthermore, harmonic function clusters from the diatonic set of C major chords are observed in the pitch class space (leftmost image in Fig. 2). In the pitch distribution space, the relations between articulated chords denote proximity from tone retention; however, the harmonic functions clusters are less represented (e.g., C major and A minor chords do not align with any common tone, as in the pitch class space). Nonetheless, parsimonious voice-leading is better expressed in this space.

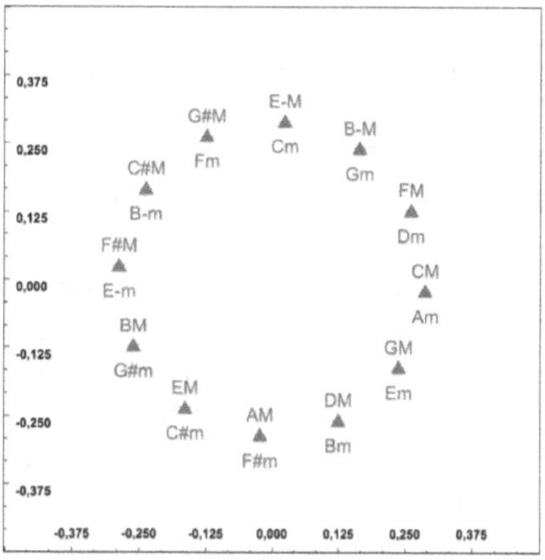

Fig. 1. Major and minor tonalities in the DFT phase spaces of pitch class and pitch distributions are visually depicted in a two-dimensional representation employing Locally Linear Embedding (LLE) for dimensionality reduction. Notably, both spaces exhibit complete alignment in their visual representation.

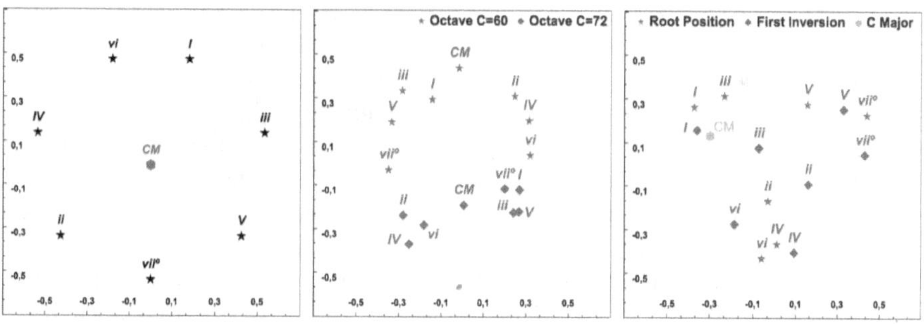

Fig. 2. A two-dimensional visualization is presented, showcasing the DFT phase spaces of pitch class (on the left) and pitch distributions for diatonic chords in the key of C major. This visualization employs Locally Linear Embedding (LLE) for dimensionality reduction. In the middle, chords in the space of pitch are represented across different octaves, while on the right, the representation includes the chords in root position and first inversion.

4 Experiment

To assess our hypothesis that VAEs' latent spaces correlate with DFT phase spaces, we utilize a benchmark of a prototypical tonal music corpus comprising

371 chorales by J.S. Bach. The corpus is sourced from the `music21` library.[1] It encompasses 195 chorales in major keys (53%) and 176 in minor keys (47%), with G Major (14%), A minor (12%), and G minor (11%) emerging as the most prevalent tonalities. On average, a chorale consists of 84 chords. Our training process incorporates 60% of the corpus chorales, while the remaining 40% are allocated for testing. The chorales used for training undergo augmentation through transposition, involving both upward and downward adjustments of the encoding across all twelve keys. This augmentation process promotes balance among tonalities, ensuring the model remains unbiased towards any particular tonalities that might be more prominent in the dataset.

To create a latent space from a MusicVAE, we adopt an architecture employing a recurrent encoder featuring a two-layer bidirectional LSTM with 1024 units. The decoder is structured hierarchically, employing a two-layer unidirectional LSTM with a hidden size of 1024 for both the conductor and decoder components. A latent size of 256 is chosen to significantly diminish input dimensionality while retaining sufficient information for effective reconstruction [17]. In the training process, we employ the Adam optimizer with an initial learning rate set to 10^{-4} and a 16 batch size to minimize the loss function and fine-tune the model parameters.

We analyze three different spaces—the latent space derived from the VAE trained on a binary 128-element vector piano roll encoding, denoting note activation per unit of time segmented as 'salami slices,'[2] and the DFT of a 12-element pitch class vector, and 128-element pitch distributions. The resulting DFT vectors with non-trivial or non-symmetrical coefficients have the same elements, i.e., 12 for a pitch class space and 128 for a pitch space. To minimize the impact of the rhythmic information, we represent every slice with equal duration. This representation can be explored at https://nadiacarvalho.github.io/Latent-Tonal-Music/.

Subsequently, we calculate distance matrices for points representing each slice in every space per chorale in the test fold, employing specific distance metrics – both Euclidean and cosine distances in the latent space, and cosine similarity for both DFT spaces to capture phase differentiations. Euclidean distances across the pitch slices from the test corpus capture both the vectors' magnitude and phase information. Conversely, cosine distances only consider the phase information. Finally, we establish correlations between pairwise distances in each latent space and the DFT phase space, computed through the Pearson (r), and Spearman (ρ) correlations. We expect high positive correlation values close to 1 (both metrics exist in the [-1,1] range), denoting linear or ranked relationships between the latent and DFT phase spaces. The code adopted to compute the evaluation procedure can be found online at https://github.com/NadiaCarvalho/MCM-Latent-Spaces.

---

[1] http://web.mit.edu/music21/, last accessed on January 10, 2024.

[2] These slices entail segmented information, capturing the addition or subtraction of pitches from the musical surface each time a change occurs. The salami slice segmentation can be computed from MIDI files by the `chordify()` function within the `music21` software package.

5 Results and Discussion

Table 1 presents the Pearson r and Spearman ρ correlation values for the corpus' latent spaces and DFT phase distances. Using cosine distance to calculate the distances in the latent space yields a better alignment with DFT phase spaces in $\{0,1\}^{12}$ and in $\{0,1\}^{128}$.[3] The robust linear and rank association in both Pearson and Spearman correlations (.65/.65 for pitch classes and .64/.61 for pitch distribution) between latent and DFT phase spaces indicates effective preservation of common-tone retention at both the pitch class and pitch spaces.

While sharing the same common-tone property, the musical interpretation of the two DFT phase spaces under analysis has different analytical interpretations, informing the properties best captured in the latent space. At the chordal level, the DFT phase space from pitch classes indicates close proximity of chords sharing the same harmonic function. The DFT phase space from pitch, on the other hand, enforces more effective common-tone retention through parsimonious voice-leading. Despite a residual higher correlation with the pitch class DFT phase space, this suggests that latent spaces are semantically driven, particularly at the harmonic function levels, without diminishing the interpretation of chord distances as reflective of parsimonious voice leading.

In Western traditions of tonal functional harmony, the intricate relationship between functional harmony and voice leading in the articulation of musical time is notable. We argue that the nuanced distinction between these two spaces introduces considerations of *style*, a concept heavily influenced by the curated corpus used to train the latent space. While being a cornerstone of functional harmony, Bach's chorales are representative of parsimonious voice leading. Vocal music for choir tends to homogenize the pitch range of the musical texture, enforcing a parsimonious horizontal conduction per voice. A corpus featuring diverse instrumentation by the same composer or an even more varied pool of composers within a tonal framework would provide better evidence for distinguishing between the pitch class and pitch DFT phase spaces in revealing stylistic aspects of the corpus and the balance between the semantic elaboration of harmonic functions and voice leading.

To further illustrate the interpretation of the latent space, whether as a common-tone functional space or a voice-leading space, we will delve into a specific example from Bach's chorales. Our chosen illustration is the opening phrase of the choral BWV 307, in Bb Major. This phrase serves as an apt example due to its functional harmony typical of a non-modulating passage in a Bach chorale. Among non-modulating chorales, this specific chorale displays the highest correlation values. Moreover, the majority of the chords in this passage are crafted through concise voice leadings between the individual voices.

[3] Unreported informal evaluations have shown poor linear alignment between DFT magnitudes, namely the L_2 norm in latent and the DFT of pitch classes and pitch distribution spaces under study. Further research ought to inspect what information is captured in the magnitudes of the latent spaces.

Table 1. Pearson r and Spearman ρ correlation values for the corpus' latent spaces and DFT phase distances.

Latent Space Correlation	DFT $\{0,1\}^{12}$		DFT $\{0,1\}^{128}$	
	Pearson (r)	Spearman (ρ)	Pearson (r)	Spearman (ρ)
Cosine	$.65 \pm .04$	$.65 \pm .05$	$.64 \pm .04$	$.61 \pm .05$
Euclidean	$.61 \pm .06$	$.59 \pm .06$	$.65 \pm .05$	$.56 \pm .07$

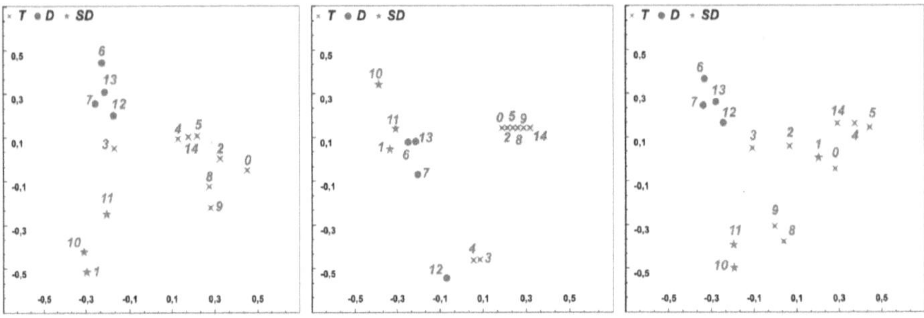

Fig. 3. Two-dimensional visualization of the latent (on the left), DFT phase of pitch classes (in the middle) and pitch (on the right) spaces of the initial phrase of the choral BWV 307, using the Locally Linear Embedding (LLE) dimensionality reduction. In the caption, the tonic (T), dominant (D), and sub-dominant (SD) functions are visually distinguished by the colours blue, red, and green, accompanied by the shapes x, circle, and star, respectively. (Color figure online)

Figure 3 showcases a two-dimensional visualization of the latent space and DFT of pitch classes and pitch distribution phase spaces utilizing Locally Linear Embedding (LLE) dimensionality reduction [20]. LLE is selected for its proficiency in preserving the local structure of the data while also capturing intricate, nonlinear relationships inherent in this scenario. These visualizations aim to illustrate and explore clusters of chords, highlighting specific spatial properties, while correlation metrics are calculated directly within the latent spaces themselves. The tonic, dominant, and sub-dominant functions are distinguished by the colours blue, red, and green, respectively. Each function is represented by unique shapes: x for tonic, circle for dominant, and star for sub-dominant.

The clusters illustrating tonal functions within the latent space exhibit clear distinctions, as depicted in Fig. 3, effectively capturing tonal functionality across the structure. Only the third chord is positioned amidst the clusters representing tonic, dominant, and sub-dominant functions. Interestingly, the separation between functional groups in the latent space appears even more distinct than in the pitch DFT space. In the latter, the first chord, assigned to the sub-dominant group, noticeably resides in the middle of the tonic group cluster.

Fig. 4. Initial phrase of the choral BWV 307, in Bb Major: Chords are ordered by 1) original chord, and order of distance (shorter to longer) to the first chord in 2) latent space, 3) DFT of pitch classes, and 4) DFT of pitch. Chords in a box share the same distance to chord 0. The count of common notes with chord 0 is annotated in red above the respective chords, while the functional analysis is indicated atop the original score. (Color figure online)

Finally, Fig. 4 depicts the chords from the initial phrase of the choral BWV 307, annotated with their tonal function. Initially, they are presented in their original sequence, and subsequently, they are arranged based on their proximity to the first chord for each space. Aligned with the corpus rank correlation findings, the order of chords organized sequentially per proximity reflects similar ranks in all spaces. This suggests that latent space distances align with the same hierarchical representations reflected in the DFT phase spaces, namely a transpositional rotational space, where the circle of fifth regional spaces and relative major minor relations are preserved. Furthermore, functional harmony clusters emerge similar to relations in the phase Fourier space or *Tonnetz*.

Despite the excerpt adhering to the principles of voice leading, as observed in Fig. 4, there is a consistent pattern where almost always one note is retained from the previous chord, and the other voices move by diatonic intervals (except for the bass which stylistically exhibits more significant jumps). The latent space, while primarily reflecting common tones, transforms the temporal sequence of the structure into a semantic space that prioritizes tonal functionality and voice leading.

An intriguing aspect emerges regarding the significance of chord transition frequencies within the latent space, influenced by the specific corpus on which the model was trained. It's worth noting that MusicVAE enables us to examine chord transitions due to its ability to learn temporal structures by using LSTMs. Illustrated in Fig. 4, it becomes apparent that the sequence of chords 7 and 11 varies compared to their arrangement in DFT spaces. Particularly noteworthy is the higher frequency of transitions from chord 0 to 7 in the training data, contrasting with the less frequent transition from chord 0 to 11. Moreover, an unanticipated discrepancy is noted in the latent space concerning the proximity of chord 1 and chord 13 in terms of voice-leading distances. This observation contrasts with expectations based on the assumption of closer intervals between notes in each voice of the chord, highlighting a nuanced aspect of the latent space dynamics. Contrary to harmonic expectations, chord 1 does not exhibit closer proximity to chord 0 in the latent space. The training data unveils a more prevalent occurrence of the transition from chord 0 to 13 (I-V) compared to 0 to 1 (I-ii7). However, examining the pitch class DFT space in Fig. 2, it becomes evident that the V chord is actually closer to the I chord than the ii chord. Notably, in this instance, the V chord should be even closer, given that the ii chord is articulated with the seventh and without the fifth note. This observation further underscores the notion that, in the context of Bach chorales, the latent space tends to prioritize functionality over voice leading, aligning with the sequence of training events.

6 Conclusions and Future Work

This paper contributes to existing research by employing advanced VAEs to uncover the structural aspects of musical harmony within a representative J.S. Bach chorale corpus. Our hypothesis posits a correlation between VAEs' latent spaces and DFT phase spaces, aligned with common-tone principles and yielding spaces where functional harmony and voice leading can be interpreted.

The results underscore the intricate interplay between pitch representations and the latent space. The DFT spaces (pitch class and pitch distributions) reveal nuanced properties, with the pitch class space indicating the proximity of chords with similar harmonic functions due to common-tone retention, while the pitch space emphasizes effective common-tone retention through parsimonious voice-leading. Despite a residual higher correlation with the pitch class space, latent spaces exhibit semantic influence, particularly at harmonic function levels, without diminishing chord distances' reflection of parsimonious voice-leading. The intricate relationship between functional harmony and voice-leading in Western tonal traditions suggests stylistic considerations influenced by a curated corpus, as seen in Bach's chorales exemplifying parsimonious voice leading with four textured voices. The latent space encapsulates elements from both the DFT common-tone space of pitch and pitch class distributions. The analysis reveals disparities in chord transition frequencies within the latent space, underscoring the impact of the training data. The latent space demonstrates a tendency to prioritize functionality over voice leading, aligning with the sequence of training events.

In future research, we aim to explore latent spaces trained on more diverse corpora, encompassing varied instrumentation and compositions by different composers within a tonal framework. This approach would yield a more comprehensive understanding of the latent space's capacity to capture stylistic nuances and enhance differentiation between pitch class and pitch spaces. Furthermore, investigating the information encapsulated in the magnitudes of latent spaces could provide insights into the observed disparities in chord proximity and transition frequencies. These findings contribute to the ongoing discourse on the interpretability and limitations of latent spaces in music analysis and cognition. Additionally, there is an intention to delve deeper into the harmonic and voice-leading properties of the latent space, with the ultimate goal of improving music generation models.

References

1. Amiot, E.: The torii of phases. In: Yust, J., Wild, J., Burgoyne, J.A. (eds.) MCM 2013. LNCS, vol. 7937, pp. 1–18. Springer, Heidelberg (2013). https://doi.org/10.1007/978-3-642-39357-0_1
2. Berkeley, I.S.N.: Density plots of hidden value unit activations reveal interpretable bands. Connect. Sci. **7**(2), 167–187 (1995). https://doi.org/10.1080/09540099550039336
3. Berkeley, I.S.N., Gunay, C.: Conducting banding analysis with trained networks of sigmoid units. Connect. Sci. **16**(2), 119–128 (2004). https://doi.org/10.1080/09540090412331282278
4. Bernardes, G., Cocharro, D., Caetano, M., Guedes, C., Davies, M.E.: A multi-level tonal interval space for modelling pitch relatedness and musical consonance. J. New Music Res. **45**(4), 281–294 (2016)
5. Bernardes, G., Cocharro, D., Guedes, C., Davies, M.E.: Harmony generation driven by a perceptually motivated tonal interval space. Comput. Entertainment (CIE) **14**(2), 1–21 (2016)
6. Bernardes, G., Davies, M.E.P., Guedes, C.: A hierarchical harmonic mixing method. In: Aramaki, M., Davies, M., Kronland-Martinet, R., Ystad, S. (eds.) CMMR 2017. LNCS, vol. 11265, pp. 151–170. Springer, Cham (2018). https://doi.org/10.1007/978-3-030-01692-0_11
7. Bernardo, G., Bernardes, G.: Leveraging compatibility and diversity in computer-aided music mashup creation. Pers. Ubiquit. Comput. **27**(5), 1793–1809 (2023)
8. Brown, A.R., Gifford, T., Davidson, R.: Techniques for generative melodies inspired by music cognition. Comput. Music. J. **39**(1), 11–26 (2015)
9. Bryan-Kinns, N., Zhang, B., Zhao, S., Banar, B.: Exploring variational autoencoder architectures, configurations, and datasets for generative music explainable AI. Mach. Intell. Res. **21**(1), 29–45 (2024). https://doi.org/10.1007/s11633-023-1457-1
10. Carvalho, N., Bernardes, G.: Exploring latent spaces of tonal music using variational autoencoders. In: The International Conference on AI and Musical Creativity (AIMC) (2023). https://doi.org/10.5281/ZENODO.8328558
11. Dawson, M.R., Perez, A., Sylvestre, S.: Artificial neural networks solve musical problems with Fourier phase spaces. Sci. Rep. **10**(1) (2020). https://doi.org/10.1038/s41598-020-64229-4
12. De Haas, W.B., Wiering, F.: Hooked on music information retrieval. Empirical Musicol. Rev. **5**(4), 176–185 (2010)
13. Ferguson, S., Parncutt, R.: Composing in the flesh: perceptually-informed harmonic syntax. In: Journées d'informatique musicale (2004)

14. Guo, Z., Kang, J., Herremans, D.: A domain-knowledge-inspired music embedding space and a novel attention mechanism for symbolic music modeling. AAAI Press (2023). https://doi.org/10.1609/aaai.v37i4.25635
15. de Haas, W., Magalhães, P., Wiering, F.: Improving audio chord transcription by exploiting harmonic and metric knowledge, pp. 295–300 (2012)
16. Perez, A., Ma, H.L., Zawaduk, S., Dawson, M.R.W.: How do artificial neural networks classify musical triads? A case study in eluding Bonini's paradox. Cogn. Sci. **47**(1), e13233 (2023). https://doi.org/10.1111/cogs.13233
17. Prang, M., Esling, P.: Signal-domain representation of symbolic music for learning embedding spaces. CoRR abs/2109.03454 (2021). https://arxiv.org/abs/2109.03454
18. Quinn, I.: General equal-tempered harmony. Perspect. New Music **44**(2), 114–48 (2006). https://doi.org/10.1353/pnm.2006.0010
19. Roberts, A., Engel, J., Raffel, C., Hawthorne, C., Eck, D.: A hierarchical latent vector model for learning long-term structure in music. In: International Conference on Machine Learning, vol. abs/1803.05428 (2018). http://arxiv.org/abs/1803.05428
20. Roweis, S.T., Saul, L.K.: Nonlinear dimensionality reduction by locally linear embedding. Science **290**(5500), 2323–2326 (2000). https://doi.org/10.1126/science.290.5500.2323
21. Turker, M., Dirik, A., Yanardag, P.: Midispace: finding linear directions in latent space for music generation. In: Proceedings of the 14th Conference on Creativity and Cognition, C&C 2022, pp. 420–427. Association for Computing Machinery, New York (2022). https://doi.org/10.1145/3527927.3532790
22. Tymoczko, D.: Set-class similarity, voice leading, and the Fourier transform. J. Music Theory **52**(2), 251–272 (2008)
23. Yust, J.: Schubert's harmonic language and Fourier phase space. J. Music Theory **59**(1), 121–181 (2015)
24. Yust, J.: Geometric generalizations of the Tonnetz and their relation to Fourier phase spaces. In: Mathematical Music Theory: Algebraic, Combinatorial, Topological and Applied Approaches to Understanding Musical Phenomena, pp. 253–278 (2018)

Fourier Qualia Wavescapes: Hierarchical Analyses of Set Class Quality and Ambiguity

Samuel Pereira[1(✉)], Giovanni Affatato[2], Gilberto Bernardes[1], and Fabian C. Moss[3]

[1] University of Porto—Faculty of Engineering—and INESC TEC, Porto, Portugal
up201709719@up.pt
[2] Politecnico di Milano, Milan, Italy
[3] Julius-Maximilians-Universität Würzburg, Würzburg, Germany

Abstract. We introduce a novel perspective on set-class analysis combining the DFT magnitudes with the music visualisation technique of wavescapes. With such a combination, we create a visual representation of a piece's multidimensional qualia, where different colours indicate saliency in chromaticity, diadicity, triadicity, octatonicity, diatonicity, and whole-tone quality. At the centre of our methods are: 1) the formal definition of the *Fourier Qualia Space* (FQS), 2) its particular ordering of DFT coefficients that delineate regions linked to different musical aesthetics, and 3) the mapping of such regions into a coloured wavescape. Furthermore, we demonstrate the intrinsic capability of the FQS to express qualia ambiguity and map it into a synopsis wavescape. Finally, we showcase the application of our methods by presenting a few analytical remarks on Bach's *Three-part Invention* BWV 795, Debussy's *Reflets dans l'eau*, and Webern's *Four Pieces for Violin and Piano*, Op. 7, No. 1, unveiling increasingly ambiguous wavescapes.

Keywords: *Fourier Qualia Space* · wavescapes · Fourier transform · music analysis · harmonic qualities

1 Intro: Shedding Light on Wavescapes and the DFT

A key question for tonal music analysis is how to connect the interrelations of tonal structures present at different timescales and how harmonic patterns, modulation plans, and key regions interact at local, regional, and global levels. Understanding the hierarchical representation of harmonic structure in Western classical music is thus a central aspect of music analysis. It provides insights into the organisation of tonal and harmonic relationships from the surface pitch

The Portuguese Foundation for Science and Technology supported this project under doctoral scholarship 2021.05059.BD, as well as the Junior Professorship for Digital Music Philology and Music Theory at Julius-Maximilians-Universität Würzburg.

© The Author(s), under exclusive license to Springer Nature Switzerland AG 2024
T. Noll et al. (Eds.): MCM 2024, LNCS 14639, pp. 317–329, 2024.
https://doi.org/10.1007/978-3-031-60638-0_25

and chordal elements [4,8,11,20]. These dynamic structural changes have been pursued computationally to foster music analysis, with a strong emphasis on empirical and corpus studies, namely to assess the tonal organisation as intuitive visual and spatial representations at multiple hierarchical levels.

1.1 Hierarchical Music Analysis with Keyscapes

Keyscapes are computational visual representations of the tonal structure of a piece of music. They commonly map maximum-likelihood estimates of a key-finding algorithm to musical segments at different temporal scales—from microelements on the musical surface to macro inference of a piece's global key [21,22]. Thus, keyscapes show the hierarchical organisation of tonality using a colour-coded scheme that indicates the degree of familiarity with a key. Its use is suitable for music analysis as it can intuitively unveil patterns of modulation, tonicisation, and key relations in a piece and highlight overall tonal coherence or diversity of musical structure. Furthermore, it fosters new methods for empirical music analysis at scale [6] and applies symbolic and audio music encodings. A generalisation of keyscapes rendering probabilistically-weighted clusterings of key estimates for each segment instead of a single maximum value has been explored in [14], which entails some tonal ambiguity concerning uncertainty in the key-finding procedure.

1.2 Analysing Pitch-Class Sets with the DFT

Independently of the development of keyscapes, recent years have seen a growth of music-theoretical and analytical approaches drawing on the capacities of the discrete Fourier transform (DFT) for music analysis based on pitch-class sets (pcsets). The existing literature has proposed scalable and music-theoretically interpretable computational methods for analysing the hierarchical structure of pitch-class relations in empirical corpus studies. In essence, the DFT, initially introduced by Lewin [13] and recently revitalised [1–3,10,18,19], decomposes pitch-class distributions into components resonating with prototypes of music-theoretically relevant structures. The DFT is agnostic to traditional tonal notions, making it suitable for analysing styles beyond common-practice tonality.

While applying the DFT to specific pcsets has been explored from various perspectives, several approaches have been used to analyse segments of pieces or entire compositions in corpus studies. A straightforward application is to compute pcsets of music segments with multiple granularity levels [10,27]. This, however, requires choosing a particular level of analysis, which is challenging to do in a principled way without intricate knowledge of a given piece that can only be obtained through detailed manual analysis. To avoid the issue of choosing specific levels of partitioning a piece, a different approach seeks *exhaustive segmentation*, where each possible contiguous segment of a piece is considered, leading to a bottom-up hierarchical structure. Moreover, the core assumption of the keyscapes method requires the existence of major and minor keys, thus strongly limiting its scope of meaningful applicability beyond tonal music.

1.3 Combining Keyscapes and the DFT: Wavescapes

To overcome these limitations, *wavescapes* [25] have been recently proposed as a novel visualisation technique that combines the advantages of keyscapes visualisation with the power and flexibility of DFT-based music analysis. Wavescapes are based on the idea that each Fourier coefficient $\hat{f}_k$ resonates with a specific prototype of music-theoretically relevant pitch-class structures [1]. The prototypes are interpreted as the following qualities: chromaticity ($\hat{f}_1$), diadicity ($\hat{f}_2$), triadicity ($\hat{f}_3$), octatonicity ($\hat{f}_4$), diatonicity ($\hat{f}_5$), and whole-tone quality ($\hat{f}_6$). Given a pitch-class distribution, the magnitude of each DFT coefficient indicates, loosely speaking, how the distribution is representative of the corresponding prototype [1,5,26,27]. Wavescapes can unveil the patterns of tonal stability and tension, modulation, and similarity within a piece or a corpus by highlighting and quantifying degrees of tonal coherence or diversity. Practically, wavescapes can be produced using the dedicated Python library wavescapes[1] or the web application midiVERTO[2] [9,16], which does not require programming skills and allows for interactive analytical exploration.

The DFT of pcsets yields a 12-dimensional vector of complex numbers. Typically, only the non-trivial dimensions, i.e., coefficients 1 through 6 are considered due to symmetry [1]. Consequently, DFT data is essentially multi-dimensional, although analysts often opt to display the dimensions separately or select only a subset. Interpreting six-dimensional (6-D) spaces can be challenging; hence, exploring methods to reduce this complexity is a logical consideration. A recent approach to reduce the dimensionality of wavescapes is the *summary wavescape*, used in [12] to study the diachronic development of Debussy's musical style. The summary wavescape only stores information on the most salient Fourier coefficient at each time scale. It can thus reduce the complexity and redundancy of the original 6-D wavescape and provide a clearer and more concise overview of the tonal hierarchy of a piece at one glance. However, this approach has no explicit notion of Fourier coefficient ambiguity as it only shows the most salient one for a given segment at a particular time scale.[3]

2 The *Fourier Qualia Space*: Definition and Properties

The *Fourier Qualia Space* (FQS) is a hexagonal plane partitioned into six regions, each corresponding to distinct harmonic qualities derived from the DFT [17]. Each unique point within this space denotes set classes, and their coordinates are intricately linked to their coefficient magnitudes. In other words,

[1] Link to the wavescapes library: https://github.com/DCMLab/wavescapes.

[2] Link to the midiVERTO application: https://dcmlab.github.io/midiVERTO/.

[3] Ambiguity has a rich history in musical composition and analysis. For example, it is associated with tonality, where identical pitches can be interpreted differently in various modes or tonalities. The qualia ambiguity we seek to discuss aligns with Temperley's [24] concept, i.e., indicating the degree to which a set class is linked with two or more harmonic qualities.

all pcsets belonging to the same set class exhibit identical geometric coordinates irrespective of the pitch-classes within each set. Below, we comprehensively explore the mathematical and geometrical underpinnings of the FQS, elucidating hierarchical aspects of harmonic qualia and degrees of qualia ambiguity.

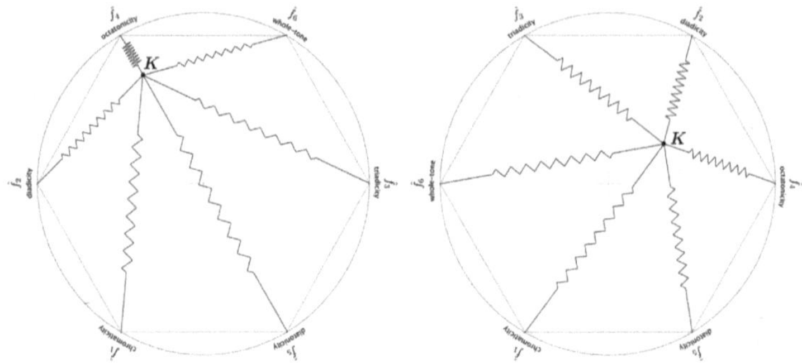

Fig. 1. Two *RadViz*-derived hexagonal depictions for the DFT coefficients. Left and right images differ in the ordering of coefficients, resulting in distinct coordinates for Stravinsky's *Petrushka* chord K (two major triads a tritone apart).

2.1 Projecting Set Classes in the FQS

The FQS uses the *RadViz* dimensionality reduction technique [7] to project the six-dimensional non-trivial DFT magnitudes of a set class onto a two-dimensional hexagon with $n = 6$ anchors positioned equidistantly, as illustrated in Fig. 1. The anchors $s_j = (x_j, y_j)$ with $j = 1, \ldots, n$ are the hexagon vertices and represent the DFT coefficients and their respective harmonic qualities. A physical model of springs is used to determine the two-dimensional coordinates of a set class. The DFT coefficient magnitudes $\left|\hat{f}_1\right|, \ldots, \left|\hat{f}_6\right|$ correspond to the stiffness of the n springs connected to the anchors S_n. Given the DFT magnitudes of a set class $K = \left(\left|\hat{f}_1\right|, \ldots, \left|\hat{f}_6\right|\right)$, the final position of K in the FQS is the two-dimensional point $P_K = (x, y)$ where the sum of the spring forces is null, such that:

$$P_K = \left(\frac{\sum_{j=1}^{n} x_j \cdot \left|\hat{f}_j\right|}{\sum_{j=1}^{n} \left|\hat{f}_j\right|}, \frac{\sum_{j=1}^{n} y_j \cdot \left|\hat{f}_j\right|}{\sum_{j=1}^{n} \left|\hat{f}_j\right|} \right). \tag{1}$$

Figure 1 illustrates how the hexagon vertices are associated with the DFT coefficients. It also displays the position of pcset K determined with equation (1)—K is the *Petrushka* chord with pitch-classes $\{0, 1, 4, 6, 7, 10\}$, belonging to set class (013679)—, exhibiting DFT magnitude values $\left|\hat{f}_1\right| = 0$, $\left|\hat{f}_2\right| = 2$, $\left|\hat{f}_3\right| = 0$, $\left|\hat{f}_4\right| \approx 3.5$, $\left|\hat{f}_5\right| = 0$, $\left|\hat{f}_6\right| = 2$, which determine the stiffness of each spring and K's position in space.

Mapping the DFT coefficient magnitudes to the anchors S_1, $S_2, \ldots,$ S_6 in the FQS is crucial to the analytical interpretation of a set class position, since distinct arrangements result in different positionings to a given set class, as Fig. 1 shows. To adress this issue, We adopt the order reported in [17], which empirically evaluated the order that best 1) promotes a wider dispersion of pcsets in the FQS and 2) maintains the highest possible correlation between pcsets' positioning in the FQS and in the six-dimensional DFT magnitude space. The permutation adopted resulted in the following order $\hat{f}_1, \hat{f}_5, \hat{f}_3, \hat{f}_4, \hat{f}_6, \hat{f}_2$.[4] Although other methods could be used to define the order of the DFT coefficients on the *RadViz* circle, the current order facilitates several analytical and interpretive properties that we detail below.

2.2 Properties: (Fuzzy) Regions, Convexity, and "Singularity"

The adopted coefficient order has proven pivotal to musical analysis. This value is understood by delineating six regions, each representing higher saliency in its respective quality, as illustrated in Fig. 2. Furthermore, this allows the identification of three meta-harmonic regions, each encompassing two harmonic qualities: 1) diatonicity ($\hat{f}_5$) and triadicity ($\hat{f}_3$); 2) octatonicity ($\hat{f}_4$) and whole-tone quality ($\hat{f}_6$); and 3) diadicity ($\hat{f}_2$) and chromaticity ($\hat{f}_1$). Meta-regions, illustrated in Fig. 2, are linked with different musical aesthetics. The first group is associated with tonal functional harmony; the second group is significant in impressionism and musical nationalism; and the third group is fundamental to 20th-century music and atonal music.[5]

However, any set class in the FQS is a convex combination of the anchors' positions. Therefore, any pcset positioned away from anchors is a convex combination of two or more DFT coefficients. This mathematical feature suggests that the regions should not be seen as confined spaces, motivating us to depict their boundaries with dashed lines, as can be seen in Fig. 2. In fact, they allow many convex combinations between the six harmonic qualities, particularly when a set class is near the space's centre. For instance, the pentachord (01468), shown in Fig. 2, is near the centre and the boundary between the $\hat{f}_4$ and $\hat{f}_6$ regions. A restricted view might classify this set class as solely octatonic. However, its position is a convex combination of all harmonic qualities, with a DFT magnitude profile of $\left|\hat{f}_1\right| = 0.52$, $\left|\hat{f}_2\right| = 1.73$, $\left|\hat{f}_3\right| = 2.23$, $\left|\hat{f}_4\right| = 1$, $\left|\hat{f}_5\right| = 1.93$, $\left|\hat{f}_6\right| = 3$. The greater saliency in $\hat{f}_6$ plots this pcset towards the whole-tone area, but the secondary saliency in $\hat{f}_3$ shifts its position towards triadicity, resulting in a final

[4] The order described was determined by computing Euclidean distances between all $\mathbb{Z}_{12}$ set classes in both the six-dimensional space and all possible anchor permutations in the FQS, correlating both matrices to find the optimal permutation. The order $\hat{f}_1, \hat{f}_5, \hat{f}_3, \hat{f}_4, \hat{f}_6, \hat{f}_2$ maximised the Pearson correlation coefficient with $r = .72$.

[5] We should also note that the adopted order presents a new analytical layer by dividing the FQS into groups of three DFT coefficients. This allows us to interpret the positioning of a set class based on the the axis that separates $\hat{f}_3, \hat{f}_5, \hat{f}_1$ vs. $\hat{f}_2, \hat{f}_6, \hat{f}_4$ indicating the presence or absence of tritones in a given set class.

position inside the $\hat{f}_4$ region. This effect is more apparent when the DFT magnitude profile of a pcset does not have a very prominent primary quality and is positioned near the centre. This property of the FQS expands existing reductionist approaches to the six DFT coefficients by encompassing their interaction, as opposed, e.g., to [12] that retain the most salient coefficient only.

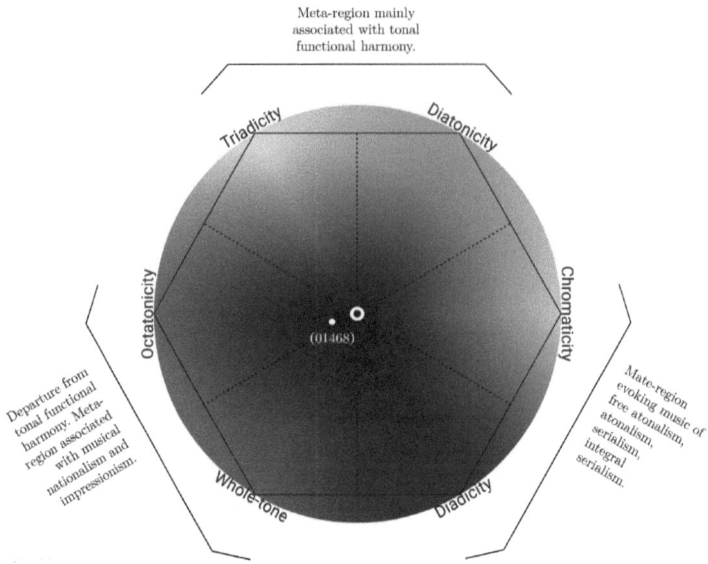

Fig. 2. Depiction of the FQS and the final arrangement of DFT coefficients [17]. The image also illustrates the three meta-regions arising from this specific order.

The pentachord (01468) exemplifies that the FQS centre is the peak of qualia ambiguity, implying a balance of forces between two or more DFT coefficients. Hence, the FQS centre and its vicinity form a "singular" zone due to its qualia ambiguity, with two or more qualities residing with equal intensity. Consequently, a set classes' proximity to the centre broadens its qualia interpretation. Figure 2 illustrates this with a gradient that darkens towards the centre, symbolising increasing qualia ambiguity. Despite its peculiarity, the ambiguous zone in space is beneficial as it allows the assessment of the recurring analytical remark of ambiguity in music. Furthermore, 2 illustrates how we coloured the FQS to map the set-class position's colour into a wavescape. In our approach, the six harmonic qualities are represented by a unique colour in a radial gradient.

3 Qualia Hierarchies in Bach, Debussy, and Webern

We aim to elucidate the harmonic qualities of three pieces by composers of different historical periods: Bach's *Three-part Invention* in F minor, BWV 795;

Debussy's *Reflets dans l'eau* (Images, Book I, No. 1); and Webern's *Four Pieces for Violin and Piano*, Op. 7, No. 1. Our analysis focus on their respective wavescapes and positioning within the FQS, illustrating the relationship between the meta-regions and the pitch-organisation across different compositional periods.

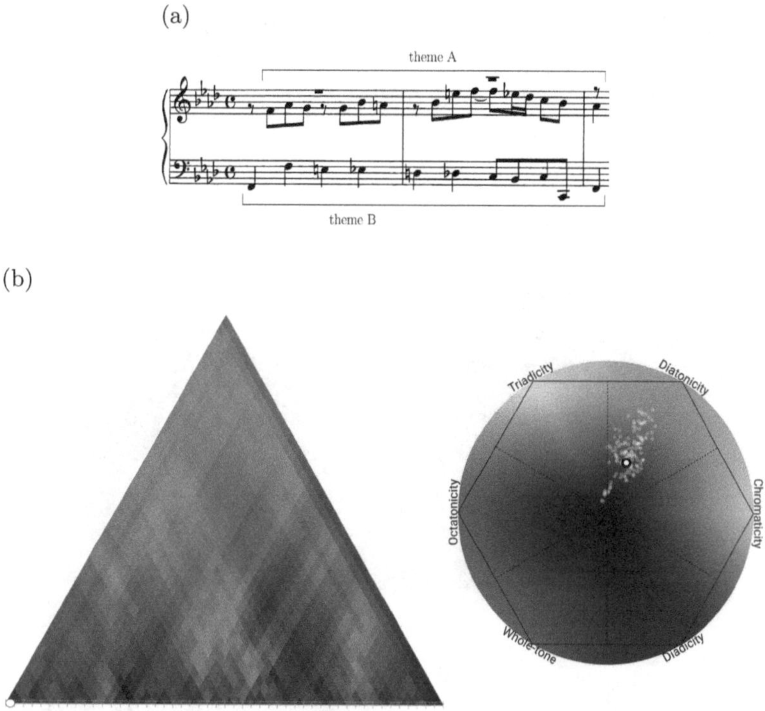

Fig. 3. In (a) are mm. 1 and 2 of Bach's BWV 795, highlighting the two structural motifs in the piece (score in public domain). In (b) one can examine both the wavescape and the FQS positioning of the piece's segments—time resolution (tr) = and window length (wl) = 26.

Bach's *Three-part Invention* BWV 795 in F minor unfolds through two themes exhibited in Fig. 3a. The first motif involves an ascending leap of a third followed by a descending second, expanded to include a tritone leap and a descent from F to A . The second motif, a *lament bass*, involves a chromatic descent from the tonic to the dominant, followed by a cadential ornamentation. These motifs undergo inversion, expansion, imitation, and combination throughout the piece. The tonal plan of the composition evolves through subtle modulations, creating a sense of tonal unpredictability with cadences in C minor, A major, E major, D major, and F minor. Despite the frequent emergence of chromatic elements, the pitch organisation favours diatonic structures, resulting in low variability in the work's qualia. This is reflected in the wavescape illustrated in

Fig. 3b, where the prevalence of diatonicity ($\hat{f}_5$) is visible, as a consequence of the musical segments' positioning in the FQS lying well within the diatonic region and meta-region of functional harmony. The overall brightness of the wavescape suggests low qualia ambiguity. However, focal points of ambiguity are associated with lower hierarchical levels of the composition. As we ascend in the wavescape, diatonicity becomes increasingly apparent. Despite the tonally ambiguous sensation imparted by the structuring use of the chromatic motif, the prevalence of diatonic pitch structures and clear cadences in various tonalities attest to diatonicity as the primary harmonic quality in the piece.

(a)

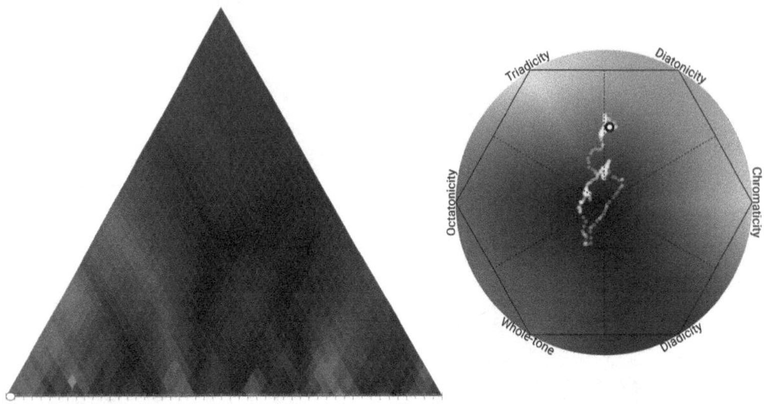

(b)

Fig. 4. In (a) is the wavescape for *Reflets dans l'eau* outlining the movement between $\hat{f}_5$ and $\hat{f}_6$ qualities, as well as its FQS positioning performing a circular trajectory in space—tr = and wl = 86. In (b) are mm. 20 and 21 with the sequence of ninth chords that create the wavescape's redish spot (public domain).

Debussy's Images, Book I, No. 1 exemplifies increased complexity and structural relevance in harmonic qualia as the result of an anchored harmonic path from opposing regions in the FQS. The wavescape of this piece exhibits pronounced colour variance, indicating the prominence of various harmonic qualities, as Fig. 4a depicts. Darker patches related to qualia ambiguity emerge at higher hierarchical levels, suggesting a more complex predominant harmonic quality. The wavescape analysis reveals a pendulum-like movement between diatonic (greenish tones) and whole-tone (purplish tonalities) qualities. This movement establishes diatonicity as the starting and ending point of the harmonic motion, with whole-tone quality as the contrasting material. This movement is embodied in the arrangement of the musical segments within the FQS, forming a clockwise motion through the meta-region encompassing $\hat{f}_4$ and $\hat{f}_6$. Two pitch organisations favour the DFT coefficient $\hat{f}_5$. The use of the major scale and pentatonic segments, closest to the anchor of $\hat{f}_5$ in the FQS, emphasises diatonicity.[6] The section beginning with the *En animant* passage at measure 44, composed of the whole-tone scale with pitch-classes $\{1, 3, 5, 7, 9, 11\}$, contributes to the prominence of $\hat{f}_6$ and the purplish patch in the wavescape. A brief octatonic section is evident in the lower tiers of the wavescape, associated with the seventh and ninth chords sequence in the *Quasi Cadenza* at measure 20 (this sequence is shown in Fig. 4b). The presence of $\hat{f}_4$, while not directly emanating from the auditory experience of the piece, links to the fact that four notes out of the five forming a major seventh and ninth chord constitute a subset of the octatonic scale, and to the fact that the motif is sequenced by minor thirds.

Webern's *Four Pieces for Violin and Piano* (Op. 7, No. 1) is a nine-measure miniature composition. The piece is divided into two sections, each characterised by different roles for the violin and piano. In the first section (mm. 1–5), the violin carries the melody, supported harmonically by the piano. This dynamic is reversed in the second section (mm. 6–9), with the piano taking the melodic lead and the violin providing an *ostinato* accompaniment. The bipartite structure of the work is reflected in its qualia wavescape (in Fig. 5a), which reveals two nested triangular structures. These structures suggest a two-fold structure (mm. 1–5 and 6–8), with a subtle cadential coda (second half of m. 8 to m. 9). The coda, marked by a *ritardando* and a thinning texture, signals the conclusion of the phrase. The greenish tones on the right side of the wavescape further emphasise the coda's individuality, indicating diatonicity as a structuring element. This diatonic character arises from the cadence's explicit reference to a subset of an E major scale (pitch-classes $\{2, 3, 7, 10\}$). The final E -major chord played by the piano reinforces this diatonic conclusion.

Moreover, the wavescape reveals that the section from mm. 6 to 8 undergoes brief, slight deviations into various harmonic qualities, displaying mostly yellow, green, and red tonalities. These hues arise from a complex texture where each note subtly alters the work's qualia. Conversely, the piece's initial part, represented on the wavescape's left side, shows less qualia variability, with a dominance of bluish tones linked to diadicity ($\hat{f}_2$) and faint purplish patches related to whole-tone quality ($\hat{f}_6$). The prevalence of $\hat{f}_2$ situates the work in the meta-region connecting $\hat{f}_2$ and $\hat{f}_1$, underscoring the first chord's importance for the work's development and the high qualia ambiguity conveyed by the wavescape. Listening to the second measure's chord depicted in Fig. 5b, we

[6] The DFT coefficient $\hat{f}_5$ also tends to exhibit high magnitude for pentatonic sets. This coefficient measures how close the analysed set is to a segment of the circle of fifths.

perceive it differently based on our attentional focus. Concentrating on the pitch-classes $\{1, 3, 5, 9\}$ highlights the whole-tone quality, while focusing on the piano's G and the pitch-classes $\{3, 8, 9\}$ accentuates the diadic quality.[7] All these elements suggest a composition of high qualia plasticity, typical of the period, swiftly transitioning between multiple harmonic universes and often highlighting two or more qualities in the same harmonic event, depending on our auditory focus.

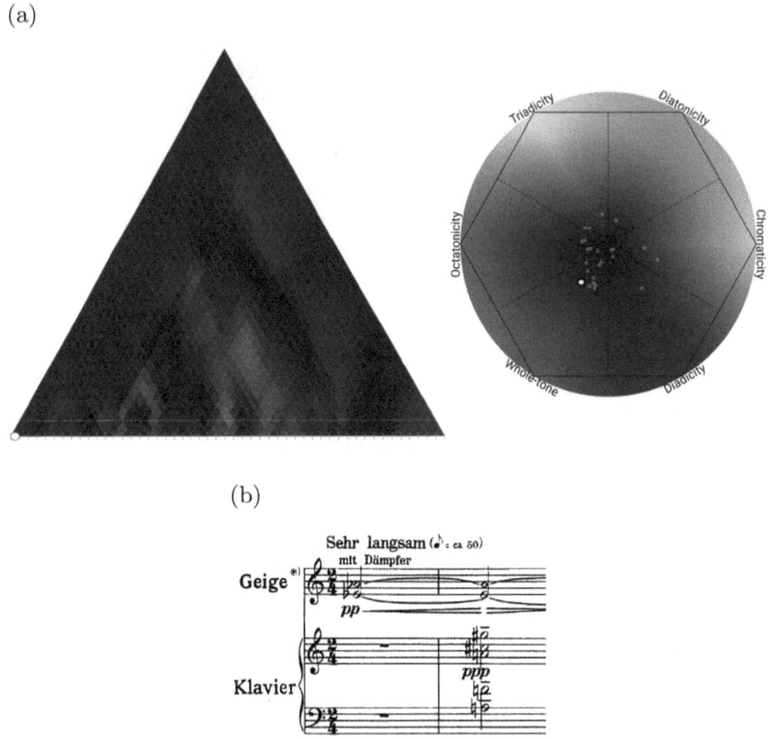

Fig. 5. In (a) is the two-triangle structure of Webern's Op.7/1 wavescape and its FQS positioning—tr = and wl = 8—, and in (b) is the piece's opening chord (score in public domain).

4 Final Remarks

We proposed a method combining the FQS—a hexagonal space for interaction of six harmonic qualities—with the hierarchical visualisation technique, wavescapes. This approach visualises a work's qualia in a triangular structure,

[7] The DFT coefficient $\hat{f}_2$ is more salient for sets arranged in tritone sequences at a minor second distance, as exemplified by the pcset $\{0, 5, 6, 11\}$.

with colours indicating salience in six harmonic qualities and higher levels offering a more global qualia view. The complementarity between the two representations is noteworthy in light of Saussure's [23] syntagmatic/paradigmatic dichotomy, which McCreless [15] employs in the context of the musical discourse: (1) the FQS provides a paradigmatic visualisation by arranging analytical information in a non-temporal way and (2) the wavescape presents the analytical information in a syntagmatic context, ordered in the time of the music itself. Integrating FQS's qualia ambiguity into the wavescape visualisation enhances its interpretation, providing insights into a work's dynamic qualia nature. This is crucial as a musical passage's perception can vary based on its interval and pitch organisation. The wavescapes of the three pieces presented, with their progressively darker tones, imply higher qualia ambiguity. Future application of our method to an extended corpus of scores could elucidate the relationship of qualia ambiguity across compositional periods. In addition, future analysis would benefit from the ability to include multiple orders of DFT coefficients in the FQS to illustrate how they affect the resulting wavescapes. Furthermore, our method allows the identification of distinct spots at high hierarchical levels in some composers, indicating movement between harmonic qualities and their structural importance. Studying these movements and the specific syntactic roles of distinct harmonic qualities could be insightful. Lastly, our analysis of Webern's Op. 7, No. 1 reveals its high qualia plasticity, signifying music in constant qualia metamorphosis, navigating various harmonic fields without a discernible destination, meriting further exploration.

References

1. Amiot, E.: Music Through Fourier Space: Discrete Fourier Transform in Music Theory. Computational Music Science, Springer, Heidelberg (2016). https://doi.org/10.1007/978-3-319-45581-5
2. Bernardes, G., Carvalho, N., Pereira, S.: FluidHarmony: defining an equal-tempered and hierarchical harmonic lexicon in the Fourier space. J. New Music Res. **51**(2–3), 142–161 (2023). https://doi.org/10.1080/09298215.2023.2202641
3. Bernardes, G., Cocharro, D., Caetano, M., Guedes, C., Davies, M.E.: A multi-level tonal interval space for modelling pitch relatedness and musical consonance. J. New Music Res. **45**(4), 281–294 (2016). https://doi.org/10.1080/09298215.2016.1182192
4. Bharucha, J., Krumhansl, C.L.: The representation of harmonic structure in music: hierarchies of stability as a function of context. Cognition **13**(1), 63–102 (1983)
5. Chiu, M.: Macroharmonic progressions through the discrete fourier transform: an analysis from Maurice Duruflé's requiem. Music Theory Online **27**(3) (2021). https://doi.org/10.30535/mto.27.3.1
6. Domínguez, A.M.: Tonal stability modeling from audio chroma features. Ph.D. thesis, Universitat Pompeu Fabra, Barcelona (2009)
7. Dzemyda, G., Kurasova, O., Žilinskas, J.: Multidimensional Data Visualization: Methods and Applications. Springer Optimization and Its Applications, vol. 75. Springer, New York (2013). https://doi.org/10.1007/978-1-4419-0236-8

8. Granroth-Wilding, M.T.: Harmonic analysis of music using combinatory categorial grammar. Ph.D. thesis, The University of Edinburgh (2013)
9. Harasim, D., Affatato, G., Moss, F.C.: midiVERTO: a web application to visualize tonality in real time. In: Montiel, M., Agustín-Aquino, O.A., Gómez, F., Kastine, J., Lluis-Puebla, E., Milam, B. (eds.) MCM 2022. LNCS, pp. 363–368. Springer, Cham (2022). https://doi.org/10.1007/978-3-031-07015-0_31
10. Harding, J.D.: Computer-aided analysis across the tonal divide: cross-stylistic applications of the discrete Fourier transform. In: Music Encoding Conference (2020)
11. Krumhansl, C.L., Kessler, E.J.: Tracing the dynamic changes in perceived tonal organization in a spatial representation of musical keys. Psychol. Rev. **89**(4), 334 (1982)
12. Laneve, S., Schaerf, L., Cecchetti, G., Hentschel, J., Rohrmeier, M.: The diachronic development of Debussy's musical style: a corpus study with Discrete Fourier Transform. Humanit. Soc. Sci. Commun. **10**(1), 1–13 (2023). https://doi.org/10.1057/s41599-023-01796-7
13. Lewin, D.: Re: intervallic relations between two collections of notes. J. Music Theory **3**(2), 298–301 (1959). https://doi.org/10.2307/842856
14. Lieck, R., Rohrmeier, M.: Modelling hierarchical key structure with pitch scapes. In: Proceedings of the 21st International Society for Music Information Retrieval Conference (ISMIR 2020), Montreal, Canada, pp. 811–818 (2020) https://doi.org/10.5281/zenodo.4245558
15. McCreless, P.: Syntagmatics and paradigmatics: some implications for the analysis of chromaticism in tonal music. Music Theory Spectr. **13**(2), 147–178 (1991). https://doi.org/10.2307/745896
16. Moss, F.C., Affatato, G., Harasim, D.: Phantom curves: scientific discovery through interactive music visualization. In: Pugin, L. (ed.) 9th International Conference on Digital Libraries for Musicology, DLfM2022, pp. 60–64. Association for Computing Machinery, New York (2022) https://doi.org/10.1145/3543882.3543886
17. Pereira, S., Bernardes, G., Martins, J.O.: Qualia motion in Fourier space: formalising linear, nondirected and contrapuntal ambiguity in Schoenberg's Op. 19, No. 1. Music Theory Spectr. (2024). (In press)
18. Quinn, I.: General equal-tempered harmony (introduction and part I). Perspect. New Music **44**(2), 114–158 (2006)
19. Quinn, I.: General equal-tempered harmony: parts 2 and 3. Perspect. New Music **45**(1), 4–63 (2007)
20. Rohrmeier, M.: Towards a generative syntax of tonal harmony. J. Math. Music **5**(1), 35–53 (2011). https://doi.org/10.1080/17459737.2011.573676
21. Sapp, C.S.: Harmonic visualizations of tonal music. In: Proceedings of the International Computer Music Conference, Havana, Cuba, vol. 1, pp. 419–422 (2001)
22. Sapp, C.S.: Visual hierarchical key analysis. Comput. Entertain. **3**(4), 1–19 (2005). https://doi.org/10.1145/1095534.1095544
23. de Saussure, F., Baskin, W., Meisel, P., Saussy, H.: Course in General Linguistics. Columbia University Press (2011). https://books.google.pt/books?id=ffzWX9LeeykC
24. Temperley, D.: The tonal properties of pitch-class sets: tonal implication, tonal ambiguity, and tonalness. Comput. Musicol. **15** (2007)
25. Viaccoz, C., Harasim, D., Moss, F.C., Rohrmeier, M.: Wavescapes: a visual hierarchical analysis of tonality using the discrete Fourier transform. Music Sci. **27**(2), 390–427 (2023). https://doi.org/10.1177/10298649211034906

26. Yust, J.: Special collections. J. Music Theory **60**(2), 213–262 (2016). https://doi.org/10.1215/00222909-3651886
27. Yust, J.: Stylistic information in pitch-class distributions. J. New Music Res. **48**(3), 217–231 (2019). https://doi.org/10.1080/09298215.2019.1606833

Similarity and Distance Measures for Music

A Fingerprinting-Based Strategy for Musical Genre Similarity

Aarón López-García(✉)

Valencian Research Institute for Artificial Intelligence (VRAIN),
Universitat Politècnica de València, Valencia, Spain
alopgar3@upv.es

Abstract. Music information retrieval is a field of signal processing that offers sophisticated tools for generating computer-assisted musical systems. Acoustic fingerprinting is a well-known application of such discipline whose purpose is to extract perceptual features of music samples. The main advantage of implementing audio fingerprinting algorithms is to facilitate the task of matching acoustic samples. For that reason, in this paper, we propose a strategy for measuring musical genre similarity based on the use of audio fingerprints. The notion of musical similarity has been approached employing the use of class-representative fingerprints, defined as an aggregation of the fingerprints of each musical genre. We have also proposed a similarity index defined for the class-representative fingerprints for which we have demonstrated the advantages of using this function as a measure of music resemblance. For the experimental study, we have chosen the GTZAN dataset. The results indicate that our strategy proves to be useful for determining musical similarity between genres. However, future work is required to evaluate and extend it to other domains or datasets.

Keywords: Audio Fingerprinting · Music Information Retrieval · Stylometry · Musical genre

1 Introduction

The rise of music streaming technologies is having a clear impact on our society. It is clear that the way people consume, discover, and share music has significantly transformed over the years. That is why industrial and technological advances in the field of music have not stopped evolving to adapt to the changes brought about by this new era. One of the main challenges of cutting-edge multimedia techniques is to be able to handle the huge volume of data offered by online platforms. The more time passes, the greater the amount of information available. Hence, it is required the use of advanced signal processing techniques in order to facilitate the handling of the musical content.

One of the fields of signal processing that is quite helpful for solving such a concern is the Music Information Retrieval (MIR) subject [3]. MIR is a multidisciplinary field that has considerably evolved over the years because of the

© The Author(s), under exclusive license to Springer Nature Switzerland AG 2024
T. Noll et al. (Eds.): MCM 2024, LNCS 14639, pp. 333–344, 2024.
https://doi.org/10.1007/978-3-031-60638-0_26

number of applications that have emerged in academic and industrial areas. For instance, MIR is behind the state-of-the-art technologies that make personalized recommendations for users that seek new or similar music [4].

Another field of MIR that supports automatic recognition tasks is musical stylometry [5]. It is composed of a set of statistical tools for analyzing the low-level features that ascertain music style [6]. Its applicability has been shown effective in determining the attribution of authorship [7–9]. It is important to emphasize that not only traditional statistical tools have been used to address musical stylometry, but also advanced machine learning techniques such as semi-supervised approaches have also been applied for such a purpose [10].

Regarding the acoustic field, audio fingerprinting is understood as an algorithm that extracts the main components taking into account the perceptual characteristics of the audio [12]. Most of the time, these techniques are applied over the spectrogram representation of the signal. Then, the pattern extraction is conducted by means of time domain [13], frequency domain [14], or a combination of both called time-frequency domain [15]. As far as implementation is concerned, there are some techniques created to this end, although they have their own advantages and limitations. The phase-based [16] and chroma-based [17] fingerprinting techniques are widely used. In regard to data transformations, wavelets have been very effective in this field [18]. Nonetheless, this is not the only feature utilized for this purpose [19].

One of the most representative audio fingerprinting strategies was developed by Wang [20] for the Shazam Entertainment company. He, together with his team, introduced the notion of constellation maps to enhance the implementation of their audio search algorithm. Although there have been several developments in audio fingerprinting and music-matching algorithms, Wang's proposal motivates many of today's systems that focus on that purpose. It should be noted that, despite Wang's work has attracted the attention of many MIR researchers, most of their framework relied on the musical theory of [21], in which it presented the processing of time-pitch musical information pairs so that they remain invariant through transformations.

Over the years more sophisticated techniques have arisen. This is the case of symbolic fingerprinting. It consists of calculating, storing, and retrieving differences among a space vector representation of musical features [23]. This approach has proven to be quite successful in identifying musical excerpts and discovering distinctive patterns in music corpora [22]. On top of all this, symbolic fingerprinting is also used for measuring plagiarism in automatic music generation algorithms [24].

The different algorithms and techniques that have been developed for audio fingerprinting can vary depending on the signal processing strategy that generates the output [25]. Among the many new developments that have emerged over the past few years, we can find Olaf [29] or PeakFP [30]. In past editions of Mathematics and Computation in Music (MCM) we have presented SpectroMap [27], an audio fingerprinting algorithm for the extraction of constellation maps (i.e. a map composed of topological prominence elements) from a given spectrogram. The search engine is based on a local peak search using either time or frequency windows.

It is worth mentioning that the implementation of audio fingerprinting techniques in acoustic artificial intelligence is very useful to reduce training costs, and so enable faster implementations. For example, we can also find recognition of activities of daily living via audio fingerprinting matching [26]. For machine-learning classification tasks, the use of fingerprints as input data for feeding models also results in a promising approach [11]. Its applicability has been proven in complex systems, such as generative adversarial networks (GANs), for speech enhancement [2]. The particular field of music recognition via audio fingerprinting has attracted lots of attention as well. The applications contain a wide variety of cases such as music plagiarism [27] or and audio synchronization [1].

In this paper, we proposed a strategy for musical genre similarly through the use of a music fingerprinting technique. The analysis is computed over the class-representative fingerprints, which are defined by means of a data fusion technique over the fingerprints of the same label. Such fingerprints are extracted utilizing the Spectromap algorithm and the experimental study has been carried out using the GTZAN dataset [31].

2 Audio Fingerprinting

2.1 Signal Processing

The methodology used in this work for audio fingerprinting extraction is based on the generation of a constellation map. As with state-of-the-art systems, we begin by calculating the short-time Fourier transform (STFT). The constellation map is obtained by a window that extracts local prominence in the neighboring values. Then, it is required to define two windows $\phi_T^{d_T}$ and $\phi_F^{d_F}$ to process the local pairwise comparisons with a respective length of d_T and d_F. Without limiting the generality of the foregoing, we can mathematically describe the time-band window mechanism with a length of $0 < d_T \leq n$ and structure $T_i = (T_i^1, ..., T_i^n)$ as:

$$\phi_T^{d_T}(T_i) = \left(\max\{T_i^k, \ldots, T_i^{k+d_T}\}\right)_{1 \leq k \leq n-d_T-1}, \quad (1)$$

per each band $i \in \{1, \ldots, n\}$.

When we group all the values we drop those elements that have an equal index to avoid duplicates. Hence, we can group the window of each band to create the set:

$$\mathbf{\Phi}_T^{d_T} = \left\{\phi_T^{d_T}(T_i)\right\}_{i=1}^{n}. \quad (2)$$

From (1), it is easy to note that even though there are $n - d_T - 1$ matches. Although the window $\phi_T^{d_T}(T_i)$ may contain a smaller number of elements whenever $d_T > 2$. Depending on how restrictive we need to be, we can proceed with just one of the bands or combine them to create a more stringent search and distortion resistance since it returns only the peaks that are prominent in both directions. Finally, the algorithm merges all the band-dependent peaks, as shown in equation (2), to give us the total number of spatial points that determine the so-called audio fingerprint. In Fig. 1, we can see a graphical example of an audio fingerprint extraction.

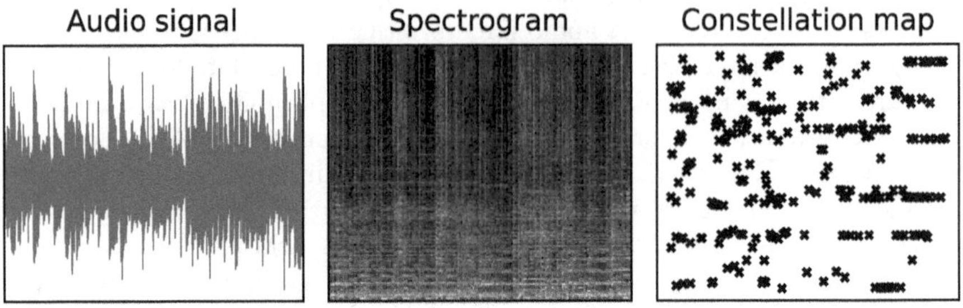

Fig. 1. Signal processing stages from which an audio excerpt is transformed into its constellation map.

2.2 SpectroMap Algorithm

SpectroMap is an algorithm for audio fingerprinting extraction from a given acoustic signal [28]. It is designed for processing the spectrogram representation of audio signals and then generating the map with its topological prominences, known as a constellation map, through a local search of peaks. The search engine has been designed for computing local comparisons with windows over time or frequency bands. When needed, the local search can also be conducted considering both bands simultaneously. In this manner, an element of the spectrogram is considered part of the constellation if it turns out to be a peak in both time and frequency patterns. Finally, the algorithm also yields two output matrices. The first one has the indexes of the constellation and the second one is the vector of the apparition of the peaks as (time, frequency, amplitude) magnitudes. In Fig. 2 we have depicted the description of the SpectroMap.

The architecture of SpectroMap has been designed to carry out the entire process required to get the fingerprint from a given audio signal. Depending on the objective of the user or the data available for the study, we can process a raw acoustic signal or an already computed spectrogram as input when initializing the SpectroMap object in Python. In case we decide to use musical signals, we can also include the required parameters for the signal-processing step that generates the spectrogram. The architecture of SpectroMap is depicted in Fig. 3.

In step Step 4 we highly recommend selecting both time and frequency bands to perform the peak detection since the output is more filtered and spatially consistent. For the remainder steps, the choice is a personal decision that depends on the scope of the research. The algorithm is published as a Python package, so it is publicly available in https://pypi.org/project/spectromap/.

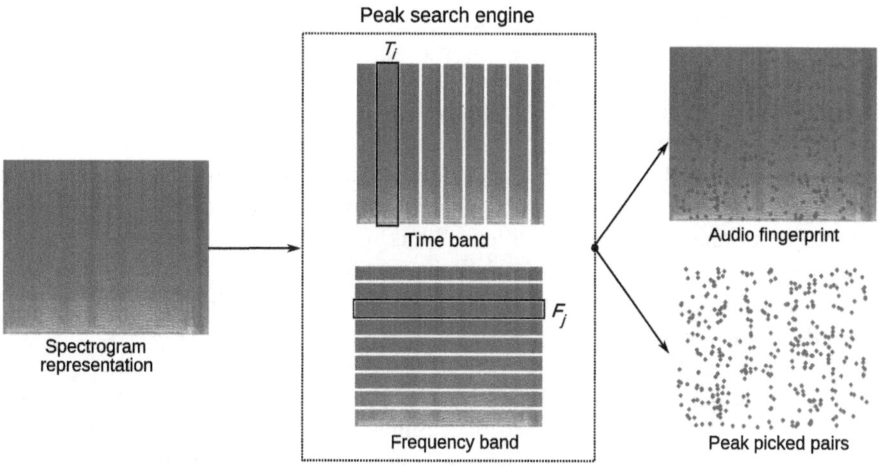

Fig. 2. Flowchart with the inner architecture of the SpectroMap to detect the peaks of a spectrogram through time and/or frequency bands.

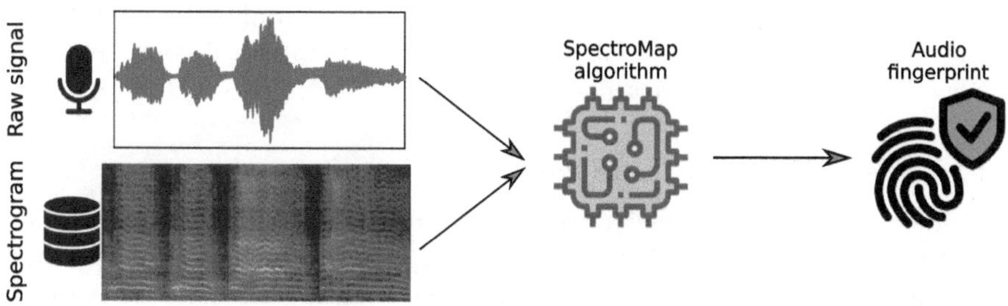

Fig. 3. Architecture of the SpectroMap algorithm with the two possible input audio samples.

Algorithm 1. SpectroMap

Input Parameters N_{FFT} and N_O.

Step 1 Read the audio file to get its amplitude vector and its sample rate.

Step 2 Compute the spectrogram through the associated Fourier transformations.

Step 3 Set a fixed window length (d_T, d_F or both) for the pairwise comparisons.

Step 4 Choose the settings to proceed with the peak detection over a selected band or a combination of both.

Step 5 Create an identification matrix consisting of a binary matrix with the same shape as the spectrogram with the position of the highlighted prominences.

Step 6 Extract such elements and create a file with the (time, frequency, amplitude) vectors.

Output Audio fingerprint $\mathbf{\Phi}$.

---

3 Musical Similarity Between Audio Fingerprints

In this section, we show our proposal for measuring musical similarity. Addressing this problem is crucial for analyzing the musical objects and cognitive/perceptual processes involved in human engagement with music within music theory [33]. One of the main concerns that arises in musical similarity tasks is to respond to the answer "*Similar with respect to what?*" [34], due to the subjective nature of the problem. Thus, musical similarity can be considered one of the most difficult and challenging areas of MIR [35]. Therefore, it is not a trivial problem and requires a lot of effort from both mathematical and computational perspectives.

Given a dataset $\mathcal{D} = \{x_i, y_i\}_{i=1}^{N}$ composed of musical excerpts $x_i \in \mathbb{R}^n$ and its annotated musical label $y_i \in \{1, \ldots, l\}$, we want to analyze the overall similarity among the l classes. For the sake of simplicity, we assume that all the audio samples have the same length, although this can be omitted since there exist techniques for handling this imbalance. Then, we compute the spectrogram of each sample and its fingerprint for a given setting. The pairwise comparisons of musical labels are conducted through the class-representative fingerprints, which are defined as a data fusion of the fingerprints of each class. Finally, we measure the similarity among musical labels by employing a statistical coefficient over the lass-representative fingerprints.

3.1 Class-Representative Fingerprint

The first stage is the extraction of the fingerprints per each audio sample so that we get the respective constellation maps $\{\mathbf{\Phi}_i\}_{i=1}^{n}$. To facilitate the mathematical notation of the following results, we consider $n_T = n - d_T - 1$ and $n_F = n - d_F - 1$ as the time and frequency lengths of the fingerprints. We have not specified whether we get them with either time or frequency bands. It is worth noting that such maps are binary matrices, that is to say, $\mathbf{\Phi}_i \in \{0, 1\}^{n_T \times n_F}$ per each $i \in \{1, \ldots, n\}$. Now, we determine a data fusion strategy that converts the different fingerprints into a reduced set of elements that keeps as much information as possible. The transformation is the following data fusion strategy:

Definition 1 (Class-representative fingerprint). *Let $\{(\mathbf{\Phi}_i, y_i)\}_{i=1}^{N}$ a set of pairs that relates the fingerprint $\mathbf{\Phi}_i$ with its label $y_i \in \{1, \ldots, l\}$, we define the k class-representative fingerprint as follows:*

$$\Xi_k = h\left(\sum_{i=1}^{N} \mathbf{1}[y_i = k] \cdot \mathbf{\Phi}_i\right), \ \forall k \in \{1, \ldots, l\}, \tag{3}$$

so that the $\mathbf{1}[\cdot]$ operator stands for the Iverson bracket and the h function is the pointwise Heaviside function defined over $]0, +\infty[$.

We note the class-representative fingerprint per each label is also a binary matrix that contains the time-frequency coordinates of the peaks of all the constellation maps of the same class. If we consider the pairs (tf) as the index of the matrices Ξ_k, i.e. the elements Ξ_k^{tf}, we can state the following proposition.

Proposition 1. *Let Ξ_k be a class-representative fingerprint of a label $k \in \{1, \ldots, l\}$. If for a given pair $(t, f) \in \{1, \ldots, n_T\} \times \{1, \ldots, n_F\}$ it is hold that $\mathbf{\Phi}_i^{tf} = 1$ with $y_i = k$, then $\Xi_k^{tf} = 1$.*

Proof. Since $\mathbf{\Phi}_i^{tf} = 1$ and $y_i = k$, we know that $\mathbf{1}[y_i = k] \cdot \mathbf{\Phi}_i^{tf} = 1$. Thus, the (t, f) element of the summation of (3) is greater than 0. As a result, the image of h for such element is equal to 1, and so, $\Xi_k^{tf} = 1$.

From the last statement, we know that each time-frequency peak of each musical fingerprint is also a peak for its class-representative element.

3.2 Fingerprint Similarity

The strategy for measuring the similarity between musical labels is conducted utilizing the set of class-representative fingerprints. As mentioned, such matrices contain all the peaks of every fingerprint per class. Hence, we can globally analyze the similarity of musical samples of a dataset by comparing the statistical similarity presented in the class-representative fingerprints.

Definition 2 (Fingerprint similarity). *Let $\{\Xi_k\}_{k=1}^l$ be the set of class-representative fingerprints and $k_1, k_2 \in \{1, \ldots, l\}$, we define the similarity between the k_1 and k_2 as follows:*

$$\mathcal{S}(\Xi_{k_1}, \Xi_{k_2}) = \frac{\sum_{t=1}^{n_T} \sum_{f=1}^{n_F} \min\left\{\Xi_{k_1}^{tf}, \Xi_{k_2}^{tf}\right\}}{\sum_{t=1}^{n_T} \sum_{f=1}^{n_F} \max\left\{\Xi_{k_1}^{tf}, \Xi_{k_2}^{tf}\right\}}. \quad (4)$$

Our similarity index can be understood as a matrix generalization of the Jaccard index, also known as Ruzicka similarity for probability vectors. The implementation of $\mathcal{S}$ over each pair of (Ξ_{k_1}, Ξ_{k_2}) has lots of benefits. In addition, the Jaccard distance has been widely used in MIR tasks (see, for instance, [32]). The following result gives us an overview of the image of the proposed similarity.

Proposition 2. *Given two indexes $k_1, k_2 \in \{1, \ldots, l\}$, the following statements hold true:*

1. $\mathcal{S}(\Xi_{k_1}, \Xi_{k_2}) \in [0, 1]$.
2. $\mathcal{S}(\Xi_{k_1}, \Xi_{k_2}) = 0 \Leftrightarrow \Xi_{k_1}^{tf} \neq \Xi_{k_2}^{tf}, \forall (t, f) \in \{1, \ldots, n_T\} \times \{1, \ldots, n_F\}$.
3. $\mathcal{S}(\Xi_{k_1}, \Xi_{k_2}) = 1 \Leftrightarrow \Xi_{k_1} = \Xi_{k_2}$.

Proof. For the first statement, $\mathcal{S}$ is defined as the quotient of two positive functions, so $\mathcal{S}(\Xi_{k_1}, \Xi_{k_2}) \geq 0$. In addition, since the minimum operator is upper dominated by the maximum operator, we get that $\mathcal{S}(\Xi_{k_1}, \Xi_{k_2}) \leq 1$.

For the second statement, to prove ($\Rightarrow$), if $\mathcal{S}(\Xi_{k_1}, \Xi_{k_2}) = 0$, then $\min\left\{\Xi_{k_1}^{tf}, \Xi_{k_2}^{tf}\right\} = 0 \;\forall (t, f) \in \{1, \ldots, n_T\} \times \{1, \ldots, n_F\}$. So $\Xi_{k_1}^{tf} \neq \Xi_{k_2}^{tf}$ for every pair (t, f). Now, to prove ($\Leftarrow$), if $\Xi_{k_1}^{tf} \neq \Xi_{k_2}^{tf}$ then $\min\left\{\Xi_{k_1}^{tf}, \Xi_{k_2}^{tf}\right\} = 0$ because they are binary matrices. So we conclude that $\mathcal{S}(\Xi_{k_1}, \Xi_{k_2}) = 0$ since the sum of the numerator is zero.

For the third statement, to prove ($\Rightarrow$), if $\mathcal{S}(\Xi_{k_1}, \Xi_{k_2}) = 1$ the sum of minimum and maximum values match, and so, all the components of Ξ_{k_1} and Ξ_{k_2} match. The proof of ($\Leftarrow$) is trivial since the value of the numerator and denominator match.

The reader must note that the previous definition is defined over the class-representative fingerprints. Nevertheless, the similarity index can be used for gauging the similarity of any two fingerprints.

4 Experimental Study

The dataset we have selected to carry out the experiments of this paper is the well-known GTZAN Genre Dataset [31]. The dataset is composed of 1000 audio files, which represent a total size of 1.3 Gb, with approximately 30-s duration. The data samples are divided into 10 different musical genres, each with 100 tracks. All the musical excerpts are stored as 22050 Hz mono 16-bit audio files in WAV format.

For the 1000 tracks of the GTZAN dataset, we have performed a data preprocessing stage in which we have selected a window of 512 frames for the fast Fourier transformation and a window overlap of 64 frames for getting the spectrogram of each sample. For the audio fingerprinting extraction with SpectroMap, we have chosen a window of the tenth length of the spectrogram in both time and frequency for a simultaneous peak search (Step 4). Then, we aggregated them as in (3) to get the class-representative fingerprints of the ten classes.

The results obtained by applying our similarity index per each label are presented in Fig. 4. We can see that the $\mathcal{S}$ index returns a value considerably lower when we apply it to different labels. In particular, the maximum similarity has been achieved when comparing the metal and rock genres, whose value is 2.251. It should be noted that the Jazz class could be considered the most different from the other classes due to the values obtained by our index. We also see that jazz and reggae classes are the most disparate since their similarity has the lowest value of 1.211.

Similarity of $100 \cdot \mathcal{S}(\Xi_{k_1}, \Xi_{k_2})$

	metal	rock	jazz	country	classical	reggae	pop	disco	hiphop	blues
metal	100.000	2.251	1.223	1.747	1.561	1.640	1.683	1.884	1.675	1.680
rock	2.251	100.000	1.277	1.888	1.692	1.677	1.738	1.875	1.866	1.660
jazz	1.223	1.277	100.000	1.378	1.294	1.211	1.284	1.263	1.307	1.315
country	1.747	1.888	1.378	100.000	1.671	1.687	1.793	1.929	1.824	1.682
classical	1.561	1.692	1.294	1.671	100.000	1.499	1.677	1.729	1.663	1.540
reggae	1.640	1.677	1.211	1.687	1.499	100.000	1.744	1.755	1.726	1.636
pop	1.683	1.738	1.284	1.793	1.677	1.744	100.000	1.863	1.794	1.707
disco	1.884	1.875	1.263	1.929	1.729	1.755	1.863	100.000	1.888	1.721
hiphop	1.675	1.866	1.307	1.824	1.663	1.726	1.794	1.888	100.000	1.708
blues	1.680	1.660	1.315	1.682	1.540	1.636	1.707	1.721	1.708	100.000

Fig. 4. Similarity index, displayed as $100 \cdot \mathcal{S}(\Xi_{k_1}, \Xi_{k_2})$, computed for the class-representative fingerprints of the GTZAN dataset.

5 Conclusions

In this paper, we have developed a method for measuring musical genre similarity. As shown in the results of the experimental study, the strategy proposed follows the human perception of genre similarity in music. The notion of class-representative fingerprints allowed us to compare the different genres by making full use of the dataset provided. In this way, we have also shown how convenient the use of fingerprints is in the field of musical computing.

The similarity index proposed for this work has proved to be suitable for computer-assisted music tasks. The application of the similarity index over the class-representative fingerprints gives us meaningful information about the similitude among classes. Apart from being able to gauge the resemblance between classes, such representative elements can also be utilized for matching musical excerpts with a particular genre.

6 Future Work

For future work, it would be useful to address the musical similarity through the class-representative fingerprints employing different perspectives. First, it is required to approach musical stylometry considering larger datasets and further musical genres. In this manner, the results will be more informative and representative. Second, additional evaluation strategies must be implemented. Then, we could implement different similarity functions and/or performance measures to test the accuracy of our proposal.

It would be convenient to consider alternative audio fingerprint algorithms than SpectroMap, as well as the settings that determine the extraction of such fingerprints. Apart from having different perspectives, we could draw conclusions about whether time or frequency bands are more representative when extracting audio fingerprints in musical tasks.

Acknowledgments. This work has been supported for the Conselleria de Innovacíon Universidades, Ciencia y Sociedad Digital Programa Ayudas PROMETEO (Ref. CIPROM/2021/077) and VAE-VADEM TED2021-131295B-C32, funded by MCIN/AEI/10.13039/501100011033 and the European Union NextGenerationEU/PRTR.

I would like to express our gratitude to the anonymous referees for their insightful comments and suggestions, which greatly improved the quality of this paper.

References

1. Yadav, T., Bidari, G., Pande, A., Surender, K.: Real time audio synchronization using audio fingerprinting techniques. In: 2022 1st International Conference on the Paradigm Shifts in Communication, Embedded Systems, Machine Learning and Signal Processing (PCEMS), pp. 16–20 (2022)
2. Faraji, F., Attabi, Y., Champagne, B., Zhu, W.: On the use of audio fingerprinting features for speech enhancement with generative adversarial network. In: 2020 IEEE Workshop on Signal Processing Systems (SiPS), pp. 1–6 (2020)
3. Schedl, M., Gómez, E., Urbano, J.: Music information retrieval: recent developments and applications. Found. Trends® Inf. Retrieval **8**, 127–261 (2014). http://dx.doi.org/10.1561/1500000042
4. Burgoyne, J., Fujinaga, I., Downie, J.: Music information retrieval. In: A New Companion to Digital Humanities, pp. 213–228 (2015)
5. Backer, E., Kranenburg, P.: On musical stylometry-a pattern recognition approach. Pattern Recogn. Lett. **26**, 299–309 (2005). In Memoriam: Azriel Rosenfeld
6. Love, H.: Attributing Authorship: An Introduction. Cambridge University Press, Cambridge (2002)
7. Whissell, C.: Traditional and emotional stylometric analysis of the songs of Beatles Paul McCartney and John Lennon. Comput. Humanit. **30**, 257–265 (1996). https://doi.org/10.1007/BF00055109
8. Bellaachia, A., Jimenez, E.: Exploring performance-based music attributes for stylometric analysis. Int. J. Comput. Inf. Eng. **3**, 1795–1797 (2009)
9. Rebora, S., Salgaro, M.: Is Felix Salten the author of the Mutzenbacher novel (1906)? Yes and no. Lang. Lit. **31**, 243–264 (2022). https://doi.org/10.1177/09639470221090384
10. Brinkman, A., Shanahan, D., Sapp, C.: Musical stylometry, machine learning and attribution studies: a semi-supervised approach to the works of josquin. In: Proceedings of the Biennial International conference on Music Perception and Cognition, pp. 91–97 (2016)
11. Kishor, K., Venkatesh, S., Koolagudi, S.: Audio fingerprinting system to detect and match audio recordings. In: Maji, P., Huang, T., Pal, N.R., Chaudhury, S., De, R.K. (eds.) PReMI 2023. LNCS, vol. 14301, pp. 683–690. Springer, Cham (2023). https://doi.org/10.1007/978-3-031-45170-6_71
12. Cano, P., Batlle, E., Gómez, E., Gomes, L.C.T., Bonnet, M.: Audio fingerprinting: concepts and bonnet. applications. In: Halgamuge, S., Wang, L. (eds.) Computational Intelligence for Modelling and Prediction. Studies in Computational Intelli-

gence, vol. 2, pp. 233–245. Springer, Heidelberg (2005). https://doi.org/10.1007/10966518_17
13. Ramalingam, A., Krishnan, S.: Gaussian mixture modeling of short-time Fourier transform features for audio fingerprinting. IEEE Trans. Inf. Forensics Secur. **1**(4), 457–463 (2006). https://doi.org/10.1109/TIFS.2006.885036
14. Seo, J.S., Jin, M., Lee, S., Jang, D., Lee, S., Yoo, C.D.: Audio fingerprinting based on normalized spectral subband moments. IEEE Signal Process. Lett. **13**(4), 209–212 (2006). https://doi.org/10.1109/LSP.2005.863678
15. Lu, C.-S.: Audio fingerprinting based on analyzing time-frequency localization of signals. In: 2002 IEEE Workshop on Multimedia Signal Processing, pp. 174–177 (2002). https://doi.org/10.1109/MMSP.2002.1203275
16. Wang, X., Gao, L., Mao, S.: PhaseFi: phase fingerprinting for indoor localization with a deep learning approach. In: 2015 IEEE Global Communications Conference (GLOBECOM), pp. 1–6 (2015)
17. Kim, S., Unal, E., Narayanan, S.: Music fingerprint extraction for classical music cover song identification. In: 2008 IEEE International Conference on Multimedia and Expo, pp. 1261–1264 (2008)
18. Baluja, S., Covell, M.: Waveprint: efficient wavelet-based audio fingerprinting. Pattern Recogn. **41**(11), 3467–3480 (2008). https://doi.org/10.1016/j.patcog.2008.05.006. ISSN 0031-3203
19. Burges, C.J.C., Platt, J.C., Jana, S.: Extracting noise-robust features from audio data. In: 2002 IEEE International Conference on Acoustics, Speech, and Signal Processing, vol. 1, pp. I-1021–I-1024 (2002). https://doi.org/10.1109/ICASSP.2002.5743968
20. Wang, A., Smith, J.: System and methods for recognizing sound and music signals in high noise and distortion. J. Acoust. Soc. Am. **121** (2007)
21. Lewin, D.: Generalized Musical Intervals and Transformations. Oxford University Press, USA (2011)
22. Collins, T., Arzt, A., Frostel, H., Widmer, G.: Using geometric symbolic fingerprinting to discover distinctive patterns in polyphonic music corpora. In: Meredith, D. (ed.) Computational Music Analysis, pp. 445–474. Springer, Cham (2016). https://doi.org/10.1007/978-3-319-25931-4_17
23. Arzt, A., Böck, S., Widmer, G.: Fast identification of piece and score position via symbolic fingerprinting. In: Proceedings of the 13th International Society for Music Information Retrieval Conference, ISMIR 2012 (2012)
24. Yin, Z., Reuben, F., Stepney, S., Collins, T.: Measuring when a music generation algorithm copies too much: the originality report, cardinality score, and symbolic fingerprinting by geometric hashing. SN Comput. Sci. **3**, 340 (2022). https://doi.org/10.1007/s42979-022-01220-y
25. Cano, P., Batlle, E.A.: Review of audio fingerprinting. J. VLSI Sig. Proc. **41**, 271–284 (2005)
26. Pires, I., et al.: Recognition of activities of daily living based on environmental analyses using audio fingerprinting techniques: a systematic review. Sensors **18** (2018). https://www.mdpi.com/1424-8220/18/1/160
27. López-García, A., Martínez-Rodríguez, B., Liern, V.A.: Proposal to compare the similarity between musical products. In: Montiel, M., Agustín-Aquino, O.A., Gómez, F., Kastine, J., Lluis-Puebla, E., Milam, B. (eds.) MCM 2022. LNCS, vol. 13267, pp. 192–204. Springer, Chma (2022). https://doi.org/10.1007/978-3-031-07015-0_16
28. López-García. SpectroMap: Peak detection algorithm for audio fingerprinting (2023)

29. Six, J.: Olaf: a lightweight, portable audio search system. J. Open Sour. Softw. **8**, 5459 (2023). https://doi.org/10.21105/joss.05459
30. Cortès, G., et al.: BAF: an audio fingerprinting dataset for broadcast monitoring (2022). https://doi.org/10.5281/zenodo.6868083
31. Tzanetakis, G., Cook, P.: Musical genre classification of audio signals. IEEE Trans. Speech Audio Process. **10**, 293–302 (2002)
32. Crawford, T., Badkobeh, G., Lewis, D.: Searching page-images of early music scanned with OMR: a scalable solution using minimal absent words. University of London, Goldsmiths (2018)
33. Anagnostopoulou, C., Chew, E., Margulis, E., Volk, A.: Music similarity: concepts, cognition and computation. J. New Music Res. **45**, 207–209 (2016)
34. Godoy, R., Haugen, M., Jensenius, A., Nymoen, K., Song, M.: Exploring sound-motion similarity in musical experience. J. New Music Res. **45**, 210–222 (2016)
35. Flexer, A., Grill, T.: The problem of limited inter-rater agreement in modelling music similarity. J. New Music Res. **45**, 239–251 (2016)

Towards Measuring the Distances of Chords of Different Cardinalities

Kristína Hurajová and Ondrej Hutník(✉)

Institute of Mathematics, Faculty of Science, Pavol Jozef Šafárik University in Košice, Jesenná 5, 040 01 Košice, Slovakia
kristina.hurajova@student.upjs.sk, ondrej.hutnik@upjs.sk

Abstract. Music can be represented by a sequence of chords. The transition between the individual chords is naturally of great importance due to the induced musical tension or the created cadence. In this work we summarize several known and less known approaches for comparing two chords of the same as well as different cardinality and calculating their distances in the corresponding chord spaces, which are used to measure transitions between individual chords.

Keywords: Chord · voice leading · n-torus · metric · pseudometric · chord space · topology

1 Introduction

There are several ways of quantifying tonal distances or affinities between chords in the literature. In general, existing approaches to qualitative chord analysis can be divided into two groups. The first involves the description of (similarity relation) functions, which assign to each pair of chords a numerical value of their similarity, easily interpretable as a kind of metric on a qualitative space. The second approach involves the construction of general classification schemes that organise chords into a hierarchical taxonomy of species, genera and possibly higher divisions. In both these distance-oriented and taxonomy-oriented modes, theorists usually appreciate the generality of the technique: rather than exhaustively asserting relationships between all possible chords, which may indeed be arbitrary, the theorist presents a relatively simple procedure that constructs an entire system.

Dmitri Tymoczko in his works, see [2,14,15], introduced chord spaces we will be working with in this paper. Chords are considered as points in the continuous space $\mathbb{R}^n$ under two equivalences. The first is the equivalence modulo a group of translations by integers $\mathbb{Z}_{12}^n$, which is musically equivalent to the equivalence modulo an octave. The resulting space is $\mathbb{R}^n/\mathbb{Z}_{12}^n$ being the n-dimensional torus $\mathbb{T}^n$. The second is the equivalence modulo the group S_n of permutations of n

This work was supported by the Slovak Research and Development Agency under the contract No. APVV-21-0468, and the grant VEGA 1/0657/22.

© The Author(s), under exclusive license to Springer Nature Switzerland AG 2024
T. Noll et al. (Eds.): MCM 2024, LNCS 14639, pp. 345–359, 2024.
https://doi.org/10.1007/978-3-031-60638-0_27

elements, which means that we're working with unordered chords. This gives us the final space $\mathcal{A}_n := \mathbb{T}^n/S_n$, see [14]. An element of $\mathcal{A}_n$ will be called a *n-tone chord.* Since the symmetric group S_n acts on the n-torus via isometries by permuting the coordinates of each point, $\mathcal{A}_n$ inherits the structure of metric space from $\mathbb{T}^n$. Moreover, since it has been obtained from a differentiable manifold through the action of a finite group, it is also an orbifold. As an OP-space, see [2], it represents voice leading relations among chords. Different theorists have favored different ways of measuring voice-leading size, and such different intuitions about voice-leading can be modeled by different choices of metric.

This paper explores the possibility of defining a distance (or, a metric) between chords of different cardinality, i.e., between chords lying in different chord spaces $\mathcal{A}_n$. We review some of the distance functions proposed in the literature from the trivial (Sect. 2) to more sophisticated ones. Among them we describe several approaches based on the Hausdorff distance between sets (Sect. 3) as well as graph theory-based distances such as the surjection distance, the fair surjection distance and the minimum link distance (Sect. 4). Finally, we mention a metric infimum method that constructs a metric from any distance functions.

2 Measuring Distances of Chords: A First Look

A metric gives us a way to measure distances in a space. Formally, a *metric* on a nonempty set X is a function $d : X \times X \to \mathbb{R}$, that possesses four properties:

(M1) (Non-negativity) $d(x, y) \geq 0$ for all $x, y \in X$;
(M2) (Identity of indiscernables) if $d(x, y) = 0$, then $x = y$;
(M3) (Symmetry) $d(x, y) = d(y, x)$;
(M4) (Triangle inequality) $d(x, z) \leq d(x, y) + d(y, z)$ for all $x, y, z \in X$.

A metric space consists of a space, along with a metric for that space. Replacing the condition (M2) by a weaker condition (M2') $d(x, x) = 0$ for all $x \in X$, we get a *pseudometric.* It means that points in a pseudometric space need not be distinguishable.

Recall that a distance in the quotient space $\mathcal{A}_n$ comes from a distance on the parent space $\mathbb{R}^n$. In general, a metric for a parent space will not necessarily give rise to a metric for the quotients of that space. It is possible only when the metric and the equivalence relation are well-matched, and one can find several results of this nature for inheriting a metric from its parent space. For our next purposes we choose the following one, see [6,9]:

Theorem 1. *Let (X, d_X) be a metric space and G be a group acting properly discontinuous and isometrically on X. Then X/G is a metric space with the metric ρ defined as $\rho([x], [y]) = \inf_{g \in G} d_X(x, gy)$.*

Recall that the action of the group G on X is *properly discontinuous* if for all $g \in G$ the mapping $x \mapsto gx$ is a homeomorphism, and for all $(x, y) \in X^2$ there exist open neighbourboods U_x and U_y such that the set $\{g \in G;\, g(U_x) \cap U_y \neq \emptyset\}$ is finite. Also, the action of G on a metric space (X, d_X) is *isometric* if $d_X(gx, gy) = d_X(x, y)$ for every $g \in G$ and $x, y \in X$. The chord space $\mathcal{A}_n$ is a concrete example of the above construction. The group of translations $G = \mathbb{Z}_{12}^n$ acts on $X = \mathbb{R}^n$ properly discontinuously, and its action is isometric, since translations preserve distances. As a consequence, the n-torus $\mathbb{T}^n$ is a metric space with

$$d_{\mathbb{T}^n}(\overline{x}, \overline{y}) = \inf_{k \in \mathbb{Z}_{12}^n} \|x - (y + k)\|_{\mathbb{R}^n}, \tag{1}$$

where $\overline{x} \in \mathbb{T}^n$ is the equivalence class of the element $x \in \mathbb{R}^n$, and $\|\cdot\|_{\mathbb{R}^n}$ is a norm on $\mathbb{R}^n$. We can choose any usual norm, but in order to find a distance that measures voice-leading size, for simplicity we consider the ℓ_1-norm $\|x\|_1 = \sum_{i=1}^n |x_i|$ for $x = (x_1, \ldots, x_n)$. The action of the symmetric group S_n on the torus $\mathbb{T}^n$ is isometric and also properly discontinuous, since S_n is finite. By Theorem 1, the space $\mathcal{A}_n = \mathbb{T}^n / S_n$ is a metric space with the metric

$$d_{\mathcal{A}_n}([x], [y]) = \inf_{\sigma \in S_n} d_{\mathbb{T}^n}(\overline{x}, \overline{\sigma(y)}),$$

where $[x]$ is the equivalence class of an element of $\overline{x} \in \mathbb{T}^n$. It is the distance measuring the smallest voice-leading size between two chords. All distances between major triads (majuscule) and minor triads (minuscule) are shown in Table 1.

Table 1. Major and minor chords in the metric $d_{\mathcal{A}_3}$

	C	$C^\sharp$	D	$D^\sharp$	E	F	$F^\sharp$	G	$G^\sharp$	A	$B^\flat$	B	c	$c^\sharp$	d	$d^\sharp$	e	f	$f^\sharp$	g	$g^\sharp$	a	$b^\flat$	b
C	0																							
$C^\sharp$	3	0																						
D	6	3	0																					
$D^\sharp$	3	6	3	0																				
E	2	3	6	3	0																			
F	3	2	3	6	3	0																		
$F^\sharp$	6	3	2	3	6	3	0																	
G	3	6	3	2	3	6	3	0																
$G^\sharp$	2	3	6	3	2	3	6	3	0															
A	3	2	3	6	3	2	3	6	3	0														
$B^\flat$	6	3	2	3	6	3	2	3	6	3	0													
B	3	6	3	2	3	6	3	2	3	6	3	0												
c	1	4	5	2	3	4	5	2	1	4	5	2	0											
$c^\sharp$	2	1	4	5	2	3	4	5	2	1	4	5	3	0										
d	5	2	1	4	5	2	3	4	5	2	1	4	6	3	0									
$d^\sharp$	4	5	2	1	4	5	2	3	4	5	2	1	3	6	3	0								
e	1	4	5	2	1	4	5	2	3	4	5	2	2	3	6	3	0							
f	2	1	4	5	2	1	4	5	2	3	4	5	3	2	3	6	3	0						
$f^\sharp$	5	2	1	4	5	2	1	4	5	2	3	4	6	3	2	3	6	3	0					
g	4	5	2	1	4	5	2	1	4	5	2	3	3	6	3	2	3	6	3	0				
$g^\sharp$	3	4	5	2	1	4	5	2	1	4	5	2	2	3	6	3	2	3	6	3	0			
a	2	3	4	5	2	1	4	5	2	1	4	5	3	2	3	6	3	2	3	6	3	0		
$b^\flat$	5	2	3	4	5	2	1	4	5	2	1	4	6	3	2	3	6	3	2	3	6	3	0	
b	4	5	2	3	4	5	2	1	4	5	2	1	3	6	3	2	3	6	3	2	3	6	3	0

All chords in $\mathcal{A}_n$ (for a fixed n) have the same cardinality, so we cannot measure distances between chords of different cardinalities directly via $d_{\mathcal{A}_n}$. From now on we consider the space of all chords.

Definition 1. *The space of all chords* is defined as $\mathcal{A} = \bigcup_{n \in \mathbb{N}} \mathcal{A}_n$.

How can a distance on $\mathcal{A}$ be defined? Firstly, in an elementary approach we can unify the sets of tones in the chords, remove the common tones, and count the number of noncommon tones to get

$$d_{\triangle}(U,V) := \#(U \triangle V) = \#(U) + \#(V) - 2\#(U \cap V), \quad U,V \in \mathcal{A},$$

where $\triangle$ and $\cap$ are the symmetric difference and the intersection of the sets, respectively. For example, with $U = [0,4,7]$ and $V = [4,7,11,2]$ we have $U \triangle V = [0,2,11]$, and hence $d_{\triangle}(U,V) = 3$. It is easy to verify that $d_{\triangle}$ is a metric on $\mathcal{A}$. Indeed, $d_{\triangle}$ can be viewed as the taxicab metric of the *characteristic vectors* of chord classes in the 12-dimensional space. More precisely, regarding a chord class as an ordered set of pitch classes, we assign to each chord class U a characteristic vector

$$\chi_U := (\chi_0(U), \chi_1(U), \ldots, \chi_{11}(U)), \text{ where } \chi_i(U) = \begin{cases} 1, & i \in U, \\ 0, & \text{otherwise.} \end{cases}$$

For instance, $\chi_{\mathrm{D}} = (0,0,1,0,0,0,1,0,0,1,0,0)$ is the characteristic vector of the chord class D, which can be interpreted as the Cartesian coordinate of D-major chord in the 12-dimensional space. Thus, the space of all chords (without multisets) is simply the set of vertices of a unit hypercube in the 12-dimensional space, and $d_{\triangle}$ is the taxicab metric in this space. This approach immediately allows the generalisation by considering a different metric instead of $d_{\triangle}$, e.g.

$$d_p(U,V) = \sqrt[p]{\sum_{i=1}^{n} |u_i - v_i|^p}, \quad p \geq 1.$$

Observe that for $p = 1$ we get the set difference metric $d_{\triangle}$ and for $p = 2$ we get the standard Euclidean metric.

Secondly, we can measure distances in $\mathcal{A}$ in the form of multisets. Easily by admitting multisets in chords, their characteristic vectors will contain numbers other than zeros and ones, and we can still compute $d_{\triangle}$. Also, consider $U \in \mathcal{A}_m$ and $V \in \mathcal{A}_n$ with $m \leq n$, and put $k = \operatorname{lcm}(m,n)$, the least common multiple of m and n, i.e., there exist $\alpha \in \mathbb{N}$ and $\beta \in \mathbb{N}$ such that $k = m\alpha = n\beta$. We embed the chords U and V into the space $\mathcal{A}_k$ by repeating each note of the chord U α-times and each note of the chord V β-times. We denote these new chords by U' and V'. The distance between the chords U and V is then defined by

$$d_0(U,V) := d_{\mathcal{A}_k}(U',V').$$

Although this distance looks mathematically correct, because it neither loses nor changes any information about the chords U and V, it is practically difficult to use because it can lead to chords of long length (consider 6-tone and 7-tone chords).

There already exist other types of distances between chords of different cardinality, see e.g. [13] for the measure *Angle*, or [11] for *cardinality transformations* between triads and tetrachords. Moreover, authors in [7] present some geometric representations comparing various distance measures without their quantification. Lastly, Hook in [8] discusses distance functions in several contexts with musical applications.

3 Second Look: Hausdorff-Based Distances

Because of computational difficulties with d_0, we will describe several more sophisticated distances in $\mathcal{A}$, for which the mathematical concept of the Hausdorff distance between arbitrary sets of a space will play a significant role, see [1]. In a metric space (X, d_X) we first introduce the distance of an element $u \in X$ from a set $V \in P(X)$ by the formula $\delta(u, V) = \inf_{v \in V} d_X(u, v)$, and then the distance between two sets $U, V \in P(X)$ as $d(U, V) = \sup_{u \in U} \delta(u, V)$. Since in general $d(U, V) \neq d(V, U)$, we introduce the Hausdorff distance between the sets U and V as the larger of these values.

Definition 2. Let (X, d_X) be a metric space. The *Hausdorff distance* between sets U and V of X is defined by

$$d_{\mathsf{H}}(U, V) := \max\left\{\sup_{u \in U} \delta(u, V), \sup_{v \in V} \delta(v, U)\right\},$$

where $\delta(u, V) = \inf\limits_{v \in V} d_X(u, v)$ and $\delta(v, U) = \inf\limits_{u \in U} d_X(u, v)$.

The Hausdorff distance can be infinite, but we will practically consider only finite sets U, V. Hence, $d_{\mathsf{H}}(U, V)$ will be finite, so we can write max and min instead of sup and inf. The Hausdorff distances between major and minor chords in $\mathcal{A}_3$ are summarised in Table 2.

Table 2. Distances of major and minor triads in d_{H}.

	C	C♯	D	D♯	E	F	F♯	G	G♯	A	B♭	B	c	c♯	d	d♯	e	f	f♯	g	g♯	a	b♭	b
C	0																							
C♯	1	0																						
D	2	1	0																					
D♯	2	2	1	0																				
E	1	2	2	1	0																			
F	2	1	2	2	1	0																		
F♯	2	2	1	2	2	1	0																	
G	2	2	2	1	2	2	1	0																
G♯	1	2	2	2	1	2	2	1	0															
A	2	1	2	2	2	1	2	2	1	0														
B♭	2	2	1	2	2	2	1	2	2	1	0													
B	1	2	2	1	2	2	2	1	2	2	1	0												
c	1	2	2	2	1	2	2	2	1	2	2	1	0											
c♯	1	1	2	2	2	1	2	2	2	1	2	2	1	0										
d	2	1	1	2	2	2	1	2	2	2	1	2	2	1	0									
d♯	2	2	1	1	2	2	2	1	2	2	2	1	2	2	1	0								
e	1	2	2	1	1	2	2	2	1	2	2	2	1	2	2	1	0							
f	1	1	2	2	1	1	2	2	2	1	2	2	2	1	2	2	1	0						
f♯	2	1	1	2	2	1	1	2	2	2	1	2	2	2	1	2	2	1	0					
g	2	2	1	1	2	2	1	1	2	2	2	1	2	2	2	1	2	2	1	0				
g♯	1	2	2	1	1	2	2	1	1	2	2	2	1	2	2	2	1	2	2	1	0			
a	2	1	2	2	1	1	2	2	1	1	2	2	2	1	2	2	2	1	2	2	1	0		
b♭	2	2	1	2	2	1	1	2	2	1	1	2	2	2	1	2	2	2	1	2	2	1	0	
b	2	2	2	1	2	2	1	1	2	2	1	1	1	2	2	1	2	2	2	1	2	2	1	0

Averaged Hausdorff Distance. Since the Hausdorff distance cannot distinguish many major and minor chords, we suggest to use the *averaged Hausdorff distance* as proposed in [12].

Definition 3. Let (X, d_X) be a metric space and $p \in \mathbb{N}$. The *p-averaged Hausdorff distance* between sets U and V of X is defined by $\Delta_p(U,V) := \max\{GD_p(U,V), IGD_p(U,V)\}$, where

$$GD_p(U,V) := \left(\frac{1}{\#(U)} \sum_{i=1}^{\#(U)} \delta(u_i, V)^p \right)^{\frac{1}{p}} = \left(\frac{1}{\#(U)} \sum_{i=1}^{\#(U)} \min_{j=1,\dots,\#(V)} \{d_X(u_i, v_j)^p\} \right)^{\frac{1}{p}}$$

and $IGD_p(U,V) := GD_p(V,U)$.

The indicator Δ_p can be viewed as a composition of slight variations of the Generational Distance (GD, see [16]) and the Inverted Generational Distance (IGD, see [3]). The definition can be extended from $p \in \mathbb{N}$ to arbitrary $p \in \mathbb{R}^* = \mathbb{R} \cup \{\pm\infty\}$ mimicking the Hölder mean when putting

$$GD_p(U,V) = \begin{cases} GD_p(U,V), & p \neq 0, \\ \left(\prod_{i=1}^{\#(U)} \delta(u_i, V) \right)^{\frac{1}{\#(U)}}, & p = 0, \\ \max_{i=1,\ldots,\#(U)} \delta(u_i, V), & p = \infty, \\ \min_{i=1,\ldots,\#(U)} \delta(u_i, V), & p = -\infty, \end{cases}$$

under the convention $GD_p(U, V) = 0$ for $p < 0$ and at least one $\delta(u_i, V) = 0$.

Example 1. For chords C $= [0, 4, 7]$ and G $= [7, 11, 2]$ we have

$$GD_2(\mathrm{C}, \mathrm{G}) = \Big(\frac{1}{3} \sum_{i=1}^{3} \min_{j=1,2,3} \{d_{\mathcal{A}_1}(u_i, v_j)^2\}\Big)^{\frac{1}{2}} = \Big(\frac{1}{3}(1+4+0)\Big)^{\frac{1}{2}} = \sqrt{\frac{5}{3}} = IGD_2(\mathrm{C}, \mathrm{G}),$$

$$\Delta_0(\mathrm{C}, \mathrm{G}) = \left(\prod_{i=1}^{3} \delta(u_i, \mathrm{G}) \right)^{\frac{1}{3}} = (1 \cdot 2 \cdot 0)^{\frac{1}{3}} = 0,$$

$$\Delta_\infty(\mathrm{C}, \mathrm{G}) = d_{\mathsf{H}}(\mathrm{C}, \mathrm{G}) = \max_{\mathrm{i}=1,2,3} \delta(\mathrm{u_i}, \mathrm{G}) = \max\{1, 2, 0\} = 2.$$

All other major and minor chord distances Δ_2 are summarized in Table 3. Note that for major and minor chords we always get the same values for GD_p and IGD_p. Also, for $p \leq p'$ we always have $\Delta_p \leq \Delta_{p'}$ making the Hausdorff distance Δ_∞ the biggest one. It has been established in [12] that Δ_p does not satisfy (M4), but a weaker version $\Delta_p(U, V) \leq N^{1/p}(\Delta_p(U, W) + \Delta_p(W, V))$, where $N \geq 1$ is a constant with $N \geq \max\{\#(U), \#(V), \#(W)\}$.

Table 3. Distances of major and minor triads in Δ_2.

	C	C♯	D	D♯	E	F	F♯	G	G♯	A	B♭	B	c	c♯	d	d♯	e	f	f♯	g	g♯	a	b♭	b
C	0																							
C♯	1	0																						
D	1.73	1	0																					
D♯	1.29	1.73	1	0																				
E	0.82	1.29	1.73	1	0																			
F	1.29	0.82	1.29	1.73	1	0																		
F♯	1.41	1.29	0.82	1.29	1.73	1	0																	
G	1.29	1.41	1.29	0.82	1.29	1.73	1	0																
G♯	0.82	1,29	1,41	1,29	0,82	1,29	1,73	1	0															
A	1.29	0.82	1.29	1.41	1.29	0.82	1.29	1.73	1	0														
B♭	1.73	1.29	0.82	1.29	1.41	1.29	0.82	1.29	1.73	1	0													
B	1	1.73	1.29	0.82	1.29	1.41	1.29	0.82	1.29	1.73	1	0												
c	0.58	1.41	1.41	1.15	1	1.63	1.41	0.82	0.58	1.41	1.73	0.82	0											
c♯	0.82	0.58	1.41	1.41	1.15	1	1.63	1.41	0.82	0.58	1.41	1.73	1	0										
d	1.73	0.82	0.58	1.41	1.41	1.15	1	1.63	1.41	0.82	0.58	1.41	1.73	1	0									
d♯	1.41	1.73	0.82	0.58	1.41	1.41	1.15	1	1.63	1.41	0.82	0.58	1.29	1.73	1	0								
e	0.58	1.41	1.73	0.82	0.58	1.41	1.41	1.15	1	1.63	1.41	0.82	0.82	1.29	1.73	1	0							
f	0.82	0.58	1.41	1.73	0.82	0.58	1.41	1.41	1.15	1	1.63	1.41	1.29	0.82	1.29	1.73	1	0						
f♯	1.41	0.82	0.58	1.41	1.73	0.82	0.58	1.41	1.41	1.15	1	1.63	1.41	1.29	0.82	1.29	1.73	1	0					
g	1.63	1.41	0.82	0.58	1.41	1.73	0.82	0.58	1.41	1.41	1.15	1	1.29	1.41	1.29	0.82	1.29	1.73	1	0				
g♯	1	1.63	1.41	0.82	0.58	1.41	1.73	0.82	0.58	1.41	1.41	1.15	0.82	1.29	1.41	1.29	0.82	1.29	1.73	1	0			
a	1.15	1	1.63	1.41	0.82	0.58	1.41	1.73	0.82	0.58	1.41	1.41	1.29	0.82	1.29	1.41	1.29	0.82	1.29	1.73	1	0		
b♭	1.41	1.15	1	1.63	1.41	0.82	0.58	1.41	1.73	0.82	0.58	1.41	1.73	1.29	0.82	1.29	1.41	1.29	0.82	1.29	1.73	1	0	
b	1.41	1.41	1.15	1	1.63	1.41	0.82	0.58	1.41	1.73	0.82	0.58	1	1.73	1.29	0.82	1.29	1.41	1.29	0.82	1.29	1.73	1	0

When comparing values from Table 2 and Table 3 we may conclude that the averaged Hausdorff distance is more sensitive, because it can distinguish between several pairs of chords which are mapped to the same value in the Hausdorff distance, e.g. $\Delta_2(\mathbf{C},\mathbf{D}) \neq \Delta_2(\mathbf{C},\mathbf{F})$ whereas $d_{\mathsf{H}}(\mathbf{C},\mathbf{D}) = d_{\mathsf{H}}(\mathbf{C},\mathbf{F})$, but still does not distinguish among $\Delta_2(\mathbf{C},\mathbf{F})$, $\Delta_2(\mathbf{C},\mathbf{G})$ and $\Delta_2(\mathbf{C},\mathbf{A})$.

Projection and Extension Distances. Next two distances use the projection/immersion of chords into lower/higher-dimensional voice-leading chord spaces. Suppose the chords $U \in \mathcal{A}_m$ and $V \in \mathcal{A}_n$ with $m \leq n$, and project them into $\mathcal{A}_k$ with $k = \min\{m,n\}$. Then we can use the distance between these projections to compute the Hausdorff distance and $d_{\mathcal{A}_k}$. This is the idea behind the projection distance, see [6].

Definition 4. The *projection distance* between sets $U \in \mathcal{A}_m$ and $V \in \mathcal{A}_n$ with $m \leq n$ is given by

$$d_{\mathsf{P}}(U,V) := \max_{S \in \mathsf{Proj}_m(V)} d_{\mathcal{A}_m}(U,S),$$

where $\mathsf{Proj}_k(V) := \{S \subset V \colon \#(S) = k\}$.

Example 2. For chords $\mathrm{C} = [0,4,7]$ and $\mathrm{G}^7 = [7,11,2,5]$ we have $m = \#(\mathrm{C}) = 3$. Then $S \in \mathsf{Proj}_3(\mathrm{G}^7) = \{[7,11,2],[7,11,5],[7,2,5],[11,2,5]\}$ and

$$d_{\mathsf{P}}(\mathrm{C},\mathrm{G}^7) = \max_{\mathrm{S} \in \mathsf{Proj}_3(\mathrm{G}^7)} \mathrm{d}_{\mathcal{A}_3}(\mathrm{C},\mathrm{S}) = \max\{3,2,3,5\} = 5.$$

It is immediate that d_{P} is symmetric, but d_{P} does not fulfill (M2). The reason is that chords U and V may not belong to the same chord space for $n \neq m$. However, when identifying the chords U and V for which $d_{\mathsf{P}}(U,V) = 0$, one can prove that the adjoint factor space is a metric space. In our case, if $d_{\mathsf{P}}(U,V) = 0$, then for every $S \in \mathsf{Proj}_m(V)$, $U = S$. For example, taking $U = [1]$ and $V = [1,1]$, we get $d_{\mathsf{P}}(U,V) = \max\{0,0\} = 0$. It follows that there exist $x \in \mathbb{R}$ such that $U = ([x],\ldots,[x]) \in \mathcal{A}_m$ and $V = ([x],\ldots,[x]) \in \mathcal{A}_n$. So we can introduce the equivalence relation

$$U \in \mathcal{A}_m \sim V \in \mathcal{A}_n \Leftrightarrow (\exists x \in \mathbb{R})\ U = ([x],\ldots,[x]) \in \mathcal{A}_m \text{ and } V = ([x],\ldots,[x]) \in \mathcal{A}_n.$$

However, even with the above factorisation, the distance d_{P} need not be a metric because it does not satisfy the triangle inequality (M4).

Example 3. Let $U = [1,3]$, $V = [0]$ and $W = [9,10]$. For $m = 1$ we have

$$d_{\mathsf{P}}(U,V) = \max_{S \in \mathsf{Proj}_1(U)} d_{\mathcal{A}_1}(S,V) = \max\{1,3\} = 3,$$
$$d_{\mathsf{P}}(V,W) = \max_{S \in \mathsf{Proj}_1(W)} d_{\mathcal{A}_1}(V,S) = \max\{3,2\} = 3.$$

Since $\#(U) = \#(W)$, we calculate the metric $d_{\mathcal{A}_2}$ and we get $d_{\mathcal{A}_2}(U,W) = 9 = d_{\mathsf{P}}(U,W)$. Therefore, $d_{\mathsf{P}}(U,V) + d_{\mathsf{P}}(V,W) < d_{\mathsf{P}}(U,W)$, and thus the triangle inequality does not hold.

From a purely mathematical point of view, another distance in $\mathcal{A}$ can be defined using a dual approach. For $V \in \mathcal{A}_m$ and $k \geq \#(V)$, define

$$\mathsf{Ext}_k(V) := \{S \supset V \colon \#(S) = k \text{ and } S \text{ contains only the notes contained in } V\}.$$

Let $U \in \mathcal{A}_m$ and $V \in \mathcal{A}_n$ with $m \leq n$, and extend U from the chord space $\mathcal{A}_m$ to the chord space $\mathcal{A}_n$ in which the chord V lies, so that we will be able to use the metric in $\mathcal{A}_n$. More specifically,

$$d_{\mathsf{E}}(U, V) := \min_{S \in \mathsf{Ext}_n(U)} d_{\mathcal{A}_n}(S, V) \tag{2}$$

is called the *extension distance* in $\mathcal{A}$. For some potential musical applications, see [6].

Table 4. Comparison of considered distances for different chords

U	V	$d_{\mathsf{P}}(U,V)$	$d_{\mathsf{E}}(U,V)$	$d_{\mathsf{H}}(U,V)$	$d_{\triangle}(U,V)$	$d_{\mathsf{sur}}(U,V)$	$d_{\mathsf{fair}}(U,V)$	$d_{\mathsf{link}}(U,V)$	$d_{\mathsf{F}}(U,V)$
[0, 4, 7]	[0, 4, 9, 7]	5	2	2	1	2	2	2	5/6
[0, 4, 7]	[0, 3, 7, 10]	6	3	2	3	3	3	3	59/30
[0, 4, 7]	[7, 11, 2, 5]	5	4	2	5	4	4	4	5/2
[0, 4, 7]	[7, 11, 5, 8]	5	3	1	5	3	3	3	16/5
[0, 4, 7]	[0, 4, 7, 11, 2]	6	4	2	2	3	3	3	19/15
[0, 4, 9, 7]	[0, 3, 7, 10]	2	2	1	4	2	2	2	13/6
[0, 4, 9, 7]	[7, 11, 2, 5]	5	5	2	6	5	5	5	75/28
[0, 4, 9, 7]	[7, 11, 5, 8]	3	3	1	6	3	3	3	71/28
[0, 3, 7, 10]	[7, 11, 2, 5]	5	5	1	6	5	5	5	73/28
[0, 3, 7, 10]	[7, 11, 5, 8]	5	5	2	6	5	5	5	41/14
[7, 11, 2, 5]	[7, 11, 5, 8]	6	6	3	2	6	6	6	3/2
[0, 4, 9, 7]	[0, 4, 7, 11, 2]	7	4	2	3	4	4	4	33/20
[0, 4, 7, 11, 2]	[0, 3, 7, 10]	5	3	1	5	3	3	3	79/35
[0, 4, 7, 11, 2]	[7, 11, 2, 5]	6	2	1	3	2	2	2	47/30
[0, 4, 7, 11, 2]	[7, 11, 5, 8]	10	9	3	5	8	8	8	73/28
[0, 1, 5, 7, 10]	[4, 6, 11]	8	6	2	8	6	7	6	16/5
[1, 3, 4, 7, 9]	[0, 5, 10]	7	6	2	8	7	8	7	47/15

4 Oddie-Based Distances: A Graph Theory Approach

In this section we describe an approach developed by Oddie [10] and described by Eiter and Mannila [4] via graph theory. We work with a general distance (not necessarily a metric) $d_X : X \times X \to [0, +\infty)$ on $X \neq \emptyset$, and we mention how to define a metric on $P(X)$. Recall that $\delta(u, V)$ always denotes the distance of an element $u \in X$ from a set $V \in P(X)$ given by $\delta(u, V) = \inf_{v \in V} d_X(u, v)$.

The Surjection Distance. Oddie [10] suggested defining the distance between two sets by considering surjections that map the larger set to the smaller set. This leads to the following distance function

$$d_{\mathsf{sur}}(U, V) := \min_{\eta} \sum_{(u,v) \in \eta} d_X(u, v), \quad U, V \in P(X),$$

where the minimum is taken over all surjections η from the larger of the sets U and V to the smaller. It is showed in [4] that computing d_{sur} can be reduced to computing the cost of a minimum weight perfect matching in a graph in polynomial time. From chords $U \in \mathcal{A}_n$ and $V \in \mathcal{A}_m$ with $m \leq n$ we construct a complete bipartite weighted graph $G = (U \cup V \cup A, E, w)$, where $A = \{a_1, \dots, a_k\}$ is a set of auxiliary vertices with $k = n - m$. The weight function w of the edge is defined as follows

$$w(e) = \begin{cases} d_X(u_i, v_j), & e = \{u_i, v_j\},\ 1 \leq i \leq n,\ 1 \leq j \leq m, \\ \delta(u_i, V), & e = \{u_i, a_j\},\ 1 \leq i \leq n,\ 1 \leq j \leq k. \end{cases}$$

Once we have assigned weights to the edges of the graph G, we look for a minimum weight perfect matching M of G, cf. [4, Lemma 4.3, Theorem 4.4]. Recall that a *matching* of an undirected weighted graph $G = (W, E, w)$ is a set of edges $M \subseteq E$ such that for each pair of distinct $e, e' \in M$ it holds that $e \cap e' = \emptyset$. The matching M is called *perfect* if $\bigcup M = W$. The weight $w(M)$ of the matching M is defined by $w(M) = \sum_{e \in M} w(e)$.

Proposition 1. *Let $U \in \mathcal{A}_n$ and $V \in \mathcal{A}_m$ with $m \leq n$. If M is a minimum weight perfect matching of $G = (V, E, w)$, then $w(M) = d_{\mathsf{sur}}(U, V)$.*

Example 4. Figure 1 shows the graph G constructed from chords $U = [0, 3, 7, 10]$ and $V = [0, 4, 7]$. Then the minimum weight perfect matching M of G is represented by bold lines, and thus $w(M) = d_{\mathsf{sur}}(U, V) = 0 + 1 + 0 + 2 = 3$.

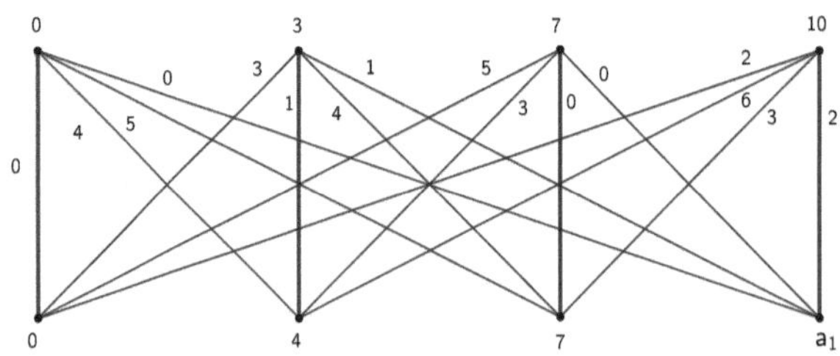

Fig. 1. Graph G for $U = [0, 3, 7, 10]$ and $U = [0, 4, 7]$.

The Fair Surjection Distance. Further Oddie introduced a variant of the surjection measure, in which admissible surjections must be fair. A surjection η between U and V is *fair*, if η maps the elements of U as evenly as possible onto the elements of V, i.e., $(\forall v, v' \in V)\, |\#(\eta^{-1}\{v\}) - \#(\eta^{-1}\{v'\})| \leq 1$, where η^{-1} denotes the relation $\{(y, x) : (x, y) \in \eta\}$. Thus,

$$d_{\mathsf{fair}}(U, V) := \min_{\eta'} \sum_{(u,v) \in \eta'} d_X(u, v),$$

where the minimum is taken over all fair surjections η'. For major and minor chords the fair surjection distance is the same as the surjection distance, since there is no surjection other than fair for 3-tone chords. Fair surjection distance can be efficiently computed by solving a network flow problem. From chords $U \in \mathcal{A}_n$ and $V \in \mathcal{A}_m$ with $m \leq n$ and two auxiliary vertices s, t we construct a directed weighted graph (a network) $N = (W, E, \kappa, c)$ with $W = U \cup V \cup \{s, t\}$ and E consisting of four types of edges e with capacity $\kappa(e)$ and cost $c(e)$ given as follows:

(i) if $e = (s, u_i)$, then $\kappa(s, u_i) = 1$ and $c(s, u_i) = 0$ for $1 \leq i \leq n$,
(ii) if $e = (u_i, v_j)$, then $\kappa(u_i, v_j) = 1$ and $c(u_i, v_j) = d_X(u_i, v_j)$ for $1 \leq i \leq n$ and $1 \leq j \leq m$,
(iii) if $e = (v_i, t)$, then $\kappa(v_i, t) = c_0$ and $c(v_i, t) = 0$ for $1 \leq i \leq m$,
(iv) if $e = (s, v_i)$, then $\kappa(s, v_i) = 1$ and $c(s, v_i) = \Omega$ for $1 \leq i \leq m$,

where $c_0 = \lceil n/m \rceil$ and Ω is an arbitrary real number greater than $\sum_{i,j} d_X(u_i, v_j)$. By [4, Lemma 4.5, Theorem 4.6] we get

Proposition 2. *Let $U \in \mathcal{A}_n$ and $V \in \mathcal{A}_m$ with $m \leq n$. If f is a maximum s-t flow of minimum cost in the network $N = (W, E, \kappa, c)$, then $c(f) = d_{\mathsf{fair}}(U, V) + (c_0 m - n)\Omega$.*

Example 5. For chords $U = [0, 3, 7, 10]$ and $V = [0, 4, 7]$ we construct the network depicted in Fig. 2. The edges are highlighted so that the flow is maximum at minimum cost, so by Proposition 2 we have $c_0 m - n = 2$ and $c(f) = d_{\mathsf{fair}}(U, V) + 2\,\Omega = 3 + 2\Omega$. Thus, $d_{\mathsf{fair}}(U, V) = 3$.

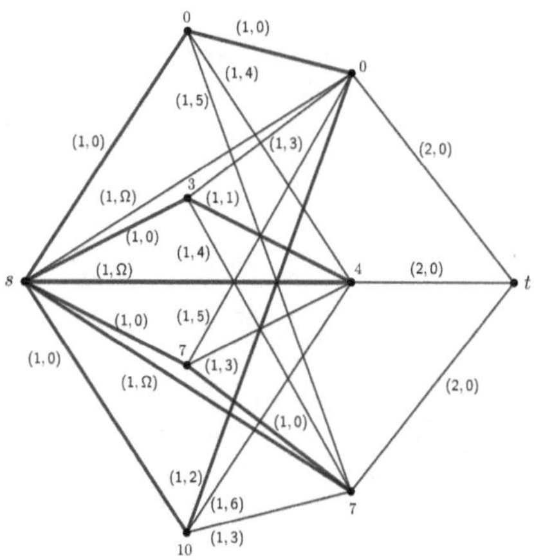

Fig. 2. Network N for $U = [0, 3, 7, 10]$ and $V = [0, 4, 7]$

The Link Distance. An alternative distance between two sets U, V is given by

$$d_{\mathsf{link}}(U, V) := \min_{R} \sum_{(u,v) \in R} d_X(u, v),$$

where the minimum is taken over all linking relations R between U and V, i.e., a relation $R \subseteq U \times V$ is a *linking*, if $(\forall u \in U)(\exists v \in V) : (u, v) \in R$ and $(\forall v \in V)(\exists u \in U) : (u, v) \in R$, see [4]. Again, as with the surjection distance, we can reduce the calculation to a perfect matching problem by defining a complete bipartite graph $G = (U \cup V, E, w)$ on the sets $U = \{u_1, \ldots, u_n\}$ and $V = \{v_1, \ldots, v_m\}$ whose edges have the weight $w(e) = d_X(u_i, v_j)$. Let $G' = (U' \cup V', E', w')$ be a zero-weighted complete bipartite weighted graph on sets $U' = \{u'_1, \ldots, u'_n\}$ and $V' = \{v'_1, \ldots, v'_m\}$ with $(U \cup V) \cap (U' \cup V') = \emptyset$ and $w(e) = 0$ for every edge. Then we take the union of G and G' by connecting all u_i, v_j to u'_i, v'_j to form the graph $G'' = (W'', E'', w'')$ with

- the set of vertices $W'' = U \cup U' \cup V \cup V'$,
- the set of edges $E'' = E \cup E' \cup \{\{u_i, u'_i\}, \{v_j, v'_j\} : u_i \in U, v_j \in V\}$,
- the weight of edges

$$w''(e) = \begin{cases} w(e), & e \in E, \\ w(e'), & e \in E', \\ \delta(u_i, V), & e = \{u_i, u'_i\},\ 1 \le i \le n, \\ \delta(v_i, U), & e = \{v_j, v'_j\},\ 1 \le j \le m. \end{cases}$$

We note that a modified version of the graph G'' provides an additional option for calculating the surjection distance. However, this calculation is more complex, cf. [4, Lemma 4.7, Theorem 4.8].

Proposition 3. *Let $U \in \mathcal{A}_n$ and $V \in \mathcal{A}_m$ with $m \le n$. If M is a minimum weight perfect matching of $G'' = (W'', E'', w'')$, then $d_{\mathsf{link}}(U, V) = w(M)$.*

Example 6. We construct the graph G'' on the sets $U = [0, 3, 7, 10]$, $V = [0, 4, 7]$, $U' = [0', 3', 7', 10']$ and $V' = [0', 4', 7']$, as depicted in Fig. 3. The perfect matching of the minimum weight is marked with bold lines, and thus we get $d_{\mathsf{link}}(U, V) = w(M) = 0 + 1 + 0 + 2 + 0 + 0 + 0 = 3$.

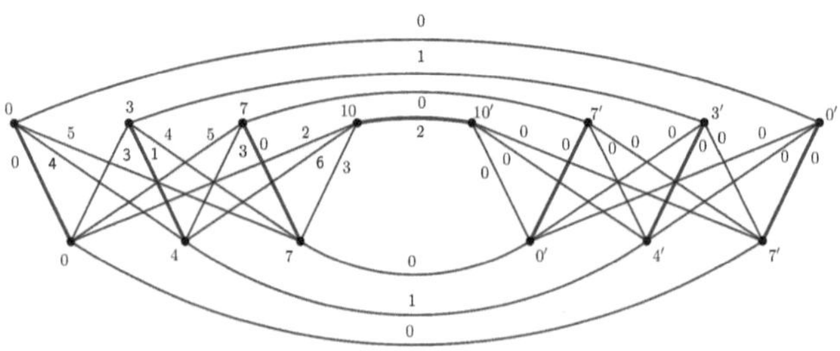

Fig. 3. Graph G'' for $U = [0, 3, 7, 10]$ and $V = [0, 4, 7]$.

The distances $d_{\mathsf{sur}}, d_{\mathsf{fair}}, d_{\mathsf{link}}$ satisfy all the properties of metric except the triangle inequality (M4). In [4] authors describe a construction (the so called *metric infimum*) that produces a metric from an arbitrary distance. It is shown [4,

Theorem 5.4] that the metric infimum of d_{sur}, d_{fair} and d_{link} collapse to a single function. However, the computation of this metric is shown to be in NP for a broad class of instances.

5 Finale: Fujita Metric

Due to its computational difficulty, Oddies' distances are not practically useful. A distance function between nonempty finite subsets based on an average of distances between their elements is provided by Fujita [5].

Definition 5. Let (X, d_X) be a metric space. The *Fujita distance* between sets U, V of X is given by

$$d_{\mathsf{F}}(U,V) = \frac{1}{\#(U \cup V)\#(U)} \sum_{u \in U} \sum_{v \in V \setminus U} d_X(u,v) + \frac{1}{\#(U \cup V)\#(V)} \sum_{u \in U \setminus V} \sum_{v \in V} d_X(u,v).$$

The Fujita distances between major and minor chords in $\mathcal{A}_3$ are summarised in Table 5. As it is proved in [5, Theorem 1], the Fujita distance is a metric on nonempty finite subsets of X.

Table 5. Distances of major and minor triads in d_{F}.

	C	C♯	D	D♯	E	F	F♯	G	G♯	A	B♭	B	c	c♯	d	d♯	e	f	f♯	g	g♯	a	b♭	b
C	0																							
C♯	3	0																						
D	3.11	3	0																					
D♯	2.6	3.11	3	0																				
E	2.4	2.6	3.11	3	0																			
F	2.47	2.4	2.6	3.11	3	0																		
F♯	3.33	2.47	2.4	2.6	3.11	3	0																	
G	2.47	3.33	2.47	2.4	2.6	3.11	3	0																
G♯	2.4	2.47	3.33	2.47	2.4	2.6	3.11	3	0															
A	2.6	2.4	2.47	3.33	2.47	2.4	2.6	3.11	3	0														
B♭	3.11	2.6	2.4	2.47	3.33	2.47	2.4	2.6	3.11	3	0													
B	3	3.11	2.6	2.4	2.47	3.33	2.47	2.4	2.6	3.11	3	0												
c	1.33	3.11	3.22	1.67	3	2.67	3.22	2.27	1.67	3.11	3	2.53	0											
c♯	2.53	1.33	3.11	3.22	1.67	3	2.67	3.22	2.27	1.67	3.11	3	3	0										
d	3	2.53	1.33	3.11	3.22	1.67	3	2.67	3.22	2.27	1.67	3.11	3.11	3	0									
d♯	3.11	3	2.53	1.33	3.11	3.22	1.67	3	2.67	3.22	2.27	1.67	2.6	3.11	3	0								
e	1.67	3.11	3	2.53	1.33	3.11	3.22	1.67	3	2.67	3.22	2.27	2.4	2.6	3.11	3	0							
f	2.27	1.67	3.11	3	2.53	1.33	3.11	3.22	1.67	3	2.67	3.22	2.47	2.4	2.6	3.11	3	0						
f♯	3.22	2.27	1.67	3.11	3	2.53	1.33	3.11	3.22	1.67	3	2.67	3.33	2.47	2.4	2.6	3.11	3	0					
g	2.67	3.22	2.27	1.67	3.11	3	2.53	1.33	3.11	3.22	1.67	3	2.47	3.33	2.47	2.4	2.6	3.11	3	0				
g♯	3	2.67	3.22	2.27	1.67	3.11	3	2.53	1.33	3.11	3.22	1.67	2.4	2.47	3.33	2.47	2.4	2.6	3.11	3	0			
a	1.67	3	2.67	3.22	2.27	1.67	3.11	3	2.53	1.33	3.11	3.22	2.6	2.4	2.47	3.33	2.47	2.4	2.6	3.11	3	0		
b♭	3.22	1.67	3	2.67	3.22	2.27	1.67	3.11	3	2.53	1.33	3.11	3.11	2.6	2.4	2.47	3.33	2.47	2.4	2.6	3.11	3	0	
b	3.11	3.22	1.67	3	2.67	3.22	2.27	1.67	3.11	3	2.53	1.33	3	3.11	2.6	2.4	2.47	3.33	2.47	2.4	2.6	3.11	3	0

Concluding Remarks

We have reviewed several approaches to measuring distances between chords in generalized voice-leading spaces. Despite several known approaches including the Hausdorff-based distances and their modifications (projection, or extension distance), we have described and exemplified less known Oddie-based distances, and the Fujita distance, see Table 4. Among all the presented distances, only the distance d_p with $p \geq 1$, the Hausdorff distance d_{H} and the Fujita distance d_{F} are metrics on $\mathcal{A}$. Comparing the last two metrics, the Fujita metric is more sensitive (compare Table 2 and Table 5). However, as a serious drawback, all the three metrics violate the property that the restriction to any $\mathcal{A}_n$ is the usual quotient metric $d_{\mathcal{A}_n}$. For instance, $d_{\mathsf{F}}(\mathrm{C}, \mathrm{c}) = 4/3 \neq 1 = \mathrm{d}_{\mathcal{A}_3}(\mathrm{C}, \mathrm{c})$, and $d_{\mathsf{H}}(\mathrm{C}, \mathrm{D}) = 2 \neq 6 = \mathrm{d}_{\mathcal{A}_3}(\mathrm{C}, \mathrm{D})$, and $d_{\triangle}(\mathrm{C}, \mathrm{e}) = 2 \neq 1 = \mathrm{d}_{\mathcal{A}_3}(\mathrm{C}, \mathrm{e})$. Thus, we are still not able to answer the question of Genuys [6, Conclusion] whether there exists a metric d on $\mathcal{A}$ whose restriction to simple cases (equal cardinality) is equal to $d_{\mathcal{A}_n}$.

Although, from a musical point of view, it does not make much sense to consider the repetition of notes in a chord (as long as we consider the octave equivalence), we cannot completely avoid their usage when considering the distance d_{O}, or d_{E}. In such cases, we are facing extensions in the form of multisets: e.g. for computing $d_{\mathsf{E}}(\mathrm{C}, \mathrm{G}^7)$ we have to complete the chord C, i.e., $\mathsf{Ext}_4(\mathrm{C}) = \{[0,0,4,7], [0,4,4,7], [0,4,7,7]\}$ being a set of multisets. On the other hand, we must handle multisets carefully because they can spoil some of the expected properties of distance. Just one example: for multisets $U = \{1,1,2\}$ and $V = \{1,2,2\}$, the Hausdorff distance is 0 because $\max_{u \in U} d_{\mathcal{A}_1}(u, V) = \max\{0,0,0\} = 0$ and $\max_{v \in V} d_{\mathcal{A}_1}(v, U) = \max\{0,0,0\} = 0$. Hence, $d_{\mathsf{H}}(U, V) = 0$, which means that d_{H} is a pseudometric on $\mathcal{A}$ when involving multisets. Consequently, d_{H} is a metric between chords of different cardinality that are not multisets. Similar problems may arise in further distances described in this paper.

References

1. Beer, G.: Topologies on Closed and Closed Convex Sets. Mathematics and Its Applications. Kluwer Academic Publisher (1993)
2. Callender, C., Quinn, I., Tymoczko, D.: Generalized voice-leading spaces. Science **320**(5874), 346–348 (2008)
3. Coello Coello, C.A., Cruz Cortés, N.: Solving multiobjective optimization problems using an artificial immune system. Genet. Program Evolvable Mach. **6**, 163–190 (2005)
4. Eiter, T., Mannila, H.: Distance measures for point sets and their computation. Acta Inform. **34**, 109–133 (1997)
5. Fujita, O.: Metrics based on average distance between sets. Jpn. J. Indust. Appl. Math. **30**, 1–19 (2013)
6. Genuys, G.: Pseudo-distances between chords of different cardinality on generalized voice-leading spaces. J. Math. Music **13**(3), 193–206 (2019)
7. Hall, R.W., Tymoczko, D.: Submajorization and the geometry of unordered collections. Amer. Math. Monthly **119**(4), 263–283 (2012)

8. Hook, J.: Exploring Musical Spaces. Oxford University Press, Oxford (2023)
9. Munkres, J.R.: Topology, 2nd edn. Pearson College Div (2000)
10. Oddie, G.: Verisimilitude and distance in logical space. Acta Philos. Fenn. **30**, 227–243 (1978)
11. Plotkin, R.: Cardinality transformations in diatonic space. In: Agon, C., Andreatta, M., Assayag, G., Amiot, E., Bresson, J., Mandereau, J. (eds.) MCM 2011. LNCS (LNAI), vol. 6726, pp. 207–219. Springer, Heidelberg (2011). https://doi.org/10.1007/978-3-642-21590-2_16
12. Schütze, O., Esquivel, X., Lara, A., Coello Coello, C.A.: Using the averaged Hausdorff distance as a performance measure in evolutionary multiobjective optimization. IEEE Trans. Evol. Comput. **16**, 504–522 (2012)
13. Scott, D., Isaacson, E.J.: The interval angle: a similarity measure for pitch-class sets. Perspect. New Music **36**(2), 107–142 (1998)
14. Tymoczko, D.: The geometry of musical chords. Science **313**(5783), 72–74 (2006)
15. Tymoczko, D.: A Geometry of Music. Oxford University Press, New York (2011)
16. Van Veldhuizen, D.A., Lamont, G.B.: Multiobjective evolutionary algorithm test suites. In: ACM Symposium on Applied Computing, pp. 351–357. ACM, New York (1999)

Assessing the Compatibility Between Musical Performance and Tuning System

Brian Martínez-Rodríguez[1(✉)] and Vicente Liern[2]

[1] Universidad Internacional de la Rioja (UNIR), Logroño, La Rioja, Spain
brian.martinez@unir.net

[2] Dep. Matemáticas para la Economía y la Empresa, Universitat de València, Valencia, Spain
vicente.liern@uv.es

Abstract. This paper we propose a novel method, based on a fuzzy approach, for assessing the compatibility of a set of experimentally measured notes with any theoretical tuning system. Through the exposition of three illustrative experiments, we demonstrate the potential of this method, where we calculate the average tuning distance between the music performance and the theoretical values, revealing insights into the suitability of different tuning systems for specific musical contexts, as well as the nuanced adjustments made by the musicians in response to varying musical scenarios.

Keywords: Tuning · Performance · Fuzzy · Compatibility

1 Introduction

The alignment between real performances and the theoretical tuning systems is an area of research that has seen substantial interest over the years. Barbour [1] provides some foundational insights into the evolution of tuning practices. The importance of tuning in musical performance cannot be overstated [2], as any specific tuning choices impact directly on musical expressiveness and authenticity [3], highlighting the discrepancies between the theoretical mathematical structure of tuning systems with the practical realities of musical performance [4]. Therefore, developing objective and quantitative tools to measure how much a performance deviates from the accepted tuning standard could be a powerful resource for performative analysis and computer-assisted musicology, offering a comprehensive analysis that bridges the gap between theoretical musings and practical applications. This paper aims to create a mathematical approach, using fuzzy numbers, to assess how much the notes obtained directly from a music performance differ from several theoretical tuning systems.

2 Related Work

Some studies related to music performance [5,6] have shown the nuanced relationship between musical ability and executive functioning, specifically inhibition, updating, and switching, across both auditory and visual modalities.

© The Author(s), under exclusive license to Springer Nature Switzerland AG 2024
T. Noll et al. (Eds.): MCM 2024, LNCS 14639, pp. 360–371, 2024.
https://doi.org/10.1007/978-3-031-60638-0_28

Findings indicate that while musical ability predicts enhanced performance in working memory updating tasks, it does not demonstrate clear associations with inhibitory control or task switching. These results suggest a modality-general yet process-specific relationship between musical ability and non-musical cognitive functions.

Aiming to expand existing knowledge on intonation in music, [7] explores the tuning performance and perception of string instrumentalists, considering factors that influence intonation accuracy, including performance tasks, frequency variations, and age groups. Results reveal significant differences in tuning preferences among high school and college/professional subjects, as well as discrepancies between verbal perception and actual tuning performance, underscoring the complexity of intonation in musical contexts.

Delving into the mechanisms for music to be expressive, [8] showcases some rules in music communication, identifying differentiation of frequency and duration categories, marking boundaries between tone groups, and ensuring timing and tuning agreement in ensembles, suggesting that expressive details in music performance are not exclusive to music but resonate with extramusical experiences of the listener.

The application of Fuzzy Logic to music entails a nuanced understanding, where blurriness and precision coexist in a complementary manner throughout the musical process [9]. This duality recognizes that blurriness is not merely an imperfection but a fundamental aspect necessary for development and evolution within music. Fuzziness in music allows for progress and innovation, challenging the notion of rigid identity and repetition.

On this direction, the idea of modelling musical notes as fuzzy sets represents a significant leap in this field [10] for a more realistic representation of music, acknowledging the inherent variability and subjectivity in musical performance. In [11] a novel approach to modeling musical notes using fuzzy numbers is presented, focusing particularly on the widely used 12-TET tuning system and establishing a theoretical framework to analyze the compatibility between theoretical notes in a score and those played by musicians, shedding light on the complex relationship between theoretical tuning systems and practical performance.

3 Theoretical Background

3.1 Tuning Systems

Based on the previous research presented in [10], we propose the following mathematical definitions aimed at generalizing the concept of a tuning system:

Definition 1. Frequency Ratio enclosed within an Octave. *Let* $r = \frac{f}{f_0}$ *with* $r \in \mathbb{R}^+$ *be the ratio of a musical interval defined by the frequencies* f_0 *and* f*, the following transformation:*

$$r^* = r \cdot 2^{-\lfloor \log_2 r \rfloor}; \quad r^* \in [1, 2[\tag{1}$$

represents the transformed ratio r^*, *enclosed within the interval* $[1,2[$, *where* $\lfloor x \rfloor$ *is the* floor function *of* x. *It is easy to verify that the number of octaves* $m \in \mathbb{Z}$ *that must be added or subtracted from* r *to limit it to* $[1,2[$ *is* $-\lfloor \log_2 r \rfloor$.

Definition 2. Tuning System. *Let* $\mathcal{B} = \{\beta_i\}_{i=1}^k$, *with* $\beta_i \in [1,2[$ *be a set of* k *frequency ratios. Let* $\mathcal{F} = \{f_i\}_{i=1}^k$ *be a set of distribution functions* $f_i \colon \mathbb{Z} \to \mathbb{Z}$, $i = \{1,2,\ldots,k\}$. *A tuning system formed by the intervals* $\mathcal{B}$ *and functions* $\mathcal{F}$ *is defined as the set*

$$\mathcal{S}_{\mathcal{B}}^{\mathcal{F}} := \left\{ r_n \,\middle|\, r_n = \left[\prod_{i=1}^{k} \beta_i^{f_i(n)} \right]^*, n \in \mathbb{Z} \right\} \tag{2}$$

where $*$ is the transformation defined in 1 and thus $r_n \in [1,2[, \forall n \in \mathbb{Z}$.

Example 1. The Open Pythagorean Tuning System. Let $\mathcal{B} = \{\beta_1 = \frac{3}{2}\}$ be the natural fifth and let $\mathcal{F} = \{f_1(n) = n\}$ be the fifth distribution function. The Open Pythagorean Tuning System is the set

$$\mathcal{S}_{pit} = \{r_n = [\beta_1^n]^*\}, \quad n \in \mathbb{Z}, \quad \beta_1 = 3/2 \tag{3}$$

Using the transformation 1 we can calculate the ratio and restrict it to the interval $[1,2[$

$$r_{pit}^*(n) = \beta_1^n \cdot 2^{-\lfloor \log_2 \beta_1^n \rfloor} \tag{4}$$

The difference between $r_{pit}^*(0)$ and $r_{pit}^*(12)$ is the *Pythagorean Comma*; the difference between $r_{pit}^*(4)$ and the natural major third (5/4) is the *Syntonic Comma.*

Example 2. Closed Pythagorean Tuning System with Wolf Fifth. Let $\mathcal{B} = \{\beta_1 = \frac{3}{2}, \beta_2 = \frac{2^{18}}{3^{11}}\}$ be the intervals of the natural fifth and the Pythagorean wolf fifth, respectively. Let the functions $\mathcal{F} = \{f_1, f_2\}$ be such that

$$f_1(n) = n - f_2(n); \qquad f_2(n) = \lfloor \frac{n+3}{12} \rfloor$$

the Closed Pythagorean Tuning System is the set

$$\mathcal{S}_{pitCerr} = \{r_n = [\beta_1^{n-f_2(n)} \cdot \beta_2^{f_2(n)}]^*\}, n \in \mathbb{Z} \tag{5}$$

Example 3. Zarlino's System. Let $\mathcal{B} = \{\beta_1 = 3/2, \beta_2 = 40/27\}$ be the intervals of the natural fifth and the syntonic fifth (natural fifth minus one Syntonic Comma), respectively. Let the functions $\mathcal{F} = \{f_1, f_2\}$ be

$$f_1(n) = n - \lfloor \frac{n+4}{7} \rfloor - \lfloor \frac{n+1}{7} \rfloor; \qquad f_2(n) = \lfloor \frac{n+4}{7} \rfloor + \lfloor \frac{n+1}{7} \rfloor$$

Zarlino's tuning system is the set

$$\mathcal{S}_{zar} = \{r_n = [\beta_1^{f_1(n)} \cdot \beta_2^{f_2(n)}]^*\}, n \in \mathbb{Z} \tag{6}$$

3.2 Notes as Fuzzy Numbers

The traditional definition of a note through the quantification of its frequency has been based on physical magnitudes like Hertz or logarithmic scales. However, the reality of music reveals a diversity of frequencies associated with the same note due to variations in execution and environmental conditions. Factors such as finger placement, changes in pressure in wind instruments, shifts in instrument placement, and temperature fluctuations introduce uncertainty in performance. This complexity does not necessarily imply detuning, but rather highlights the dynamic and unique nature of each musical interpretation.

Consider a violinist who during the course of their concert plays a note B_4 in the first position of the second string of the violin, resulting in a frequency $f_1 = 495.84$ Hz. A few bars later, they repeat the execution of the same note B_4, using the same string, position, attack, and intensity but this time resulting in a frequency $f_2 = 496.575$ Hz. Will the change in the tuning of the note be perceptible? If we calculate the number of cents in the interval formed by these two frequencies, we get that

$$r_{cents}(f_1, f_2) = 1200 \times \left| \log_2 \left(\frac{495.84}{496.57} \right) \right| = 2.547\,\text{cents}$$

There is a difference of approximately 2.5 cents between the two executions of the same note. The minimum tuning difference that a well-trained human ear is capable of distinguishing is approximately 4 cents [12]. Which of the two notes will be correct? According to the 12-ET over an reference A_4 at 442 Hz, the frequency of a B_4 is 496.128 Hz, therefore neither of the two B_4 notes coincides exactly with the theoretical value; however, both will be considered as correctly tuned.

Based on the expression 1, we can define a transformation to convert any frequency f into a pitch expressed in cents, contained between $[0, 1200[$, taking as a reference any arbitrary frequency f_0 (Table 1):

$$f^* = 1200 \left(log_2 \left(\frac{f}{f_0} \right) - \left\lfloor log_2 \left(\frac{f}{f_0} \right) \right\rfloor \right) \tag{7}$$

Table 1. Example of different intervals calculated with the transformation.

Interval	f(Hz)	ratio	f^*(cents)	Interval	f(Hz)	ratio	f^*(cents)
octave	884.00	2: 1	0.00	nat. P5 (15ma)	2652.00	6: 1	701.96
fifteenth	1768.00	4: 1	0.00	nat. P4	589.33	4: 3	498.04
nat. P5	663.00	3: 2	701.96	nat. P4 (8vb)	294.67	4: 6	498.04
nat. P5 (8vb)	331.50	3: 4	701.96	nat. P4 (15ma)	2357.33	16: 3	498.04

Definition 3. Fuzzy Degree (f-degree). *Let $\delta \in [1, 50]$ be and arbitrary level of tolerance expressed in cents. Let f_D be the theoretical frequency of an specific*

musical degree calculated using any tuning system. A Fuzzy Degree $\tilde{D}$ *is defined as the triangular fuzzy number:*

$$\tilde{D} := (f_D^*, f_D^* - \delta, f_D^* + \delta) = (f_D^*, \delta)$$

whose membership function is

$$\mu_{\tilde{D}}(x) = \begin{cases} 1 - \dfrac{|f_D^* - x^*|}{\delta} & \text{if } |f_D^* - x^*| < \delta \\ 0 & \text{if } |f_D^* - x^*| \geq \delta \end{cases}$$

Definition 4. Fuzzy Tuning System (f-tuning system). *A Fuzzy Tuning System is defined as a sequence of m distinct f-degrees:*

$$\tilde{S}(\delta) = \{\tilde{D}_1, \tilde{D}_2, \ldots, \tilde{D}_m\} = \{\tilde{D}_j\}_{j=1}^m$$

Definition 5. Fuzzy Note (f-note). *Let $\tilde{S}(\delta)$ be a* f-tuning system *of m distinct f-degrees. A Fuzzy Note is defined as*

$$\tilde{N} = \left(f_N^*; \{\mu_{\tilde{D}_j}(f_N)\}_{j=1}^m\right) \tag{8}$$

where each $\mu_{\tilde{D}}(f_N)$ indicates the degree of belonging of the experimentally measured note's frequency to the *f-degree* $\tilde{D}$.

By consequence, a *f-note* might belong to several *f-degrees* of a *f-tuning system*; therefore, the definition of the musical note requires a tuple of values formed by its measured pitch, and the set of belonging coefficients of that pitch to each of the *f-degrees* of the *f-tuning system.*

Example 4. Let $\tilde{S}^Z(\delta)$ be a *f-tuning system* based on Zarlino's system of 19° per octave. Obtain all the *f-notes* corresponding to the frequency $f_N = 842.51$ Hz that have at least one membership coefficient different from zero. Take as reference frequency $f_0 = 262.81$ Hz, and $\delta = 50$ as tolerance level.

Note	Cents	Note	Cents	Note	Cents	Note	Cents
C	0	E	316	G	631	La	955
C	71	E	386	G	702	B	1018
D	133	E	457	G	773	B	1088
D	204	F	498	A	814	Si	1159
D	275	F	569	A	884	C	1200

Continuing with the process the requested frequency should be transformed into cents: $f_N^*(842.51\,\text{Hz}) = 816.81$ cents. In the interval $(816.81-50, 816.81+50)$ there are two possible f-degrees: A and G. The membership coefficients are:

$$\mu_{\tilde{A}\flat} = 1 - \frac{|814 - 816.81|}{50} = 0.9438; \quad \mu_{\tilde{G}\flat} = 1 - \frac{|773 - 816.81|}{50} = 0.1238$$

Therefore, the f-note is configured as follows, disregarding the membership coefficients with a value of zero:

$$\tilde{N} = \left(816.81; \mu_{\tilde{A\flat}} = 0.9438; \mu_{\tilde{G\flat}} = 0.1238\right)$$

3.3 Compatibility of a Series of Notes with a Tuning System

Suppose we have a set of n experimentally measured frequencies. How could we determine their average deviation from the expected theoretical values? We need to define a measure of similarity between our set of experimentally measured frequencies and the set of degrees belonging to the tuning system that allows us to determine, among all the possibilities, what is the final deviation. According to prior definitions explained on [15] we can construct the following definitions.

Definition 6. Average Tuning Distance. *Let* $F = \{f_1, f_2, \ldots, f_n\}$ *be a set of* n *experimentally observed frequencies. Let* $\tilde{S}(\delta) = \{\tilde{D}_j\}_{j=1}^{m}$ *be a reference* f-tuning system *consisting of* m *distinct* f-degrees. *Let* $\{\tilde{N}_1, \tilde{N}_2, \ldots, \tilde{N}_n\}$ *be the set of* n f-notes *constructed from* $\tilde{S}(\delta)$, *the expression 8, and each of the frequencies existing in the set* F. *The average tuning distance is*

$$d(F, \tilde{S}) = \frac{1}{n \cdot m} \sum_{i=1}^{n} \sum_{j=1}^{m} \mu_{\tilde{D}_j}(f_{N_i}) \cdot |f^*_{N_i} - f^*_{D_j}|$$

We can again construct the membership matrix

$$\mathbf{U} = \begin{pmatrix} \mu_{\tilde{D}_1}(f_{N_1}) & \cdots & \mu_{\tilde{D}_m}(f_{N_1}) \\ \vdots & \ddots & \vdots \\ \mu_{\tilde{D}_1}(f_{N_n}) & \cdots & \mu_{\tilde{D}_m}(f_{N_n}) \end{pmatrix}$$

The sum of the membership coefficients of an *f-note* to all the *f-degrees* must be equal to 1, therefore:

$$\sum_{j=1}^{m} \mu_{\tilde{D}_j}(f_{N_i}) = 1, \quad 1 \leq i \leq n.$$

where u_{ij} is the element (i, j) of the membership matrix $\mathbf{U}$. We can therefore define new normalized membership coefficients $\hat{u}_{ij}$ as follows:

$$\hat{u}_{ij} = \frac{\mu_{\tilde{D}_j}(f_{N_i})}{\sum\limits_{k=1}^{m} \mu_{\tilde{D}_k}(f_{N_i})} \tag{9}$$

Finally, rewriting the expression 6 with the normalized coefficients will yield the expression for the normalized distance

$$\hat{d}(F, \tilde{S}) = \frac{1}{n \cdot m} \sum_{i=1}^{n} \sum_{j=1}^{m} \hat{u}_{ij} \cdot |f^*_{N_i} - f^*_{D_j}| \tag{10}$$

4 Experiments

4.1 Tuning System of a Recorded Wind Instrument

Consider a saxophone playing a specific melody. By analyzing a digital recording of this performance, and using the method proposed in 10, we will calculate the average distance of the real performance with several tuning systems: 12-ET, Meantone 1/4, Zarlino, Kellner-Bach. The reference frequency is set on $A_4 = 440\,\text{Hz}$ and the arbitrary tolerance level $\delta = 50$.

First, we will determine the fundamental frequencies of each of the notes played by the instrument. For this, we will divide the digital recording into temporal windows, composed of 2^{14} samples, applying a *Hamming* type window function to each. Subsequently, we will obtain the energy spectrum of each window by applying the FFT algorithm. We will calculate the fundamental frequency of each window using the *Harmonic Product Spectrum* method ([13] and [14]): by summing several times the original spectrum with compressions of ratio $1 : k$ of the same spectrum, we manage to cancel all spectral components that are not harmonics. The fundamental frequency to be determined f_0 will be the first of the harmonics that have survived the HPS process (Fig. 1).

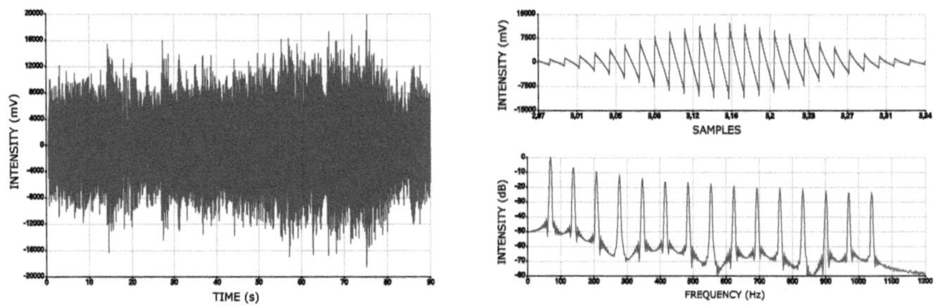

Fig. 1. a) Recording of the melody; b) A temporal window; c) Energy Spectrum.

Table 2. Histogram of occurrences of the measured notes.

f_0(Hz)	Nº	f_0(Hz)	Nº	f_0(Hz)	Nº	f_0(Hz)	Nº	f_0(Hz)	Nº	f_0(Hz)	Nº	f_0(Hz)	Nº	f_0(Hz)	Nº
262.53	4	273.37	4	276.06	10	278.63	6	289.45	5	292.14	10	294.68	6	308.80	1
326.97	10	328.30	5	332.30	5	343.03	5	347.02	5	348.42	10	353.69	1	364.48	5
369.78	5	385.93	5	389.96	4	391.09	10	407.23	1	412.67	10	413.87	5	417.95	4
439.49	9	460.81	6	463.48	10	466.16	5	490.47	4	491.61	5	493.03	10	514.56	6
309.51	4	310.77	15	367.01	5	368.50	5	436.71	5	437.97	4	519.94	10	521.14	5

Using the tolerance level $\delta = 50$, we can calculate the membership coefficient $\mu_{\tilde{D}_j}(f_{N_i})$ of each *f-note* f_{N_i} to each *f-degree* $\tilde{D}_j$. Using expression 10, we can calculate the normalized average tuning distance of the set of notes from Table 2 with various tuning system (Fig. 2).

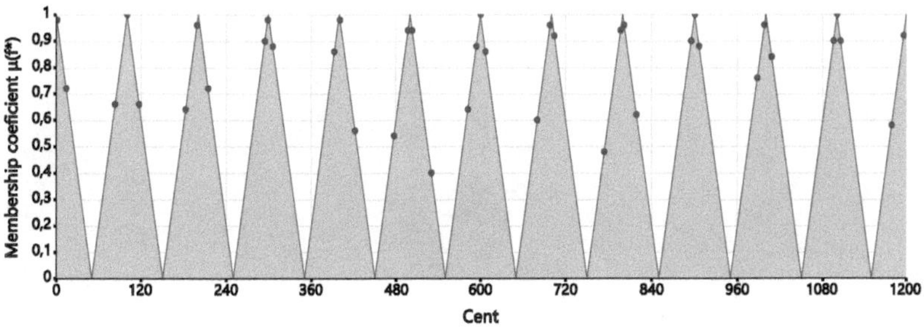

Fig. 2. Superimposition of measured notes on the 12-ET.

The computational results obtained for the normalized average tuning distance between the set of notes and the different tuning systems are as follows (Table 3):

Table 3. Normalized average tuning distance results.

Tuning System	Difference
Meantone 1/4	1.8113336683
Zarlino	1.2553954370
Kellner-Bach	1.2089606993
12-ET	0.8495630336

The findings of this first experiment strongly indicate that the saxophone player's intonation aligns more closely with the 12-ET tuning system compared to other tuning systems. The observed deviations between the theoretical values of the notes and the measured ones can be attributed to the inherent nature of wind instruments, such as the saxophone, and the nuanced adjustments made by musicians in response to varying musical contexts.

4.2 Tuning System of a Vocal Line

In this example we will analyse the original vocal line of the song *Billie Jean*, by Michael Jackson. Using the same method, we will calculate the average distance of the vocal performance with several tuning systems: 12-ET, Meantone 1/4, Zarlino, Kellner-Bach. We will take as a reference frequency of $A_4 = 440\,\text{Hz}$ and tolerance level $\delta = 50$. The FFT analysis will be made with temporal windows composed of 2^{13} samples and a *Blackman-Harris* window function. The notes of each window will be calculated using *HPS* method. Setting a tolerance level $\delta = 50$, we will calculate the membership coefficient $\mu_{\tilde{D}_j}(f_{N_i})$ of each *f-note* f_{N_i} to each *f-degree* $\tilde{D}_j$ with expression 10. The results found are shown on the Table 4 (Fig. 3).

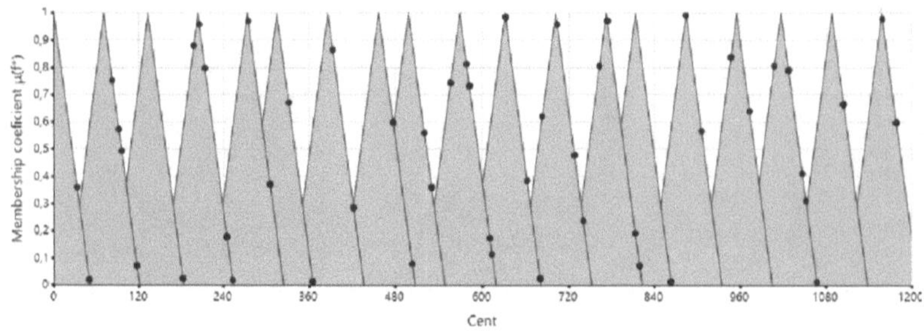

Fig. 3. Superimposition of measured notes on the Zarlino system.

Table 4. Normalized average tuning distance results.

Tuning System	Difference
Meantone 1/4	2.1915144356
Zarlino	1.1862056198
Kellner-Bach	2.6134632491
12-ET	2.7803101878

The second experiment reveals a notable departure of Michael Jackson's vocal line in *Billie Jean* from the conventional 12-ET tuning system, with a closer alignment observed towards Zarlino's 19-note-per-octave system. However, it is crucial to exercise caution in categorically attributing the analyzed vocal line to Zarlino's system, as such a classification would be anachronistic and overlooks the complexities of musical interpretation and performance practice.

Instead, our focus should shift towards a comprehensive analysis of the song itself, delving into the nuances of Michael Jackson's vocal delivery and the expressive elements inherent within the song. An examination of *Billie Jean* reveals a dynamic interplay between vocal technique and stylistic considerations characteristic of pop, rock, and R&B genres. These genres often entail a diverse range of vocal inflections, articulations, and emotive nuances, resulting in greater frequency fluctuation within the emitted pitches. As such, the observed affinity towards a 19-note-per-octave system may stem from the inherent expressiveness and stylistic idiosyncrasies employed by Jackson in his performance.

Furthermore, the multifaceted nature of musical interpretation and performance, wherein deviations from traditional tuning systems are not merely technical anomalies but rather expressive manifestations reflective of stylistic conventions and artistic intent, contextualizes our findings within the broader musical landscape and delves into the intricacies of performance practice, gaining a deeper appreciation for the dynamic interplay between intonation, style, and musical meaning in Michael Jackson's performance.

4.3 Tuning System of a Synthesised Melody

In this example we will analyse an original melody digitally synthesised using *Meantone 1/4 tunning system.* We will take as a reference frequency of $A_4 =$ 440 Hz and tolerance level $\delta = 50$. The FFT analysis will be made with temporal windows composed of 2^{13} samples and a *Blackman-Harris* window function. *HPS* method will be used. Tolerance level $\delta = 50$. The results found for the normalized average tuning distance of the melody is shown on the Table 5.

Table 5. Normalized average tuning distance results.

Tuning System	Difference
Meantone 1/4	0.9019778812
Zarlino	1.1271079067
Kellner-Bach	1.8718559679
12-ET	2.1283800555

As expected, the results of our third experiment prove the extracted notes of the melody to be more aligned with Meantone 1/4 tuning.

5 Conclusions

By embracing a fuzzy approach and extend findings of prior work [15] to the field of music performance and tuning, we have developed a novel method to quantify the compatibility of a music performance with any theoretical tuning system. This approach led to the calculation of the average tuning distance, providing a metric to evaluate the alignment of experimental measured frequencies with theoretical tuning models. We have successfully tested the method with three different examples. The findings offer insights into the suitability of each tuning system for specific musical contexts.

Future research stemming from this study could take several promising directions like expanding the application of fuzzyness to other aspects of music theory, such as agogics, dynamics, articulation or rhythm; enhancing signal processing algorithms to improve the precision in analyzing musical notes; extending this methodology to a broader range of musical styles and instruments, and integrating machine learning techniques for enhanced tuning analysis.

By embracing the concepts of fuzzyness, our research bridges the gap between theoretical musings and the practical realities of musical expression. The fuzzy approach developed here has proved to be not only innovative but also flexible and adaptable, offering potential applications in various fields related to music technology and acoustics.

A Appendix

Calculations based on Definition 2 for several tuning systems used on this research (Tables 6, 7 and 8).

Table 6. Calculation of the 19 notes of the Zarlino Tuning System.

Note	n	r^*	cents	Note	n	r^*	cents
C	0	$\beta_1^0 \cdot \beta_2^0 \cdot 2^0$	0.00	G	-6	$\beta_1^{-4} \cdot \beta_2^{-2} \cdot 2^4$	631.28
C	7	$\beta_1^5 \cdot \beta_2^2 \cdot 2^{-4}$	70.67	G	1	$\beta_1^1 \cdot \beta_2^0 \cdot 2^0$	701.96
D	-5	$\beta_1^{-3} \cdot \beta_2^{-2} \cdot 2^3$	133.24	G	8	$\beta_1^6 \cdot \beta_2^2 \cdot 2^{-4}$	772.63
D	2	$\beta_1^2 \cdot \beta_2^0 \cdot 2^{-1}$	203.91	A	-4	$\beta_1^{-3} \cdot \beta_2^{-1} \cdot 2^3$	813.69
D	9	$\beta_1^7 \cdot \beta_2^2 \cdot 2^{-5}$	274.58	A	3	$\beta_1^2 \cdot \beta_2^1 \cdot 2^{-1}$	884.36
E	-3	$\beta_1^{-2} \cdot \beta_2^{-1} \cdot 2^2$	315.64	A	10	$\beta_1^7 \cdot \beta_2^3 \cdot 2^{-5}$	955.03
E	4	$\beta_1^3 \cdot \beta_2^1 \cdot 2^{-2}$	386.31	B	-2	$\beta_1^{-1} \cdot \beta_2^{-1} \cdot 2^2$	1017.60
E	11	$\beta_1^8 \cdot \beta_2^3 \cdot 2^{-6}$	456.99	B	5	$\beta_1^4 \cdot \beta_2^1 \cdot 2^{-2}$	1088.27
F	-1	$\beta_1^{-1} \cdot \beta_2^0 \cdot 2^1$	498.04	B	12	$\beta_1^9 \cdot \beta_2^3 \cdot 2^{-6}$	1158.94
F	6	$\beta_1^4 \cdot \beta_2^2 \cdot 2^{-3}$	568.72				

Table 7. Calculation of the 12 notes of Meantone 1/4.

Note	n	r^*	cents	Note	n	r^*	cents
C	0	$\beta_1^0 \cdot 2^0$	0	F	6	$\beta_1^6 \cdot 2^{-3}$	579.47
D	7	$\beta_1^7 \cdot 2^{-4}$	76.05	G	1	$\beta_1^1 \cdot 2^0$	696.58
D	2	$\beta_1^2 \cdot 2^{-1}$	193.16	G	8	$\beta_1^8 \cdot 2^{-4}$	772.63
E	-3	$\beta_1^{-3} \cdot 2^2$	310.26	A	3	$\beta_1^3 \cdot 2^{-1}$	889.74
E	4	$\beta_1^4 \cdot 2^{-2}$	386.31	B	-2	$\beta_1^{-2} \cdot 2^2$	1006.84
F	-1	$\beta_1^{-1} \cdot 2^1$	503.42	B	5	$\beta_1^5 \cdot 2^{-2}$	1082.89

Table 8. Calculation of the 12 notes of the Kellner-Bach temperament.

Note	n	r^*	cents	Note	n	r^*	cents
C	0	$\beta_1^0 \cdot \beta_2^0 \cdot 2^0$	0	F	6	$\beta_1^1 \cdot \beta_2^5 \cdot 2^{-3}$	588.27
C	7	$\beta_1^2 \cdot \beta_2^5 \cdot 2^{-4}$	90.22	G	1	$\beta_1^0 \cdot \beta_2^1 \cdot 2^0$	697.26
D	2	$\beta_1^0 \cdot \beta_2^2 \cdot 2^{-1}$	194.53	G	8	$\beta_1^3 \cdot \beta_2^5 \cdot 2^{-4}$	792.18
E	-3	$\beta_1^{-3} \cdot \beta_2^0 \cdot 2^2$	294.13	A	3	$\beta_1^0 \cdot \beta_2^3 \cdot 2^{-1}$	891.79
E	4	$\beta_1^0 \cdot \beta_2^4 \cdot 2^{-2}$	389.05	B	-2	$\beta_1^{-2} \cdot \beta_2^0 \cdot 2^2$	996.09
F	-1	$\beta_1^{-1} \cdot \beta_2^0 \cdot 2^1$	498,04	B	5	$\beta_1^1 \cdot \beta_2^4 \cdot 2^{-2}$	1091.01

References

1. Barbour, J. M.: Tuning and Temperament: A Historical Survey. Courier Corporation (2004)
2. Rahkonen, C.: How Equal Temperament Ruined Harmony (and Why You Should Care) (2008)
3. Halewood, M.: On equal temperament: tuning, modernity and compromise. Hist. Hum. Sci. **28**(3), 3–21 (2015)
4. Milne, A., Sethares, W., Plamondon, J.: Tuning continua and keyboard layouts. J. Math. Music **2**(1), 1–19 (2008)
5. Slevc, L.R., Davey, N.S., Buschkuehl, M., Jaeggi, S.M.: Tuning the mind: exploring the connections between musical ability and executive functions. Cognition **152**, 199–211 (2016)
6. Sloboda, J.A. : The acquisition of musical performance expertise: deconstructing the talent account of individual differences in musical expressivity. In: The Road to Excellence, pp. 107–126 (2014)
7. Geringer, J., Witt, A.: An investigation of tuning performance and perception of string instrumentalists. Bull. Counc. Res. Music. Educ. **85**, 90–101 (1985)
8. Sundberg, J.: How can music be expressive? Speech Commun. **13**(1), 239–253 (1993)
9. Heister, H.W.: Music and Fuzzy Logic: The Dialectics of Idea and Realizations in the Artwork Process. Springer, Heidelberg (2021). https://doi.org/10.1007/978-3-662-62907-9
10. Liern, V.: Fuzzy tuning systems: the mathematics of musicians. Fuzzy Sets Syst. **150**(1), 35–52 (2005)
11. León, T., Liern, V.: A fuzzy framework to explain musical tuning in practice. Fuzzy Sets Syst. **214**, 51–64 (2013)
12. Wood, A.: The Physics of Music. Davies Press, London (2008)
13. André, H., Khelf, I., Leclere, Q.: Harmonic product spectrum revisited and adapted for rotating machine monitoring based on IAS. In: Proceedings of the 9th International Conference Surveillance, Fes, Morocco. Institut International des Sciences Appliquées (2017)
14. Noll, A.M.: Pitch determination of human speech by the harmonic product spectrum, the harmonic sum spectrum, and a maximum likelihood estimate. In: Proceedings of Symposium on Computer Processing in Communication, vol. 19, pp. 779–797. University of Brooklyn Press, New York (1970)
15. Martínez, B., Liern, V.: A fuzzy-clustering based approach for measuring similarity between melodies. In: Agustín-Aquino, O.A., Lluis-Puebla, E., Montiel, M. (eds.) MCM 2017. LNCS (LNAI), vol. 10527, pp. 279–290. Springer, Cham (2017). https://doi.org/10.1007/978-3-319-71827-9_21

Short Papers

Advanced Visualization Techniques for Music Theory

Gilles Baroin[1] and Richard Cohn[2](✉)

[1] University of Toulouse, Toulouse, France
[2] Yale University, New Haven, USA
richard.cohn@yale.edu

Abstract. Authors such as Hook, Tymozcko, and Cohn have used still imagery to model relation systematic pitch correlation and convey new analytical insights. Here we explore the pedagogical advantages of using industry-level graphic software to produce animations in a visual language to which many users are habituated. This paper describes some of the technological resources and the advantages that they afford. It then presents fixed-image representations, and describes animated ones, of analyses of short cyclic progressions of harmonies in songs by Richard Rodgers (as performed by Ella Fitzgerald) and Nick Cave (as adapted and performed by Johnny Cash), as well as a passage of Schubert.

Keywords: Mathemusic · Visualization · Music Theory

1 Introduction

This short paper introduces a forthcoming book, *Mathematical Models of Musical Pitch and Time* (MMTP) [1], on which the authors have collaborated. Similarly to Julian Hook's *Exploring Musical Spaces* [2], *MMTP* is not primarily a presentation of new research, but rather a pedagogically oriented synthesis. *MMTP* is distinct from Hook's extraordinary book in three ways. First, where *EMS* builds on a university-level training in music theory, *MMTP* presupposes readers who only know scales, chords, and basic notation. Second, it presents a mathematical approach to rhythm as well as to pitch, in the form of a sustained textbook-style treatment of beat-class set theory. (That aspect of the book is not featured here.) Third, whereas *EMS* incorporates beautifully rendered black and white still images, MMTP uses color images in the print book, and for a supplementary website to circulate multimedia animations.

The images created for the supplementary website take advantage of the most advanced visualization software currently available, with the intention of meeting students, most of whom have been exposed from an early age to visualizations created for advertising, film, and gaming, at the high level that they have been habituated to expect. The images are designed using virtual reality and CGI (computer-generated imagery), and have several special features, including inverse kinematic calculations, hierarchical joint systems, and the ability to respond to physical forces such as gravity and magnetism. Programs and scripts allow the developer to establish control in real time, altering or

© The Author(s), under exclusive license to Springer Nature Switzerland AG 2024
T. Noll et al. (Eds.): MCM 2024, LNCS 14639, pp. 375–380, 2024.
https://doi.org/10.1007/978-3-031-60638-0_29

reorienting texts, modifying colors and textures, depending on the position of the viewer and the nature of their interactions with the images. Images are generated with Virtual Reality software, which can easily be exported to more conventional formats, including 360-degree movies, 3D movies for flat screens, and 3D printed still images. The exported image can be made available to viewers using a variety of platforms, including virtual reality headsets, personal computers, interactive web pages, phones, and tablets.

Some of the images pertain to topics that are already represented in some undergraduate and post-graduate courses, including atonal pitch-class theory, theory of musical tuning, and neo-Riemann theory. However a priority of this project is to help convince the community that, with appropriate multimedia support, topics often considered too advanced for the undergraduate classroom, including beat-class set theory and the voice-leading spaces introduced by Callender, Quinn, and Tymoczko [3], can be taught, even to students with no university-level training in either music or mathematics.

2 Three Examples

Here we present images that model three short pieces of music, across a variety of styles, including jazz (Ella Fitzgerald), classical (Franz Schubert), and popular (Johnny Cash). The compositions or performances were selected in part because their pace of harmonic change is both slow and consistent, and in part because the progressions are cyclic or sequentially patterned. The slow and consistent speed makes for effective coordination between music and graphics, in such a way that user cognitive processing is not overwhelmed, and the cyclic repetition encourages habituation that supports development of cognitive and performative grasp of the musical relations.

2.1 Richard Rodgers, Blue Moon, as Sung by Ella Fitzgerald

First, we introduce three graphic models of "Blue Moon," a 1934 standard by Richard Rodgers sung by Ella Fitzgerald and recorded in 1957 [4]. It is presented here in Example 1. The entirely diatonic song cycles a standard jazz turn-around progression.

Example 1. "Blue Moon", first line

2.1.1 A Triadic Circle-of-Thirds Graph

Figure 1, on the left, places the progression on a circle of diatonic triads whose roots are ordered by diatonic thirds. Chords adjacent on the circle share maximum pitch classes. The path is legible in the still image, but becomes particularly clear when the cyclical motion is animated in conjunction with the music. The progression is a maximally even division of the diatonic circle, emphasizing the primacy of maximal evenness in various dimensions of musical structure.

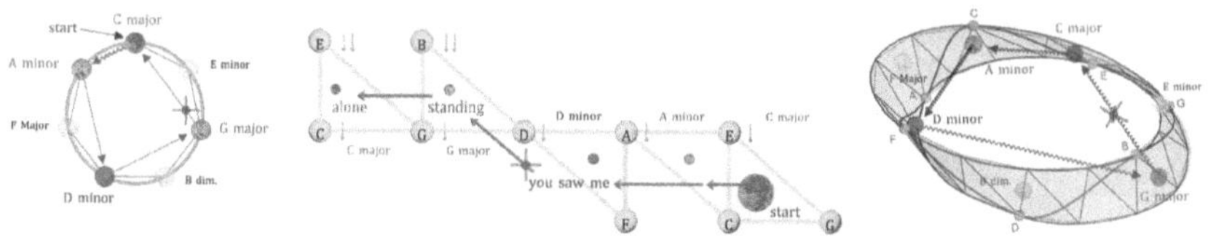

Fig. 1. Three representations of the chords progression in Blue Moon.

2.1.2 An Eulerian *Tonnetz*

The center drawing in Fig. 1 models the same progression on a subset of the Eulerian *Tonnetz*. Horizontal axes are generated by Pythagorean fifths, vertical axes by just major thirds, and triads are represented as right triangles joining their constituent pitch classes. "Blue Moon" is modelled as a right-to-left motion as shown. The upward transfer of levels models the syntonic comma that would arise if the accompanying instruments were not tempered, and emphasizes the difficulty of making music in just intonation, even for such a simple diatonic progression. Again, the progression is legible on the still image, but becomes vivid with animation plus audio.

2.1.3 A Diatonic Möbius Strip

The right part of Fig. 1 models the same progression on a Möbius strip, whose single edge contains seven equally distributed points, representing the white-note pitch classes arranged by diatonic fifths. Connections between pitch classes related by diatonic third, across the interior of the strip, partition its surface into seven triangles of equal size. (The triangles are inevitably distorted, given the non-linear topology of the shape.) A model of "Blue Moon" on a still image of this figure is not particularly instructive, but graphic animations in three dimensions, with changes in orientation together with rotations of the object, allow the viewer to be admitted inside the space; we believe that this provides viewers a unique and profound insight into the nature of diatonic space as it projects a progression that divides that space maximally evenly.

2.2 A Passage from a Schubert Sonata

Example 2 provides a passage from the final movement of Franz Schubert's Piano Sonata in Bb major, D. 960, composed in 1828. The passage, which begins in D minor and ends in F major, consists of a motion through the 1-flat diatonic scale's cycle of thirds, but with two chromatic alternations (indicated with the grey circles) that first substitute Eb for E, and then reverse the substitution, in both cases to avoid diminished triads.

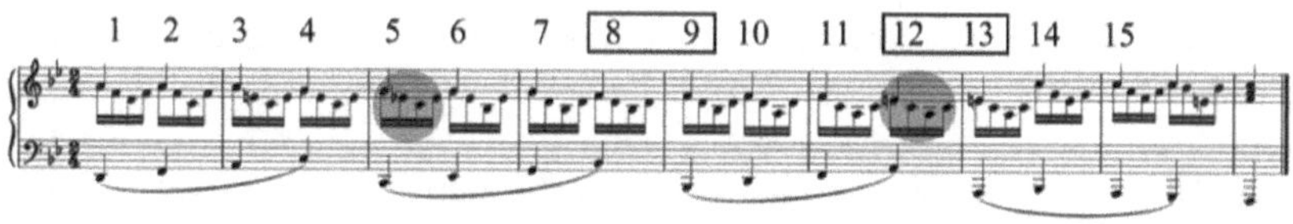

Example 2. Selection from Schubert. D. 960 Finale

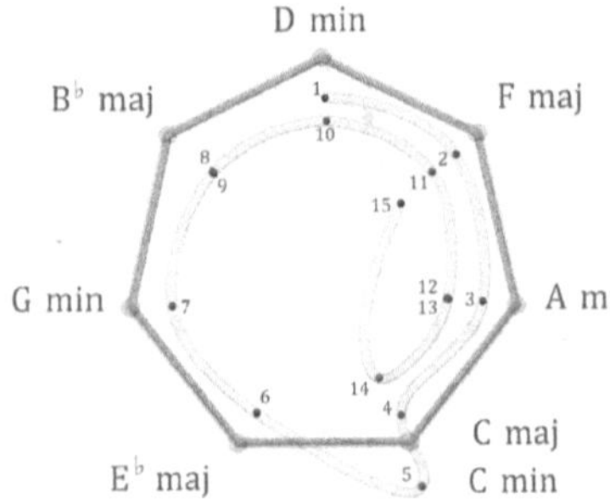

Fig. 2. .

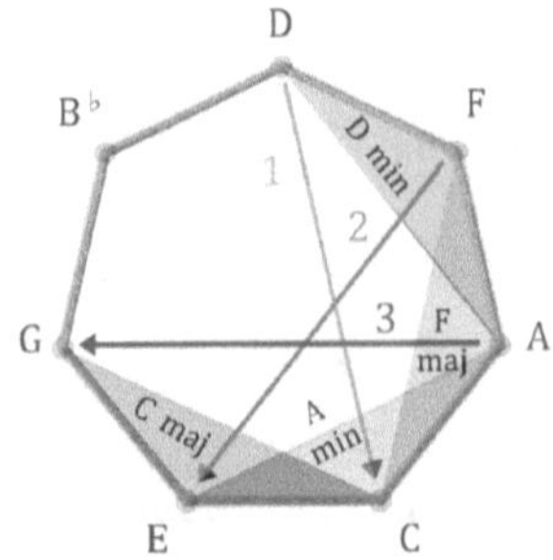

Fig. 3. .

2.2.1 Circle-of_Thirds Models of Schubert

Figure 2 presents the path of Example 2 on a circle of diatonic triads ordered by thirds (the same space as Fig. 1). Figure 3 models the beginning of the progression using a circle of diatonic **pitch classes** ordered by thirds. Here, triads are composed of three consecutive pitch classes on the circle, and triadic progressions are modelled successively by the arrows. Although as a still image Fig. 3 is not completely legible, and becomes quickly cluttered, an animated version of this progression, and Schubert's extension of it, on this space is fully intelligible, highlighting the parsimonious voice leading that is the focus of neo-Riemannian models of triadic chromaticism such as Cohn's *Audacious Euphony* [5].

2.2.2 A Hypertoroidal Tonnetz

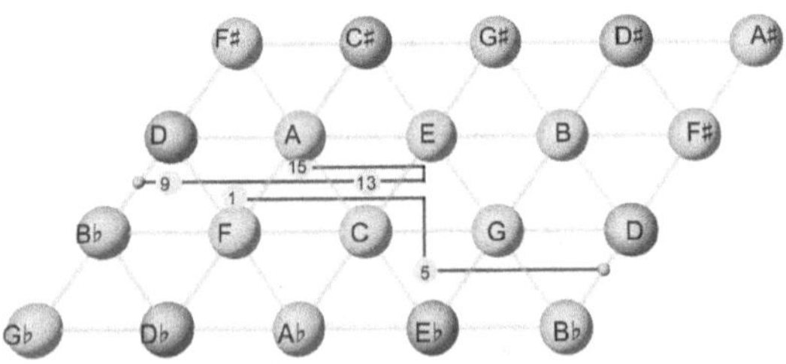

Fig. 4. The Hypertoroidal Tonnetz.

Figure 4 presents the same progression on a hypertoroidal *Tonnetz*, which, unlike the Eulerian *Tonnetz*, assumes equal temperament. Duplications of pitch classes at opposite edges are identified by matched colors. The model again highlights the parsimonious voice leading of diatonic thirds, but also emphasizes that the Eb for E substitution, which transfers down one level, is necessary in order for Schubert to return to his starting region, rather than exiting it to the sharp side by transferring up one level. The reverse of direction also highlights the "braking" effect necessary to transition out of maximally parsimonious sequential logic and back into diatonic framework in preparation for a theme well-grounded in F major.

2.3 Nick Cave, "The Mercy Seat," Performed by Johnny Cash

The final example is from Johnny Cash's 2000 cover of a 1988 song by Nick Cave, "The Mercy Seat," whose lyrics record the perseverations of a guilty prisoner preparing to die by electrocution [6]. The example is presented in Fig. 5; this seven-bar phrase is cycled approximately a dozen times in the second half of the song. The representation beneath the transcription models the passage on Cube Dance. To the right, the model is "unclasped," yielding a version of Cube Dance that serves as the core of three-note voice-leading space introduced in the work of Callender, Quinn, and Tymoczko [3], and explained in greater detail in Tymoczko's *A Geometry of Music* [7] (see also Hook, *Exploring Musical Spaces* [2], chapter 11).

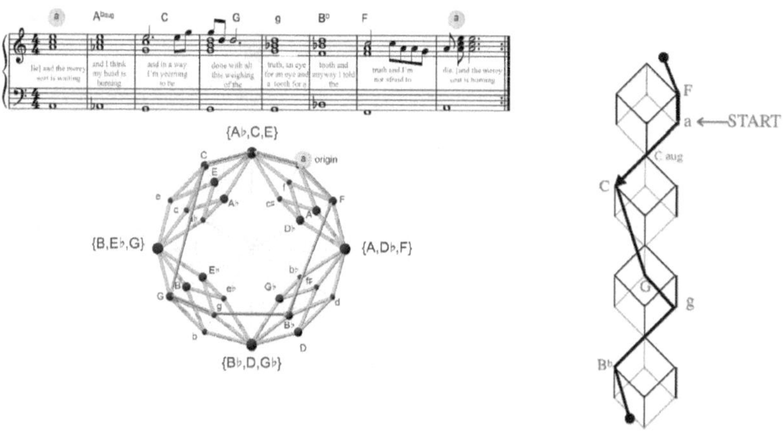

Fig. 5. The cyclic progression of "The Mercy Seat" modelled by two representations of Cube Dance.

3 Summary

The images exhibited here, and throughout this short paper, serve as the basis for animations that take advantage of the affordances of the new technologies, in their most sophisticated and flexible corporate iterations. In addition to allowing music theorists and students the opportunity to more flexibly inspect complex topological musical models, in greater detail than is afforded by still graphics, we hope that students and other young musicians will develop interest in mathematical music theory as a result of being intrigued by advanced-technology graphics and compelling music.

References

1. Cohn, R.: Mathematical Models of Musical Pitch and Time, with Graphics by Gilles Baroin. MIT Press (forthcoming)
2. Hook, J.: Exploring Musical Spaces. Oxford University Press (2022)

3. Callender, C., Quinn, I., Tymoczko, D.: Generalized voice-leading spaces. Science **320**(5874), 346–348 (2008)
4. Cohn, R.: Audacious Euphony: Chromaticism and the Triad's Second Nature. Oxford University Press (2012)
5. https://www.youtube.com/watch?v=T3FFOju3VM0
6. https://www.youtube.com/watch?v=XoJ0bAp4Eo8
7. Tymoczko, D.: A Geometry of Music, chap. 3. Oxford University Press (2011)

Recurrence Relations Rhythm

Michael Cohen(✉) and Yasuyuki Kachi

University of Aizu, Aizu-Wakamatsu, Fukushima 965-0001, Japan
xilehence@gmail.com, kachi@u-aizu.ac.jp

Abstract. Recurrence relations have analogies to 1D tessellations, including rhythmic patterns. We elaborate the first 16 binary recurrence relations to consider such extensions of ordered partition, exploring musical instances of such analogies. The nth term of the sequence defined by an arbitrary binary (i.e., with coefficients 0 or 1) homogeneous (i.e., with no constant term) recurrence relation counts the number of ways of partitioning (a space of size) n as an ordered sum (accumulation) of elements of a fixed set of positive integers (coalescences of unit intervals), corresponding to the non-zero precursive ("historical") terms of the recurrence relation. Rhythmically, such recurrence relation-generated sequences count the number of ways of arranging percussive (tone-oblivious) events with particular durations determined by their respective recurrence relations' precursive terms (antecedent summands). We take the opportunity to mention some characteristics of several of these recurrence relations and our corresponding neologisms: The 2nd binary recurrence relation, and the first nontrivial one, the Fibonacci, is embedded within the 13th recurrence, which we dub the "metaFibonacci." The 5th, Narayana's cows, is embedded within the 14th, so we call it "meta-Narayana's cows." The 6th, the Padovan, is embedded in the 11th, which we call "metaPadovan."

Keywords: Recurrence relations · Number partition · Rhythm

1 Introduction

A sequence is an ordered list of members (terms, elements, items). Sequences can be defined recursively. In the class of linear recurrence relations with constant binary coefficients and no constant term, the first non-trivial instance is the Hemachandra or Fibonacci sequence [4,9], #3 in Table 1. Its recursive definition, expressed by well-known recurrence relation $d_{3,n+1} = d_{3,n} + d_{3,n-1}$, has combinatorial interpretation, namely the number of ways a natural number can be partitioned order-sensitively as a sum of 1 s and 2 s. A rhythmic application is the number of ways a musical phrase with integer length (such as a 4-beat bar) can be orderedly partitioned using only single- and double-unit elements (constituent elements), counting the ways of making a rhythm of fixed length using only single- and double-length pulses [6]. The induction is that for each successive dilation of the pattern, a new pattern of length n will start (WLOG,

© The Author(s), under exclusive license to Springer Nature Switzerland AG 2024
T. Noll et al. (Eds.): MCM 2024, LNCS 14639, pp. 381–386, 2024.
https://doi.org/10.1007/978-3-031-60638-0_30

without loss of generality) with either a single- or double-unit prefix extension, followed by any of the foregoing patterns of length $n-1$ or $n-2$.

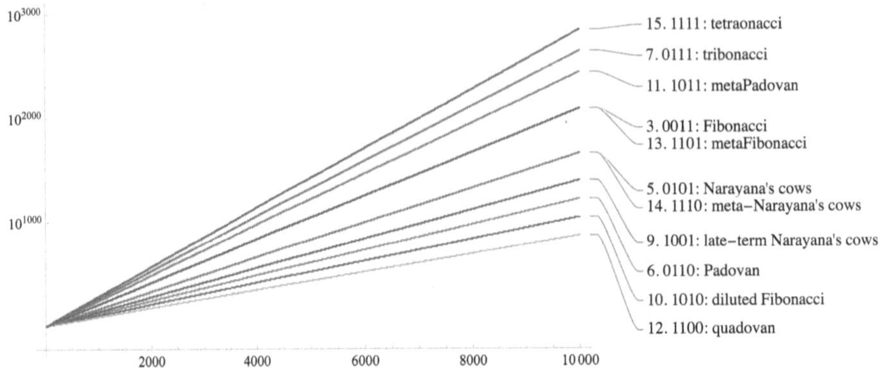

Fig. 1. Log plot of recurrences, indicating asymptotic exponential growth

In Fig. 1, the #3 & #13 sequences appear conflated, as do the #5 & #14 sequences. Even though the distances between respective elements are astronomical, the differences are small compared to the absolute values. In fact, the sequence #13 embeds the #3 Fibonacci sequence within it, since $d_{13,n} + d_{13,n-2} = d_{3,n}$. Therefore the 13th sequence in some sense has more information than the Fibonacci, and we dub it [7] "metaFibonacci." The #9 sequence has structural similarity to #5, Narayana's cows, so we call it "late-term Narayana's cows." #5 Narayana's cows is also embedded in #14, since $d_{14,n} + d_{14,n-1} = d_{5,n}$, so we call #14 "meta-Narayana's cows." #11 embeds the #6 Padovan sequence, since $d_{11,n} + d_{11,n+1}$ yields the odd terms of d_6, and $d_{11,n+2} - d_{11,n}$ yields the even terms of that sequence, so we dub it the "meta-Padovan," although those sequences are not conflated. We are unaware of any embedding by the #12 sequence, and call it "quadovan" after its structural similarity to the Padovan generation.

The so-called "tribonacci" number sequence, #7 in Table 1, corresponds to solutions to the recurrence relation $d_{7,n} = d_{7,n-1} + d_{7,n-2} + d_{7,n-3}$. In the rhythmic metaphor explained for the Fibonacci numbers, a tribonacci number n counts the number of ways a musical passage can be filled using only single-, double-, and triple-unit elements.

The Padovan sequence, #36 in Table 1, is like the Fibonacci sequence, but each of the taps is delayed one step, as defined by the recurrence relation $d_{6,n} = d_{6,n-2} + d_{6,n-3}$, summing the 2nd- and 3rd-last elements (not the immediately previous and 2nd-last), counting the number of ways to partition a 1D structure using only double- and triple-unit elements.

For instance, musically, one can respectively identify elements of length 1 with a quarter note (♩), 2 with a half note (𝅗𝅥), 3 with a dotted half note (𝅗𝅥.) and 4 with a whole note (𝅝). In 4/4 ("common") meter, we elaborate the ways that drum patterns of lengths 1, 2, 3, & 4 can fit a 4-beat bar. (As shown in the

Table 1. Rhythmic partition of 4 by first 16 binary recurrence relations

	Elements: Corresponding Notes	Name	OEIS tag	(4) 𝅝	(3 1) 𝅗𝅥.♩	(2 2) 𝅗𝅥𝅗𝅥	(2 1 1) 𝅗𝅥♩♩	(1 3) ♩𝅗𝅥.	(1 2 1) ♩𝅗𝅥♩	(1 1 2) ♩♩𝅗𝅥	(1 1 1 1) ♩♩♩♩	Recurrence Relation	Characteristic Eqn.	Largest Real Root
0.	{}	(degenerate)	A000004									$d_{0,n} = 0$		
1.	{1}: ♩	(constant)	A000012								♩♩♩♩	$d_{1,n} = d_{n-1}$	$\alpha = 1$	1
2.	{2}: 𝅗𝅥	alternating (period 2)	A000035			𝅗𝅥𝅗𝅥						$d_{2,n} = d_{n-2}$	$\alpha^2 = 1$	1
3.	{2, 1}: 𝅗𝅥♩	Hermechandra, Fibonacci	A000045			𝅗𝅥𝅗𝅥	𝅗𝅥♩♩		♩𝅗𝅥♩	♩♩𝅗𝅥	♩♩♩♩	$d_{3,n} = d_{n-1} + d_{n-2}$	$\alpha^2 = \alpha + 1$	1.62
4.	{3}: 𝅗𝅥.	periodic (period 3)	A079978									$d_{4,n} = d_{n-3}$	$\alpha^3 = 1$	1
5.	{3, 1}: 𝅗𝅥., ♩	Narayana's cows	A000930		𝅗𝅥.♩			♩𝅗𝅥.			♩♩♩♩	$d_{5,n} = d_{n-1} + d_{n-3}$	$\alpha^3 = \alpha^2 + 1$	1.47
6.	{3, 2}: 𝅗𝅥., 𝅗𝅥	Padovan	A000931			𝅗𝅥𝅗𝅥						$d_{6,n} = d_{n-2} + d_{n-3}$	$\alpha^3 = \alpha + 1$	1.32
7.	{3, 2, 1}: 𝅗𝅥., 𝅗𝅥, ♩	tribonacci	A000073		𝅗𝅥.♩	𝅗𝅥𝅗𝅥	𝅗𝅥♩♩	♩𝅗𝅥.	♩𝅗𝅥♩	♩♩𝅗𝅥	♩♩♩♩	$d_{7,n} = d_{n-1} + d_{n-2} + d_{n-3}$	$\alpha^3 = \alpha^2 + \alpha + 1$	1.84
8.	{4}: 𝅝	periodic (period 4)	A121262	𝅝								$d_{8,n} = d_{n-4}$	$\alpha^4 = 1$	1
9.	{4, 1}: 𝅝, ♩	"late-term Narayana's cows"	A003269	𝅝							♩♩♩♩	$d_{9,n} = d_{n-1} + d_{n-4}$	$\alpha^4 = \alpha^3 + 1$	1.38
10.	{4, 2}: 𝅝, 𝅗𝅥	"diluted Fibonacci"	A079977	𝅝		𝅗𝅥𝅗𝅥						$d_{10,n} = d_{n-2} + d_{n-4}$	$\alpha^4 = \alpha^2 + 1$	1.27
11.	{4, 2, 1}: 𝅝, 𝅗𝅥, ♩	"metaPadovan"	A060945	𝅝		𝅗𝅥𝅗𝅥	𝅗𝅥♩♩		♩𝅗𝅥♩	♩♩𝅗𝅥	♩♩♩♩	$d_{11,n} = d_{n-1} + d_{n-2} + d_{n-4}$	$\alpha^4 = \alpha^3 + \alpha^2 + 1$	1.75
12.	{4, 3}: 𝅝, 𝅗𝅥.	"quadovan"	A017817	𝅝								$d_{12,n} = d_{n-3} + d_{n-4}$	$\alpha^4 = \alpha + 1$	1.22
13.	{4, 3, 1}: 𝅝, 𝅗𝅥., ♩	"metaFibonacci"	A006498	𝅝	𝅗𝅥.♩			♩𝅗𝅥.			♩♩♩♩	$d_{13,n} = d_{n-1} + d_{n-3} + d_{n-4}$	$\alpha^4 = \alpha^3 + \alpha + 1$	1.62
14.	{4, 3, 2}: 𝅝, 𝅗𝅥., 𝅗𝅥	meta-Narayana's cows	A013979	𝅝		𝅗𝅥𝅗𝅥						$d_{14,n} = d_{n-2} + d_{n-3} + d_{n-4}$	$\alpha^4 = \alpha^2 + \alpha + 1$	1.47
15.	{4, 3, 2, 1}: 𝅝, 𝅗𝅥., 𝅗𝅥, ♩	tetranacci	A000078	𝅝	𝅗𝅥.♩	𝅗𝅥𝅗𝅥	𝅗𝅥♩♩	♩𝅗𝅥.	♩𝅗𝅥♩	♩♩𝅗𝅥	♩♩♩♩	$d_{15,n} = d_{n-1} + d_{n-2} + d_{n-3} + d_{n-4}$	$\alpha^4 = \alpha^3 + \alpha^2 + \alpha + 1$	1.93

last row of Table 1, there happen to be eight ways in total: four uninterrupted beats, three uninterrupted beats followed by a single beat, two uninterrupted beats followed then two single beats, and so on.) Such analogy can be extended to any binary recurrence relation [3], which implicitly restricts the corresponding elements that can be used for the temporal tiling.

Any linear recurrence relation can be solved to generate its sequence using matrix exponentiation and multiplication by an initial seed. For instance, using a square matrix with size 3 for recurrences with order-3 retrospection, (such Narayana's cows and tribonacci sequences), the Padovan matrix expression is

$$\begin{bmatrix} d_{6,n} \\ d_{6,n-1} \\ d_{6,n-2} \end{bmatrix} = \begin{bmatrix} 0\,1\,1 \\ 1\,0\,0 \\ 0\,1\,0 \end{bmatrix} \begin{bmatrix} d_{6,n-1} \\ d_{6,n-2} \\ d_{6,n-3} \end{bmatrix}, \tag{1}$$

and the generation of the sequence values can be expressed as

$$\begin{bmatrix} d_{6,n} \\ d_{6,n-1} \\ d_{6,n-2} \end{bmatrix} = \begin{bmatrix} 0\,1\,1 \\ 1\,0\,0 \\ 0\,1\,0 \end{bmatrix}^n \begin{bmatrix} d_{6,0} \\ d_{6,-1} \\ d_{6,-2} \end{bmatrix}. \tag{2}$$

Eigenvalues of the matrix are the roots of the characteristic equation, which determine the ratio of adjacent terms of the sequence, which combination is normally dominated in the limit by the largest real root, shown as the right-most column in Table 1. Since the sequence growth is approximately exponential, the slope of each plot line in the semi-log plot of Fig. 1 is the log of the asymptotic ratio $r = d_{n+1}/d_n$ for large n (since $y = r^x \Rightarrow \log y = x \log r$, $(\log y)/x = \log r$). For instance, the ratio of successive terms of the Fibonacci sequence approaches Φ, the so-called "Golden Ratio," because of its appearance in nature [5] and art.

2 DTM (desktop music) Rendering

Musical interpretation of the notion of recurrence relations is a creative undertaking on its own. Many possible mathematical expressions arisen from recurrence relations can be reified musically, including modulation of timbre, intensity, pitch, and spatial position (such as stereophonic panning). In this paper, we use their simplest rhythmic expression, impulsive drum beats (Fig. 2).

These ideas have been explored in DTM (desktop music) DAWs (digital audio workstations), as well as with mathematical symbolic computation. Simple patterns use a bass (kick) drum on the down beats (natural numbers), a snare drum on the backbeats ("&") to indicate continuation (so not between subsequence partition elements), and a clap on termination of each subsequences.

A non-unit integer coefficient corresponds to allowing alternative variations of the same length, for instance by changing the timbre or dynamics of the newly interpolated element. For example, $d_n = 2d_{n-1}$ means that 2 kinds of elements can be introduced (quiet & loud, bass kick & snare, left & right, open & closed cymbal, etc.).

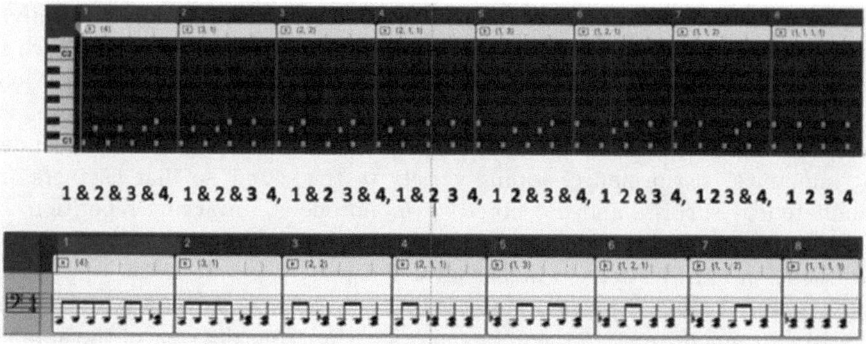

Fig. 2. Elaborations of partitions of 4 in contemporary DAW: GarageBand piano-roll notation, rhythmic counting, and musical score notation

For instance, if a coefficient is 2 instead of 1, one musical equivalence (there are many) would be to double the dynamic range of the intensity or amplitude or volume or velocity loudness say by adding a bit to its encoding. (These attributes are all related but different, but for the purposes of this example, any will do.) For example the possible drum pattern elements remain the same but with twice as many dynamics options, such as *piano* and *forte*. Normal MIDI encoding uses 7 bits (0–127), so such interpretation could be realized for the first seven levels. A recent release of the MIDI standard, MIDI 2.0, allows Voice Channel Messages to be parameterized by velocity of 16 bits [1,2], so the extended dynamic range (higher resolution, finer granularity) is 64K.

3 Discussion

In this contribution, we have explored simple single-instrument rhythmic patterns of fixed length, such as using just a snare drum to express note durations, experimenting with changing the ways that elements of various lengths can pack (exactly fill) a fixed interval, but this consideration applies to any elements of those lengths, including more complicated drumming or melodies. Such analysis applies to musical forms that have a certain rhythmic consistency—such as popular music, rock, jazz, blues—but not so much to longer forms with richer variation such as operas and symphonies. Most popular songs have a duration of about 3 min, such as might be scored on around 30 4-beat bars—with an intro, verse-chorus repetition twice or thrice with a bridge, and outro—and a typical tempo of 120 bpm, so such a "radio-friendly" song, would have a "unrolled" duration of around 100 bars. Since the shortest note in ordinary music is the 64th 𝅘𝅥𝅱 (or rarely the 128th), that is the quantum of temporal resolution, and the total number of such notes would be on the order of 10,000 elements, so that is the upper limit of the domain used in Fig. 1.

The assumption motivating this research, that one might choose to hobble one's musical composition by precluding certain rhythmic elements, is more like writing a novel without certain vowels [10] than it is like painting with only a fixed color palette. Contemporary computerized music composition tools such as DAWs for DTM have many parameters and creative options and settings. Modern sequencers, music programming applications, allow nondeterminism, pseudorandom fluctuation in otherwise repeated patterns. Such variation is intended to "humanize" performance, adding variety to the sound, so that listeners are stimulated by surprise and not bored by monotonous, unvarying repetition. In a general purpose sequencer, such variation could be intentional "jitter," timing variations, or substitutions of higher level elements.

The conceit underlying our inquiry is that musical expressiveness can be characterized by an information theoretic quantification, consideration of the number of possible ways of expressing a particular idea, such as a pattern captured by an underlying rhythm. It is something of a stretch to grant that, for instance, a rhythm composed only of 2- and 3-beat elements (as corresponding to a Padovan recurrence) is somehow richer than one made of 1- and 4-beats (corresponding to late-term Narayana's cows recurrence) because there are more possible ways to arrange the respective elements. It could be that ultimately, such ideas are more interesting in the abstract than in actual listening to real songs, that musical enjoyment defies entropic analysis, and that the essence of transcendence in musical experience is in the "intangibles."

Acknowledgements. We thank Masanobu Kaneko for inspiring original musings about recurrence relations and number theory.

References

1. Details about MIDI 2.0, MIDI-CI, Profiles and Property Exchange (2023). https://www.midi.org/midi-articles/details-about-midi-2-0-midi-ci-profiles-and-property-exchange
2. What Musicians & Artists need to know about MIDI 2.0 (2023). https://www.midi.org/midi-articles/what-musicians-and-artists-need-to-know-about-midi-2-0
3. Cohen, M., Kachi, Y.: Full and partial ordered number partitioning: recurrence relations and their analogy to rhythmic pattern variety. In: Roy, D., Fragulis, G.F. (eds.) Proceedings ETLTC-ICETM International Conference on ICT Integration in Technical Education in Conjunction with International Conference on Entertainment Technology & Management. Aizu-Wakamatsu, Japan (2024)
4. Dickau, R.: Fibonacci numbers (2023). http://www.prairienet.org/~pops/fibboard.html
5. Livio, M.: The Golden Ratio: The Story of Phi, the World's Most Astonishing Number. Broadway Books, New York (2002)
6. Richter-Gebert, J.: Musica! — Music meets Math. science-to-touch (2021). https://apps.apple.com/app/musica-math-meets-music/id1466301336
7. Sloane, N.J.A.: The on-line encyclopedia of integer sequences (1989). https://oeis.org
8. Weisstein, E.W.: Binary Bracketing. MathWorld–A Wolfram Web Resource (2023). https://mathworld.wolfram.com/BinaryBracketing.html
9. Weisstein, E.W.: Fibonacci Number. MathWorld–A Wolfram Web Resource (2023). https://mathworld.wolfram.com/FibonacciNumber.html
10. Wright, E.V.: Gadsby. Wetzel Publishing Co., Los Angeles (1939)

Mining Significant Sequential Contrast Patterns

Darrell Conklin[1,2(✉)]

[1] Department of Computer Science and Artificial Intelligence, University of the Basque Country UPV/EHU, San Sebastian, Spain
darrell.conklin@ehu.es
[2] IKERBASQUE, Basque Foundation for Science, Bilbao, Spain

Abstract. Contrast pattern mining is the process of finding patterns whose frequency differs significantly from one group of data to another. This paper presents an anti-monotonic constraint that is also a necessary condition for pattern statistical significance. This reduces the search space and enables statistical significance to be used as the primary measure of pattern interest. The effectiveness of the method is evaluated on a dataset of folk tune families.

Keywords: sequential pattern mining · contrast pattern mining · significant patterns · data mining

1 Introduction

Pattern mining is a large field of study concerned with finding regularities in data, hopefully those that have some functional significance and extrapolate to future data. The paradigmatic application of pattern mining is *association rule mining*, aiming to find patterns that have high support in the data, and high confidence when oriented into antecedent/consequent rules. Contrast data mining [4], also referred to as *emerging pattern mining* [5,7] and *supervised descriptive rule mining* [10], is a variant of rule mining where only one fixed attribute is used in the consequent of rules. The goal is to discover patterns that have significantly differing support in the positive and negative datasets, and these need not have high support or confidence. Several researchers have adapted *sequential pattern mining* [6] for finding contrast patterns in sequential data [1–3,9].

In sequential pattern mining, algorithmic issues and computation time are paramount, due to the potentially large search spaces involved: from F features one can form F^{ℓ} possible patterns of length ℓ. Pattern mining methods manage this complexity with *anti-monotonic constraints* [11] to restrict both the search space and the number of patterns reported. The standard constraint is *minimum support* (any specialization of an infrequent pattern will be infrequent), but this will be specific to a particular dataset and pattern representation, and support alone is not a good measure of pattern interestingness. Statistical measures are

© The Author(s), under exclusive license to Springer Nature Switzerland AG 2024
T. Noll et al. (Eds.): MCM 2024, LNCS 14639, pp. 387–392, 2024.
https://doi.org/10.1007/978-3-031-60638-0_31

preferable, but are generally not anti-monotonic. This paper presents an anti-monotonic constraint that is also a necessary condition for pattern statistical significance. The method is illustrated on Dutch folk tune families [12].

2 Methods

This section first presents the notation used for contrast pattern mining. Then the anti-monotonic p-value constraint will be introduced. This method follows the p-value bounding method previously introduced in the context of association rule mining [8]. Here this bound is rephrased as an anti-monotonic constraint and is adapted for contrast data mining for sequential patterns.

2.1 Definitions and Notation

Feature. A *feature* of an event is an attribute/value pair: for example, the pitch of a musical note or its interval from the previous note.

Sequence. A sequence is a list of features.

Dataset. A dataset $\mathcal{D}$ is a set of sequences, assumed to be partitioned into a positive set γ, called a *corpus*, and a negative set $\gamma' = \mathcal{D}\setminus\gamma$, called an *anticorpus*.

Sequential Pattern. A sequential pattern is a list of features. A pattern ϕ *occurs* in a sequence or another pattern ψ if it is contiguous subsequence of ψ. This relation is written $\phi \sqsubseteq \psi$ and it is said that ψ is a *specialization* of ϕ (if it is a pattern) or an *instance* of ϕ (if it is a sequence). The *extension* of a pattern ϕ in a dataset $\mathcal{D}$ is all instances of the pattern, and is written $\mathcal{E}_\mathcal{D}[\phi]$. The *empty pattern* Λ has the extension $\mathcal{E}_\mathcal{D}[\Lambda] = \mathcal{D}$.

Counting. To count pattern instances, the following notation is used: $n_{\gamma\phi} = |\mathcal{E}_\mathcal{D}[\phi] \cap \gamma|$, the number of positive instances of the pattern; $n_\phi = |\mathcal{E}_\mathcal{D}[\phi]|$, the number of instances of the pattern in the entire dataset; $n_\gamma = |\gamma|$, the number of sequences in the corpus; and $n_\mathcal{D} = |\mathcal{D}|$, the number of sequences in the dataset.

Contrast Pattern. A contrast pattern is one that occurs in significantly more sequences than would be expected. Here we consider only patterns that are *over-represented* in the corpus (see Fig. 1).

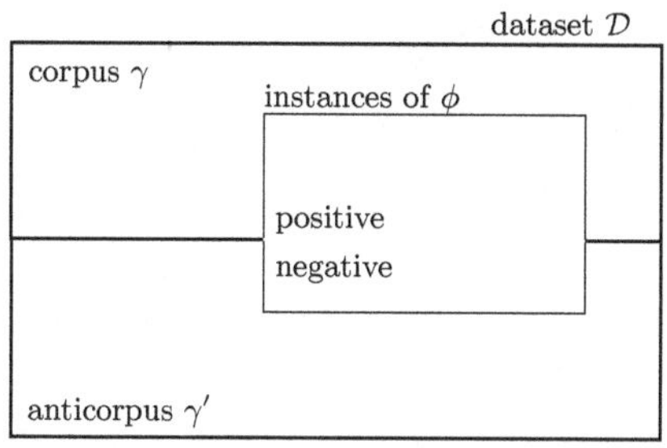

Fig. 1. Schematic view of contrast pattern mining. The goal of contrast pattern mining is to find over-represented patterns ϕ: where the proportion of positive instances in γ is greater than the proportion of negative instances in γ'.

2.2 Pattern Significance

Fisher's exact test can be used to evaluate whether a contrast pattern is significantly over-represented [8]. This test uses the hypergeometric distribution as the null distribution. A random variable X has a hypergeometric distribution, written $X \sim \mathrm{Hyp}(n, m, N)$, if it has the probability mass function:

$$P(X = k) = h(k, n, m, N) = \frac{\binom{m}{k}\binom{N-m}{n-k}}{\binom{N}{n}} \ , \tag{1}$$

which gives the probability of k successes in a sample of size n, when there are m possible successes in the population of size N.

The upper cumulative distribution of $X \sim \mathrm{Hyp}(n, m, N)$ gives the *p-value*: the probability of k *or more successes* in the sample:

$$P(X \geq k) = 1 - \sum_{i=0}^{k-1} P(X = i) \ . \tag{2}$$

To apply this to contrast pattern mining, we consider the random variable $X_{\phi}^{\gamma} \sim \mathrm{Hyp}(n_{\phi}, n_{\gamma}, n_{\mathcal{D}})$ modelling the number of positive instances (in γ) when drawing a sample of n_{ϕ} instances. A pattern ϕ is considered *significant*, written $i(\phi)$, if it is unlikely that $n_{\gamma\phi}$ or more positive instances could arise from random choice: that $P(X_{\phi}^{\gamma} \geq n_{\gamma\phi}) \leq \alpha$ for a specified significance level α.

As an example of a significance calculation, consider the following melodic interval contrast pattern and example instantiation:

interval : $[-1, -2, 0, -2, 0, 2]$

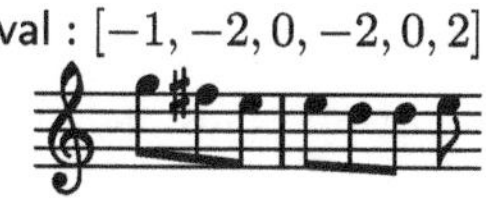

This pattern ϕ is characteristic and discriminative of a tune family γ of Dutch songs (see Results), with $n_{\gamma} = 17$, $n_{\phi} = 16$, $n_{\gamma\phi} = 15$, and $n_{\mathcal{D}} = 360$. It is the most significant pattern found in this group: according to (2) it has a low p-value of 1.7e-23.

2.3 Anti-monotonic Constraints

Sequential pattern mining algorithms search a tree of patterns from general (shortest) to specific (longest) starting at the empty pattern. For contrast patterns, the group γ is specified in advance. Before testing a pattern for significance and specializing it by appending all possible features, it is evaluated whether the search needs to be continued. This is done by checking whether the pattern satisfies *anti-monotonic* constraints [11]:

Definition 1. *An* anti-monotonic constraint *is a predicate C with the property that for all patterns $\phi \sqsubseteq \phi'$, $\neg C(\phi) \Rightarrow \neg C(\phi')$.*

By definition, if a pattern does not meet an anti-monotonic constraint, then neither can any more specific pattern. This can enable the pruning of entire branches of the search space.

2.4 Necessary Conditions on Pattern Significance

Statistical significance $i(\phi)$ alone is not anti-monotonic (a specialization of a pattern could have either a higher or lower p-value), but we can define a constraint that is both anti-monotonic and also a necessary condition for pattern significance:

$$C(\phi) = h(n_{\gamma\phi}, n_{\gamma\phi}, n_\gamma, n_\mathcal{D}) \leq \alpha \ . \tag{3}$$

Theorem 1. *C is an anti-monotonic constraint, that is, $\neg C(\phi) \Rightarrow \neg C(\phi')$, for any specialization ϕ' of ϕ.*

Proof. From (1) we see that $h(k', k', m, N) \geq h(k, k, m, N)$ when $k' \leq k$. Since $\mathcal{E}_\mathcal{D}[\phi'] \subseteq \mathcal{E}_\mathcal{D}[\phi]$, we have $h(n_{\gamma\phi'}, n_{\gamma\phi'}, n_\gamma, n_\mathcal{D}) \geq h(n_{\gamma\phi}, n_{\gamma\phi}, n_\gamma, n_\mathcal{D})$, therefore $\neg C(\phi) \Rightarrow \neg C(\phi')$. □

Theorem 2. *C is a necessary condition for pattern significance, that is, $i(\phi) \Rightarrow C(\phi)$.*

Proof. From (1) we see that $h(k, n, m, N) \leq h(k, n', m, N)$ when $n \leq n'$. Since $n_{\gamma\phi} \leq n_\phi$, we have $h(n_{\gamma\phi}, n_{\gamma\phi}, n_\gamma, n_\mathcal{D}) \leq h(n_{\gamma\phi}, n_\phi, n_\gamma, n_\mathcal{D})$. So if $\neg C(\phi)$, then from (2) it follows that $P(X_\phi^\gamma \geq n_{\gamma\phi}) > \alpha$, so $\neg C(\phi) \Rightarrow \neg i(\phi)$. □

Putting these two theorems together, we have:

Corollary 1. *If $\neg C(\phi)$, then no specialization ϕ' of ϕ can be significant.*

Proof. From Theorem 1 we have $\neg C(\phi) \Rightarrow \neg C(\phi')$ and by Theorem 2 we have $\neg C(\phi') \Rightarrow \neg i(\phi')$, so by implication chaining we have $\neg C(\phi) \Rightarrow \neg i(\phi')$. □

It also follows immediately from Corollary 1 that certain features can be entirely removed from consideration from the search:

Corollary 2. *For a feature f: if $\neg C(f)$, then f cannot participate in a significant pattern.*

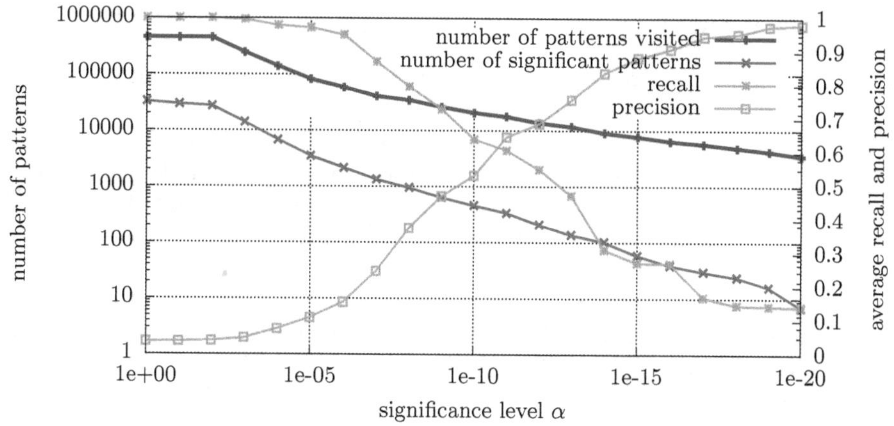

Fig. 2. Left axis: number of significant patterns, and number of patterns visited. Right axis: the recall and precision of the pattern set averaged over all tune families.

3 Results

This section presents some results of the significant pattern mining method on the Meertens Tune Collections [12]: selected subsets of a larger collection of Dutch songs from the Middle Ages through the 20th century. A subset of 360 tunes in this collection was annotated and grouped into 26 tune families according to melodic similarity, and was given an extensive encoding into melodic features [12]. Here we use the melodic interval attribute which represents the distance, in semitones, between the note and its previous note. There are 23 distinct interval values present in the dataset. All patterns were required to have a minimum of two positive instances.

The effectiveness of the method can be measured by the total number of patterns visited (tested for statistical significance) during the search. This gives a good indication of the relative runtime of the performance under varying levels of α. For each α, every tune family was used as a corpus, with the remaining sequences as the anticorpus. The number of patterns visited, and the total number of significant patterns that exist at that α, were summed for all families. Additionally, for each family the recall (proportion of distinct family members covered by all significant contrast patterns for that family) and precision (proportion of sequences covered that are family members) were computed. The mean of the precision and recall over all families is computed for each α.

Figure 2 shows the results. It can be seen that as α decreases, more search branches are pruned, with two orders of magnitude (100x) fewer patterns visited for low α than for $\alpha = 1$ (which has no pruning). At low α most branches can be pruned and relatively few patterns are visited. As expected, the recall and precision display an inverse relationship. As expected, the total number of significant patterns decreases as α decreases.

4 Discussion

Mining for sequential patterns is difficult without some anti-monotonic constraint on the search space. This paper presented a way to evaluate the significance of sequential contrast patterns, and more importantly, showed how a lower bound of significance, defined earlier for association rule mining [8] can be used within an anti-monotonic constraint for sequential pattern mining. A significant reduction can be achieved in the number of patterns considered during sequential pattern mining.

An advantage of this method is that it circumvents the need to specify a minimum acceptable support for patterns. The single parameter α is not as sensitive as minimum support to a particular dataset, the same α can be used across different datasets, and patterns can be ranked by statistical significance.

Disclosure of Interests. The author has no competing interests to declare that are relevant to the content of this article.

References

1. Chan, S., Kao, B., Yip, C.L., Tang, M.: Mining emerging substrings. In: Proceedings of the Eighth International Conference on Database Systems for Advanced Applications, pp. 119–126. Washington, DC, USA (2003)
2. Conklin, D.: Discovery of distinctive patterns in music. Intell. Data Anal. **14**(5), 547–554 (2010)
3. Deng, K., Zaïane, O.R.: An occurrence based approach to mine emerging sequences. In: 12th International Conference on Data Warehousing and Knowledge Discovery, Bilbao, Spain (2010)
4. Dong, G., Bailey, J. (eds.): Contrast Data Mining: Concepts, Algorithms, and Applications. Chapman and Hall/CRC, Boca Raton (2012)
5. Dong, G., Li, J.: Efficient mining of emerging patterns: discovering trends and differences. In: Proceedings of the Fifth ACM SIGKDD International Conference on Knowledge Discovery and Data Mining, KDD '99, San Diego, pp. 43–52 (1999)
6. Fournier-Viger, P., Lin, J.C.W., Kiran, R.U., Koh, Y.S., Thomas, R.: A survey of sequential pattern mining. Data Sci. Pattern Recogn. **1**(1), 54–77 (2017)
7. García-Vico, A., Carmona, C., Martín, D., García-Borroto, M., del Jesus, M.: An overview of emerging pattern mining in supervised descriptive rule discovery: taxonomy, empirical study, trends, and prospects. WIREs Data Min. Knowl. Discovery **8**(1), e1231 (2018)
8. Hämäläinen, W.: Kingfisher: an efficient algorithm for searching for both positive and negative dependency rules with statistical significance measures. Knowl. Inf. Syst.: Int. J. (KAIS) **32**, 383–414 (2012)
9. Ji, X., Bailey, J., Dong, G.: Mining minimal distinguishing subsequence patterns with gap constraints. Knowl. Inf. Syst. **11**(3), 259–296 (2007)
10. Novak, P.K., Lavrač, N., Webb, G.I.: Supervised descriptive rule discovery: a unifying survey of contrast set, emerging pattern and subgroup mining. J. Mach. Learn. Res. **10**, 377–403 (2009)
11. Pei, J., Han, J., Wang, W.: Constraint-based sequential pattern mining: the pattern-growth methods. J. Intell. Inf. Syst. **28**(2), 133–160 (2007)
12. Van Kranenburg, P.: Melodic Features for Meertens Tune Collections and ESSEN Folksong Collection (2019). https://doi.org/10.5281/zenodo.3551003

Bits and Beats: Computing Rhythmic Information as Bitwise Operations Optimized for Machine Learning

Fernando Gualda(✉)

UFRGS, Porto Alegre, RS 90020-180, Brazil
gualda@ufrgs.br

Abstract. Effective Machine Learning algorithms require meticulous data preparation. In contemporary microprocessors, ensuring predictable and efficient memory access is crucial to mitigate cache misses. Therefore, compact data structures with high information density are crucial for optimal computation. This paper discusses the interplay of musical meter and microtiming, as long-term expectation of pattern repetition and short-term deviations. A review of musicological, cognitive and computational models is presented. The literature regards meter perception of rhythmic patterns as a manifestation of human attentive behavior, which is also culturally dependent. Additionally, the paper discusses dissimilarity measures of rhythmic patters. The author introduces two algorithms featuring normalized binary representations of rhythmic patterns of inter-onset intervals, activation functions for computing linear, and logarithmic time differences. Bitwise operations are implemented for the bit-sequences of rhythmic patterns. The author offers an implementation of mutual information calculation as bitwise operations.

Keywords: Rhythm · Machine Learning · Vector Embedding · Activation Functions · Binary Neural Networks · Music Information Retrieval

1 Introduction

This paper proposes representing rhythmic patterns as binary data for machine learning applications. Timing in rhythmic patterns can be represented either *symbolically*, typically as integer proportions in a musical score, or *empirically* estimated from performances, often measured in milliseconds. Although both representations of timing can be converted into binary data, their treatment differs: performances that differ in their onset timings in milliseconds might still correspond to the same symbolic timing representation.

An emergent property of rhythmic patterns is the metric structure: some onsets might be (or might be felt) more prominent than others, giving rise to a sense of hierarchy within those onsets, known as meter. Beats are another

© The Author(s), under exclusive license to Springer Nature Switzerland AG 2024
T. Noll et al. (Eds.): MCM 2024, LNCS 14639, pp. 393–404, 2024.
https://doi.org/10.1007/978-3-031-60638-0_32

emergent property, experienced by humans while listening to music. A common experience is the urge to tap along or move the body with the music. In symbolic representations, beats are defined by metric division, while in music performances, beats can only be inferred.

2 Rhythmic Patters, (Micro)Timing and Meter

Rhythm, defined as a pattern of inter-onset-intervals (IOI), occurs across a wide range of time periods. According to [1] musical time scales can be divided into *Supra* (days, years), *Macro* (minutes, hours), *Meso* (infrasonic[1] periods), *Sound objects*, and *Micro*.[2] In this paper, we consider sound objects as any acoustic phenomenon, the experience of which listeners may be able to describe.

Microtiming, though perceptible, might remain an indiscernible phenomenon at the level of individual onsets. Their combinations are related to expressiveness, entrainment, and musicians' technical abilities at the *meso* structure level (e.g. *swing*, *rubato*, *accelerando*, *melody lead*). Some aspects of this meso structure level are represented as meter and have been the focus of much research on rhythm in musical scores. Other aspects, however, are performance practices specific to certain musical eras, genres, or styles, containing irregularities not entirely possible to represent symbolically.

Timing microstructures typically occur in performances. In the 1990s, [2] discusses *normative* and *individual* aspects of microtiming. Normative timing microstructures refer to the expected timing structure to be communicated by the performer (e.g. *rallentando* should not accelerate), whereas the individual aspect refers to patterns that differ from most other performers. The literature often refers to microtiming as *expressive timing*, thus assuming intentionality, as in [3], who states that in regard to structural and expressive properties of musical rhythm, "the listener separates this temporal information into rhythmic categories and expressive timing" (p. 343).

The relationship between rhythmic structure, cognition and representation is incredibly rich and complex. The scientific literature is extensive, and authors vary significantly in their approaches to research. This diversity is emphasized by [4], whose author concludes that "meters and rhythms are among the most fluffy concepts in musicology." This poignant criticism might seem rightful, though. Within an interdisciplinary field of mathematics, computation, and music, it might appear essential to refrain from ill-defined problems. Because, in order

[1] Sound, infrasound and ultrasound refer to the range humans perceive acoustic waves. Sound occurs between roughly 20 Hz to 20 kHz; whereas infrasounds, below 20 Hz, and ultrasounds, above 20 kHz. Melodic sequences of musical notes always occur within infrasonic periods. Even very fast melodies do note present more than 20 notes per second (50 ms. IOIs).

[2] [1] proposes a time scale from *Infinite* as "the ideal time span of mathematical durations such as the infinite sine waves of classical Fourier analysis", *Sample* as "the atomic level of digital audio systems", up to *Infinitesimal* as "the ideal time span of mathematical durations such as the infinitely brief delta functions" (p. 4).

to be computable, musical concepts need to be sufficiently defined to a point at which data can be effectively processed by algorithms. On the other, within the broader perspective of digital humanities, the purpose of computation is to provide humans with more powerful tools to better understand music making, artistic expression, affective communication, meaning attribution, etc. All inherently ill-defined problems that, nevertheless, need to be addressed.

2.1 Brief Musicological Perspective

One of the most cited musicological approaches to music research has been proposed by [5], a seminal work that makes a distinction between intrinsic and extrinsic rhythmic structures referring to the sound objects themselves and their metric structure. Software for identifying metric structures has been developed by [4]. Nevertheless, not always there is a score into which the data can be fit (e.g. improvisation); many musical works present more than one published score, and during practicing sessions, musicians may simplify or change some excerpts according to their learning habits.

Moreover, many works of Western (classical) musical repertoire might exhibit hypermetric structures [6]. [7] provides discussions on the differences of additive rhythm and divisive rhythm with emphasis on superimposed two and three subdivisions, referred to as *African Hemiola Style*. Also, from an ethnomusicological perspective, rhythms cannot be divorced from their collective meaning; social norms and ways of life within a social group are integral to the concept of rhythm [8]. Additionally, in numerous musical genres, *microtiming* variations are an intrinsic part of the notion of rhythm itself, as evident in genres like Jazz [9,10] and in [11].

Toussaint [12] offers a thorough examination of rhythms using diverse corpora that include both Western and non-Western music. Reviewers [13], [14] commend his significant contributions to the intersection of mathematics and music. Though, [13] criticizes several flaws in [12], specially in regard to the inappropriate application of mathematical tools to describe musical phenomena. [15] presents an extensive review of cognitive theories, some of which incorporate algorithms. The musicology of the 21st century may increasingly rely on mathematics and computation, given the rapid technological advancements in recent decades. Nevertheless, the human capacity to create and appreciate music is profoundly influenced by the interplay of human biology and culture [16], which evolves at a much slower pace.

2.2 Brief Cognitive Perspective

The experience of rhythm that emerges in the human mind [17] is very complex, modulated by cumulative effects [18]. Despite millennia of discussion and study on rhythm, the literature often emphasizes the limited understanding of its complexity. Studies in music cognition in the late 20th century [19,20] often propose theories rooted in statistical inference drawn from listening experiments. [21]

views the perception of meter in rhythmic patterns as a manifestation of human attentive behavior.

Studies within Systematic Musicology [17,22] rely on interdisciplinary research based on the interplay of acoustic and psychoacoustics studies, digital signal processing, and neuroscientific evidence, such as Neural Resonance Theory [23]. A theory for expectation in music listening, presented in [24], outlines five stages: Imagination, Tension, Prediction, Reaction, and Appraisal (ITPRA), merging aesthetic considerations with information flow [25]. Recently proposed theories, like IDyOM and IDyOT, postulate information dynamics in the study of cognition [25] of music structures.

2.3 Representations of Meter, Beat, and Micro (Expressive) Timing

Most representations of rhythm focus on meter using cyclic timelines (necklace notation [26]). Meter is commonly described as a metric hierarchy. A thorough review on *just noticeable differences* (JND) for distinguishing rhythmic patterns is provided by [21]. The thresholds vary across different studies, with the range of 200 to 250 ms emerging as a significant threshold. The distinction between short (200 to 300 ms) and long (450 to 900 ms) durations suggests the grouping (chunking) of faster onsets. [21] concludes that "one of the functions of musical meter is to mediate among these qualitatively different durations by integrating them into a coherent attentional framework." [27] discusses how different authors define metric complexity, usually as the interplay of integer divisions of bars into regular onsets (between 300 to 900 ms).

The emergence of the *beat* (or *pulse*), as a salient, psychological effect, happens regardless of recurrent occurrences of onsets on downbeats. Additionally, the grouping within sequences of identical stimuli into beats, already described as *subjective rhythmization* since the late 19th century [21], suggests that the experience of beat is an expectation of regularity, not a perception of an actual physical phenomenon. Though it arises from the experience of rhythm, it is not driven by direct stimuli.

Microtiming is typically described as small variations from the score, also referred to as *expressive timing*, when intentional. During music performances, onsets that were expected to happen simultaneously, according to the score, rarely occur precisely at the same moment. Despite this, listeners may perceive them as if they did, as the time difference between these onsets is so minimal that it becomes indiscernible. The JNDs for recognition of small deviations within IOIs in isochronous rhythms has been reported as 5.2% at 100 ms (5 ms), up to 2.6% at 1000 ms (26 ms) [28]. Microtiming also contribute to the emergence of *rubato* (horizontal temporal fluctuations) and *melody lead* (vertical temporal fluctuations) as demonstrated by [29].

The normalized Pairwise-Variability Index (nPVI), was first applied to music examples by [30]. Since then, numerous academic papers have been published using nPVI as a dissimilarity measure in both music and speech. Notably, there are distinctions in the data between these domains: music corpora typically contain score representations without *microtiming*, while speech naturally involves irregular rhythmic variations. The nPVI has received criticism for not describing "more concrete empirical descriptions of specific rhythmic features" [31]. Nevertheless, it has provided effective results in identifying meaningful patterns in large corpora [32].

Since the nPVI does not depend on any prior assumptions about rhythmic structure, it solely computes IOIs of musical excerpts. Similarly, phase space plots of consecutive IOIs produce geometrical visualizations of rhythmic patterns "without making any *a priori* assumptions concerning isochrony, beat induction, or metrical hierarchies." [33] This approach can depict 3 consecutive IOIs as 2D images, provided that their total duration is constant, which is seldom the case when microtiming is present.

3 Bitwise Rhythmic Representations

Rather than assuming *a priori* metric representations, including bar measures and beats, this paper approaches rhythmic representation simply as irregular patterns, the structure of which should be inferred from the data itself. This is because algorithms for rhythmic analysis could identify multiple rhythmic structures within the same dataset.

Two rhythmic aspects are relevant for this paper. Firstly, since the metric structures can be hypermetrical [6], hierarchical (even presenting fractal subsymmetries [21]), and either additive or divisive, many periodic repetitions might be identifiable by listeners. Consequently, all these competing, concurrent expected repetitions are possible *beats* and need to be representable. Secondly, *microtiming* is an invaluable aspect of music performance, and should not be underestimated. Therefore, research on rhythm should model both expected and dynamically changing structures.

In the field of Machine Learning, Convolutional Neural Networks have found widespread application across diverse industries, significantly transforming society. However, processing multiple layers of data using 32-bit floating-point arithmetic with millions of operations is rather expensive. Binary Neural Networks (BNN) is a recent development that reduces the requirements of memory usage and matrix operations. Its hidden layers are represented with just 1-bit operations [34–37].

Given that the cyclic timeline (necklace notation) is a commonly utilized rhythmic representation, the proposed bitwise representations for rhythmic patterns could potentially be adopted within the music research community. Cyclic timelines divide the circumference into multiples of two or tree, depending on the meter. Therefore, a bitwise representation of 48 or 60 bits should cover the majority of divisive meters. Fortunately, modern computers, including most smartphones and edge devices, inherently process 64-bits, providing an efficient platform for these representations in BNNs.

3.1 Activation Functions and Binarization

Rectified Linear Units (ReLU) are examples of continuous piecewise linear functions, alongside Gaussian Error Linear Unit (GeLU), logistic sigmoid, hyperbolic tangent and Softmax [38,39]. Activation Functions (AF) introduce non-linearities into a neural network, enabling it to learn (approximate) complex patterns within the data. Equation 1 presents the ReLU AF, the image of which, is always constant below a threshold, in this case: zero. Values below a threshold do not activate deeper layers in the neural network, hence the name "activation function."

$$f(x) = max(0, x) \tag{1}$$

In the specific case of rhythmic patterns represented as binary digits, each bit set to one holds musical significance. Consequently, there is no necessity to further binarize the input data. When the bit length matches that of the first layer of the network, it can be seamlessly incorporated into the network as its initial layer. This approach achieves optimal efficiency as there is no additional processing required.

In BNNs, both weights and activations are represented as binary digits. Consequently, the binarization process is typically applied to both within the internal layers. The binarization of input data, however, ensures that the data aligns with the binary representation used throughout the network, facilitating efficient computations and reducing memory requirements. In addition to deterministic binarization, which uses a *Sign*( ) function to convert input data into +1 or −1, [34] introduced a new AF that incorporates stochastic binarization. In Eq. 3, σ(x) is the *hard sigmoid function*. In Eq. 2, x^b represents a binarized variable:[3]

$$x^b = \begin{cases} +1, & p = \sigma(x) \\ -1, & 1 - p \end{cases} \tag{2}$$

$$\sigma(x) = max(0, min(1, (x + 1)/2))) \tag{3}$$

3.2 Binary Representations of Rhythmic Patterns

Rhythmic patterns may represent two different musical data: score and performance. Corpora containing score representations provide clearly defined rhythmic divisions. In this case, each bit could represent the smallest division presented in the data, or a minimal time resolution within which the score could be quantized - provided the addition of quantization error is considered. For most of the repertoire written as score, discretizing it into binary digits can result in regular differences between bits, resembling the necklace representation. However, instead of a single circle, the entire musical excerpt becomes a stream of *ones* and *zeros*. The cycles need to be inferred.

[3] superscript b is not an actual exponent, but the notation used in [34] to represent x as a single binary digit.

In the case of data representing music performance, however, *microtiming*, *rubato*, and *melody lead* might be the meaning within the data that needs to be approximated. Therefore, it becomes essential to represent the data within multivariate time frames. When the conveyed information about onsets to the BNN are presented within the JNDs for rhythmic perception, the classifications that result from the BNN might more meaningful for human interpretation than just isolated numbers, such as changes in musical *tempo* in *beats per minute* (BPM) or timing changes (Δt) in ms.

Concerning data acquisition, regular sampling of the rhythms into a bit-stream can processed as if it were a digital signal. Then the BNN operated like a digital filter, regularly providing output results. While this method ensures predictable time, it may become computationally expensive, despite the inherent efficiency of BNNs. Conversely, sampling each discrete onset can produce a rhythm pattern in a frame: either a time window (e.g. 1000 ms) or a fixed-cardinality set (e.g. every n onsets).

3.3 Activation Functions for the Binarization of Rhythmic Patterns

Now, let us consider three potential representations for rhythms: (1) regularly sampled bits, (2) exponential time, and (3) step functions. Figure 1 provides three possible mappings of onsets into sequences of binary numbers. They are examples of activation functions that binarize the onsets into specific bits. In this figure, the time-window of each bit is depicted in the vertical axis, offering a visualization of the first difference. Regularly sampled windows at a constant rate provide a linear-time representation of a local rhythm time-window in the form of a bit-sequence. It can represent a single *beat* (or *bar*) of the score. It contains onsets that are locally related to each other given the metric structure. Table 1 provides and example of two sequences of 12 onsets:

Table 1. Example of bitwise operations in rhythmic patterns within a 64-bit sequence.

Bitwise	64-bit binary sequence representing rhythmic patterns	bitcount
X	1010000000001100000000010010000000000100000000001010000000000101	11
Y	1100000000001000000000000011000000000101000000000100010000000001	10
X ∨ Y	1110000000001100000000010011000000000101000000001110010000000001	15
X ∧ Y	1000000000001000000000000010000000000100000000000000000000000001	5
X ∧ ¬Y	0010000000000100000000010000000000000000000000001010000000000101	7
Y ∧ ¬X	0100000000000000000000000001000000000001000000000100010000000000	5

Table 1 presents a sequence of 64 bits representing the two versions of 12 onsets of a two part isorhythmic example with microtiming. Boolean algebra operators OR, AND and NOT are applied to demonstrate processing of the rhythmic data. Bits set to *one* represent a least one onset, since concurrent onset occur at the same time. Bits set to *zero* are not counted. By representing a rhythmic pattern as a single memory element allows the computation of each comparison with a single assembly instruction using 128 bits instead of a loop with 12 comparisons between two 32-bit floating-point data (768 bits). Thus it runs at least 100 times faster and uses 6 times less memory.

4 Implementation, Application and Conclusion

The binarization of rhythmic patterns can be achieved by assigning each bit to a short-time window (Δt). When all the bits represent the same Δt, it is just a division (e.g. 1000 ms ÷ 64 bits ≈ 16 ms).

In order represent *microtiming* around the vicinity of the current time, Eqs. 4 and 5 calculate an exponentially time-stretched representation of local rhythm, resembling a *fisheye-lens* view. In these equations, c is the bit representing the current time, and b is the number of bits used in for zooming into the time window. In Fig. 1, the solid, red line is achieved by applying Eq. 4, centered at bit 32 ($c = 32$), with 6 bits of resolution ($b = 6$). All information contained in the 13-bit corresponds to the same time interval of two bits of the other 51, albeit at lower resolution.

$$f[x, b, c] = \begin{cases} min(2^{x-c},\ 2^b) & \textit{if } x > c \wedge x < 64 \\ 0 & \textit{if } x = c \\ min(2^{c-x},\ 2^b) & \textit{if } x \geq 0 \wedge x < c \end{cases} \tag{4}$$

$$f[x, b, c] = \begin{cases} min(2^{x-c},\ 2^b) & \textit{if } x \geq c \wedge x < 64 \\ min(2^{c-x},\ 2^b) & \textit{if } x \geq 0 \wedge x < c \end{cases} \tag{5}$$

The step function operates in the opposite direction, compressing onsets into half of the bits in a *bird's-eye* view. In doing so, it retains specific information at a global level for input into the BNN. It can be achieved by reducing the time resolution further away from the bits that potentially represent the beat of interest. In Fig. 1, in the dashed line, bits 26 to 38, represents the *fisheye-lens* view. The total time from bits 26 to 32 is equivalent to the Δt of a single bit from bits 11 to 25. Thus, within bits 11 to 53 there is a 32-bit representation of onsets with the same Δt. Bits 0 to 10 and 54 to 63 offer a compressed representation with half of the time resolution ($2\Delta t$ per bit), providing more rhythmic information at the cost of lowering the time resolution.

In the linear time representation (small dotted line), each of the 64 bits equally represent the same time period. In the piecewise linear representation (red, solid line), 13 bits zoom into the current time in order to represent almost concurrent onsets as a *fisheye-lens*. In the stepped version of the piecewise linear function (dashed line), in addition to the zoom into the current time, there are step functions with smaller time resolution. All time-windows represent a segment of the data that can refer to a musical excerpt, rhythmic exercise, a free improvisation, etc. Reference time is presented at bit 32. Past onsets are represented from bit 0 to 31; future onsets, from bit 33 to 63.

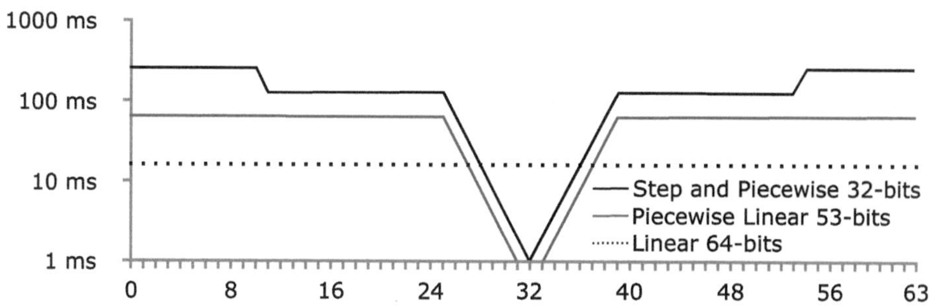

Fig. 1. Relative time difference in ms (vertical axis) represented bit-by-bit in logarithmic scale.

The window-size of 1000 ms seems to accommodate most of the literature on the cognition of beat. In fact, since bits are quantized, within a window-size of 1024 integer numbers, each of the 64 bits would accommodate 16 possible representations. For instance, if 1000 ms onsets are projected into this quantized representation, they would occupy exactly 62.5 bits.

The function *countBits*() counts the number of bits set to *one* within a 64-bit unsigned integer (uint64_t) using either an efficient, single instruction *popcount*, or by implementing a loop, in the unlikely case it is not present in the instruction set architecture.

By superimposing a meter on this 64-bit representation (using a bitwise mask), it is possible to compare any rhythmic structure within this 64-bit representation using mutual information calculation. Mutual Information also provides a better metric for comparing binarized rhythms than the Kullback-Leibler divergence (D_{KL}), for MI presents some properties (e.g. symmetry) that D_{KL} lacks.

Results might provide more musically meaningful information, one of the few complaints with the nPVI, for it retains more information than only a pair of onsets. The proposed representations are BNN optimized. Thus, there might be a computationally efficient way to classify rhythms, even with the presence of *microtiming*. It might work with short-time predictions used in score following and free improvisation with autonomous agents, in keeping with the ITPRA theory by [24].

The author has no competing interests to declare that are relevant to the content of this article.

A Appendix

```
int countBits(uint64_t value) {
   #ifdef __POPCNT__
      return __builtin_popcountll(value);
   #else
```

```
      int count = 0;
      while (value) {
         count += value & 1;
         value >>= 1;
      }
      return count;
   #endif
}

int countBitsArray(uint64_t * array, size_t size) {
   int totalSetBits = 0;
   for (size_t i = 0; i < size; i++)
      totalSetBits += countBits(array[i]);
   return totalSetBits;
}

float H(float p[], int size) {
   float h = 0.0f;
   for (int i = 0; i < size; i++)
      if (p[i] > 0)
         h += -p[i] * log2(p[i]);
   return h;
}

float MutualInfo(uint64_t X, uint64_t Y) {
   float Ixy = 0.0f;
   if ( X | Y ) {
      float bits= countBits( X | Y );
      float px0 = countBits(~X & Y ) / bits; // p(X=0)
      float px1 = countBits( X     ) / bits; // p(X=1)
      float py0 = countBits( X &~Y ) / bits; // p(Y=0)
      float py1 = countBits(       Y ) / bits; // p(Y=1)
      float pxy = countBits( X & Y ) / bits; // p(X=1, Y=1)
      float Hx  = H((float[])px0, px1, 2);
      float Hy  = H((float[])py0, py1, 2);
      float Hxy = H((float[])pxy, px0, py0, 3);
      Ixy = Hx + Hy - Hxy;
   }
   return Ixy;
}
```

References

1. Roads, C.: Microsound. MIT Press, Cambridge (2001)
2. Repp, B.H.: Diversity and commonality in music performance: an analysis of timing microstructure in Schumann's "Träumerei.". J. Acoust. Soc. Am. **92**, 2546–2568 (1992)

3. Desain, P., Honing, H.: The formation of rhythmic categories and metric priming. Perception **32**(3), 341–365 (2003)
4. Mazzola, G.: The Topos of Music: Geometric Logic of Concepts, Theory, and Performance. Springer, Basel (2002)
5. Lerdahl, F., Jackendoff, R.: A Generative Theory of Tonal Music. MIT Press, Cambridge (1983)
6. Bosworth, W. T.: Accentual counterpoint and metrical narrative in the music of Brahms. PhD Dissertation. University of Cambridge (2017)
7. Fuhr, J.: Experiencing Rhythm: Contemporary Malagasy Music and Identity. Cambridge Scholars Publishing, Newcastle (2013)
8. Agawu, K.: African Rhythm: A Northern Ewe Perspective. Cambridge University Press, Cambridge (1995)
9. Marchand, U., Peeters, G.: Swing ratio estimation. Digital Audio Effects 2015 (DAFx15), Trondheim, Norway (2015)
10. Nelias, C., Sturm, E.M., Albrecht, T., et al.: Downbeat delays are a key component of swing in jazz. Commun. Phys. **5**, 237 (2022)
11. Mendes, J.J.F., Gualda, F.: Identificação e Registro de Estruturas Sonoras da Viola Machete do Recôncavo Baiano. In: Proceedings of the Brazilian Association of Ethnomusicology, Maceió - Brazil (2008)
12. Toussaint, G.T.: The Geometry of Musical Rhythm. CRC Press, Boca Raton (2020)
13. Gómez-Martín, F.: A review of Godfried Toussaint's the geometry of musical rhythm. J. Math. Music **16**(2), 239–247 (2022)
14. Yust, J., White, C.W., VanHandel, L.: Rhythm, mathematics, and Godfried Toussaint. J. Math. Music **16**(2), 133–137 (2022)
15. Temperley, D.: The Cognition of Basic Musical Structures. MIT Press, Cambridge (2001)
16. Cross, I.: Music, cognition, culture, and evolution. An. N. Y. Acad. Sci. **930**, 28–42 (2001)
17. Hermes, D.J.: The perceptual structure of sound. In: Bader, R., Leman, M., Godoy, R.I. (eds.) Current Research in Systematic Musicology, vol. 11. Springer, Cham (2023)
18. Bregman, A.S.: Auditory Scene Analysis: The Perceptual Organization of Sound. MIT Press, Cambridge (1990)
19. Howell, P., Cross, I., West, R.: Musical Structure and Cognition. Academic Press, London (1985)
20. Leman, M.: Music and Schema Theory: Cognitive Foundations of Systematic Musicology. Springer Verlag, Heidelberg (1995)
21. London, J.: Hearing in Time: Psychological Aspects of Musical Meter. Oxford University Press, New York (2004)
22. Bader, R.: Springer Handbook of Systematic Musicology. Springer, Cham (2018)
23. Parker, T., Chul, K.J., Large, E.W.: A dynamical, radically embodied, and ecological theory of rhythm development. Front. Psychol. **13**, 653696 (2022)
24. Huron, D.: Sweet Anticipation: Music and the Psychology of Expectation. MIT Press, Cambridge (2006)
25. Forth, J., Agres, K., Purver, M., Wiggins, G.A.: Entraining IDyOT: timing in the information dynamics of thinking. Front. Psychol. **7**, 196169 (2016)
26. Sethares, W.A., Budney, R.: Topology of musical data. J. Math. Music **8**(1), 73–92 (2014)
27. Nguyen, T., Gibbings, A., Grahn, J.: Rhythm and Beat Perception. In: Bader, R. (ed.) Springer Handbook of Systematic Musicology, pp. 507–522. Springer, Cham (2018). https://doi.org/10.1007/978-3-662-55004-5_27

28. Friberg, A., Sundberg, J.: Time discrimination in a monotonic, isochronous sequence. J. Acoust. Soc. Am. **98** (5, Pt 1), 2524-2531 (1995)
29. Large, E.W., Palmer, C.: Perceiving temporal regularity in music. Cogn. Sci. **26**, 1–37 (2002)
30. Patel, A.D., Daniele, J.: An empirical comparison of rhythm in language and music. Cognition **87**(1), B35-45 (2003)
31. Condit-Schultz, N.: Deconstructing the nPVI: a methodological critique of the normalized pairwise variability index as applied to music. Music. Percept. **36**(3), 300–313 (2019)
32. Daniele, J.R., Patel, A.D.: Stability and change in rhythmic patterning across a composer's lifetime: a study of four famous composers using the nPVI equation. Music. Percept. **33**(2), 255–265 (2015)
33. Ravignani, A.: Visualizing and interpreting rhythmic patterns using phase space plots. Music. Percept. **34**(5), 557–568 (2017)
34. Courbariaux, M., Hubara, I. Soundry, D., El-Yaniv, R.: Binarized neural networks: training deep neural networks with weights and activations constrained to +1 or -1. arXiv preprint: arXiv:1602.02830 (2016)
35. Yuan, C., Agaian, S.S.: A comprehensive review of binary neural network. Artif. Intell. Rev. **56**, 12949–13013 (2023)
36. Sayed, R., Azmi, H., Shawkey, H., Khalil, A.H., Refky, M.: A systematic literature review on binary neural networks. IEEE Access **11**, 27546–27578 (2023)
37. Deisenroth, M.P., Faisal, A.A., Ong, C.S.: Mathematics for Machine Learning. Cambridge University Press, Cambridge (2023)
38. Apicella, A., Donnarumma, F., Isgrò, F., Prevete, R.: A survey on modern trainable activation functions. Neural Netw. **138**, 14–32 (2021)
39. Ding, B., Qian, H., Zhou, J.: Activation functions and their characteristics in deep neural networks. In: CCDC 2018, Shenyang, pp. 1836-1841 (2018)

Regular Temperament Theory: Exploring the Landscape Between JI and ETs with Linear Algebra

Dave Keenan[1] and Douglas Blumeyer[2](✉)

[1] Brisbane, Australia
[2] San Francisco, USA
douglas.blumeyer@gmail.com
https://dkeenan.com, https://douglasblumeyer.com

Abstract. Dissonance is easy. Consonance is rare. We want scales and tunings that give different flavors of consonance, without being too complex, and with small errors. Regular Temperament Theory (RTT) is a powerful tool, reducing the problem to approximating a few small prime numbers, and generating tunings by stacking a few intervals called generators. RTT opens a middle ground between just intonation and equal temperaments that goes far beyond meantone, applying linear algebra to discover new temperaments that balance complexity and error in different ways while providing harmonies that do not exist in conventional tunings. We introduce the theory and list some open problems.

Keywords: Microtonal · Temperament · Just Intonation

1 Introduction

This paper describes an application of linear algebra to an area of music theory called Regular Temperament Theory (RTT), whose purpose is to access novel harmonies based on ratios of small whole numbers up to 13 or so, without just intonation's complexity in number of notes, number of different step sizes, and difficulty of modulation. This comes at a cost in tuning accuracy. Various tradeoffs are possible.

Its development began around 1998 in a public online forum known as *the tuning list* [1], facilitated by Mills College, Oakland, California. The first author was involved from the beginning of that collaboration. Other major contributors to the theory were Paul Erlich [2], Graham Breed [3], Gene Ward Smith [4] and Mike Battaglia [5]. The theory has roots going back to the 1970s and earlier, with George Secor [6], Erv Wilson [7] and Adriaan Fokker [8]. Some composers who enriched this collaboration were Joe Monzo [9], Herman Miller, Margo Schulter, Dan Stearns, and Joseph Pehrson.

In January 2021 we set out to make the field more accessible. After surveying the existing resources, many including abstract algebra (group theory) and exterior algebra (wedge products), we laid out an approach covering all practical needs with linear algebra alone, giving examples in Wolfram Language. We replaced jargon and eponyms

© The Author(s), under exclusive license to Springer Nature Switzerland AG 2024
T. Noll et al. (Eds.): MCM 2024, LNCS 14639, pp. 405–411, 2024.
https://doi.org/10.1007/978-3-031-60638-0_33

with consistent descriptive terms, and standardized unique variable letters and their styling. Finally, we reframed the basics of tuning optimization in favor of flexibility and real-world musical considerations, rather than mathematical purity or computational efficiency. This paper is a summary of the first 6 of 10 chapters in the resulting online textbook [10].

Milne et al. summarized a part of this theory to generate keyboard layouts [11].

2 Mapping Just Primes to Temperament Generators

RTT allows us to generate new pitch systems (tunings) prior to scale generation and composition. The input vectors of this theory represent justly-intoned intervals modeled as frequency-ratios whose numerator and denominator are small positive integers. We represent them as *prime-count vectors* whose entries are integers that give the counts (or exponents) of successive primes in the prime factorization of the ratio, with negative counts for the denominator. For example, the interval of a just major third is a frequency ratio of $5/4 = 5/(2 \times 2) = 2^{-2} \times 3^{0} \times 5^{1}$ which is represented as the column vector $\mathbf{i} = [-2\ 0\ 1]^{\mathrm{T}}$. We can call them "PC-vectors", but to avoid confusion with "pitch class vectors" we can simply call them "vectors" in RTT, where it is conventional to use a variant of Dirac's bra-ket notation. $\mathbf{i} = [\,-2\ 0\ 1\rangle$.

A row vector is a linear map. For example, the *just(-prime) tuning map* $\boldsymbol{j} = 1200 \times \log_2 (\langle 2\,3\,5]) = \langle\, 1200.0\ \ 1902.0\ \ 2786.3]$ can be used to obtain the size of a just interval in cents. e.g. for the just major third, $\boldsymbol{j}\mathbf{i} = \langle\, 1200.0\ \ 1902.0\ \ 2786.3]\ \ [-2\ 0\ 1\rangle = 386.3$ ¢.

A regular temperament is a temperament that can be represented by a rectangular integer matrix that corresponds to a linear mapping from prime-count vectors to integer vectors of lower dimensionality called *generator-count vectors* (GC-vectors). This matrix is called the temperament mapping matrix, or simply the *mapping*.

For example, 12-tone equal temperament (12-ET) is represented by the single-row mapping $M = [12\ 19\ 28]$, which tells us that there is a single generator and that prime 2 is approximated by 12 generators, prime 3 by 19 generators, and prime 5 by 28 generators. From this we can deduce that the generator is an approximate semitone. This is confirmed when we map the just major third: $\mathbf{y} = \boldsymbol{M}\mathbf{i} = [12\ \ 19\ \ 28]\ [-2\ 0\ 1\rangle = [4\rangle$. We use a curved angle bracket to distinguish a GC-vector from a (PC-)vector. The single entry of 4 means the just major third is approximated by 4 generators, in this case 4 semitones.

We can think of the vector entries as having units of primes **p** and the GC-vector entries as having units of generators **g**, and so the mapping entries have units of generators per prime **g**/**p**.

We said the generator was an *approximate* semitone because this way of defining temperaments leaves open the exact tuning of the generators. That belongs to a separate optimization phase. All we need to know at this stage is that the generator is 1/12 of an approximate octave. We do not assume pure octaves or octave-equivalence.

Equal temperaments have mapping matrices with rank $r = 1$, and so are called *rank-1 temperaments*. So far, our input vectors have dimensionality $d = 3$ and are known as *5-limit* vectors (after the highest prime). A mapping matrix is always full-rank and so it is an $r \times d$ matrix.

Meantone is a well-known *rank-2 temperament*, dating from the early 1500s. It is conventionally described as having a generator that is a slightly-narrow perfect fifth, an approximate 3/2 ratio (~3/2). But from RTT's point of view, meantone has two generators, the first one being an approximate octave (~2/1). It is well known that meantone approximates a just major third (~5/4) using a stack of four fifths, octave-reduced. So it approximates prime 5 (2786.3 ¢) by a stack of four fifths (4 × ~700 = ~2800 ¢).

We can write the corresponding meantone mapping as $M = \begin{bmatrix} 1 & 1 & 0 \\ 0 & 1 & 4 \end{bmatrix}$.

Looking at each column in turn, we see that prime 2 is approximated by one of the first generator, prime 3 by one of the first generator plus one of the second generator, and prime 5 by four of the second generator.

With more than one row, our mapping matrix is no longer unique. We can perform elementary integer row operations to find alternative mappings for meantone, associated with different generators. For example, we could use an octave and a perfect fourth (~4/3) as the generators. Since the fourth is the octave minus the fifth, we can perform the inverse operation on the rows of the mapping, adding the second row to the first and negating the second, to obtain $M = \begin{bmatrix} 1 & 2 & 4 \\ 0 & -1 & -4 \end{bmatrix}$.

This shows the need for a canonical form for mapping matrices, so we can recognize when different procedures give the same temperament. A first pass at a canonical form is to use Hermite normal form (HNF), integer analog of RREF. In Wolfram Language:

```
hnf[a_]:= Last[HermiteDecomposition[a]].
```

And we obtain $M = \begin{bmatrix} 1 & 0 & -4 \\ 0 & 1 & 4 \end{bmatrix}$.

We can see that the first generator approximates prime 2 on its own, as before, so it is an approximate octave. The second generator approximates prime 3 on its own, so it is an approximate perfect twelfth (~3/1).

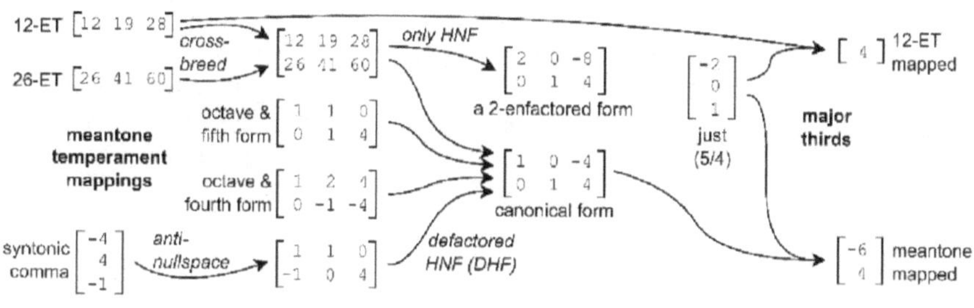

Fig. 1. Mapping just primes to temperament generators

We have so far obtained the meantone mapping by considering the generators. Another way is by "cross-breeding" equal temperaments. Both 12-ET and 26-ET can be considered as supporting (rather extreme) tunings of meantone. The 5-limit mapping for 26-ET is [26 41 60]. If we stack the two ET mappings, we have $M = \begin{bmatrix} 12 & 19 & 28 \\ 26 & 41 & 60 \end{bmatrix}$.

Taking the HNF we obtain $M = \begin{bmatrix} 2 & 0 & -8 \\ 0 & 1 & 4 \end{bmatrix}$.

The latter *would* be the canonical meantone mapping except for a common factor of 2 in the top row. We say the matrix is *enfactored*. This is musically wasteful as half of the generator count vectors would not correspond to the tempering of any interval. Note that the common factor is hidden in the original matrix. The HNF may not reveal the common factor either. It may only become apparent in some far-from-obvious linear combination of the rows, using coprime integer multipliers (if not coprime, they would be *introducing* rather than *revealing* a common factor). A simple defactoring (or saturation) algorithm, which works even in these hidden cases, consists in performing a column-wise Hermite decomposition, keeping only the unimodular matrix, inverting it and taking the top r rows, where r is the rank. Taking the HNF of that yields our final canonical form which we call *defactored Hermite form* (DHF). In Wolfram Language:

```
dhf[m_]:= hnf[Take[Inverse[Transpose[First[
HermiteDecomposition[Transpose[m]]]]], MatrixRank[m]]]
```

Meantone equates the tones 10/9 and 9/8, hence the name. This means that their quotient, the syntonic comma 81/80 = $[-4\ 4\ -1\rangle$, is in the nullspace of the mapping, and since nullity is dimensionality minus rank, $n = d - r = 1$, this comma alone forms a basis for the nullspace. So the mapping matrix can be obtained as the *dual* of the comma basis matrix $[[-4\ 4\ -1\rangle]$, which in this case has only one column (vector). If this was a 7 limit (4D) rank-2 temperament, a second comma would vanish, such as 225/224, and the comma basis would have two columns $[[-4\ 4\ -1\ 0\rangle\ [5\ -2\ -2\ 1\rangle]$. In Wolfram Language:

```
dualMapping[c_]:= dhf[NullSpace[c]]
```
(NullSpace is its own inverse)

When we apply this to the comma basis consisting only of the syntonic comma, we again obtain the canonical meantone mapping.

If we apply the canonical meantone mapping to the vector for the just major third we obtain the GC-vector $\mathbf{y} = M\mathbf{i} = [\text{-}6\ 4\rangle$. This says that the meantone-mapped major third consists of 4 of the approximate twelfth generators minus 6 of the approximate octave generators. The concepts covered in this section are summarized in Fig. 1.

3 Tuning Optimization

Now for the optimization phase. We seek a *generator tuning map* $\boldsymbol{g} = \langle\!\langle\sim1200\ \sim1900]$ whose entries are optimal generator sizes. With $\boldsymbol{g}$ we could obtain the size of any tempered interval in cents as $\boldsymbol{g}\mathbf{y} = \boldsymbol{g}M\mathbf{i}$ analogous to obtaining the size of a just interval as $\boldsymbol{j}\mathbf{i}$. Therefore, the *error* in cents of a tempered interval is $\mathrm{e} = \boldsymbol{g}M\mathbf{i} - \boldsymbol{j}\mathbf{i} = (\boldsymbol{g}M - \boldsymbol{j})\mathbf{i}$.

An optimal $\boldsymbol{g}$ would minimize the audible harm done to the just intervals we care about. For a single interval we model this harm as the *damage* d, defined as the absolute value of the error multiplied by some positive *weight* w, so $\mathrm{d} = |\mathrm{e}|w$. The weight is typically one of three things: unity (unweighted) $w = 1$, or some measure of the *complexity* of the interval $w = c$, or the reciprocal of a complexity, called a *simplicity*, $w = 1/c$. The interval complexity measure is typically some pre-scaled power-norm $c = \|X\mathbf{i}\|_q$ where X is a diagonal (scaling) matrix. The most common complexity measure scales each entry by the base-2 log of its prime, then takes the taxicab norm: $X = \text{diag}(\log_2([2\ 3\ 5\ \ldots]))$, $q = 1$. This is called the *log product* (lp) complexity as it is equal to the log of the product of the numerator and denominator of the corresponding ratio.

The *target-interval list* T is the list of intervals whose damage we care about; the fewer, the less damaged each can be. It may be chosen from a type of music, or the intervals playable on an instrument, but a useful default is the *truncated integer-limit triangle* (TILT). This begins as a triangular table of the ratios > 1 between the positive integers less than the prime after the temperament's prime limit; for the 5-prime-limit this is the 6-integer-limit and for the 7-prime-limit it is the 10-integer-limit. We then remove intervals smaller than 15/13 (247.7 ¢) and larger than 13/4 (2040.5 ¢). See [10].

So for 5-limit meantone we suggest the 6-TILT, which consists of the eight intervals 2/1, 3/1, 3/2, 4/3, 5/2, 5/3, 5/4, 6/5, converted to prime-count vectors and placed side-by-side to form a target-interval list T, a $d \times k$ matrix where k is the number of target intervals. See Fig. 2.

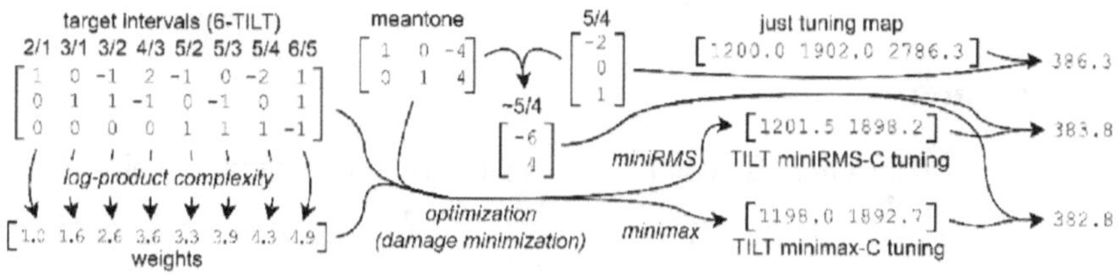

Fig. 2. Tuning optimization

We then compute the *target-interval error list* as $\mathbf{e} = (\boldsymbol{g}M - \boldsymbol{j})\mathrm{T}$, and for this run we'll compute the *target-interval weight list* $\boldsymbol{w}$ as the log product complexities of the target intervals. We then turn the weight list into a diagonal weight matrix $W = \text{diag}(\boldsymbol{w})$ and compute the *target-interval damage list* as $\mathbf{d} = |\mathbf{e}|W$, where $|\cdot|$ is entry-wise absolute value. In terms of the generator tuning map $\boldsymbol{g}$, the damage list s $\mathbf{d} = |(\boldsymbol{g}M - \boldsymbol{j})\mathrm{T}W|$.

Popular statistics for the overall damage are the maximum damage $⟪\mathbf{d}⟫_\infty$ and the RMS damage $⟪\mathbf{d}⟫_2$, and therefore the popular optimization procedures are minimax and miniRMS (least squares). The double-angle-brackets are our notation for power-means, by analogy with power-norms.

MiniRMS has the advantage of a simple closed-form solution for $\boldsymbol{g}$ involving the Moore-Penrose inverse, $\boldsymbol{g} = \boldsymbol{j}\mathrm{T}W(M\mathrm{T}W)^+$. In Wolfram Language:

```
g = j.t.w.PseudoInverse[m.t.w]
```

The result is $\boldsymbol{g} = ⟨1201.5\ 1898.2]$. We call this the TILT miniRMS-C tuning for meantone. The "C" stands for "complexity weight". To obtain the size of the more conventional generator, the fifth, we can subtract the first generator from the second to obtain 696.7 ¢. For reference, $\mathbf{e} = [1.5\ {-3.8}\ {-5.3}\ 6.8\ {-1.2}\ 4.1\ {-2.6}\ {-2.6}]$ ¢.

A minimax value for $\boldsymbol{g}$ can be found using: `MaxIterations->100;`

```
NMinimize[ Max[Abs[(g.m-j).t.w]], g∈Vectors[2,Reals] ]
```

And we find that the TILT minimax-C tuning for meantone is $\boldsymbol{g} = \langle 1198.0\ 1892.7]$. So the fifth is 694.7 ¢, and $\mathbf{e} = [-2.0\ {-9.3}\ {-7.2}\ 5.2\ {-5.4}\ 1.8\ {-3.4}\ {-3.8}]$ ¢.

The minimax tuning found by this method is often not unique. In such cases, what we really want is the limit of the $\boldsymbol{g}$s that minimize the p-mean of the damages $\langle\!\langle \mathbf{d} \rangle\!\rangle_p$ as p approaches infinity. We can use the following Wolfram Language code, replacing the "2" with successive powers of 2, until the generators change by less than, say, 0.1 ¢. Numeric precision may need to be increased to avoid spurious results.

```
NMinimize[ Total[((g.m-j).t.w)^2], g∈Vectors[2,Reals] ]
```

We can also include constraints, such as holding some intervals unchanged by the tuning—typically the octave. A list of vectors can be assembled into a *held-interval basis* H, applying the constraint that $\boldsymbol{g}\mathrm{MH} = \boldsymbol{j}\mathrm{H}$. In Wolfram Language:

```
NMinimize[ { Total[((g.m-j).t.w)^2], g.m.h==j.h }, …]
```

4 Conclusion

Although we used meantone as a familiar example, an extraordinary example is George Secor's *miracle* temperament [6], $M = \begin{bmatrix} 1 & 1 & 3 & 3 & 2 \\ 0 & 6 & -7 & -2 & 15 \end{bmatrix}$. Its held-octave 12-TILT minimax-U generators are $\boldsymbol{g} = \langle 1200.0\ 116.72]$ ¢ with a maximum unweighted damage of 3.32 ¢. A 21-note-per-octave miracle scale has only two step sizes, contains many melodic subset scales, and supports a network of 11-(prime)-limit harmony that would require many times that number of notes in JI. Its rediscovery in 2001 started a "gold rush" that led to the development of the tools described above, and many others. Searches were conducted both by combining commas and by combining ETs.

Open problems regarding temperaments include: staff notation, family classification, anti-JI intervals [12], refining search criteria (including complexity × damage = badness measures), and psychoacoustic experiments to validate them.

Many temperaments have been cataloged in the Xenharmonic Wiki, where you will also find many details that were omitted from this paper due to space limitations, in our extended exposition entitled *D&D's Guide to RTT* [10].

We thank Paul Erlich, Graham Breed, Mike Battaglia, Keenan Pepper, Sintel, Kite Giedraitis, Flora Canou, and the two anonymous reviewers for improving this paper.

References

1. Blumeyer, D., Keenan, D. (eds.): Yahoo Tuning Groups Ultimate Backup (2020). https://yahootuninggroupsultimatebackup.github.io. Accessed January 2024

2. Sethares, W. (ed.): Papers by Paul Erlich. https://sethares.engr.wisc.edu/papers/erlich.html
3. Breed, G.: Temperament Finding Scripts (2010–2021). http://x31eq.com/temper
4. Smith, G.W.: Xenharmony. http://lumma.org/tuning/gws/home.htm. Accessed January 2024
5. Battaglia, M.: Mike's lectures on regular temperament theory (2012). Accessed January 2024. https://en.xen.wiki/w/Mike%27s_lectures_on_regular_temperament_theory
6. Secor, G.: A New Look at the Partch Monophonic Fabric. Xenharmonikôn 3 (1975). https://www.anaphoria.com/secor.pdf
7. Narushima, T.: Microtonality and the Tuning Systems of Erv Wilson. Routledge (2018)
8. Fokker, A.: Les mathématiques et la musique. Trois conférences, ch. IV, Archives du Musée Teyler 10, Martinus Nijhoff, Den Haag (1947). https://huygens-fokker.org/docs/mm4.html
9. Monzo, J.: Tonalsoft. http://tonalsoft.com/enc/encyclopedia.aspx. Accessed January 2024
10. Keenan, D., Blumeyer, D.: Dave Keenan & Douglas Blumeyer's Guide to RTT (2023). https://en.xen.wiki/w/Dave_Keenan_%26_Douglas_Blumeyer%27s_guide_to_RTT
11. Milne, A., Sethares, W., Plamondon, J.: Tuning continua and keyboard layouts. J. Math. Music **2**(1), 1–19 (2008)
12. Keenan, D.: Noble frequency ratios as prime-count vectors in $\mathbb{Q}(\sqrt{5})$: the Feudal Manifesto (2022). https://forum.sagittal.org/viewtopic.php?t=557

Exploring Mode Identification in Irish Folk Music with Unsupervised Machine Learning and Template-Based Techniques

Juan José Navarro-Cáceres[1(✉)], Nádia Carvalho[2], Gilberto Bernardes[2], Diego M. Jiménez-Bravo[1], and María Navarro-Cáceres[1]

[1] Expert Systems and Applications Lab, Faculty of Science, University of Salamanca, Plaza de los Caídos s/n, 37008 Salamanca, Spain
juanjo_810@usal.es

[2] Faculty of Engineering & INESC TEC, University of Porto, Porto, Portugal

Abstract. Extensive computational research has been dedicated to detecting keys and modes in tonal Western music within the major and minor modes. Little research has been dedicated to other modes and musical expressions, such as folk or non-Western music. This paper tackles this limitation by comparing traditional template-based with unsupervised machine-learning methods for diatonic mode detection within folk music. Template-based methods are grounded in music theory and cognition and use predefined profiles from which we compare a musical piece. Unsupervised machine learning autonomously discovers patterns embedded in the data. As a case study, the authors apply the methods to a dataset of Irish folk music called The Session on four diatonic modes: Ionian, Dorian, Mixolydian, and Aeolian. Our evaluation assesses the performance of template-based and unsupervised methods, reaching an average accuracy of about 80%. We discuss the applicability of the methods, namely the potential of unsupervised learning to process unknown musical sources beyond modes with predefined templates.

Keywords: Mode Detection · Folk music · Unsupervised learning · Template-based method

1 Introduction

Western folk music is characterized by its varied composition, including elements such as rhythm, instrumentation, tempo, and compositional patterns, such as profiles and mode. The mode, meaning the scale in which the folk tune is based on, provide a framework that evoke specific emotions or atmospheres. Additionally, the choice of mode can vary depending on the cultural background and regional influences. Overall, the mode attribute plays a significant role in shaping the character of Western folk music compositions.

This research has been funded by the European Union's Horizon Europe research and innovation program under the Marie Sklodowska-Curie grant agreement nº 101086338.

© The Author(s), under exclusive license to Springer Nature Switzerland AG 2024
T. Noll et al. (Eds.): MCM 2024, LNCS 14639, pp. 412–420, 2024.
https://doi.org/10.1007/978-3-031-60638-0_34

Within the field of music computing research, mode detection remains an underexplored domain. Most of the methods to detect mode have focused on the major and minor modes, which are the most common in Western classical and commercial music. In modal folk music tradition, modes other than major (Ioninan) and minor (Aeolian) – Dorian, Phrygian, Lydian, Mixolydian, and Locrian – require thorough research.

However, similar to this, key detection has been much explored within the context of classical Western music. The concept of key or tonality holds significant importance in classic Western music [18]. Nonetheless, in the case of folk music usually lacks significance, as plenty of vocal melodies are sung in different keys, according to the vocal features of the singer, and instrumental tunes can be transported according to the instruments that interpret them.

However, next paragraphs will delineate the existing literature pertaining to key detection, serving as an approach for mode detection in folk music analysis.

Computational efforts in automatic key detection have supported various content-driven applications within musicology and music theory, notably facilitating large-scale empirical analysis [4,20].

Key-detection methods often utilize templates derived from key profiles [12, 14,15,23]. Extensive experimental studies [13,14], statistical analyses [1], and probabilistic or statistical models [2,23] have been employed in creating key profiles.

Supervised machine learning algorithms have also been extensively investigated and applied to key detection. The use of labeled datasets and explicit guidance during the training phase allowed us to get accurate predictions in various musical contexts. [11,16] makes an approach using different supervised algorithms, such as KNN, SVM and CNN. Finally [10] tries to compare different Artificial Neural Networks (ANN) to detect minor and major modes. However, the exploration of unsupervised learning approaches in key detection remains relatively limited in comparison.

For unsupervised techniques, we can mention [9] that proposes an unsupervised approach to detect major and minor modes based on Manhattan distance among the features of the data. From a folk music perspective, unsupervised techniques hold relevance due to the presence of unidentified traditions and modes within folk music. Using unsupervised learning methods may unveil previously unknown modes, thus expanding our understanding of folk music traditions. This fact, as well as the scarcity of research in this area, underscores a need for a more comprehensive exploration of unsupervised learning techniques.

In this paper, we present a novel approach for mode detection, which aims to identify the mode of a symbolic encoded modal music piece within folk music tradition, specifically an Irish folk music dataset called The Session. Our work adopts and compares traditional template-based approaches with unsupervised machine-learning strategies.

2 Methods

The work proposed follows three stages, illustrated in Fig. 1. First of all, we selected the dataset and extracted the scores in ABC notation, and the tag related to mode detection. Secondly, the data was processed to obtain the PCP. Two PCP are computed: the simple PCP, which only considers pitch occurrences, and the weighted PCP, which also considers pitch duration and bar position. Finally, we detect mode by applying two different paradigms. On the one hand, we use a statistical approach called template-based mode detection. On the other hand, we use a dimensionality reduction to optimize the performance of four non-supervised classical algorithms: DBScan, K-means, MeanShift and Self-organizing map (SOM).

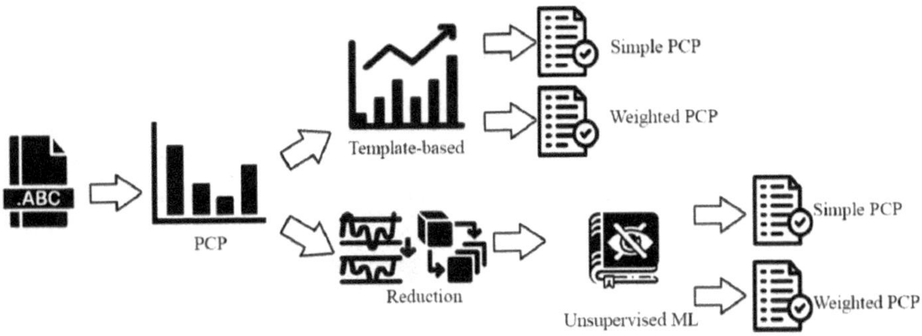

Fig. 1. Process workflow.

2.1 Dataset and Data Pre-processing

Our experimental approach employs "The Session" dataset [21] as the empirical ground truth data. We utilize the dataset pre-processed and revised by Sturm et al. [22], which predominantly retains monophonic songs and incorporates ABC fields with musical surface information and formal annotations, such as the mode. They have reviewed the dataset fields twice to verify any misclassification in the annotations. Additionally, all songs are transposed to the tonic C. In total, the dataset encompasses 23,636 folk music songs. The distribution per mode is as follows: Ionian (15861), Dorian (2971), Mixolydian (1620), and Aeolian (3184).

To obtain successful results, the data usually needs to be preprocessed. In this case we have used the Pitch Class Profile (PCP) extraction, based on the K-K Profiles by Krumhansl and Kessler [13,14]. In this technique, each mode can be represented by a pitch profile composed by the number of occurrences of the 12 notes of the temperament music. Therefore, we parse each folk song in the dataset using the Python library `music21` to compute PCP. Two distinct PCP templates are generated, labeled as *simple* and *weighted*. The *simple* template results from converting pitch content into pitch classes, accumulating them on a 12-element pitch class histogram, where each note in the song carries a unity value. The *weighted* template, inspired by previous research [7], introduces a nuanced approach by assigning weights to pitch classes based on their duration and metrical position within the bar. This approach assumes that longer notes in strong metrical positions hold higher hierarchical salience and stability. To implement this bias, each note is weighted by its duration relative to a quarter note, while the metrical position is ranked in the [0, 1] range to five metrical pulses: the measure, half-measure, beat, division, and subdivision pulse. These measure positions are assigned to the following beat strength weights: 1, .5, .25, .125, and .0625. The beat strength weights are computed using the `Music21Object.beatStrength` object from `music21` Python library [8].

The final result of a PCP is a 12-dimension vector, one element per musical note. Therefore, to illustrate the dataset distribution, we need to reduce the dimensionality using the UMAP algorithm. Figure 2 depicts the visualization of the data. On the left, the representation of the simple Pitch Class Profile (PCP) is presented, and on the right, the weighted PCP.

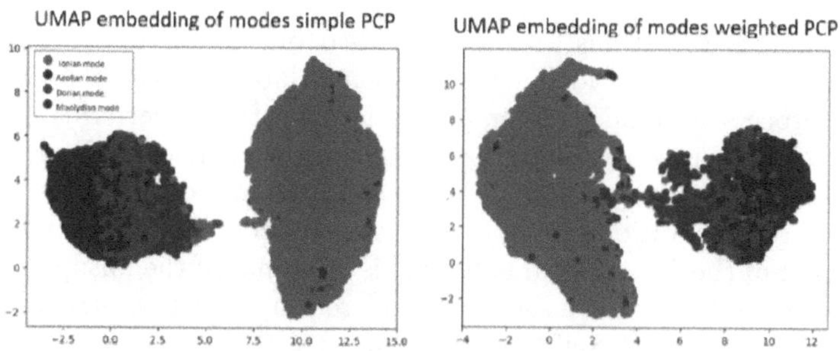

Fig. 2. UMAP dataset visualization (Color figure online)

The visual map reveals two predominant groups, one comprising the Ionian mode (denoted in red) and the other encompassing the remaining modes. In consequence, the expected outcomes are to detect at least these two big groups. The biggest challenge is the detection of the three groups of Mixolidian, Dorian, and Aeolian modes.

2.2 Template-Based Mode Detection

Seminal to template-based mode detection is adopting key profiles that denote the relative stability of one of the twelve pitch classes within a key and mode. To this end, we employed K-Fold Cross Validation to assess the performance of our approach, specifically 4 folds. Each fold was alternately designated as the training set and the testing set, with a 75/25% split ratio utilized for training and testing, respectively. During each iteration of the cross-validation procedure, were computed from both *simple* and *weighted* PCP using the training data. This approach allowed us to evaluate the effectiveness of our mode detection method across multiple train-test splits and assess its robustness.

Given the template-based mode profiles, to infer the mode of an unknown folk music song, we compute their respective *simple* and *weighted* PCP, followed by the distance or alignment to the four respective mode profiles. Common metrics adopted in the latter step include cosine, Euclidean, and Pearson correlation, whose efficacy we assess in the evaluation of our work, whose results are shown in Sect. 3.

2.3 Unsupervised Mode Detection

In our approach to unsupervised learning for musical mode detection, we employ a systematic method encompassing feature selection, normalization, dimensionality reduction, and clustering algorithms.

First, we select the features of the songs to infer the mode of a folk music song. The features selected are, on the one hand, the *simple* PCP and, on the other hand, the *weighted* one. Then, we normalize the features using the MinMax Scale [19].

Unsupervised algorithms are usually based on distances between features so they are not optimized for high dimensional data. Therefore, a dimensional reduction process is applied to improve the algorithm's performance. In this case, the technique used to reduce the dimensionality is the UMAP technique [3], which effectively captures complex relationships within the data, providing a low-dimensional representation that preserves both local and global structures [17].

The result of the dimensional reduction is the input of the unsupervised algorithms. In this work, we have selected 4 different approaches, namely, DBScan, KMeans, MeanShift and Self-Organized Maps (SOM), as classical clustering algorithms [5,6,24].

Each algorithm has been configured with specific parameters. For the DBScan algorithm, the parameters epsilon (eps) and minimum samples ($min\_samples$) have been set to 0.3 and 5, respectively. In the case of KMeans, the algorithm has been configured to identify four clusters. MeanShift has been assigned a bandwidth of 2.5 to enhance its capacity to recognize more than two clusters. Lastly, the Self-Organized Maps (SOM) algorithm has been configured with four centroids.

3 Results and Discussion

The goal of the section is to demonstrate that unsupervised ML effectively produces accurate results in mode detection in our case study, Irish folk music. In the next subsection, we will show the results obtained by applying a classical approach (template-based method). In the last section, we will discuss the performance of four classical unsupervised algorithms after data preprocessing.

3.1 Results of Template-Based Method

Table 1 presents the classification results for the template-based method in identifying the four modes studied on The Session dataset. The percentage of accurate mode classifications is reported for both *simple* and *weighted* PCP, using three metrics (cosine and Euclidean distances and Pearson correlation coefficient). Surprisingly, the results from simple PCP demonstrate superiority in mode classification by a residual percentage points margin. Additionally, Pearson correlation outperforms in detecting the Dorian, Mixolydian, and Aeolian modes. Therefore, the best-performing method in terms of accuracy for the template-based method adopts *simple* PCP.

Table 1. Percentages of correct mode classification using three measures – cosine, Euclidean distances, and Pearson correlation – and two pitch class distribution – simple and weighted – in the four Ionian, Dorian, Mixolydian, and Aeolian diatonic modes. SOM best results are also included in the table.

Measures	Cosine		Euclidean		Correlation		SOM
PCP	Simple	Weighted	Simple	Weighted	Simple	Weighted	Simple
Ionian	97.69	95.78	97.65	95.73	**97.75**	95.78	**98.91**
Dorian	77.72	69.24	77.52	69.30	**77.89**	69.10	45.88
Mixolydian	73.70	66.79	**73.77**	66.91	73.70	66.85	**79.88**
Aeolian	78.61	76.91	78.70	76.72	**78.73**	77.04	63.65

Table 2 provides the confusion matrix for the four modes under evaluation in The Session. An expected misclassification between Ionian and Mixolydian modes and between Dorian and Aeolian modes is observed as they align as major and minor $\hat{3}$ degree scales, respectively.

Table 2. Confusion matrix for the mode classification task using the template-based method and Pearson correlation coefficient, and the SOM classifier. Classification results are expressed in percentages.

	Pearson				SOM			
	Ionian	Dorian	Mixolydian	Aeolian	Ionian	Dorian	Mixolydian	Aeolian
Ionian	97.75	0.02	2.22	0.02	98.91	.02	1.03	.04
Dorian	0.03	77.89	2.63	19.46	.27	45.88	43.45	10.40
Mixolydian	13.21	13.09	73.70	0.00	16.67	3.33	79.88	.12
Aeolian	0.07	20.96	0.26	78.73	.06	31.54	4.74	63.65

3.2 Unsupervised Algorithms

Figure 3 illustrates the outcomes of the applied clustering algorithms. Each map displays the result of the clustering performed by the respective algorithm using both PCP features.

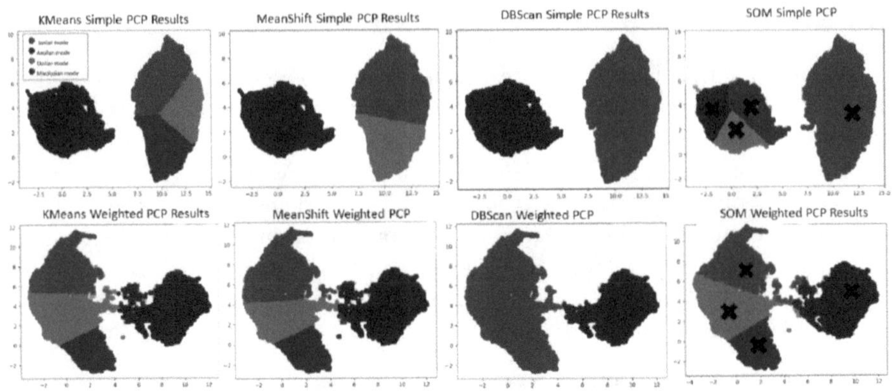

Fig. 3. Algorithms results visualization

These visualizations indicate an inclination to form a cluster for the Aeolian, Mixolydian, and Dorian modes, while also dividing the Ionian group -except in the case of the Self-Organized Maps (SOM) algorithm using the simple PCP. This outcome aligns logically with the dataset's Ionian group size: algorithms tend to subdivide larger groups. In terms of clustering evaluation, the silhouette evaluation would be very good in the case of DBScan, as it splits perfectly the two big groups, but in the case of mode detection the performance is not, as it doesn't detect the Aeolian, Mixolydian and Dorian modes. Remarkably, the SOM algorithm has achieved a visualization closely resembling the original dataset, signifying superior performance. The corresponding confusion matrix is presented in Table 2.

The confusion matrix shows the algorithm's ease in distinguishing the Ionian mode. However, outcomes for the Dorian, Mixolydian, and Aeolian modes suggest challenges in pattern recognition.

A noteworthy limitation of the SOM algorithm lies in its reliance on a predefined number of clusters. Consequently, it may encounter difficulties adapting to new modes. This limitation is shared with the KMeans algorithm. However, the MeanShift and DBScan algorithms exhibit autonomous group discovery capabilities, although they have performed worse outcomes.

3.3 Discussion

The obtained results show higher performance of the unsupervised learning in Ionian and Mixolydian modes detection compared to the template-based method. On the other side, the Aeolian and Dorian mode have presented worse results in comparison to the traditional method. The comparative analysis is presented in Table 1. However, this outcome aligns with expectations, considering the high degree of similarity between the PCP of Aeolian and Dorian modes, differing by just one note. In addition, it is noteworthy that unsupervised ML introduces a more versatile and adaptive method accommodating the inclusion of new modes, while the template-based method needs a large quantity of data from each mode for training and testing. Therefore, this unsupervised learning presents very optimistic results for advancing research endeavors in mode detection.

4 Conclusions

In this study, our main focus was on exploring a relatively unexplored area: detecting modes in folk music, with a specific focus on Irish folk music. Unlike more mainstream music genres, folk music presents a unique challenge because of its diverse scales and pitch relationships, which are not as straightforward as the major and minor modes commonly found in Western music. In addition, unsupervised learning techniques have been applied which hold greater relevance due to the presence of unidentified traditions and modes within folk music. Our study sheds new light on the potential of using unsupervised learning techniques in folk music research. By delving into this lesser-known aspect of music theory, we hope to contribute to a better understanding of the rich and varied traditions found in folk music.

Our exploration of unsupervised machine learning methods stemmed from their inherent flexibility in detecting patterns, particularly in scenarios where predefined groups might be elusive. We compared four distinct algorithms against a template-based method, which utilized classical distances to compare PCPs. The results demonstrated that the unsupervised approach not only provided good accuracy but also outperformed the template-based method, showing the efficacy of these algorithms in mode detection in the case of Irish folk music (using a dataset from TheSession.org).

Future work will include exploring new datasets of digital score files that include different folk cultures. Additionally, we aim to explore supervised approaches to compare the results with unsupervised ML. Also, the clustering and dimensionality reduction algorithms will be wider explored to improve the performance. Finally, a deeper musicological analysis can be done to understand the idiosyncrasies of the modal music within the folk context.

References

1. Aarden, B.: Dynamic Melodic Expectancy. Ph.D. thesis, Ohio State University, Columbus, OH, USA (2003)
2. Albrecht, J., Shanahan, D.: The use of large corpora to train a new type of key-finding algorithm: an improved treatment of the minor mode. Music Percept. Interdiscip. J. **31**, 59–67 (2013)
3. Allaoui, M., Kherfi, M.L., Cheriet, A.: Considerably improving clustering algorithms using UMAP dimensionality reduction technique: a comparative study. In: El Moataz, A., Mammass, D., Mansouri, A., Nouboud, F. (eds.) ICISP 2020. LNCS, vol. 12119, pp. 317–325. Springer, Cham (2020). https://doi.org/10.1007/978-3-030-51935-3_34

4. Bernardes, G., Davies, M.E., Guedes, C.: Automatic musical key estimation with adaptive mode bias. In: 2017 IEEE International Conference on Acoustics, Speech and Signal Processing (ICASSP), pp. 316–320. IEEE (2017)
5. Cao, L., Zhao, Z., Wang, D.: Clustering algorithms. In: Target Recognition and Tracking for Millimeter Wave Radar in Intelligent Transportation, pp. 97–122. Springer, Singapore (2023). https://doi.org/10.1007/978-981-99-1533-0_5
6. Carreira-Perpinán, M.A.: A review of mean-shift algorithms for clustering. arXiv preprint arXiv:1503.00687 (2015)
7. Carvalho, N., Diogo, D., Bernardes, G.: Computational similarity of Portuguese folk melodies using hierarchical reduction. In: Proceedings of the 10th International Conference on Digital Libraries for Musicology, pp. 22–29. DLfM '23, Association for Computing Machinery, New York, NY, USA (2023). https://doi.org/10.1145/3625135.3625152
8. Cuthbert, M.S., Ariza, C.: Music21: a toolkit for computer-aided musicology and symbolic music data (2010)
9. Finley, M., Razi, A.: Musical key estimation with unsupervised pattern recognition. In: 2019 IEEE 9th Annual Computing and Communication Workshop and Conference (CCWC), pp. 0401–0408. IEEE (2019)
10. Garg, M., et al.: Comparative analysis of deep learning architectures and vision transformers for musical key estimation. Information **14**(10), 527 (2023)
11. George, A., Mary, X.A., George, S.T.: Development of an intelligent model for musical key estimation using machine learning techniques. Multimed. Tools Appl. **81**(14), 19945–19964 (2022)
12. Krumhansl, C.L.: Cognitive Foundations of Musical Pitch. Oxford University Press, New York, NY, USA (1990)
13. Krumhansl, C.L.: Cognitive Foundations of Musical Pitch, vol. 17. Oxford University Press, Oxford (2001)
14. Krumhansl, C.L., Kessler, E.J.: Tracing the dynamic changes in perceived tonal organization in a spatial representation of musical keys. Psychol. Rev. **89**(4), 334 (1982)
15. Longuet-Higgins, H.: Letter to a musical friend. Music Rev. **23**, 244 (1962)
16. Mahieu, R.: Detecting musical key with supervised learning (2017)
17. McInnes, L., Healy, J., Melville, J.: UMAP: uniform manifold approximation and projection for dimension reduction. arXiv preprint arXiv:1802.03426 (2018)
18. Milne, A.: A computational model of the cognition of tonality. Open University (United Kingdom) (2013)
19. Pedregosa, F., et al.: Scikit-learn: machine learning in Python. J. Mach. Learn. Res. **12**, 2825–2830 (2011)
20. Quinn, I., White, C.W.: Corpus-derived key profiles are not transpositionally equivalent. Music Percept. Interdiscip. J. **34**(5), 531–540 (2017)
21. Session, T.: Traditional Irish music on the session (2021). https://thesession.org/
22. Sturm, B., Santos, J.F., Korshunova, I.: Folk music style modelling by recurrent neural networks with long short term memory units. In: 16th International Society for Music Information Retrieval Conference (2015)
23. Temperley, D.: Bayesian models of musical structure and cognition. Music Sci. **8**, 175–205 (2004)
24. Venkatkumar, I.A., Shardaben, S.J.K.: Comparative study of data mining clustering algorithms. In: 2016 International Conference on Data Science and Engineering (ICDSE), pp. 1–7 (2016). https://doi.org/10.1109/ICDSE.2016.7823946

Communication-Performances

Of All Interval Tetrachords and Octatonic Scales

Emmanuel Amiot(✉) and Tom Johnson

LAMPS, Perpignan, France
manu.amiot@free.fr

Abstract. Serendipity is another name for method. When order appears unexpectedly in music, the explanation is so often mathematical. Here, a nice feature of a hand-made picture revolves about the Sylow subgroups of the affine group modulo 12 (and of its maximal normal subgroup $\boldsymbol{T} / \boldsymbol{I}$), explaining the surprise occurrence of octatonic scales when trying to organize all-interval tetrachords by common dyads.

Keywords: all-interval tetrachords · stabilizer · Sylow · octatonic scale

Notations

- The ambient space is the cyclic group $\mathbb{Z}_{12}$ of pitch-classes (pcs) with 12 elements.
- T_k is the transposition $x \mapsto x + k$ in $\mathbb{Z}_{12}$. All transpositions form a cyclic group, $\boldsymbol{T}$.
- I_k is the inversion $x \mapsto -x + k$ in $\mathbb{Z}_{12}$. Inversions and transpositions form the dihedral group $\boldsymbol{T} / \boldsymbol{I}$.
- M_k is the multiplication $x \mapsto x \times k$ in $\mathbb{Z}_{12}$. The value of k is restricted to 1, 5, 7 or 11 (alias -1). The *affine maps* are the $T_j \circ M_k$. See [5].
- $\mathcal{O}_k$ is the octatonic scale that does NOT contain $k \in \mathbb{Z}_{12}$.

An *all-interval tetrachord*, or AIT, is a transposition or inversion of one of the two seminal forms $\{0, 1, 4, 6\}$ or $\{0, 1, 3, 7\}$, the only 48 pitch-class tetrachords admitting all possible intervals in their interval vector (4-Z15 and 4-Z29 in Forte's classification).

1 The Seminal Case

1.1 The Initial Picture

In an old drawing, the second author regrouped the 48 all-interval tetrachords as pairs of connected dyads (for compositional purposes) and found three connected subgraph components, see Fig. 1.

© The Author(s), under exclusive license to Springer Nature Switzerland AG 2024
T. Noll et al. (Eds.): MCM 2024, LNCS 14639, pp. 423–435, 2024.
https://doi.org/10.1007/978-3-031-60638-0_35

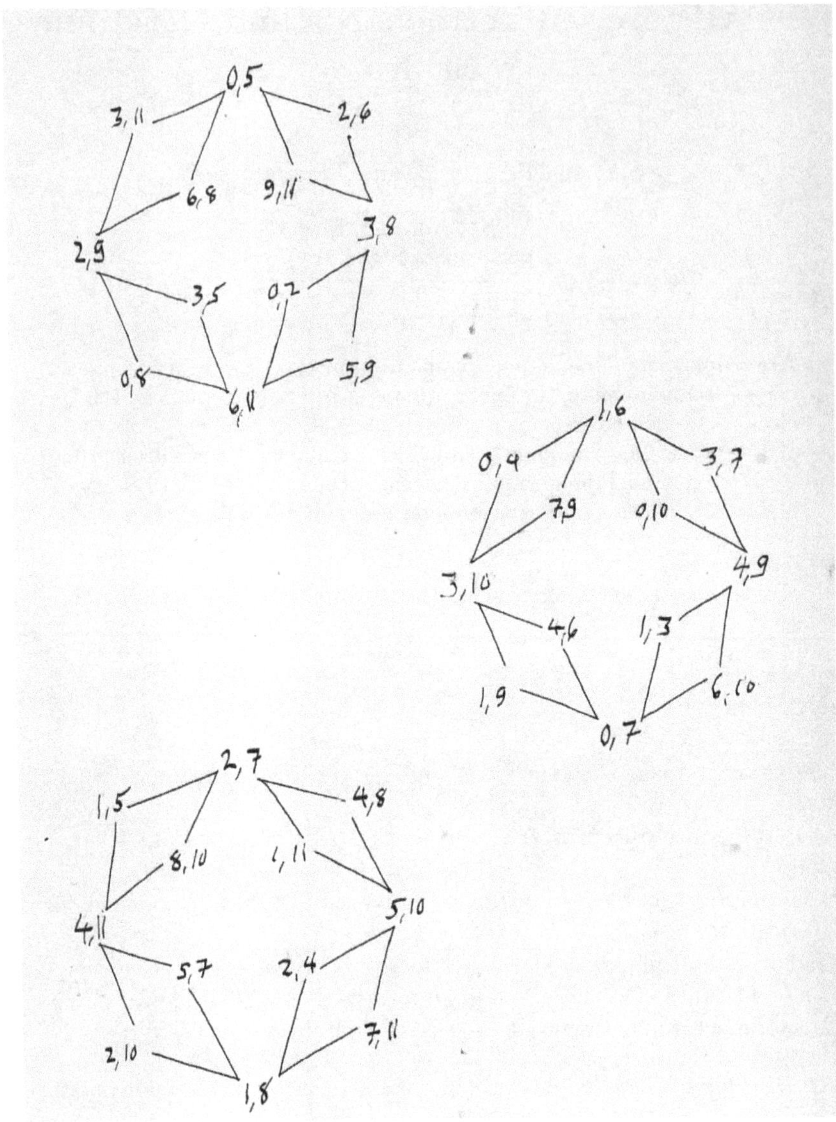

Fig. 1. The seminal picture

Several years afterwards, he noticed a funny fact (do you? Have a good look again at the pictures before carrying on with your reading, or listen to [3]: each of the 3 graphs is made of the tones in one of the three octatonic scales! For instance the left one has tones 1, 2, 4, 5, 7, 8, 10, 11, the complement of the diminished seventh 0, 3, 6, 9. This octatonic scale can be heard several times quite clearly in the lower voice, for instance in the piece *Minor seconds.* Why is this? True, any given all-interval tetrachord is included in one and only one octatonic scale, by why should its neighbours in the diagram belong to the same scale? It brings up several interesting points, perhaps not ground-breaking, but possibly suggestive for composers, theoreticians or teachers alike.

First let us make clear how the picture was constructed: the line (say) in the first picture connecting dyads/pairs 1,5 and 2,7 results in the tetrachord 1, 2, 5, 7, which is a transposition of one of the canonical shapes for all-interval tetrachords (or **AIT** for short thereafter), 0, 1, 4, 6. As we can see at a glance, after we start (say) from 1, 5, the graph is more or less determined; however, in this picture some pairs have two and some have four neighbours. More specifically, not all possible intervals appear, tritones, minor thirds and semi-tones being excluded, or more properly said, out of the picture – so far. Here, fourths (or fifths, the pairs being orderless) admit four completions in all-interval tetrachords, whereas whole tones and major thirds have two only. As we will see there was an alternative for the composer. But we need to do some maths first.

1.2 The Stabilizer of an Octatonic Scale

There is a nice Galois-like correspondence between the 48 AIT and an algebraic group, the group G of affine transformations of $\mathbb{Z}_{12}$ which contains all maps $x \mapsto ax + b$ wherein b is any element of $\mathbb{Z}_{12}$ and a is invertible in the ring, that is to say it is either 1, 5, 7 or 11 (notice that modulo 12, 11 = -1). G contains the famous $\boldsymbol{T}\,/\,\boldsymbol{I}$ group of transpositions/inversions as a subgroup (take $a = \pm 1$) and it is a normal subgroup. So is the sub-subgroup of transpositions, $\boldsymbol{T}$: by a direct computation, any conjugate of a transposition (resp. inversion) is a transposition (resp. inversion). Each of these three groups corresponds to a more or less precise definition of the shape of pc-sets: there are four orbits under $\boldsymbol{T}$, which are the four classes of AIT under transposition, only two under $\boldsymbol{T}\,/\,\boldsymbol{I}$, and one under G.

G has exactly 48 elements, the same cardinality as the set of AIT, and it is no coincidence since each transformation of G maps an AIT onto another AIT; and conversely, for any pair of AIT $(\mathcal{T}_1, \mathcal{T}_2)$, there exists exactly one affine map g in G such that $g(\mathcal{T}_1) = \mathcal{T}_2$. Depending on taste, you can call it a simply transitive group action, or a Generalized Interval System, see [2] – perhaps the latter is more appropriate in this context (Generalized *All*-Interval System?). So G exchanges AIT in the nicest possible way[1]. What of octatonic scales, like $\mathcal{O}_0 = \{1, 2, 4, 5, 7, 8, 10, 11\}$, which appear in the picture? Some affine maps exchange them (like transposition T_1 or inversion I_2), some do not (like transposition T_3, inversion I_3 or multiplication M_5, for $\mathcal{O}_0$). One basic fact in algebra is that all elements in a group G preserving a subset, say this octatonic scale $\mathcal{O}_0$, constitute a subgroup that we shall call here G_0. A slightly less simple result is that the ratio of their cardinalities, $|G|/|G_0|$, is the number of distinct octatonic scales, namely 3. This helpfully proves that there are 16 elements in G_0. As a shortcut and a vindication of this notation, let us notice that preserving $\mathcal{O}_0$ is the same

[1] The reader may be already familiar with the simpler case of $\boldsymbol{T}\,/\,\boldsymbol{I}$ acting on the set of major and minor triads, or the trivial case of $\boldsymbol{T}$ acting on major triads.

as preserving its complement, $\{0, 3, 6, 9\}$ which is the diminished seventh 7_0 beginning with 0. In other words, elements of G_0 are those affine maps who send a multiple of 3 onto a multiple of 3. When put that way, they are easily checked and listed:

- The four transpositions T_k for $k = 0, 3, 6, 9$.
- The four inversions I_k for $k = 0, 3, 6, 9$.
- The eight affine maps $T_k \circ M_5$ or $T_k \circ M_7$ for $k = 0, 3, 6, 9$, like $x \mapsto 7x + 9$.

Summing up, these are the transformations $x \mapsto ax + k, a = 1, 5, 7, 11, k = 0, 3, 6, 9$. This is a 16-element subgroup of G. Though its size lies between $\boldsymbol{T}$ and $\boldsymbol{T} / \boldsymbol{I}$, it is *not* normal, for interesting reasons that we will now explore.

What is characteristic in the above analysis is that there is a reciprocal: any pc-set invariant by G_0 is either $\mathcal{O}_0$ or its complement 7_0 (barring the empty set and the entire $\mathbb{Z}_{12}$). This goes a long way to explain this occurrence of octatonic scales in AIT landscape: the group acting on the AIT has non-trivial subgroups stabilizing the octatonic scales. Let us characterize these subgroups more precisely.

1.3 The 2-Sylows of G

If we do the same for the two other octatonic scales, the results are similar but not identical: the transpositions are the same, but for instance (working by hand or with a computer) G_1 contains the four inversions I_k for $k = 2, 5, 8, 11$ and the affine maps $x \mapsto 5x+k, k = 2, 5, 8, 11$ and $x \mapsto 7x+k, k = 0, 3, 6, 9$ – as we notice here, the first case was deceptively simple. However, there is a straightforward and satisfying relationship between the three stabilizers G_0, G_1, G_2: they are *conjugate*. Although this is a theoretical and general result, it is worthwhile to compute it by hand.

Let us consider for instance some $g \in G_0$: we have $g(7_0) = 7_0$ (remember this is the complement of the octatonic $\mathcal{O}_0$). But since $7_0 = T_{-1}(7_1)$, we get $g(T_{-1}(7_1)) = 7_0$ or more conveniently $T_1(g(T_{-1}(7_1)) = 7_1$, meaning exactly that $T_1 \circ g \circ T_{-1}$ is in G_1. In short, $G_1 = T_1 \circ G_0 \circ T_{-1}$. Similarly, $G_2 = T_2 \circ G_0 \circ T_{-2}$. This is a nice example illustrating the useful (and famous) theorems of Ludwig Sylow concerning subgroups whose order is a maximal power of a prime – here, the maximal power of the prime 2 dividing the cardinality of G, 48, is 16. The Sylow theorems state, first and foremost, that such subgroups exist (this partial reciprocal to Lagrange's theorem is much less obvious than it seems), that they are conjugate, and express some conditions on how many there are (here, 3). See any textbook, for instance [6], for comprehensive statements and proofs. Here we have three conjugates, hence they are not normal. Of course, the musical meaning of these aspirin-requiring results is that a piece featuring AIT inside an octatonic scale can be transposed while preserving its musical qualities. Some of these qualities are the potential symmetries of the piece, and most can be heard in [3].

1.4 Geometrical Embedding

Let us try and follow G_0 when we apply it to the picture of the AIT pertaining to the first octatonic scale $\mathcal{O}_0$, isolated for convenience on Fig. 2. Some of the transformations have a straightforward geometrical interpretation.

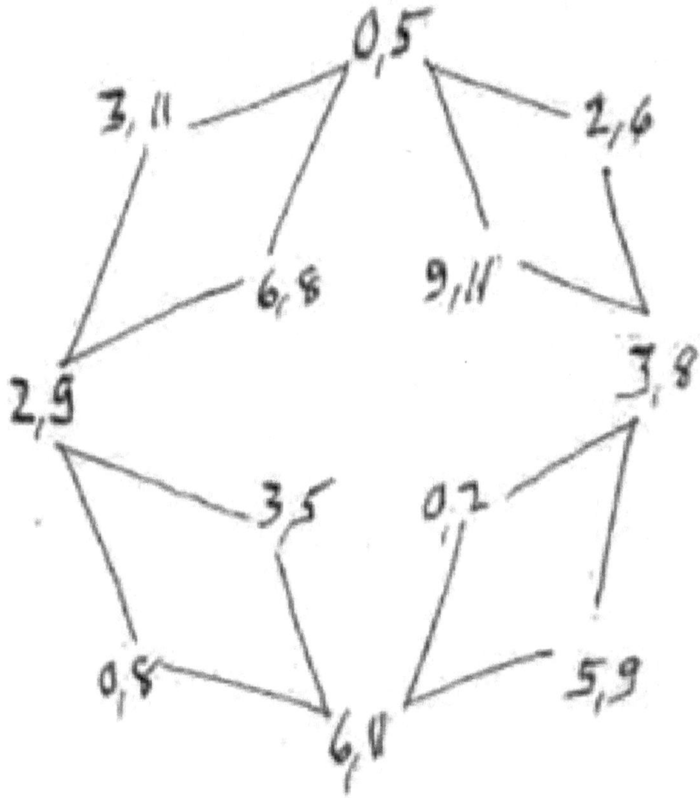

Fig. 2. The first connected component corresponding to $\mathcal{O}_0$

The transpositions appear on this diagram as rotations. For instance, the tritone transposition T_6 is a half-turn, while T_3 maps 4, 11 to 7, 2 (written as 2, 7), 1, 5 to 4, 8, etc.: a quarter-turn clockwise. Inversions, just as in a clock diagram of pitch-classes, are axial symmetries. For instance, I_9 exchanges 8, 10 and 1, 11 around a vertical axis, likewise 1, 5 and 4, 8, while preserving 2, 7 (2 goes to 7), etc. Together they form the classical, geometrical dihedral (sub)group with eight elements, the symmetries of a square, which acts on the sixteen AIT lying in the scale – I_9 turns 1, 2, 5, 7 into 2, 4, 7, 8. This dihedral subgroup is one of the 2-Sylows of $\boldsymbol{T}\,/\,\boldsymbol{I}$, which could be denoted as $\boldsymbol{T}\,/\,\boldsymbol{I}_0$, say. Other affine transformations preserving the octatonic, like M_5, do not preserve the diagram since they turn fifths into semi-tones, which are conspicuously absent on the picture. They do act on AIT though, for instance M_5 turns 1, 2, 5, 7 into 1, 5, 10, 11 with the basic shape being changed from 0, 1, 4, 6 to the other one, 0, 1, 3, 7 (²).

1.5 The Dual AIT Picture

Is this "a bug or a feature"? Definitely the latter! applying (say) M_5 to all elements of the pictures draws an entirely different though closely related graph[3], which can be considered as its twin. It also has three connected components, one for each octatonic scale. In Fig. 3 we draw the image of the component corresponding to $\mathcal{O}_0$ (the elements of the connected subgraph reduce to the set $\mathcal{O}_0$).

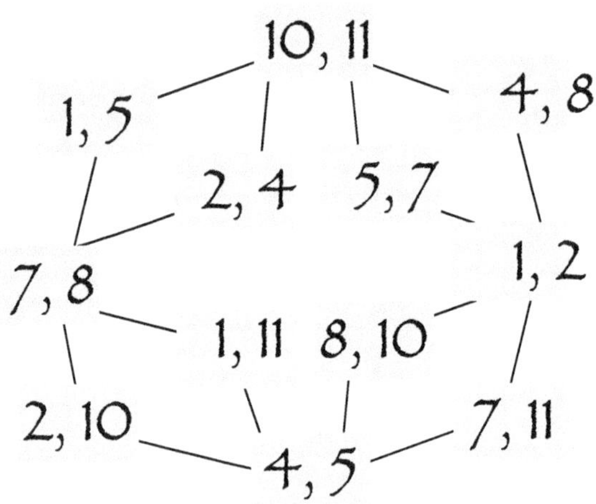

Fig. 3. First component of the dual graph

Doing the same on the two remaining components is left as a treat to the reader (there is a rewarding surprise).

1.6 Are Such Designs Unique?

A cold mathematical analysis yields a more complicated graph[4].

---

[2] These are non-isometric transformations for the natural metric on $\mathbb{Z}_{12}$. They are however isometrical for the torus metric on $\mathbb{Z}_3 \times \mathbb{Z}_4$, as first noticed by Mazzola in [4].

[3] Actually it appeared in another sketch by the second author.

[4] Actually the constructions in the first pictures are reminiscent of *hypergraphs*, whose "edges" are arbitrary subsets, but it is not quite the case since the same pitch-classes occur in several different locations.

If one wants to complete (say) a fourth/fifth into an AIT, then as mentioned previously there are exactly four possible complements, which are two major thirds and two seconds. How to complete a given pc-set – say, a pair – into some definite pc-set *shape* in general would be a worthwhile field of study. But here we can use the very special specificity of an AIT:

Lemma 1. *Any interval* a, b *except the tritone belongs to exactly four AIT.*

This is because any AIT contains one and only type of interval by its very definition – barring the tritone which comes twice, being its own inversion –, and there are four distinct shapes up to transposition, hence for each of the 4 types there is one transposition exactly containing a given interval. For instance, 0, **1**, 4, **6** contains the fourth 1, 6 which is completed by 0, 4, a major third. The same happens with the only inversion of the AIT exchanging 1 and 6: **1**, 3, **6**, 7 complements its fifth by 3, 7. But a different complement of the pair occurs with the non-isometric form **1**, **6**, 7, 9 where the complement is a second (same for its inversion 0, **1**, **6**, 10).

Start with the fourth/fifth 4, 11 in Fig. 2 for instance: it can be completed only by two (major) thirds in the first type, or two seconds in the other type. These are exactly its neighbours on the picture.

However, the lemma above promises four neighbours again for each of these major thirds complements. Consider any of these thirds, say 1, 5: it can be completed by fifths 4, 11 or 2, 7, as in the original picture, which build up 11, 1, 4, 5 and 1, 2, 5, 7 which are transposed and inverted of 0, 1, 4, 6. But one could have chosen to complete also (or alternatively) to the other form by adding a *minor* second: 10, 11 or 7, 8. These do not appear on the original picture Fig. 2, but are featured as neighbours of 1, 5 on the twin picture Fig. 3.

Carrying on the process, one obtains a more complicated graph, whose vertices all have *degree 4* (they are all connected to four neighbours) and can be fourths, major and minor seconds, or major thirds. It is the merging of Fig. 2 and 3. It is featured on Fig. 4 – which shows well why the initial drawing skipped some possible edges for the sake of neatness. On the right side, we tried to draw the graph in 3D for better legibility. Arguably, it could be even simpler in 4D, where it looks a lot like a hypercube. Notice that we still get a connected component whose underlying set is an octatonic scale, a new musical object combining AIT by common pairs in a maximal way. It can be viewed as a glueing together of cubes (see right picture). Alternatively, it is a combinatorial object where tetrachords are connected whenever their intersection is a pair, and more specifically this pair is any interval except minor thirds or tritones.

What does **not** work too well is starting with one of the missing intervals, a tritone or a minor third: any tritone can be completed into no less than *eight* distinct AIT – always by a minor third[5], and each minor third has four tritone

[5] The doubling from four to eight is easily explained by the fact that every tritone is two intervals, say 5, 11 is also 11, 5. Besides, the complement pair's *shape*, a minor third, is invariant by multiplication by 5 (or 7), since $5 \times 3 = 3 \mod 12$ and it remains the same in all four forms of AIT.

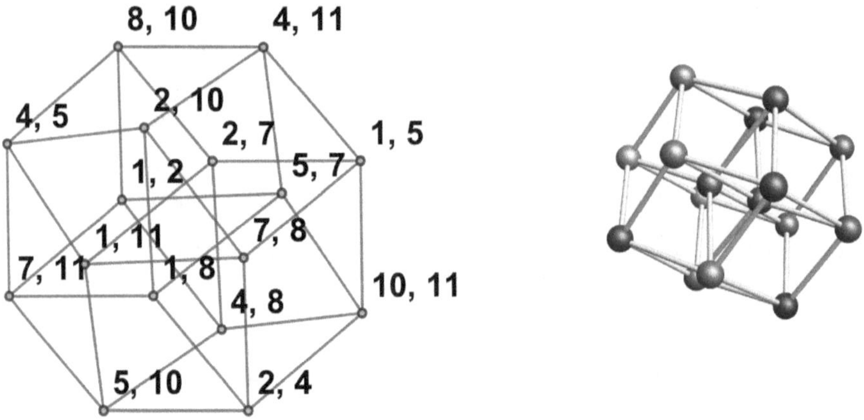

Fig. 4. The seminal graph combined with its multiple by 5

completions, following the general rule in the Lemma above. All tetrachords (in a given octatonic scale) are obtained in these decompositions, but in the less than legible graph on Fig. 5.

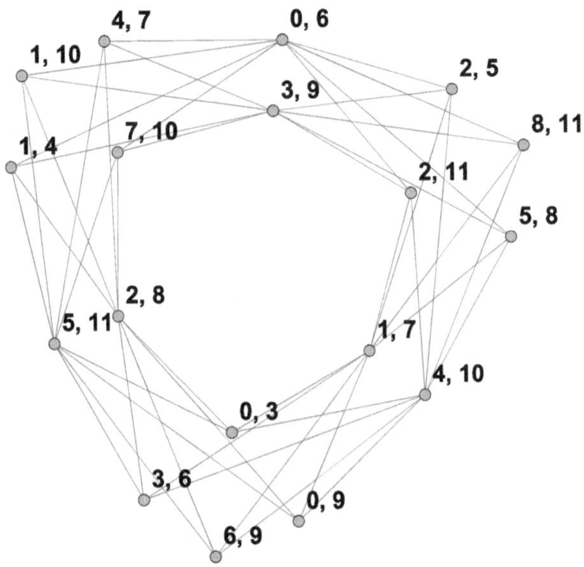

Fig. 5. Completing tritones by minor thirds

2 Variants

This does not work very well either if we try to use major/minor triads instead of AIT, because, to begin with, G is not acting anymore on these chords: only

$\boldsymbol{T} / \boldsymbol{I}$ does. Still, similar graphs are possible, recalling the majestic major chord progressions written by Messiaen when he used his second Mode M2 (the name he gave to the octatonic scale, among his "modes à transpositions limitées"). In this graph, vertices are connected by the neo-Riemannian transformations P and R alternating, or alternatively with elements of $\boldsymbol{T} / \boldsymbol{I}$. We have lost the nice decomposition in pairs, obviously. But again we get three connected orbits, each stable by one of the three 2-Sylows $\boldsymbol{T} / \boldsymbol{I}_i, i = 0, 1, 2$. They are octatonic cycles of triads in the Tonnetz and each one covers one of the three octatonic scales! These octatonic cycles (in more senses than one) will be familiar to many readers. We present one of them on Fig. 6. One could also add their images by multiplication by 5, which are transpositions of 0, 1, 4 and its inverse 0, 3, 4 but are less familiar in tonal music (though subsets of *minor* scales, for instance, see the famous theme of Bach's *Fugue in C♯minor* in WTC 1). Pairs of corresponding cycles could be connected by those triads sharing a major third. We do not produce the corresponding graph, which is decidedly not as legible as the other ones.

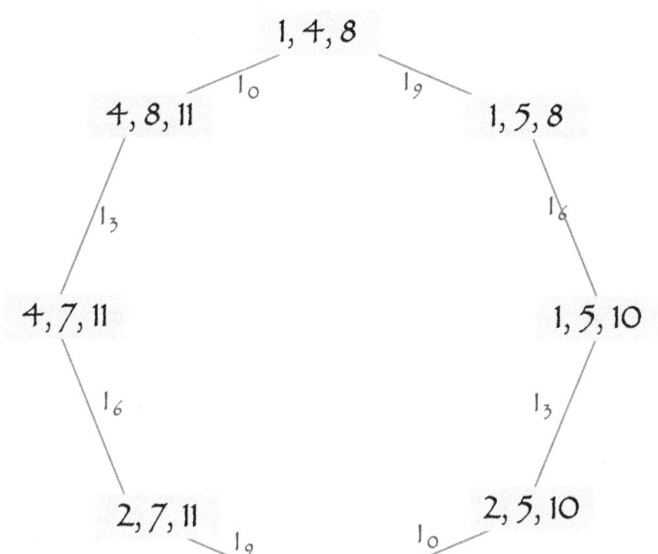

Fig. 6. Octatonic cycle of triads

Let us revert to tetrachords: similar pictures can be drawn with dominant sevenths and their inverses, Tristan chords, which have almost flat interval distribution and in this regard are more similar to AIT than, say, diminished sevenths. Again, connecting tetrachords in case of common pairs, the pc-sets can be grouped into three octatonic components, which are rich in connections, each vertex being connected to all others but one (the opposite vertex in the cube, see Fig. 7, each edge corresponding to a common pair, with an exchanging symmetry in $\boldsymbol{T} / \boldsymbol{I}$.

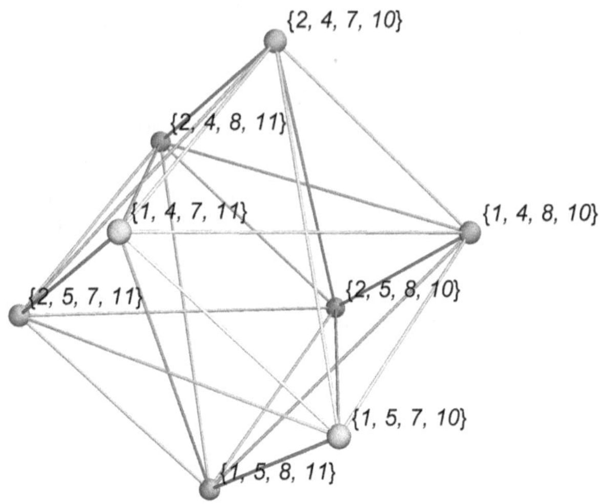

Fig. 7. A cubic graph of dominant sevenths and their inverses in $\mathcal{O}_0$

Among these symmetries is the inversion in Tristan's first leitmotif that exchanges the Tristan chord $\{3, 5, 8, 11\}$ and the following Dominant seventh $\{2, 4, 8, 11\}$, cf. [1] among many other references.

It is also quite possible to draw a graph more akin to the seminal one, which would connect complementary pairs (the vertices) into sevenths (the edges). But the resulting graph is either unreadable (when all possible edges are drawn) or barren (when limiting the possible pcs to an octatonic scale): for instance, a major third, like 7, 11 admits only two complements in $\mathcal{O}_0$ which are 1, 4 and 2, 5. These two minor thirds are only complemented by 7, 11 and 8, 10 and the picture is a simple square. The 24 pairs are partitioned into 6 similar squares, see Fig. 8. On the other hand, trying to connect sevenths without first segregating them into ambient octatonic scales leads to very messy graphs, which have been omitted here.

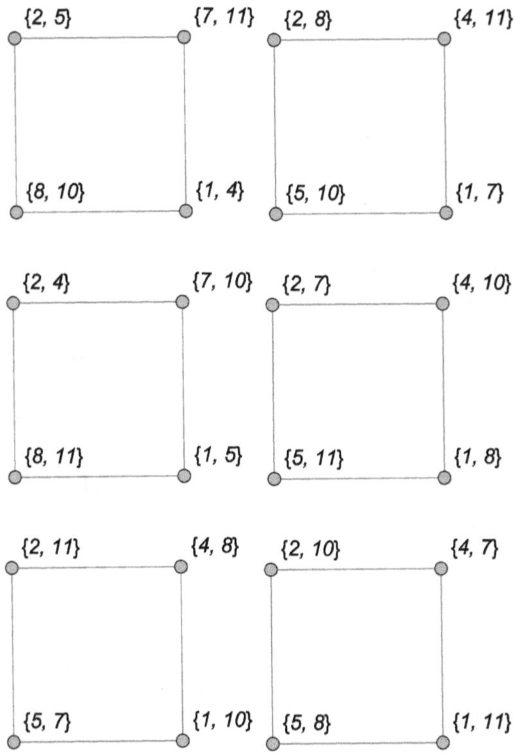

Fig. 8. Dominant sevenths and Tristan chords grouped inside octatonic scales

Finally, limiting oneself to four dominant sevenths (or four Tristan chords) provides sequences that have been seen time and again in actual compositions, invariant by T_3 and generating an octatonic context. Ironically, the fist author wrote the excerpt on Fig. 9 in a song, featuring such a sequence of Tristan chords, a few months before hearing about the octatonic scale of the seminal picture by the second author.

Fig. 9. Composition using an orbit of Tristan chords under T_3

I am grateful to an anonymous reviewer for suggesting checking other moduli. Again it confirms the miracle of all-interval tetrachords in 12-TeT:

- Mod 7, for the famous Fano (projective) plane which can be seen as trichords (1 2 4) in a diatonic universe (7 notes), there are exactly 2 trichords sharing any given note, two trichords never share a pair ("there is only one line passing through two points"); it is difficult to imagine a graph similar to the ones in this paper.
- Raising the modulo instead, one can find all-interval *pentachords* modulo 21, like $0, 1, 4, 14, 16$. There are 42 of them, less that AIT modulo 12 and indeed they intersect too sparsely for beautiful graphs, with as a rule only *one* other pentachord sharing a given pair, though ten share any given pc. Perhaps it does not work as well as the seminal case because here the affine group works transitively but not simply transitively, though $\boldsymbol{T} / \boldsymbol{I}$ does – there are only two "shapes" for the pentachords.
- Other experiments (for instance with collections of 64 homometric hexachords modulo 16) also fail to recover the beauty and simplicity of the seminal situation.

Conclusion

We did not achieve a comprehensive survey of all interesting geometrical or algebraic features of the seminal graph: the reader may find some other points not devoid of appeal. Still, we feel that listing the mathematical concepts involved may be useful for a better understanding of the relationships between these chords (pc-sets) and help in both analysis and composition. We hope that we have made clear how and why the initial picture is, in many ways, the best possible, keeping a balance between meaning and usefulness unequaled by the other logical possibilities. The actions of 2-Sylows of $\boldsymbol{T} / \boldsymbol{I}$ also explain how maximally *non*-symmetrical objects like the AIT give rise to the elegant Fig. 1.

If any musician happens to remember a couple of the ideas presented in this paper in their practice, then our purpose will be fulfilled. Alternatively, an algebra course could use these graphs and groups as telling examples, less barren that the ones provided in textbooks as a rule.

For instance, to mention a trend popular nowadays, one could search for Hamiltonian paths (passing exactly once through each pair) or Eulerian ones (passing through all edges, i.e. realizing once and only once each AIT in the chosen octatonic scale). We leave it as a treat for the reader, in the simplest graphs (wherein there are few, if any, solutions). The same problem for the complete octatonic graph on Fig. 4 requires a computer with quite a bit of memory.

References

1. Amiot, E.: Pour en finir avec le désir, Revue d'Analyse Musicale, 22 February 1991
2. Lewin, D.: Generalized Musical Intervals and Transformations. Yale University Press, New Haven (1987)

3. Johnson, T.: Intervals, for any two instruments, Editions 75, 75 rue de la Roquette, Paris, France (2013). https://www.youtube.com/watch?v=2ps49ceQe8
4. Mazzola, G., ii, et al.: Topos of Music. Birkhauser, Zurich (2002)
5. Rahn, J.: Basic Atonal Theory. Longman, New York (1980)
6. Sylow, Ludwig, 3 Theorems. https://en.wikipedia.org/wiki/Sylow_theorems

Stages in My Vuza Rhythmic Canons

Violeta Dinescu(✉)

Universität Oldenburg, Fakultä III – Institut für Musik, 26111 Oldenburg, Germany
violeta.dinescu@uni-oldenburg.de

Abstract. Several stages in understanding and using the Vuza rhythmic canon structure are enumerated. The starting idea was to compose **Vuza canons of moments** instead of Vuza canons of single notes. The next idea was to integrate such Vuza canons in more complex structures: polyphony of a Vuza canon and one or more melodies, polyphony of Vuza canons, polyphony of Vuza canons and other musical structures. Another stage was that of **Vuza canons of attacks**, allowing overlapping of voices, but not of attacks. This musical form is in the spirit of classical canons.

Keywords: rhythmic canon · Vuza rhythmic canon · attack · polyphony

1 Introduction

I made my first Vuza canons in the time of the great covid crisis. It happened that, before the covid restrictions in teaching, I planned to give composition courses at Oldenburg University about different types of canons, including Vuza rhytmic canons. Due to the interest of the students, in spite of the frame of online meetings, this topic was retaken another two semesters. Along analysing classical canons, the students wrote their own canons. Among these, there were many Vuza canons, in the spirit of the numerous existing pieces available online, written by composers from all over the world.

In the following I describe the stages of my path in the realm of Vuza rhythmic canons. The first five stages are related to my double CD "Vuza Canons". The next natural challenge was to write Vuza canons for live performances. I wrote two pieces, with a new (for me) understanding of the Vuza canon structure: "Cronicari" (2023) after the well known poem by URMUZ and "Exalta vocem tuam" (2024) after Isaiah 58:1.

2 Stage 1: Vuza Canon of Disjoint Soloistic Moments

My first Vuza canon were realized patching together compatible chunks of certain of my own compositions, for which I had professional recordings, played by outstanding musicians. The Vuza canon structure was interpreted in the way described in cf. [4]:

© The Author(s), under exclusive license to Springer Nature Switzerland AG 2024
T. Noll et al. (Eds.): MCM 2024, LNCS 14639, pp. 436–441, 2024.
https://doi.org/10.1007/978-3-031-60638-0_36

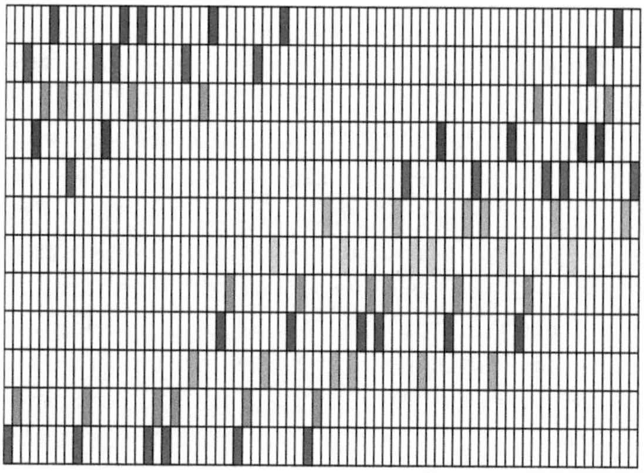

In the following, making abstraction of other contents, we consider a musical work (or a part of it) as consisting of M *events*, called also *moments* or *entries*, or *interventions*, for V voices.

Dan Tudor Vuza has introduced, in his works [2], a type of rhythmic canon which is defined by the requierment to fulfil the properties 1–4 below (in the logic of his works, it was named "complementary regular canon of maximal category"). These canons were later called *Vuza canons.*

We denote the successive events by numbers from 0 to (M-1), for reasons which will become obvious in the course of the explanation.

Conditions:

1. *Each moment is assigned to a voice and only to one; so, we have no overlapping voices.*

   Convention. The first voice (called voice 0) is the one that starts the piece, i.e. it manifests at time 0. Let $(M_0 = 0, M_1, \ldots, M_{L-1})$ be the moments of this voice.
2. *All other entries follow the same pattern, i.e., there is a set of numbers* $(d_0 = 0, d_1, \ldots, d_{V-1})$ *such that:*

   Voice 1 has intervening moments $(M_0 + d_1, M_1 + d_1, \ldots, M_{L-1} + d_1)$

   Voice 2 has interventions at times $(M_0 + d_2, M_1 + d_2, \ldots, M_{L-1} + d_2)$

   etc.

   It follows that each voice has L entries and $M = L \times V$*.*
3. *The sequence of numbers* $(M_0, M_1, \ldots, M_{L-1})$ *is not periodic, i.e. no voice has periodic interventions.*
4. *The sequence of numbers* $(d_0 = 0, d_1, \ldots, d_{V-1})$*, which describes translations, is not periodic.*

All the numbers $M_0, M_1, \ldots, M_{L-1}$ and $d_0 = 0, d_1, \ldots, d_{V-1}$ are taken in the set of natural numbers $\{0, 1, \ldots, M-1\}$ and in additions $M_i + d_j$ the results are "brought" to be all in the set $\{0, 1, \ldots, M-1\}$, by substracting, if necessary, M from the sum obtained (example: if $M = 72$ then $50 + 33$ will be considered $83 - 72 = 11$). With this rule of addition "modulo M", a sequence $\{M_0, M_1, \ldots, M_{L-1}\}$ is called periodic if there is a natural number p with the property that the sequence obtained by translation with p, i.e. $\{M_0 + p, M_1 + p, \ldots, M_{L-1} + p\}$ is a permutation of the initial sequence. To give an example: the sequence 0, 3, 12, 15, 24, 27, 36, 39, 48, 51, 60, 63 modulo 72 is periodic because it has a symmetry of translation (by $p = 12$ modulo 72).

We call the set of numbers $(M_0, M_1, \ldots, M_{L-1})$ the inner rhythm and the set of numbers $(d_0, d_1, \ldots, d_{V-1})$ the outer rhythm of the canon.

Given a Vuza canon structure, one obtains a new Vuza canon structure (called *the dual* of the given one) taking $(d_0, d_1, \ldots, d_{V-1})$ as the inner rhythm and $(M_0, M_1, \ldots, M_{L-1})$ as the outer rhythm.

Without a theoretical study it is difficult to find Vuza canon structures, i.e. to find $(M_0 = 0, M_1, \ldots, M_{L-1})$ and $(d_0 = 0, d_1, \ldots, d_{V-1})$ that meet the above requirements.

One of the surprising results of Dan Tudor Vuza is that the number of voices in a Vuza canon has at least two prime factors (so M also has at least two prime factors); hence we have no Vuza canons with 2, 3, 4, 5, 7, 8, 9, 11, 13, 16 ... voices and no Vuza canon of length M=64 ($= 2^6$) or M=81($= 3^4$) ....

Note. In general, the composers who authored Vuza canons have interpreted the *event* as a single musical note. ..."

A first practical step was to consider the event as a fixed length time interval (in my pieces of 7 or 8 s). A Vuza canon of moments is a piece with disjoint (not overlapping) voices.

This stage is exemplified by the piece "Round Table I" for oboe, clarinet, saxophone, violine, viola and piano (Vuza canon of internal rhythm (0, 1, 21, 24, 25, 30, 36, 45, 49, 60, 66, 69) and external rhythm (0, 8, 16, 18, 26, 34).

3 Stage 2: Vuza Canon of Disjoint Multiple Voices Moments

The next step was to allow a voice of the canon to consist in a group of instruments, from duo to an orchestra. This stage is represented, for instance, by "Round Tables II, III and IV". The groups of instruments are: two guitars, clarinet and viola, voice and trombone, trio of recorders, trio of flute, percussion and piano, horn quartet, wind quintet and string orchestra.

4 Stage 3: Polyphony of Strict Vuza Canon and Melody

An example of this kind is "Transparency I" (Two times Vuza Canon for 6 voices with solo violin).

5 Stage 4: Polyphony of Strict Vuza Canons with or Without Other Melodies

This can be seen in "Transparencies III. Three Vuza Canons for 18 Voices with Mezzosopran Solo and Clarinet Solo".

6 Stage 5: Polyphony of Strict Vuza Canons and Flexible Traditional Canons

An example of this kind is "Transparency II. Three Times Vuza Canon for 6 Voices with a Sequence in Canon of Other Voices" and "1989–1990. Vuza Canon for 12 Voices with 3 Celli in Canon".

7 Stage 6: Vuza Canon of Attacks in Nondisjoint Moments, with Some Incidental Attacks

In writing the piece "Cronicari", after the first part, which is a strict 6 voices Vuza rhythmic canon modulo 72, of 6 voices (here three voices are actually DUOs) with internal rhythm (0, 19, 23, 24, 30, 36, 43, 47, 60, 66, 67, 71) and external rhythm (0, 22, 38, 40, 54, 56), I added two other parts in which I allowed some voices to sing/play supplementary. This way the moments of intervention of the voices are no more disjoint, but the structure of Vuza canon remains at the level of attacks, as a subset of the set of all attacks. The "new attacks" are not synchronous with the "old" ones. Cf. https://www.youtube.com/watch?v=1ugbFdW3mKY for the premiere on 24th of May 2023, The 32nd International Week of New Music (SIMN-EUTOPIA), Bucharest and https://www.youtube.com/watch?v=4KSiboW_MjM at 40:04 for the premiere of the variant in three parts on 10th of November 2023, UrMuzică și Literatură, Bucharest.

8 Stage 7: Vuza Canon of Attacks in Nondisjoint Moments

The composition "Exalta vocem tuam", after Isaiah 58:1 has a first part which follows strictly the structure of a Vuza canon **at the level of attacks**. It has module 108, 6 voices, internal rhythm (0, 28, 29, 30, 36, 37, 38, 46, 47, 66, 72, 82, 83, 91, 92, 100, 101, 102) and external rhytme (0, 33, 57, 60, 81, 84) and the overlapping of voices brings no other attacks. This structure is a kind of purification of that from **Stage 6**. I learned incidentally about this natural interpretation being present when Dan Tudor Vuza spoke about it in a small group of interested people.

I would like to mention two further Vuza canons of the Romanian composers Laurenţiu Beldean ("Stupore II. Hoquetus on a Vuza Canon". Text: URMUZ for 7 Singers, Percussion, two "Hesitant Performers", a Slavic Spectator and Electroacoustic Sounds; premiere in Bucharest, 10th of November 2023, UrMuzică și

Literatură, Bucharest), Laura Manolache ("Perspective" for 6 Percussions and a Violin on a structure of Vuza Canon, 2024), and Mihaela Vosganian ("Balinese Interferences" for Voices, Gamelan Ensemble and Electronics). Their pieces are also essentially Vuza rhytmic canons of several seconds long moments.

Acknowledgement. Some time after the performance of "Cronicari", I was very fortunate to attend an explanation given by Dan Tudor Vuza about the understanding of Vuza canons as canons of attacks of maybe overlapping voices.

References

1. Vuza, D.T.: Sur le rythme périodique. Revue Roumaine de Linguistique - Cahiers de Linguistique Théorique et Appliquée **22**(1), 73–103 (1985). republished in Musikometrika I (editor : M.G. Boroda), p. 83 - 126. Studienverlag Dr. N. Brockmeyer, Bochum 1988
2. Vuza, D.T.: Supplementary sets and regular complementary unending canons I - IV. Perspectives of New Music, **29**, 2 (1991), pp. 22–49; **30**, 1 (1992), pp. 184–207; **30**, 2 (1992), pp. 102–125; **31**, 1 (1993), pp. 270–305
3. Vuza, D.T.: Supplementary sets - theory and algorithms. Muzica **1**, 75–99 (1995)
4. ***, Naive Presentation of Vuza Rhythmic Canons, booklet of the double CD Vuza Canons **gutingi 262**, **charisma Musikproduktion**, January 2023

Configurations of Disjoint Augmentation Canons

Jeremy Kastine(✉)

Life University, Marietta, GA 30060, USA
jeremy.kastine@life.edu

Abstract. This paper introduces the concept of configurations of disjoint augmentation canons. A simple necessary and sufficient condition is given for the feasibility of any given 2-part n-note configuration.

Keywords: Rhythmic Canon · Augmentation

1 Introduction

Consider the following rhythmic canon:

Fig. 1. An example of a disjoint augmentation canon.

The rhythm in Part 2 is the result of augmenting that of Part 1 by a factor of $\frac{2}{3}$ (a diminution, really) and then translating it 2 beats to the right. (For simplicity, the augmentation is applied to the onset times of the notes, and not the durations.) In so doing, the onset times of Part 2 are disjoint from those of Part 1. Different pitches were chosen for the two parts so that we can write the canon on a single clef as follows.

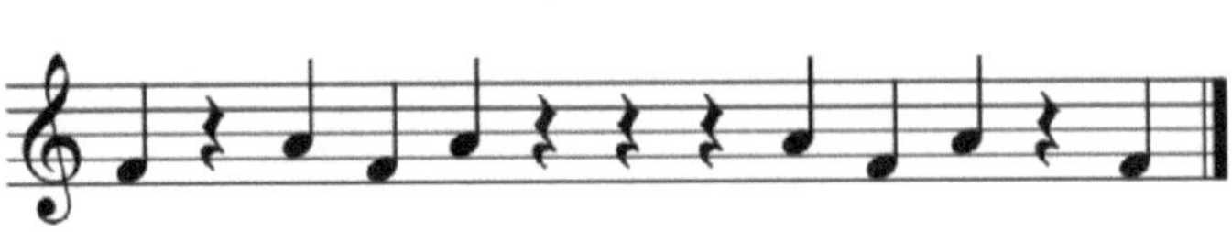

Fig. 2. Another way to display the canon from Fig. 1.

Definition 1. Let $C = (T, A, R)$ where

© The Author(s), under exclusive license to Springer Nature Switzerland AG 2024
T. Noll et al. (Eds.): MCM 2024, LNCS 14639, pp. 442–446, 2024.
https://doi.org/10.1007/978-3-031-60638-0_37

- $T = (t_1, t_2, \ldots, t_p) \subseteq \mathbb{R}$ is strictly increasing,
- $A = (a_1, a_2, \ldots, a_p) \subseteq (0, \infty)$, and
- $R = (r_1, r_2, \ldots, r_n) \subseteq \mathbb{R}$ is strictly increasing.

We will call C a p-part n-note augmentation canon, where T gives the translations of the parts, A gives the augmentation factors of the parts, and R is the rhythmic theme.

We will use the notation $[a, b]_{\mathbb{Z}}$ for the set of all integer values between a and b inclusive.

Define $f_C : [1, p]_{\mathbb{Z}} \times [1, n]_{\mathbb{Z}} \to \mathbb{R}$ by $f_C(i, j) = t_i + a_i r_j$. If f_C is one-to-one, then we will say that C is a disjoint augmentation canon. In this case, we let $<_C$ be the strict total order on $[1, p]_{\mathbb{Z}} \times [1, n]_{\mathbb{Z}}$ such that $(i_1, j_1) <_C (i_2, j_2)$ if and only if $f_C(i_1, j_1) < f_C(i_2, j_2)$. We call $<_C$ the configuration of C.

Example 2. The canon given in Figs. 1 and 2 can be represented by $C = ((0, 2), (3, 2), (0, 1, 3, 4))$.

Definition 3. Let $\prec$ be a strict total order on $[1, p]_{\mathbb{Z}} \times [1, n]_{\mathbb{Z}}$ such that.

- $(i, 1) \prec (i, 2) \prec \cdots \prec (i, n)$ for each $i \in [1, p]_{\mathbb{Z}}$ and
- $(1, 1) \prec (2, 1) \prec \cdots \prec (p, 1)$.

We will call $\prec$ a p-part n-note disjoint rhythmic configuration, where each ordered pair (i, j) represents the j-th note of the i-th part If there exists a p-part n-note disjoint augmentation canon C such that $<_C$ is equivalent to $\prec$, then we will say that C is a solution of $\prec$ and that $\prec$ is feasible.

Example 4. Consider the 3-part 3-note disjoint rhythmic configuration defined as follows.

$$(1, 1) \prec (1, 2) \prec (2, 1) \prec (3, 1) \prec (2, 2) \prec (2, 3) \prec (1, 3) \prec (3, 2) \prec (3, 3)$$

Musically, the task of searching for a solution of $\prec$ amounts to that of inserting rests into the following template in such a way that the three rhythms become proportional. (Part 1 = F, Part 2 = A, Part 3 = C.) (Fig. 3)

Fig. 3. The disjoint rhythmic configuration given by $\prec$.

A solution is given in Fig. 4, which corresponds to the disjoint augmentation canon $C = ((0, 7, 8), (6, 2, 5), (0, 1, 2))$.

Thus, $\prec$ is feasible.

Example 5. Consider the 3-part 3-note configuration defined by.

$$(1, 1) \prec (2, 1) \prec (3, 1) \prec (1, 2) \prec (3, 2) \prec (2, 2) \prec (1, 3) \prec (2, 3) \prec (3, 3).$$

Fig. 4. The solution C of $\prec$.

This configuration is infeasible. The reader is encouraged to draw out the corresponding template as in the example above and try to see why no solution can exist before reading the explanation in the following paragraph.

Since we have $(2, 1) \prec (3, 1) \prec (3, 2) \prec (2, 2)$, the augmentation factor of Part 2 must be strictly greater than that of Part 3. But we also have $(3, 2) \prec (2, 2) \prec (2, 3) \prec (3, 3)$, which implies that the augmentation factor of Part 3 must be strictly greater than that of Part 2. This contradiction makes it impossible to find a solution of $\prec$.

2 Theoretical Results

Proposition 6. Let $\prec$ be an p-part n-note disjoint rhythmic configuration. If there exist $i_2 > i_1 \geq 1$ and $j_2 > j_1 > 1$, such that $(i_2, j_1) \prec (i_1, j_1) \prec (i_1, j_2) \prec (i_2, j_2)$, then $\prec$ is infeasible.

Proof. Suppose to the contrary that $\prec$ is feasible and that $C = (R, A, T)$ is a solution. Then we have.

$$a_{i_1}\left(r_{j_2} - r_{j_1}\right) = f_C(i_1, j_2) - f_C(i_1, j_1) < f_C(i_2, j_2) - f_C(i_2, j_1) = a_{i_2}\left(r_{j_2} - r_{j_1}\right),$$

which implies that $a_{i_1} < a_{i_2}$. But by the definition of disjoint rhythmic configurations, we also have

$$(i_1, 1) \prec (i_2, 1) \prec (i_2, j_1) \prec (i_1, j_1),$$

which (in the same way as above) implies that $a_{i_2} < a_{i_1}$. This contradiction indicates that $\prec$ is infeasible.

Proposition 7. Let $\prec$ be a 2-part n-note disjoint rhythmic configuration with $n \geq 2$. If there is no $i \in [1, n-1]_{\cap\mathbb{Z}}$ for which $(2, i) \prec (1, i) \prec (1, i+1) \prec (2, i+1)$, then $\prec$ has a solution $C = (T, (a_1, a_2), R)$ where $a_1 > a_2$.

Proof. We will prove the proposition by induction. First consider the base case of $n = 2$. Below, we examine the three possible 2-part 2-note configurations and find that each of them has a solution with $a_1 > a_2$.

- If $\prec$ is defined by $(1, 1) \prec (1, 2) \prec (2, 1) \prec (2, 2)$, then $C = ((0, 3), (2, 1), (0, 1))$ is a solution.
- If $\prec$ is defined by $(1, 1) \prec (2, 1) \prec (1, 2) \prec (2, 2)$, then $C = ((0, 2), (3, 2), (0, 1))$ is a solution.
- If $\prec$ is defined by $(1, 1) \prec (2, 1) \prec (2, 2) \prec (1, 2)$, then $C = ((0, 1), (3, 1), (0, 1))$ is a solution.

Now suppose that the proposition holds for a fixed value of $n = k$ where $k \geq 2$. Let $\prec$ be a 2-part $(k+1)$-note configuration and suppose that there is no $i \in [1, k]_{\cap\mathbb{Z}}$ for which $(2, i) \prec (1, i) \prec (1, i+1) \prec (2, i+1)$. Let $\prec_0$ be the 2-part k-note configuration obtained by removing $(1, k+1)$ and $(2, k+1)$ from $\prec$. Then by the inductive hypothesis, $\prec_0$ has a solution $C_0 = (T, (a_1, a_2), R_0)$ where $a_1 > a_2$. Consider the following four cases.

1. Suppose that $(1, k) \prec (2, k)$ and $(1, k+1) \prec (2, k+1)$. Let (i_1, j_1) be the element that immediately precedes $(1, k+1)$ in $\prec$ and let (i_2, j_2) be the element that immediately follows $(1, k+1)$ in $\prec$. Choose a value of y strictly between $f_{C_0}(i_1, j_1)$ and $f_{C_0}(i_2, j_2)$. Let $r_{k+1} = \frac{y-t_1}{a_1}$ and append r_{k+1} to the end of R_0 to form R. Then $C = (T, (a_1, a_2), R)$ is a solution of $\prec$ with $a_1 > a_2$.
2. Suppose that $(1, k) \prec (2, k)$ and $(2, k+1) \prec (1, k+1)$. Choose $r_{k+1} > \max\left(r_k, \frac{t_2-t_1}{a_1-a_2}\right)$, append r_{k+1} to the end of R_0 to form R, and let $C = (T, (a_1, a_2), R)$. Note that

$$\begin{aligned} &f_C(1, k+1) - f_C(2, k+1) \\ &= t_1 - t_2 + (a_1 - a_2)r_{k+1} \\ &> t_1 - t_2 + (a_1 - a_2)\left(\frac{t_2-t_1}{a_1-a_2}\right) \\ &= 0. \end{aligned}$$

   So $f_C(2, k+1) < f_C(1, k+1)$, which means that C is a solution of $\prec$.
3. We need not consider the case that $(2, k) \prec (1, k)$ and $(1, k+1) \prec (2, k+1)$, since it would contradict Proposition 6.
4. The case in which $(2, k) \prec (1, k)$ and $(2, k+1) \prec (1, k+1)$ can be proved in the same way as Case 1.

Theorem 8. Let $\prec$ be a 2-part n-note disjoint rhythmic configuration. Then $\prec$ is feasible if and only if there is no $i \in [1, n-1]_{\cap\mathbb{Z}}$ for which $(2, i) \prec (1, i) \prec (1, i+1) \prec (2, i+1)$.

Proof. Follows directly from the previous two propositions.

3 Next Steps

Given the simple feasibility test of Theorem 8, we can easily enumerate feasible 2-part n-note configurations by a straight-forward recursive search procedure. The results are as follows (Table 1).

Table 1. Counting feasible 2-part n-note configurations.

Notes	1	2	3	4	5	6	7	8
Solutions	1	3	9	28	90	297	1001	3432

This sequence seems to match A000245 from the Online Encyclopedia of Integer Sequences (OEIS), the formula of which is

$$a(n) = \frac{3(2n)!}{(n+2)!(n-1)!}$$

I hope to prove that the sequence of the number of feasible 2-part n-note configurations is indeed given by the formula above for all n. If so, I would like to see if any of the applications in which this sequence arises (as given by the OEIS) is related to the topic of this paper in any useful way.

I will also investigate whether the feasibility criteria given in this paper can be extended, either exactly or in some modified form, to configurations of three or more parts.

The proof of Proposition 7 provides a way to construct a solution to any feasible 2-part n-note configuration. But not all solutions may be equally desirable from a musical point of view, and (so far, at least) we do not have a feasibility test or construction for configurations with three or more parts. Therefore, I would like to develop reliable numerical methods for determining feasibility and finding musically desirable solutions. Unfortunately, these problems are non-convex in nature. But luckily, they can be framed in terms of low rank and/or rank-constrained optimization problems which recent research seems to suggest are relatively well-behaved.

Sonification of Wigner Functions: Case Study of Intense Light-Matter Interactions

Reiko Yamada[1(✉)], Antoine Reserbat-Plantey[3], Eloy Piñol[1], and Maciej Lewenstein[1,2]

[1] ICFO – Institute of Photonic Science, Barcelona Institute of Science and Technology, Av. C.F. Gauss, 3, 08860 Castelldefels, Barcelona, Spain
reiko.yamada@icfo.eu
[2] ICREA Lluis Companys, 23, 08010 Barcelona, Spain
[3] Université Côte d'Azur, CNRS, CRHEA. Valbonne, Sophia, Antipolis, France

Abstract. In quantum mechanics, the Wigner function $\rho_W(\mathbf{r}, \mathbf{p})$ serves as a phase-space representation, capturing information about both the position $\mathbf{r}$ and momentum $\mathbf{p}$ of a quantum system. The Wigner function facilitates the calculation of expectation values of observables, examination of quantum system dynamics, and analysis of coherence and correlations. Therefore, it might serve as a tool to express quantum systems intuitively, for example, by using sonification techniques. This paper summarizes the experimental strategies employed in a previous project and delineates a new approach based on its outcomes. Emphasizing the attribution of specific Wigner functions to their underlying quantum states, dynamics, and sources; our proposed methodology seeks to refine the sonification and scoring process, aiming to enhance intuitive understanding and interpretation of quantum phenomena.

Keywords: quantum mechanics · sonification · Wigner function · electro-acoustic composition

The development of atomic and molecular optics [12], ultra-fast optics [8], nanophotonics [7] or quantum transport - to name a few - has led to experimental situations involving genuine and complex quantum mechanics phenomena. While quantum mechanics textbooks often refer to the building blocks such as quantum harmonic oscillator, single particle in a box, or single spin states, modern physics has to deal with systems with physical properties that evolve on a very short length scale, often smaller than the De Broglie wavelength [3]. Consequently, quantum phenomena such as entanglement, cluster states, mesoscopic quantum state superposition, or strongly correlated states appear. Their description and modeling sometimes require specific mathematical tools, such as Wigner functions [4]. Sonification of quantum phenomena has often revolved around simple physical systems, such as a Gaussian wavepacket in a potential well [5], or quantum harmonic oscillators [6,11]. Motivations for sonifying quantum phenomena vary, from enhancing understanding of quantum behavior [5] to

© The Author(s), under exclusive license to Springer Nature Switzerland AG 2024
T. Noll et al. (Eds.): MCM 2024, LNCS 14639, pp. 447–455, 2024.
https://doi.org/10.1007/978-3-031-60638-0_38

developing sound-based protocols for quantum information storage [11] or even creating tools for sorting and categorizing features within quantum systems [2]. Campo's 2005 study [2] elegantly demonstrated the sonification of mass spectra of subatomic particles like baryons. Some approaches also emphasize the aesthetic aspect of sonifying quantum phenomena, exemplified by Rodney Du Plessis' piece *Psi* (2021), which utilizes the *QHOSYN* sound emulator to simulate quantum harmonic oscillators' superposition [6]. In such a manner, this tool enables the simulation and generation of quantum-based sound. Another illustration comes from the sonification of many-body systems - which are situations often challenging to analyze -, as presented in Bob Sturm's piece *50 Particles* (2001), where musical gestures map inter-particles coupling, and the system dynamics drive the piece development [14]. Alternative approaches are based on mapping momentum with musical transposition and position with musical modulations as in a two-dimensional representation of tonality, such as the Regener space [9,10].

As this article will detail, Wigner functions are powerful tools for describing and computing massively quantum systems. We attempted to sonify the Wigner function describing a quantum superposition of optical states - known as Schrödinger cat state - using various methodologies, and we anticipate that sonifying its quantum nature may yield unexpected manifestations beyond preconceived paradigms. Therefore, our objective is to explore various strategies for sonifying quantum phenomena, to achieve a multi-perspective understanding, exemplified by optical Schrödinger cat states [8] through auditory information.

Simultaneously, we are interested in providing an auditory perspective on the Wigner function that faithfully describes the system as it exists rather than concentrating solely on specific studies of it. In some respects, our approach echoes the sentiment expressed by John Cage [1]: "I have nothing to say, and I am saying it, and that is poetry as I need it".

We will begin by examining Wigner functions' utilization to depict highly non-classical states. Subsequently, we will describe the methodology for obtaining datasets within a specific experimental context: intense light-matter interactions aimed at generating Schrödinger cat states. Finally, we will introduce various sonification strategies and scoring processes employed in our approach.

1 Wigner Distributions

1.1 Classical Distributions and Quasi-distributions

In classical mechanics, a particle's position r and momentum p can be determined with arbitrary precision. If the particle's state is defined statistically, it can be described by a distribution function, $\rho_{cl}(r,p)$. This function is always positive, must be normalized to unity, and may include correlations between the particle's position and velocity.

In quantum mechanics, these two representations - position and momentum space (connected through Fourier transformation) - are also commonly used. In contrast to the classical situation, these representations are exclusive. All information regarding the particle's momentum is lost in the position representation, while the converse is also true. A necessary consequence is that no information about correlations between the particle's position and momentum can be extracted.

The latest quantum mechanical description is often not practical, and there is a need for an intermediate description that maintains information about r and p - with a specific limitation in the precision - while respecting the laws of quantum mechanics. Such a tool was introduced by Wigner in 1932 and consists of a quantum mechanical function - named $\rho_W(r,p)$ - that allows computing average values, or more accurately, expectation values, in a similar way to what can be done classically. Indeed, in a classical distribution $\rho_{cl}(r,p)$, if one wants to know the average value of a given operator $\mathcal{A}$, the distribution shall be multiplied by this operator and summed over all phase space (r,p). Wigner distributions - or Wigner functions - are thus helpful in computing any average values of observables. From a practical standpoint, the Wigner function contains all information about a quantum system and its evolution. This article considers sonifying Wigner functions to gain a more intuitive perception of a quantum system and its dynamics. Therefore, we shall first describe the key differences and similarities between a classical statistical distribution function and a Wigner quasi-distribution function.

1.2 Negative Values in Wigner Functions

While classical distributions are always positive or zero and normalized to 1, the description of a quantum system can be more nuanced. On the one hand, there are cases where the Wigner function $\rho_W(r,p)$ is always positive and thus behaves like a classical distribution. Here, such Wigner function can be seen as a probability distribution and is used to describe systems having quantities that vary in space over large distances (i.e. on a length scale much larger than the de Broglie wavelength $\lambda = h/\|\mathbf{p}\|$), as found in quantum electronic transport. Notably, to fulfill the Heisenberg uncertainty principle, the width of the Wigner function around a given value - i.e. $p_0(r)$ - is of the order of $\hbar/d$, where d is the coherence length of the wavefunction. In systems where physical quantities vary at a length scale comparable to λ, quantum effects become important. Surprisingly, the Wigner distribution can take negative values. As a result, it is no longer possible to refer to it as a probability distribution, and the more appropriate term is quasi-distribution. It is often said that negative values of the Wigner function carry the quantumness of a physical system. We will derive the Wigner function for a case study and explain the emergence and signification of negative values. We consider the first excited state $\Psi_1(t)$ of a 1D quantum harmonic oscillator. The envelope of this odd wavefunction is similar to the first harmonics of a string, having a node at its center. Its Wigner function at

$(r,p) = (0,0)$ is defined by [4]:

$$\rho_W(r=0,p=0) = \frac{1}{2\pi\hbar}\int dy\, \Psi_1\left(\frac{y}{2}\right)\Psi_1^*\left(-\frac{y}{2}\right)$$
$$= -\frac{1}{2\pi\hbar}\int dy\, |\Psi_1\left(\frac{y}{2}\right)|^2 < 0 \tag{1}$$

The negative value highlights that a Wigner function cannot be seen as a probability distribution but rather a quasi-distribution. Negative values for ρ_W are often observed in various situations, ranging from simple interference experiments like Young's slits (tamasic or dark rays) to complex quantum states resulting from intense laser-atom interactions [8].

2 Data from Intense Light-Matter Interactions Experiments

Our data is obtained through the collaboration between theoreticians and experimentalists who study exotic quantum states that originate from the interaction of intense laser fields with atomic clouds [8]. Especially for our second sonification approach described in the methodology section, a comprehensive understanding of the experimental conditions [8,13] is imperative. A coherent laser state initially interacts with a cloud of N Xenon atoms in their ground state. This interaction produces higher harmonic states and amplitude-shifted coherent states. The quantum superposition of both states can produce an optical Schrödinger cat state - in this case, kitten state -, similar to the quantum superposition of at least two different macroscopic or mesoscopic states, such as $|cat\ alive\rangle$ and $|cat\ dead\rangle$. Their Wigner function has the form:

$$\rho_W^{Cat}(\beta) = \frac{2}{\pi(1-e^{(-|\delta\alpha|^2)})}[e^{(-2|\beta-\alpha-\delta\alpha|^2)}$$
$$+ e^{(-|\delta\alpha|^2)}e^{(-2|\beta-\alpha|^2)} - e^{(-|\delta\alpha|^2)}e^{(-2|\beta-\alpha|^2)}$$
$$(e^{(2(\beta-\alpha)\delta\alpha^*)} + e^{(2(\beta-\alpha)^*\delta\alpha)})]. \tag{2}$$

Here, α and β are complex numbers related to position and momentum through the expression: $\beta - \alpha = r + ip$. The value $\delta\alpha$ reflects the shift in amplitude for the fundamental state. This shift is due to correlations between the fundamental and the generated harmonics and is central to understanding system evolution. When this shift is large, e.g. $\delta\alpha < -3$, the resulting state is a coherent state with a Gaussian profile, almost analog to a purely classical state. However, when $\delta\alpha$ goes from -1 and tends to 0, the state evolves from a Schrödinger cat state (Eq. 2) to a quantum state with a well-defined number of photons, also known as Fock state. The Wigner function describing such Fock state is:

$$\rho_W^{F_m}(r,p) = \frac{(-1)^m}{\pi}e^{-(r^2+p^2)}\mathcal{L}_m\left(2(r^2+p^2)\right)$$
$$\rho_W^{F_1}(r,p) = -\frac{1}{\pi}e^{-(r^2+p^2)}\left(1-2(r^2+p^2)\right) \tag{3}$$

where $\mathcal{L}_m$ denotes the mth Laguerre polynomial. Both Wigner functions for the Schrödinger cat state (Eq. 2) and the first Fock state (Eq. 3) are shown in Fig. 1a.

Fig. 1. Sonification of Wigner functions. a: Phase space (r, p) representation of the Wigner functions $\rho_W(r, p)$ of a Schrödinger kitten state for $\delta\alpha = -1$ (top, Eq. 2), and the first Fock state (bottom, Eq. 3) are plotted. **b**: Excerpt of the score for string quartet using the method II. **c**: Excerpt of the score for string quartet using the method III. Both methods II and III are described in Sect. 3.2.

3 Sonification Approaches to Wigner Functions

3.1 Mesh Discretization Method of Wigner Quasi-distribution

In our previous trials [15], we explored direct sound generation from Wigner functions, employing a mapping method that discretizes its arguments and values into grids, ensuring 99 % coverage of the distribution. This mesh discretization strongly depends on the type of grid used. Therefore, we considered two approaches: regular grids with all the lattice sites equidistant and non-regular grids based on Gaussian intervals, as detailed previously [15]. The latter phase-space discretization accurately pictures the variations of the Wigner function as it reduces the binning size to gain sampling resolution.

3.2 Sound Generation

These meshes were subsequently translated into auditory representations through various techniques: (I) mapping grid points to phase, frequency, and amplitude of waves, generating up to 900 concurrent sinusoidal waves; (II) directly mapping minimum and maximum Wigner function values to frequencies; (III) segmenting Wigner function values in four sections and correlating with frequency intensity; and (IV) mapping mathematical moments of the Wigner function to a set of harmonic oscillators distributed over a Gaussian envelope.

The neutral sinusoidal waves and triangular pulses controlled by Wigner function data were used for acousmatic rendering. For acoustic instruments, frequencies were rounded to the nearest quarter division of 12-note equal-tempered pitches, and different playing techniques were employed to control timbres, reflecting the negativity of the Wigner function (shown in Fig. 1b-c).

In the method IV, we compute the central value r_0 and the standard deviation σ_r of $\rho_W(r, p)$. These parameters were then associated with the central frequency f_0 and the spectral filter bandwidth (often denoted as Q-value in EQ treatment) of the generated sound event. Each event was produced from a Gaussian distribution centered at $f_0(r_0)$ of N different harmonic oscillators. The envelope of this Gaussian distribution has a bandwidth mapped by $Q(\sigma_r)$. The resulting sonogram (or Fourier space representation of the generated sound) is displayed in Fig. 2a. Here, one can distinguish a Gaussian envelope for a set of $N = 21$ oscillators (appearing as a bright set of lines at low frequencies) from upper harmonics. The central value of the Gaussian distribution increases linearly as the Wigner function second moment $\sigma_r(\delta\alpha)$ changes, according to our mapping protocol. Notably, the individual oscillator bandwidth remains constant throughout this process.

The resulting audio events in every case exhibited a high degree of dissonance, characterized by transformations in the nuances of dissonance and a rich, sustained texture, generating a plethora of overtones. This auditory experience starkly contrasts with the prevalent visual representation of the Wigner function. Depending on the mapping variables employed, massively quantum states (generally understood as when ρ_W displays negative values) can be distinguished from classical states through sound, albeit discernible only to those informed about the mapping approach to these variables.

3.3 Integrating Wigner Function's Evolution into Electro-Acoustic Medium

In our latest methodology, we augment the diverse expressions of the Wigner function with a focused consideration of its experimental conditions. We integrate the method IV, as described in the previous section, using single-reed wind instrument multiphonics to articulate the superposition of fundamental and high harmonics in our experiments. Specifically, in this latest method, we combine the functional elements derived from the quasi-distribution ρ_W - namely, the first (r_0) and second (σ_r) moments -, with the multiphonics of wind instruments (Fig. 2b). This integration gives a dual perception of ρ_W and its corresponding experimental situation. Indeed, on the one hand, one can track the evolution of the Wigner function ρ_W as a function of $\delta\alpha$, transiting from a Fock state ($\delta\alpha \rightarrow 0$) to a kitten state ($\delta\alpha = -1$), and finally to an almost classical coherent state ($\delta\alpha = -3$). On the other hand, multiphonic sound generation in single-reed wind instruments maps the experimental process of quantum superposition of higher harmonic states and amplitude-shifted coherent laser states. Thus, this integration enables us to simultaneously and audibly track the system's evolution. Furthermore, the position and momentum information encoded in the Wigner function $\rho_W(r,p)$ can be directly utilized for multi-channel spatialization in live performance settings.

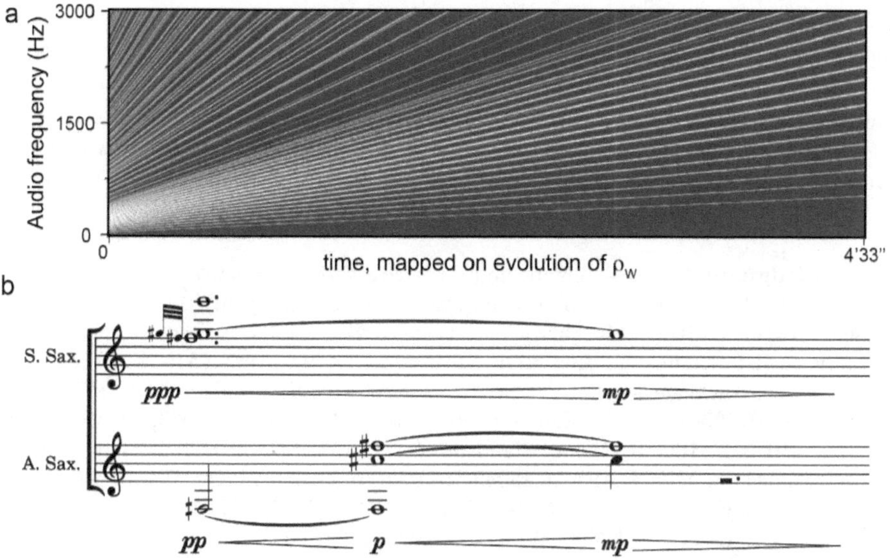

Fig. 2. Spectral analysis of the rendition IV. a: Audio sonogram representation (reduced to monophonic) of the transition from the first Fock state to a kitten state and then to a coherent state in 4'33". **b**: Excerpt from the score for soprano and alto saxophones using multiphonics.

4 Conclusion and Outlook

Highly non-classical states, such as the ones studied here and originating from intense light-matter interactions, require mathematical tools such as Wigner functions for description and computation. While attributing physical significance to the Wigner function must be cautious, they prove invaluable for computing average values of quantum observables and representing the physical properties of a given system [4]. In this work, we introduced various strategies to sonify Wigner functions to offer an auditory perspective. In particular, we proposed a mesh discretization method for accurately mapping Wigner functions and sonifying ρ_W from four perspectives based on direct parameter mapping (I-III). Furthermore, in some cases, the scoring (e.g. the use of *sul ponticello* or ricochet when $\rho_W < 0$) highlighted the negative values of the Wigner function.

The latest sonifying method IV, based on extracting the moments of ρ_W, offers a statistical analysis of the Wigner quasi-distribution. Our study was limited to computing the two first moments (mean value and standard deviation). Additionally, extracting the third and fourth moments - skewness and kurtosis, respectively - could provide further insights into the asymmetric behavior (in (r, p) space) and the tail extension of the Wigner quasi-distribution ρ_W.

One exciting area of research involves tackling interference phenomena, such as the Young double-slit experiment. In this experiment, the wavefunction of a particle (such as a photon) is split into two coherent wavepackets situated at

$r = -a$ and $r = a$ after passing through the slits. The diagonal components of the corresponding Wigner function describe the two wavepackets. In contrast, the off-diagonal components are non-zero at $r = 0$, where the probability of finding a particle is zero. These off-diagonal components, also known as a *ghost* [4] components, explain the observed interference pattern when the wavepackets overlap. This scenario presents opportunities for sonification, as the system's observable can be precisely calculated using Wigner functions.

Acknowledgments. The academic and artistic research mentioned in this article was made possible by ICFO and Phonos Foundation (Barcelona, Spain). The authors thank Philipp Stammer (ICFO) for his generous help with topics related to the Wigner function. A.R-P thanks funding from ANR (23-CE47-0015) NEAR-2D. R.Y., E.P. and M.L. acknowledge support from: ERC AdG NOQIA; MCIN/AEI (PGC2018-0910.13039/501100011033, CEX2019-000910-S 10.13039/501100011033, Plan National FIDEUA PID2019-106901GB-I00, Plan National STAMEENA PID2022-139099NB-I00 project funded by MCIN/AEI/10.13039 501100011033 and by the "European Union NextGenerationEU/PRTR" (PRTR-C17.I1), FPI); QUANTERA MAQS PCI2019 111828-2); QUANTERA DYNAMITE PCI2022 132919 (QuantERA II Programme co-funded by European Union's Horizon 2020 program under Grant Agreement No 101017733), Ministry for Digital Transformation and of Civil Service of the Spanish Government through the QUANTUM ENIA project call - Quantum Spain project, and by the European Union through the Recovery, Transformation and Resilience Plan - NextGenerationEU within the framework of the Digital Spain 2026 Agenda; Fundació Cellex; Fundació Mir-Puig; Generalitat de Catalunya (European Social Fund FEDER and CERCA program, AGAUR Grant No. 2021 SGR 01452, QuantumCAT/U16-011424, co-funded by ERDF Operational Program of Catalonia 2014–2020); Barcelona Supercomputing Center MareNostrum (FI-2023-1-0013); EU Quan-tum Flagship (PASQuanS2.1, 101113690) EU Horizon 2020 FET-OPEN OPTO-logic (Grant No 899794); EU Horizon Europe Program (Grant Agreement 101080086 - NeQST), results incorporated in this standard have received funding from the European Innovation Coun-cil and SMEs Executive Agency under the European Union's Hori-zon Europe programme), ICFO Internal "QuantumGaudi" project; European Union's Horizon 2020 program under the Marie Sklodow-ska-Curie grant agreement No 847648; La Caixa Junior Leaders fellowships, La Caixa Foundation (ID 100010434): CF/BQ/PR23/11980043. Views and opinions expressed are, however, those of the author(s) only and do not necessarily reflect those of the European Union, European Commission, European Climate, Infra-structure and Environment Executive Agency (CINEA), or any other granting authority. Neither the European Union nor any granting authority can be held responsible for them. E.P. has been supported by Ayuda (PRE2021-098926) funded by MCIN/AEI/ 10.13039/501100011033 and by FSE+.

References

1. Cage, J.: Silence: Lectures and Writings, 50th Anniversary Edition. Wesleyan University Press (2012). google-Books-ID: MUvYNgbo39IC
2. Campo, A.D., Frauenberger, C., Höldrich, R., Melde, T., Plessas, W., Sengl, B.: SONIFICATION OF QUANTUM SPECTRA. Limerick, Ireland (2005). https://www.semanticscholar.org/paper/SONIFICATION-OF-QUANTUM-SPECTRA-Campo-Frauenberger/4b088fc452cae3e863876a7f19766c54cf260514

3. Cohen-Tannoudji, C., Diu, B., Laloë, F., Hemley, S.R., Ostrowsky, N., Ostrowsky, D.B.: Quantum Mechanics, Volume 1: Basic Concepts, Tools, and Applications, 2nd Edition. Wiley-VCH Verlag GmbH & Co, Weinheim, Germany, second edition edn. (1997)
4. Cohen-Tannoudji, C., Diu, B., Laloë, F., Hemley, S.R., Ostrowsky, N., Ostrowsky, D.B.: Quantum Mechanics, Volume 3: Fermions, Bosons, Photons, Correlations, and Entanglement. Wiley-VCH Verlag GmbH & Co, Weinheim, Germany, second edition edn. (2020)
5. Cádiz, R.F., Ramos, J.: Sound synthesis of a gaussian quantum particle in an infinite square well. Comput. Music. J. **38**(4), 53–67 (2014)
6. DuPlessis, R.: rodneydup/QHOSYN (2023). https://github.com/rodneydup/QHOSYN. original-date: 2021-02-26T20:32:09Z
7. Hensen, B., et al.: Loophole-free Bell inequality violation using electron spins separated by 1.3 kilometres. Nature **526**(7575), 682–686 (2015). https://doi.org/10.1038/nature15759, https://www.nature.com/articles/nature15759
8. Lewenstein, M., Ciappina, M.F., Pisanty, E., Rivera-Dean, J., Lamprou, T., Tzallas, P.: Generation of optical Schrödinger cat states in intense laser-matter interactions. Nature Phys. **17**, 1104–1108 (2021). https://doi.org/10.1038/s41567-021-01317-w
9. Noll, T., Graben, P.B.: Quantum-musical explorations on Zn. In: Montiel, M., Agustín-Aquino, O.A., Gómez, F., Kastine, J., Lluis-Puebla, E., Milam, B. (eds.) MCM 2022. LNCS, vol. 13267, pp. 369–375. Springer, Heidelberg (2022)
10. Regener, E.: Review of pitch notation and equal temperament: a formal study. J. Music Theory **19**(1), 161–171 (1975). https://doi.org/10.2307/843755, https://www.jstor.org/stable/843755
11. Saranti, A., Eckel, G., Pirrò, D.: Quantum harmonic oscillator sonification. In: Ystad, S., Aramaki, M., Kronland-Martinet, R., Jensen, K. (eds.) CMMR/ICAD -2009. LNCS, vol. 5954, pp. 184–201. Springer, Heidelberg (2010). https://doi.org/10.1007/978-3-642-12439-6_10
12. Srakaew, K., et al.: A subwavelength atomic array switched by a single Rydberg atom. Nat. Phys. (2023). https://doi.org/10.1038/s41567-023-01959-y, https://www.nature.com/articles/s41567-023-01959-y
13. Stammer, P., et al.: High photon number entangled states and coherent state superposition from the extreme ultraviolet to the far infrared. Phys. Rev. Lett. **128**, 123603 (2022)
14. Sturm, B.L.: Composing for an ensemble of atoms: the metamorphosis of scientific experiment into music. Organised Sound **6**(2), 131–145 (2001). https://doi.org/10.1017/S1355771801002102
15. Yamada, R., Piñol, E., Grandi, S., Zakrzewski, J., Lewenstein, M.: Towards the intuitive understanding of quantum world: sonification of rabi oscillations, wigner functions, and quantum simulators (2023). https://doi.org/10.48550/arXiv.2311.13313, http://arxiv.org/abs/2311.13313, arXiv:2311.13313 [quant-ph]

Tribute to Yves Hellegouarch

A Tribute to Yves Hellegouarch

Wynne Hellegouarch and Franck Jedrzejewski(✉)

CEA Paris Saclay University, Saclay, France
franckjed@gmail.com

1 Introduction

Honorary founder of the *Journal of Mathematics and Music*, Yves Hellegouarch was a French mathematician. He spent most of his career at the University of Caen. A former student of the École Normale Supérieure (ENS), he was a student at the National Conservatory of Music in Paris, where he studied cello with Maurice Maréchal and chamber music with Maurice Hewitt. In 1954, he decided to pursue a career in mathematics. He specialized in number theory, and all his life worked on Fermat's last theorem. He introduced special elliptic curves, which are now called Frey curves, in honor of the German mathematician Gerhard Frey. The latter discovered that one of these curves, introduced by Yves Hellegouarch, was a potential counter-example to the Taniyama-Shimura conjecture. This result was proved by Jean-Pierre Serre and Ken Ribet, and this was the starting point for the proof of Fermat's conjecture obtained by Andrew Wiles in 1994. Three years later, Yves Hellegouarch presented all of this work in a well-known book *Invitation aux mathématiques de Fermat-Wiles.*

Although a university mathematician, he never gave up practicing the cello and performed in concerts when the opportunity arose. He liked just intonation and justified it in several passages. In an unpublished letter, dated 11/19/2015, he writes:

> When I was a professional cello student, I studied Ravel's Piano trio (with M. Hewitt) and discovered that the last note of the first movement of this trio was clashing with the piano: it was always too low! At that time I thought it was just a problem of gut strings or bad coaching, or bad student; and cello tuning was not part of the curriculum! But a few decades later, having some mathematical knowledge, I discovered the real reasons were actually mathematics had something to do with the tuning of Pythagorean fifths and published a new equal temperament scale that solved this problem and which I call the Rachmaninov-Ravel (or RR for short) temperament and I also discovered that it was very good at improving the main resonances of my two celli.

© The Author(s), under exclusive license to Springer Nature Switzerland AG 2024
T. Noll et al. (Eds.): MCM 2024, LNCS 14639, pp. 459–463, 2024.
https://doi.org/10.1007/978-3-031-60638-0_39

Yves Hellegouarch wrote 14 articles on mathematics and music (see references [1–14]), mainly on questions of tuning and temperament. He became interested in Leonhard Euler's *Tentamen novae theoriae musicae.* His main article is the first he published, *Scales*, in the *Proceedings of the Canadian Academy of Sciences* [1]. This article was updated and republished in two parts in the *Gazette des mathématiques*, organ of the *French Mathematical Society* [9]. These last texts are dedicated to the Ravel-Rachmaninoff temperament. He had asked me to format his last article which remained unpublished and which we propose below. It was note really Yves' intention to invent and study a new temperament, although he called it the RR Temperament, but it was his intention to adjust the frequencies of certain notes on the cello in order to make certain inflections more beautiful. The method proposed here fixes the four strings of a cello (or of the 4 instruments of a quartet) and induces an equal temperament by defining a new and unique semitone.

The first article – *Sketch of a comparative study of the mathematical resonances of the advent of perspective (in painting) and equal temperament (in music)* – that we publish are notes that Yves probably wrote in preparation for a conference "Destines of art, designs of science" which was held at the University of Caen in 1986. Certain elements of this note are included in the text which was published in 1992 [7]. Eventually, it was used as preparatory notes for the MaMuPhi Seminar helds in Paris 12th March 2005, as we can see on the video[1] These notes are interesting because of the relationships between the birth of perspective in painting and that of equal temperament in music. Although these two events occur at different times in history since perspective was born with Filippo Brunelleschi in 1425 and equal temperament was born in the writings of the mathematician Simon Stevin (1548–1620), but will not reach its full development until with Andreas Werckmeister (1645–1706), the text is of interest in the consequences that each of this two essential events had. We have translated this note into English.

The second article – *Some common sense remarks on the tuning of string trios or quartet instruments* – is an article that Yves had written in English and which he would have liked to publish as he saw in these inflections of Ravel and Rachmaninov's string quartet a call for just intonation.

2 A Biographical Sketch by Wynne Hellegouarch

Yves was born on September 18 1936 in Angers. Both his parents were primary school teachers and amateur musicians. Yves started cello lessons aged 7 and, as he showed exceptional promise, his father decided that he and his elder sister (a pianist) should attend the Paris Conservatoire. Yves had been accepted as a pupil by Maurice Maréchal.

He attended school normally until the age of ten and, from then on, after the family moved to Paris (which he hated) concentrated on his musical studies until he obtained his First Prize in cello and "Première médaille" in chamber

[1] See the MaMuPhi Seminar https://www.youtube.com/watch?v=-hu7oJ3vM2k.

music. During this time he had no formal education, although the flat was full of books and he did benefit from a few English lessons (his parents thinking that this would be useful for his future career !).

After this he decided, to his parents' great displeasure, that he wanted to study maths – his parents explained to him that this was not feasible, his lack of education making it impossible for him to pass the baccalauréat. However he did so (First part in September 1954) and, after passing, was entitled to enter a lycée – the Lycée Jacques Decour being the only one which would accept him at this late date. He stayed there for the two years of "classes préparatoires" and was admitted to the ENS Ulm in 1957. One of his close friends was Jean-Claude Risset.

He practically gave up the cello, apart from playing occasionally for friends at the ENS and performing a concerto in a University concert at the Sorbonne.

After the Agrégation, he was appointed as Assistant at Orsay but moved to Besançon at the request of his thesis director. Finding Besançon too far removed from Anjou (and, coincidentally from Wales, my native land) we moved to Caen in 1970 and remained there until his retirement in 2000. After the first year or so, his teaching was mainly limited to Agrégation and DEA [PhD]. It was soon realised that he was not cut out for teaching first year students!

His thesis (Besançon 1972) was entitled *Courbes elliptiques et équation de Fermat.*

He had already spoken about elliptic curves at a colloquium in Bordeaux a few years before but this had not attracted much attention; also in 1975. The curves were "re-invented" by Gerhard Frey in 1985, leading eventually to Wiles'demonstration in 1994. The curves, first called "Frey curves" later became known as "F-H (or H-F) curves".

He wrote a large number of articles, mainly on Number Theory but, already in the 1970s, he became interested in Scales. He took up the cello again in about 1974, when a colleague, who was leaving Caen, left a place vacant in a amateur string quartet. After a break of more than 15 years, Yves started to play again and, a few years later, agreed to take over the University orchestra. He did not find this particularly rewarding but it did spark his interest in scales (first publication [1])

It did not take him long to get back to his previous standard on the cello and, from 1981, after meeting some professional musicians, thanks to a summer school in England, he started playing in recitals in France and Britain. One of his partners was the pianist Jacques Delannoy, a mathematician who had studied at ENS Ulm before deciding to devote himself to music professionally. Yves played concertos with the university orchestras of Würzburg (twinned with Caen) and Rennes. He also set up a series of chamber music concerts at the University, in the hope of sparking some interest among the students.

After his retirement he spent much more time on music – concerts, recitals – but also became more and more interested in scales and tuning string instruments.

Unfortunately, he was not only totally incompetent but also extremely hostile concerning anything to do with "informatique" (although he did, much against his better judgment) contribute to the creation of a "Licence d'informatique" [BSc in computer science] at the University. It was only when I showed him that he could listen to classical music (exclusively, of course) all day on YouTube that he realised that there might, after all, be some advantage to these contraptions. He was prepared to listen to a few singers like Brassens, Brel, Joan Baez, Leonard Cohen. Incidentally he was very interested in the system of "*tonic sol-fa*", widely used for choral singing in Britain, particularly in Wales – for working-class singers, as most were, who could not read music.

In 2013 he was diagnosed with Alzheimer or similar (in fact, apparently, Levy bodies). For a few years the symptoms were limited. As the neurologist said, "Il a de bonnes réserves cognitives" ! He drove – only short distances, with me, a non-driver, supervising, until 2018, when he fell and broke his wrist. He reeducated himself (it was his left wrist) by playing scales ! After that, things became more difficult, though he remained quite happy. He went into care mid-2020 and, though apparently quite cheerful, immediately ceased to recognise family members.

He died February 5, 2022 (Fig. 1).

Fig. 1. Yves Hellegouarch, ca. 2010. © Photo Alain de Trébons.

References

1. Hellegouarch, Y.: Scales, Comptes rendus Académie des Sciences, Société royale du Canada, 4 (5), 1982 et Erratum 5 (2) 1983
2. Hellegouarch, Y.: Gammes naturelles. Musique et mathématiques, Publications de l'APMEP, éd. par Barnard Parzysz, pp. 127–158 (1984)
3. Hellegouarch, Y.: À la recherche de l'arithmétique qui se cache dans la musique, Publications de l'Institut de recherche mathématiques de Rennes, fascicule 2 « Science, histoire et société », pp. 49–59 (1986)
4. Hellegouarch, Y.: Survey d'un arithméticien. Singularités **2**(8), 17–22 (1991)
5. Hellegouarch, Y.: Musique et arithmétique, in Marsyas. Revue de pédagogie musicale et chorégraphique **22**, 69–76 (1992)
6. Hellegouarch, Y.: Esquisse d'uneétude comparée des résonances mathématiques de l'avènement de la perspective (en peinture) et de celui du tempérament égal (en musique), manuscrit, non daté
7. Hellegouarch, Y.: Essai d'une nouvelle théorie de la musique de L. Euler. Destins de l'art, desseins de la science, Actes du colloque de Caen, 1986, IREM de Basse Normandie, Caen (1992)
8. Hellegouarch, Y.: De la musique des sphères aux problèmes actuels de l'arithmétique, Rapporteurs F. Mercelot et F. Deves, Atelier A63, Caen, s.d. (postérieur à) (1992)
9. Hellegouarch, Y.: Gammes naturelles, Nouvelle version dans la Gazette des mathématiciens, Société mathématique de France, 1999, 25-39 et 1999, 13-26
10. Hellegouarch, Y.: A mathematical interpretation of expressive interaction, in Mathematics and Art, édité par Claude Bruter, pp. 141–148. Springer, Berlin (2002)
11. Hellegouarch, Y.: Outils diophantiens pour la définition d'une topologie harmonique, Paris, IRCAM (2002)
12. Hellegouarch, Y.: Description d'un tempérament égal non classique. In: À la lumière des mathématiques et à l'ombre de la philosophie, edited by Moreno Andreatta and François Nicolas, Sampzon: Delatour, pp. 97–100 (2012)
13. Hellegouarch, Y.: Annexe : deux aides pratiques pour la contribution de Franck Jedrzejewski. In: À la lumière des mathématiques et à l'ombre de la philosophie. edited by Moreno Andreatta and François Nicolas, Sampzon: Delatour, p. 105 (2012)
14. Hellegouarch, Y.: L'art de se servir d'un instrument à cordes. In: Hacher, X., Papadopoulos, A. (eds.) Leonhard Euler, pp. 207–226. CNRS éditions, Paris (2015)

Sketch of a Comparative Study of the Mathematical Resonances of the Advent of Perspective (in Painting) and Equal Temperament (in Music)

Yves Hellegouarch(✉)

Caen University, Caen, France
franckjed@gmail.com

1 Comments on the Title

1) I should say "perspective in drawing" rather than "perspective in painting" and equal temperament in "architectural (or abstract) music". I'm not referring here to the relationships between perspective, colors and values ("aerial perspective"), which are in a way the "musical" aspect of painting, nor to the relationships between equal temperament and the timbre of musical sounds (often referred to as their "color").
2) A comparative study seems to me to be justified by the fact that perspective and temperament have in common that they lie at the crossroads of certain branches of science – (mathematics and optics), respectively (mathematics and acoustics) – of humanist culture and certain theorizable artistic praxis (of the Western world) See (Fig. 1).

2 Quick History

2.1 Perspective

The abstract problem painters were trying to solve was that of representing three-dimensional space on a two-dimensional canvas in as "realistic" a manner as possible. Experts dispute whether the Greeks were capable of solving this problem, but on the scientific side they had at least the beginnings of the necessary tools.

But it was in the Quattrocento that the desire to free oneself from the old techniques of representation to achieve greater realism (as in Piero della Francesca's *Resurrection*) became most apparent, prepared underground by the research of 13th-century university opticians.

Filippo Brunelleschi seems to have laid the foundations of the new practice, which was codified around 1435 in L.B. Alberti's *Trattato della pittura*.

It wasn't until over a century later that this practice attracted the interest of a scientist, the polygrapher Guidobaldo del Monte, who gave a Euclid-like interpretation of Alberti's recipes.

© The Author(s), under exclusive license to Springer Nature Switzerland AG 2024
T. Noll et al. (Eds.): MCM 2024, LNCS 14639, pp. 464–468, 2024.
https://doi.org/10.1007/978-3-031-60638-0_40

But it was in 1636 that Girard Desargues' new perspective method was published in Paris [4], followed in 1639 by the first work on projective geometry based on synthetic methods ([5], see also [9]). This represented an epistemological breakthrough, a second paradigm of which would be Blaise Pascal's *Essay on Conics* in 1640 (a disciple of Desargues...).

Despite the interest shown in Desargues' work by Descartes and Fermat, it (like Pascal's *Essay on Conics*) could have sunk into oblivion if geometry had not had the good sense to seduce the British scholars of the 18th century.

But it wasn't until the end of that century, and especially the beginning of the 19th century, that geometry literally exploded into what it is today, and much more besides, since it contains models of non-Euclidean geometries of all dimensions.

Should we attribute any influence on the birth of Cubism to these geometries (in particular, the Poincaré-Minkowski geometry of special relativity)? It's a vast question, and beyond our scope.

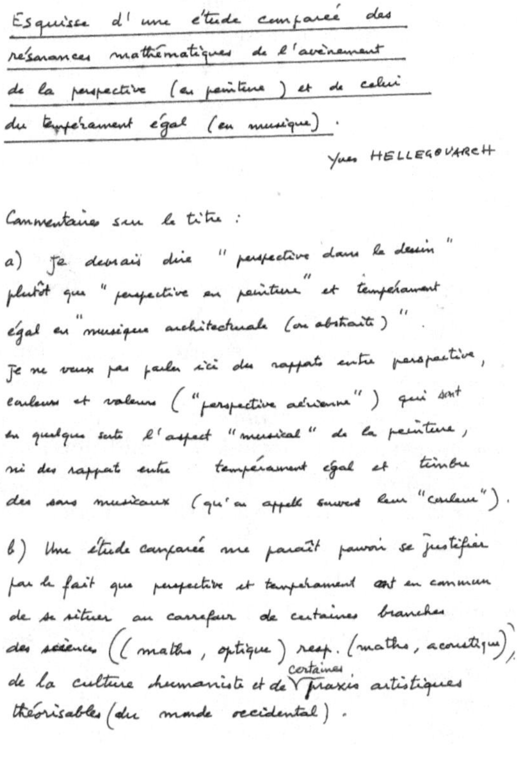
Esquisse d'une étude comparée des résonances mathématiques de l'avènement de la perspective (en peinture) et de celui du tempérament égal (en musique).

Yves HELLEGOUARCH

Commentaires sur le titre :

a) Je devrais dire "perspective dans le dessin" plutôt que "perspective en peinture" et tempérament égal en "musique architecturale (ou abstraite)".

Je ne veux pas parler ici des rapports entre perspective, couleurs et valeurs ("perspective aérienne") qui sont en quelque sorte l'aspect "musical" de la peinture, ni des rapports entre tempérament égal et timbre des sons musicaux (qu'on appelle souvent leur "couleur").

b) Une étude comparée me paraît pouvoir se justifier par le fait que perspective et tempérament ont en commun de se situer au carrefour de certaines branches des sciences ((maths, optique) resp. (maths, acoustique)), de la culture humaniste et de certaines praxis artistiques théorisables (du monde occidental).

Fig. 1. Yves Hellegouarch, Preparatory notes (in French).

2.2 Temperament

Comparatively much more is known about the relationship between mathematics and music in Greek times. Mathematically speaking, they possessed a practically complete theory of what we call positive rational fractions, as seen in Euclid, and there's no doubt that they correctly associated fractions with the main musical intervals thanks to the theory of the monochord [1].

The fact that the perception of intervals for a relativistic ear (i.e. one that has no notion of the absolute pitch of musical sounds) is projective in nature is not only a pleasant nod to the first part of this text, but the origin of a possible definition of interval addition in the mode of the addition of segments, which the Greeks mastered perfectly. In other words, in our modern language, the Greeks' unconscious could not have been unaware that all intervals (necessarily rationals for Pythagoras) formed an additive group isomorphic to the multiplicative group of positive rationals.

The main intervals were identified by their names, and the Greeks had to make their way through a virgin forest of esoteric words.

The Middle Ages were a period of consolidation of this knowledge, above all with Boethius citeBoe1492, whose aim was to set out the group of four liberal arts of a mathematical nature known as the *quadrivium* (arithmetic, astronomy, geometry, music), and with Guido of Arezzo, much later, who began to notate the intervals of the diatonic genre (the genre handed down to us by plainsong) on an outline staff, introducing clefs and giving names to the notes (manifesting a kind of desire to find the set of integers $\mathbb{Z}$).

In addition, pressure from secular music led to successive extensions of harmony (reduced to intervals of the form 2^n, $n \in \mathbb{Z}$, in Greek times). The development of fundamental musical scales (temperament) became a major preoccupation (dare I say it!) of theorists from the 16th century, and arithmetical calculations in decimal numbers for the multiplicative group of positive reals, were powerfully aided from 1614 by Neper's invention of original logarithms and Briggs' introduction of common logarithms (of base 10).

Such a revolution in calculation technique could not leave music theorists indifferent, as they were always on the lookout for the "greatest common divisor" of all usable intervals; in other words, "the" generator of the (immanent) group of all musical intervals. It was probably under Neper's influence that Werckmeister's concept of equal temperament [10] and Holder's mathematically contradictory concept of comma were developed [8].

As Christopher Bunting explains in [2]: If your idea of scales and tuning is based on piano tuning, you're dealing in trickery, to put it bluntly! This deception was partially approved by J.S. Bach and fully supported by his son C.P.E. Bach, but I don't think the virtue of that famous name alone preserves it from criticism.

At the beginning of the 19th century, when no one was interested in the problems of temperament any more, a geologist by the name of John Farey [6] introduced a brand-new idea into mathematics, which instantly began to interest mathematicians (Cauchy, [3]) and to flourish (to tell the truth, this idea had already been presented a short time before by a certain Haros with less impact, [7]).

Since then, the fascination exerted by Farey sequences has not ceased to stimulate mathematicians (Ford circles, Hardy-Ramanujan method, hyperbolic geometry, approximation using Coxeter groups, etc.), and here we see the appearance of a chapter of maths stemming from "musical" considerations (in the sense of Boethius).

Conversely, the "modern" algebra of the 20th century has made it possible to explain the immanent aspects of the scales of Pythagoras and Zarlino [11], and to answer the questions from the uninitiated without, however, completely lifting the veil on the mysteries of music theory.

3 Similarities, Differences

As we can see from this brief history, the differences outweigh the similarities, and we're going to try and clarify what these differences consist of.

We have seen that the desire for realism in painting led to a theory of perspective that is in fairly good agreement with Euclid's laws of optics, and that was first expressed in the language of Euclid's geometry.

The scientific transformation brought about by Desargues was the result of a language directly derived from the praxis of technicians and a pertinent use of the Cartesian "clean slate" that enabled the nascent projective geometry to emancipate itself from the Euclidean framework, at its peril, by asserting its own axiomatics. What we see here is an advance in geometry, but one whose continuity passes through the plastic arts, and whose re-emergence in Desargues provokes an apparent discontinuity. [The manuscript is interrupted here.]

References

1. Boethius. De institutione musica, Venice 1492. English translation Fundamentals of Music by Calvin M. Bower, edited by Claude V. Palisca, Yale University Press (1989)
2. Bunting, C.: Essay on the craft of Cello playing, t. 2. Cambridge University Press, Cambridge (1982)
3. Cauchy, A.L.: Démonstration d'un Théorème Curieux sur Les Nombres. Bulletin des Sciences, par la Société Philomatique de Paris **3**(3), 133–135 (1816)
4. Desargues, G.: Exemple de l'une des manières universelles du SGDL [Sieur Girard Desargues Lyonnois] touchant la practique de la perspective sans emploier aucun tiers-point, Paris (1636)
5. Desargues, G.: Brouillon-project d'une atteinte aux evenemens des rencontres du cone avec un plan, Paris (1639)
6. Farey, J.: On a curious property of vulgar fractions. Philos. Mag. J. **47**(3), 385–386 (1816)
7. Haros, C.: Tables pour évaluer une fraction ordinaire avec autant de decimals qu'on voudra; et pour trouver la fraction ordinaire la plus simple, et qui approche sensiblement d'une fraction décimale. Journal de l'École Polytechnique **6**(11), 364–368 (1801)
8. Holder, W.: Treatise on the Natural Grounds and Principles of Harmony, facsimile of the 1694 London edition, Broude Brothers (1967)

9. Taton, R.: L'œuvre mathématique de G. Desargues: Textes publiés et commentés avec une introduction biographique et historique. Presses universitaires de France, Paris (1951)
10. Werckmeister, A.: Musicalische Temperatur, Quedlinburg, 1691, Edited by. Rudolf Rasch. The Diapason Press, Utrecht (1983)
11. Zarlino, G.: Le Istitutioni harmoniche, Venice (1558)

Some Common Sense Remarks on the Tuning of String Trios or Quartet Instruments

Yves Hellegouarch(✉)

Caen University, Caen, France
franckjed@gmail.com

By "common sense" I mean the "scientific" common sense of physicists or chemists, not particularly the one of musicians which, nevertheless, is so important in quartet playing.

As everybody knows, since J. Haydn, the instruments of the string quartet are the violin, the viola and the cello, but what I am trying to say also applies to the five-stringed cello.

As soon as inspiration has fixed the desired "A" of the first violin, the rest of my story is just a question of common sense (everybody has the right to judge if I am right or wrong). Let us remark that fixing the "A" implies fixing all its harmonics: for example, to hear them we move a finger of the left hand, *as lightly as we can*, on the open string and press a little the bow near the bridge. What we hear then is "tout un monde lointain" of sounds which are *not* as mysterious as one may think because they are linked by precise mathematical relations between their heights: this is part of the formula of vibrating strings (the oldest part which was known since immemorial times!). For example if your left hand finger is just at the middle of the string (the vibrating part of it) you hear a sound which is twice as high as the open string, if your finger is at the third (or two thirds) part of the string you hear the Pythagorean fifth of this string, etc., and if your finger is at one of the two reduced fractions p/q or $(q-p)/q$ you hear the *same* sound whose frequency is q times the frequency of the frequency of the open string.

This is a marvelous thing which was probably known to the great Pythagoras, a well-known Greek living a long time ago.

Examples:

1. if $p/q = 1/2$, the two fractions coincide and we have the second harmonic of the string which frequency is two times the frequency of the open string.
2. if $p/q = 1/3$, the two fractions are different, they are $1/3$ and $2/3$, but we hear the same third harmonic of the open string.

1 Definition of Three Different "equal" Tunings

In fact we are only interested here in equal temperaments so the tuning of the instruments implies the temperament in an obvious way (we mean string instruments *not* pianos: see [2,3] for pianos).

© The Author(s), under exclusive license to Springer Nature Switzerland AG 2024
T. Noll et al. (Eds.): MCM 2024, LNCS 14639, pp. 469–473, 2024.
https://doi.org/10.1007/978-3-031-60638-0_41

The most obvious tuning is to take four pure (or "Pythagorean") fifths (i.e. 3/2). So we have (in frequencies):

$$\frac{3}{2} = \frac{\text{E}}{\text{A}} = \frac{\text{A}}{\text{D}} = \frac{\text{D}}{\text{G}} = \frac{\text{G}}{\text{C}}$$

It is particularly easy to do those things in a string quartet since it simply means that the upper note ("numerator") is just the third harmonic of the lower note ("denominator"). For a piano, you need a whole book to describe it! see Cordier's account in [3], this is the reason why we limit ourselves to string quartets or trios[1].

Of course, you need an "A" to begin with, and the first violin is in charge of providing an inspired one to the rest of the players! For the choice of the "A", see the entry "diapason" in [1].

So, with open strings, a quartet can realize (in frequencies)

$$\frac{\text{E}}{\text{A}} = \frac{\text{A}}{\text{D}} = \frac{\text{D}}{\text{G}} = \frac{\text{G}}{\text{C}} = \frac{3}{2}$$

the Pythagorean fifth 3/2 being here just a number without "dimension".

As a result, this tuning is extremely brilliant (see [3]), but it has a *defect*: if a viola or cello is tuned in this way the third harmonic of the A is not equal to the tenth harmonic of the C (the latter is too low). In fact, we have:

$$\frac{\text{E}}{10\ \text{C}} = \frac{3\ \text{A}}{10\ \text{C}} = \frac{3}{10}\left(\frac{3}{2}\right)^3 = \frac{81}{80} = 1.0125$$

which is called the comma of Didymus (see [1]) and is quite *audible*! This is the reason why the last note in the cello part of the first movement of the celebrated Ravel trio is *too low* in this tuning!

The "official" way of tuning instruments (see [1] at "diapason") is by choosing the fifth equal to $2^{7/12}$. Then we have:

$$\frac{3}{10}\frac{\text{A}}{\text{C}} = \frac{3}{10}\left(2^{7/12}\right)^3 = \frac{3}{10}2^{7/4} \cong 1.0090757...$$

which is *much better.*

This is the reason why I proposed to change the usual tuning of the instruments which was impoverishing their resonance whatever efforts the players could make.

[Editor's Note: The tenth harmonic of C which is E

$$C = C_{-1}\ C_1\ G_1\ C_2\ E_2\ G_2\ B_2^\flat\ C_3\ D_3\ E_3 = E$$

is compared with the third harmonic E' of A_1:

$$A_1\ A_2\ E_3' = E'$$

[1] All this can be realized without any mechanical help, but tighteners might come in handy!

A_1 is 3 fifths of size Q above C

$$C = C_{-1}\ G_{-1}\ D_1\ A_1\ A_2\ E'_3 = E'$$

The frequency of E is 10 times the frequency of C

$$\text{freq}(E) = 10\text{freq}(C)$$

The frequency of E' is 3 times the frequency of A_1 and the frequency of A_1 is three fifths Q^3 times the frequency of C

$$\text{freq}(E') = 3\text{freq}(A_1) = 3Q^3\text{freq}(C)$$

Thus

$$\frac{\text{freq}(E')}{\text{freq}(E)} = \frac{3Q^3\text{freq}(C)}{10\text{freq}(C)} = \frac{3Q^3}{10}$$

Writing $E = \text{freq}(E')$ and $\text{freq}(E) = 10C$ for short, we get Hellegouarch's formula

$$\frac{E}{10C} = \frac{\text{freq}(E')}{\text{freq}(E)} = \frac{3Q^3}{10}$$

which is computed for $Q = 3/2$ and $Q = 2^{7/12}$.]

My starting point was to choose differently the fifths while tuning, and replace the pure fifths 3/2 by smaller fifths in order to liberate the resonance of the whole. So I proposed to call "temperament RR" the equal temperament in which all the fifths had to be equal to $(10/3)^{1/3} = 1.4938016$. For example, you tune your cello as if all the fifths (between the strings) were not 3/2, but the smaller $\left(\frac{10}{3}\right)^{1/3}$.

So you tune your cello (for example) *first* in pure fifths, *then* reduce all the fifths to $\left(\frac{10}{3}\right)^{1/3}$ which means (for a cello) to tighten up the D by $(10/3)^{1/3}$, the G by $(10/3)^{2/3}$, and the C by (10/3) (hence the need for tighteners![2]). When you have done this, you should hear that the whole cello is "pleased" and Ravel's last harmonic in the first movement of his *Piano trio* is accepted by other instruments! (see [4, 5] for more details).

Remark 1. If it does not work well then you correct $\frac{10}{3}$ to $\frac{10}{3}\varepsilon$ with an ε your common sense agrees with, and do it again if you are not pleased.

Remark 2. For the first tuning (with pure fifths), we have:

$$\frac{3\text{ A}}{10\text{ C}} \cong 1.0125 = \frac{81}{80} = \text{Didymus's comma}$$

For the second tuning (with tempered fifths), we have:

$$\frac{3\text{ A}}{10\text{ C}} \cong 1.0090757...$$

and

$$\frac{2\text{ E}}{10\text{ C}} = \text{Didymus's comma} = \frac{81}{80}$$

[2] The peg, which is handled by hand, only allows a slightly rough adjustment of the tension and therefore the tuning. The player is often obliged to refine the result using tighteners: he can then lower or raise the note without difficulty. Editor's Note.

2 The Temperament RR

Experience taught me that the first E in the cello part of the *Oriental Dance* by Rachmaninov could have a particularly nice resonance if your cello were well-tuned (particularly the C-string) and, on another side, I knew that the natural tenth harmonic of the open C-string of the cello in the first movement of Ravel's *Piano trio* was always too low and precarious (in the best cases): clearly the cello was not happy with it and the other strings were protesting!

Finally, it was by studying Schubert's *Rosamunde quartet* that I learned the importance of the resonance (in quartet playing) of the four instruments (between themselves) for the quality of the global sound. So, in the RR-temperament, all the fifths are equal to $\left(\frac{10}{3}\right)^{1/3} = 1.4938016$. The semitones are $\left(\frac{10}{3}\right)^{1/21} = 99$ cents and the octave is less than the usual octave $\left(\frac{10}{3}\right)^{4/7} = 1191$ cents.

Now, we can make the following table (in cents):

Pure fifth	Tempered fifth	RR fifth
702	700	695

Summing up: RR fifths are smaller than the pure fifths and tempered fifths, but the former ones give a correct tenth harmonic of the C-string. So the quartetist tuned in RR fifths might find this tuning smoother than the official one when playing pieces like *Rosamunde quartet.*

2.1 Tuning String Trios or Quartets the RR's Way

First, we begin by describing the ambitus of the cello, viola and violin and represent them in a holistic way by the following drawing (in pure fifths):

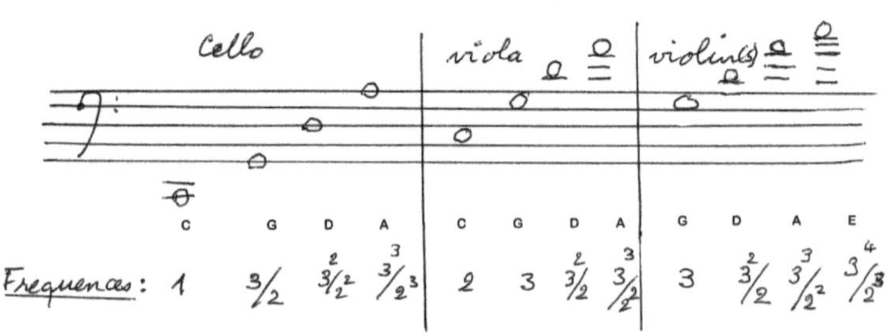

All this can be considered in the RR-temperament. So, for example, the C of the viola is almost the second harmonic of the C-string of the cello, and its G is almost the third of the same string. Also, the G of the violins is almost the third harmonic of the same string, and the D and the A are almost the same as those of the viola.

Finally, the E of the violins is almost the pure fifth of their A strings.

As a conclusion, we see that this tuning fits very well with the string quartet or trio.

Acknowledgement. *Thanks to Eric Reyssat, Yves Tastet and Franck Jedrzejewski for the interest they showed for this work.*

References

1. de Candé, R.: Dictionnaire de la musique. Seuil, Paris (1961)
2. Cordier, S.: L'accordage des instruments à claviers. Bulletin du Groupe Acoustique Musique **75** (1974)
3. Cordier, S.: Piano bien tempéré et justesse orchestrale. Buchet Chastel, Paris (1982)
4. Hellegouarch, Y.: Description d'un tempérament égal non classique. In: À la lumière des mathématiques et à l'ombre de la philosophie, edited by Moreno Andreatta and François Nicolas, pp. 97–100. Delatour, Sampzon (2012)
5. Jedrzejewski, F.: Quelques remarques sur les tempéraments. In: À la lumière des mathématiques et à l'ombre de la philosophie, edited by Moreno Andreatta and François Nicolas, pp. 101–104. Delatour, Sampzon (2012)

Author Index

© The Editor(s) (if applicable) and The Author(s), under exclusive license
to Springer Nature Switzerland AG 2024

T. Noll et al. (Eds.): MCM 2024, LNCS 14639, pp. 475–476, 2024.
https://doi.org/10.1007/978-3-031-60638-0

SPRINGER NATURE

GPSR Compliance

The European Union's (EU) General Product Safety Regulation (GPSR) is a set of rules that requires consumer products to be safe and our obligations to ensure this.

If you have any concerns about our products, you can contact us on ProductSafety@springernature.com

In case Publisher is established outside the EU, the EU authorized representative is:

Springer Nature Customer Service Center GmbH
Europaplatz 3
69115 Heidelberg, Germany

FSC
www.fsc.org
MIX
Papier | Fördert gute Waldnutzung
FSC® C083411

Kolibri 360 GmbH
Ferdinand-Jühlke-Straße 7
99095 Erfurt, Deutschland
produktsicherheit@kolibri360.de